Natural Remedies for Pest, Disease and Weed Control

Natural Remedies for Pest, Disease and Weed Control

Edited by

CHUKWUEBUKA EGBUNA, BSc, MSc BIOCHEMISTRY
Research Biochemist
Department of Biochemistry
Faculty of Natural Sciences
Chukwuemeka Odumegwu Ojukwu University
Nigeria

BARBARA SAWICKA, PHD
Full professor
Department of Plant Production Technology and Commodities Science
Faculty of Agrobioengineering
University of Life Sciences
Poland

ELSEVIER

ACADEMIC PRESS
An imprint of Elsevier

Publisher: Charlotte Cockle
Acquisition Editor: Charlotte Cockle
Editorial Project Manager: Laura Okidi
Production Project Manager: Kiruthika Govindaraju
Cover Designer: Alan Studholme

List of Contributors

Yusuf Abubakar
Department of Biochemistry
Natural Product Research Laboratory
Bauchi State University
Gadau, Bauchi State, Nigeria

Charles Oluwaseun Adetunji, PhD
Applied Microbiology
Biotechnology and Nanotechnology Laboratory
Department of Microbiology
Edo University Iyamho
Auchi, Edo State, Nigeria

Dickson Adom, PhD
Department of Educational Innovations in Science and
    Technology
Kwame Nkrumah University of Science and
    Technology
Kumasi, Ashanti, Ghana

Amrish Agarwal
Formulation Division
Institute of Pesticide Formulation Technology (IPFT)
Gurugram, Haryana, India

Muhammad Akram, PhD
Department of Eastern Medicine
Directorate of Medical Sciences
Government College University Faisalabad
Faisalabad, Punjab, Pakistan

Hakiye Aslan, PhD
Bingöl University
Engineering Faculty
Food Engineering
Turkey

İmran Aslan, Assoc. Prof.
Bingöl University
Health Sciences Faculty
Occupational Health and Safety Department
Turkey

Temitope Banjo, PhD
Department of Biological Sciences
Crawford University
Igbesa, Ogun State, Nigeria
Department of Biological Sciences
Covenant University
Ota, Ogun State, Nigeria
Institute for Human Resources Development
Federal University of Agriculture
Abeokuta, Ogun State, Nigeria

Madhvi Chawan, MSc
Assistant Professor
Haffkines Biopharmaceuticals
Mumbai, Maharashtra, India

Intan Soraya Che Sulaiman, BSc, MSc, PhD
National Defence University of Malaysia
Centre of Research & Innovation Management
Kem Sungai Besi
Kuala Lumpur, Malaysia

Muhammad Daniyal
TCM and Ethnomedicine Innovation & Development
    International Laboratory
Innovative Materia Medica Research Institute
School of Pharmacy
Hunan University of Chinese Medicine
Changsha, Hunan, P.R. China

Dominika Skiba, PhD
Department of Plant Production Technology and
    Commodities Science
University of Life Science in Lublin
Lublin, Poland

Chukwuebuka Egbuna, BSc, MSc
Research Biochemist
Department of Biochemistry
Faculty of Natural Sciences
Chukwuemeka Odumegwu Ojukwu University
Uli, Anambra State, Nigeria
Nutritional Biochemistry and Toxicology Unit
World Bank Africa Centre of Excellence
Centre for Public Health and Toxicological Research
    (PUTOR)
University of Port-Harcourt
Port Harcourt, Rivers State, Nigeria

Ayten Ekinci, PhD
Cumhuriyet University
Gemerek Vocational High School
Chemistry and Chemical Processing Technologies
    Department
Sivas, Turkey

Shahira M. Ezzat, PhD
Professor
Department of Pharmacognosy
Faculty of Pharmacy
Cairo University
Cairo, Egypt
Department of Pharmacognosy
Faculty of Pharmacy
October University for Modern Sciences and Arts
    (MSA)
Cairo, Egypt

Ajay Kumar Gautam, PhD
Department of Plant Sciences
Central University of Punjab
Bathinda, Punjab, India

Rohan Gavankar, MSc, BEd, PhD
Assistant Professor
Department of Botany
VIVA College
Virar, Maharashtra, India

S. Zafar Haider, PhD
Centre for Aromatic Plants (CAP)
Dehradun, Uttarakhand, India

Jonathan C. Ifemeje, PhD
Department of Biochemistry
Faculty of Natural Sciences
Chukwuemeka Odumegwu Ojukwu University
Uli, Anambra State, Nigeria

Agnieszka Jamiołkowska, PhD
Professor
Department of Plant Protection
Faculty of Horticulture and Landscape Architecture
University of Life Sciences in Lublin
Lublin, Poland

Jaison Jeevanandam, PhD
Department of Chemical Engineering
Curtin University Malaysia
Sarawak, Malaysia

Smriti Kala, PhD
Formulation Division
Institute of Pesticide Formulation Technology (IPFT)
Gurugram, Haryana, India
Center for Rural Development Technology (CRDT)
Indian Institute of Technology (IIT)
Delhi, India

Marek Kopacki, PhD
Habilitated Doctor
Department of Plant Protection
Faculty of Horticulture and Landscape Architecture
University of Life Sciences in Lublin
Lublin, Poland

Toskë L. Kryeziu
Department of Pharmacy
Faculty of Medicine
University Hasan Prishtina
Pristina, Kosovo

Jitendra Kumar
Formulation Division
Institute of Pesticide Formulation Technology (IPFT)
Gurugram, Haryana, India

Shashank Kumar, PhD, MSc, BSc
Assistant Professor
School of Basic and Applied Sciences
Department of Biochemistry and Microbial Sciences
Central University of Punjab
Bathinda, Punjab, India

Colin B. Lukong, PhD
Department of Biochemistry
Faculty of Natural Sciences
Chukwuemeka Odumegwu Ojukwu University
Uli, Anambra State, Nigeria

Mohammad Mehdizadeh, PhD
Weed Science
Department of Agronomy and Plant Breeding
University of Mohaghegh Ardabili
Ardabil, Iran

Rana M. Merghany
Department of Pharmacognosy
National Research Centre
Giza, Egypt

Azham Mohamad, BSc, MS
Lecturer
Centre of Foundation Studies for Agricultural Science
Universiti Putra Malaysia
UPM Serdang
Selangor, Malaysia

Waseem Mushtaq, PhD
Allelopathy Laboratory
Department of Botany
Aligarh Muslim University
Aligarh, Uttar Pradesh, India

S.N. Naik
Center for Rural Development Technology (CRDT)
Indian Institute of Technology (IIT)
Delhi, India

Keshavi Nalla
Jawaharlal Nehru Technological University
Kakinada, Andhra Pradesh, India

Ashish Kumar Nayak, PhD
Department of Microbial Genomics and Diagnostic
   Laboratory
Regional Plant Resource Centre
Bhubaneshwar, Odisha, India

Jaweria Nisar
Department of Eastern Medicine
Directorate of Medical Sciences
Government College University Faisalabad
Faisalabad, Punjab, Pakistan

Onyeka Kingsley Nwosu, BSc, MSc, PGDE
National Biosafety Management Agency
Abuja, FCT, Nigeria

Alloysius Chibuike Ogodo, BSc, MSc, PhD
Department of Microbiology
Faculty of Pure and Applied Sciences
Federal University Wukari
Wukari, Taraba State, Nigeria

Valentina V. Onipko, Doctor of Sc.
Professor
Poltava National Pedagogical University named after
   V.G. Korolenko
Poltava, Ukraine

Olumayowa Vincent Oriyomi, Msc
Institute of Ecology and Environmental Studies
Obafemi Awolowo University
Ile-Ife, Osun State, Nigeria

Chandan Kumar Panda, PhD
Assistant Professor-cum-Junior Scientist
Department of Extension Education
Bihar Agricultural University, Sabour
Bhagalpur, Bihar, India

Sunil Pareek, PhD
Department of Agriculture and Environmental Sciences
National Institute of Food Technology
   Entrepreneurship and Management
Sonepat, Haryana, India

P.K. Patanjali
Formulation Division
Institute of Pesticide Formulation Technology (IPFT)
Gurugram, Haryana, India

Kingsley C. Patrick-Iwuanyanwu, PhD
Department of Biochemistry (Toxicology unit)
University of Port Harcourt
Port Harcourt, Rivers State, Nigeria
Nutritional Biochemistry and Toxicology Unit
World Bank Africa Centre of Excellence
Centre for Public Health and Toxicological Research
   (PUTOR)
University of Port-Harcourt
Port Harcourt, Rivers State, Nigeria

Sergey V. Pospelov, PhD
Professor
Chair Agriculture & Agrochemistry
Poltava State Agrarian Academy
Poltava, Ukraine

Anna D. Pospelova, PhD
Associated Professor
Chair Plant Protection
Poltava State Agrarian Academy
Poltava, Ukraine

Seshu Vardhan Pothabathula, PhD
Assistant Professor
School of Basic and Applied Sciences
Department of Biochemistry and Microbial Sciences
Central University of Punjab
Bathinda, Punjab, India

Narashans Alok Sagar, PhD
Department of Agriculture and Environmental Sciences
National Institute of Food Technology
    Entrepreneurship and Management
Sonepat, Haryana, India

Arvind Saroj, PhD
Department of Plant Pathology
Central Institute of Medicinal and Aromatic Plants
Lucknow, Uttar Pradesh, India

Barbara Sawicka, PhD
Full Professor
Department of Plant Production Technology and
    Commodities Science
Faculty of Agrobioengeenering
University of Life Sciences
Lublin, Poland

Maxim V. Semenko
Bachelor of Biology
Poltava National Pedagogical University named after
    V.G. Korolenko
Poltava, Ukraine

Aamir Sharif
Department of Pathology
University of Sargodha
Sargodha, Punjab, Pakistan

Nisha Sogan, PhD
National Institute of Malaria Research (NIMR)
Delhi, India

Laith Khalil Tawfeeq Al-Ani
School of Biology Science
Universiti Sains Malaysia
Minden, Malaysia

Nikhil Teli, MSc, PhD
Department of Botany
VIVA College
Virar, Maharashtra, India

Habibu Tijjani, PhD
Department of Biochemistry
Natural Product Research Laboratory
Bauchi State University
Gadau, Bauchi State, Nigeria

Kingsley Ikechukwu Ubaoji, PhD
Department of Applied Biochemistry
Nnamdi Azikiwe University
Awka, Anambra State, Nigeria

Krishnan Umachandran, BE, MS, MBA, PhD
General Manager
Organization Development
Nelcast Ltd.
Chennai, Tamil Nadu, India

Deepa Verma, MSc, BEd, M.A, M.PhiL, MBA, PhD
Head of Department
Assistant Professor
Department of Botany
VIVA College
Virar, Maharashtra, India

Parisa Ziarati, PhD
Assistant Prof. of Chemistry
Nutrition & Food Sciences Research Center
Tehran Medical Sciences
Islamic Azad University
Tehran, Iran

Contents

PART I
GREEN APPROACH TO PEST AND DISEASE
CONTROL

1 **Pests of Agricultural Crops and Control Measures**, 1
Barbara Sawicka, PhD and Chukwuebuka Egbuna, BSc, MSc

2 **Plant Diseases, Pathogens and Diagnosis**, 17
Barbara Sawicka, PhD, Chukwuebuka Egbuna, BSc, MSc, Ashish Kumar Nayak, PhD and Smriti Kala, PhD

3 **Pesticides, History, and Classification**, 29
Yusuf Abubakar, Habibu Tijjani, PhD, Chukwuebuka Egbuna, BSc, MSc, Charles Oluwaseun Adetunji, PhD, Smriti Kala, PhD, Toskë L. Kryeziu, Jonathan C. Ifemeje, PhD and Kingsley C. Patrick-Iwuanyanwu, PhD

4 **Biopesticides, Safety Issues and Market Trends**, 43
Chukwuebuka Egbuna, BSc, MSc, Barbara Sawicka, PhD, Habibu Tijjani, PhD, Toskë L. Kryeziu, Jonathan C. Ifemeje, PhD, Dominika Skiba, PhD and Colin B. Lukong, PhD

5 **Natural Compounds Against Plant Pests and Pathogens**, 55
Agnieszka Jamiołkowska, PhD and Marek Kopacki, PhD

6 **Phytochemicals of Plant-Derived Essential Oils: A Novel Green Approach Against Pests**, 65
Arvind Saroj, PhD, Olumayowa Vincent Oriyomi, MSc, Ashish Kumar Nayak, PhD and S. Zafar Haider, PhD

7 **Semiochemicals: A Green Approach to Pest and Disease Control**, 81
Shahira M. Ezzat, PhD, Jaison Jeevanandam, PhD, Chukwuebuka Egbuna, BSc, MSc, Rana M. Merghany, Muhammad Akram, PhD, Muhammad Daniyal, Jaweria Nisar and Aamir Sharif

8 **Fungistatic Properties of Lectin-Containing Extracts of Medicinal Plants**, 91
Sergey V. Pospelov, PhD, Anna D. Pospelova, PhD, Valentina V. Onipko, Doctor of Sc. and Maxim V. Semenko

PART II
BIOLOGICAL CONTROL MEASURES

9 **Biological Control of Weeds by Allelopathic Compounds From Different Plants: A BioHerbicide Approach**, 107
Mohammad Mehdizadeh, PhD and Waseem Mushtaq, PhD

10 **Microbial Control of Pests and Weeds**, 119
Deepa Verma, MSc, B.ED, M.A, M.PHIL, MBA, PhD, Temitope Banjo, PhD, Madhvi Chawan, MSc, Nikhil Teli, MSc, PhD and Rohan Gavankar, MSc, B.ED, PhD

11 **Biological Control of Plant Pests by Endophytic Microorganisms**, *127*
Alloysius Chibuike Ogodo, BSc, MSc, PhD

PART III
TECHNIQUES, BIOTECHNOLOGICAL AND
COMPUTATIONAL APPROACH

12 **Techniques for the Detection, Identification, and Diagnosis of Agricultural Pathogens and Diseases**, *135*
*Ajay Kumar Gautam, PhD and
Shashank Kumar, PhD, MSc, BSc*

13 **Nucleic Acid—Based Methods in the Detection of Foodborne Pathogens**, *143*
*Hakiye Aslan, PhD, Ayten Ekinci, PhD and
İmran Aslan, Assoc. Prof.*

14 **Plant Biotechnology in Food Security**, *163*
*Barbara Sawicka, PhD, Krishnan
Umachandran, BE, MS, MBA, PhD,
Dominika Skiba, PhD and
Parisa Ziarati, PhD*

15 **Homemade Preparations of Natural Biopesticides and Applications**, *179*
*Charles Oluwaseun Adetunji, PhD,
Chukwuebuka Egbuna, BSc, MSc,
Habibu Tijjani, PhD, Dickson Adom, PhD,
Laith Khalil Tawfeeq Al-Ani and
Kingsley C. Patrick-Iwuanyanwu, PhD*

16 **The Use of Vermiwash and Vermicompost Extract in Plant Disease and Pest Control**, *187*
*Intan Soraya Che Sulaiman, BSc, MSc, PhD
and Azham Mohamad, BSc, MS*

17 **Genetic Modification as a Control Mechanism to Plant Pest Attack**, *203*
*Onyeka Kingsley Nwosu, BSc, MSc, PGDE and
Kingsley Ikechukwu Ubaoji, PhD*

18 **Biopesticides: Formulations and Delivery Techniques**, *209*
*Smriti Kala, PhD, Nisha Sogan, PhD, Amrish
Agarwal, S.N. Naik, P.K. Patanjali and
Jitendra Kumar*

19 **Safe Storage and Preservation Techniques in Commercialized Agriculture**, *221*
*Narashans Alok Sagar, PhD and Sunil Pareek,
PhD*

20 **Advances in Application of ICT in Crop Pest and Disease Management**, *235*
Chandan Kumar Panda, PhD

21 **Applications of Computational Methods in Plant Pathology**, *243*
*Keshavi Nalla, Seshu Vardhan Pothabathula,
PhD and Shashank Kumar, PhD, MSc, BSc*

INDEX, *251*

CHAPTER 1

Pests of Agricultural Crops and Control Measures

BARBARA SAWICKA, PHD • CHUKWUEBUKA EGBUNA, BSC, MSC

INTRODUCTION

A pest is any specie considered undesirable. The term "pest" is very subjective. Each specie (individual) is competing with people for food and shelter, carrying pathogens, feeding on humans, or otherwise threatening its health, well-being, and welfare (this definition also includes weeds). One of the most important pest characteristics is the level to which their numbers are regulated by natural enemies. Pests are often species that have become out of control of their natural enemies (e.g., as a result of being moved to other regions of the world, or as a result of extirpation of these natural enemies by humans) [1–4]. Pests are organisms that cause losses in agricultural, forestry, and storage activities. It is estimated that losses caused by pests in arable crops account for about 10%–15% of the total yield value on a global scale [5,6].

Pests also include animals damaging cultivated plants and crops of these plants stored in warehouses, granaries, or processing plants. Furthermore, these pests damage trees, shrubs, wood-based materials, food products, etc. Pests include some species of nematodes, snails, vines, mites, and insects (the most numerous group) and mainly rodents from vertebrates. Pests sometimes also transmit pathogens—harmful viruses, bacteria, or fungi [5,6].

Among all the biotic factors, pests are considered the most important and cause the largest losses in crops (30%–40%) [7–11]. Some of them result from improper storage, and some are as a result of the settlement, development, and feeding of various storage pests, including harmful mites and insects. These organisms can be the cause of losses, both quantitative, resulting from the loss of mass of raw material because of feeding, as well as qualitative losses. For example, in Kenya, maize losses due to insects reach up to 50% [8,12]. According to the economic model of Compton et al. [13], every percent of insect invasion causes a decrease in the value of maize by 0.6%–1% [7,8,10,13]. According to Patol [5], pests are responsible for 80%–90% of stored seed loss. In Cameroon, losses caused by insects feeding in corn seeds are estimated at 12%–44% [14]. The mite is a very severe, though invisible to the naked eye. Maize grain in storage depicts various species of storage mites, including fine mites and flour mites. Because of life processes carried out in humid places, the temperature and humidity of the stored raw material may gradually increase. As a result, its slow deterioration takes place. The feeding mites alone mainly damage the embryos that are the most moistened part of the kernel, and the microscopic size of the damage later reduces the seed sowing value. The mites not only damage and pollute the raw material that settles with their excrements, secretions, louts, and dead individuals. With a very large population, the grain becomes unpleasant, which reduces the quality of the raw material [15,16].

The stored cereal grain can feed many other species of insects, both beetles and storage butterflies, such as the giant hood, the hood of the grain, coffee cobbler, flour miller, short breader, half-wing, *Carpophilus flavipes*, *Carpophilus mutilatus*, small flatten, scrubber, grain skins, *Trogoderma variabile*, *Streptochaeta spicata*, pungent trifle, sugar beetle, *Typhaea stercorea*, *Moorish hide*, corn weevil, rice weevil, grain weevil, *Sitophilus oryzae*, bread crust, and even rodents. The butterflies are larger than mites, more mobile, and less sensitive to unfavorable environmental conditions. Grain spaces, especially in stored maize, allow the insects to freely penetrate the entire profile of the grain heap. They gather more often in more humid places, as well as in periods of unfavorable temperature. Their presence in a given place and ongoing life processes (mainly breathing) locally increase the humidity and temperature of the raw material. Insects, as well as mites, start feeding from embryos, as well as seeds grown, broken, and those covered by the fruit and seed (shell) have been damaged [16–20]. Because of the intensive feeding of insects, large amounts of dust are generated, which falls clogging gradually into the spaces between the grains, as a result of which the free flow of air through the grain prism is limited. The processes of heating and humidifying the grain in the occluded layer are then intensified. The bacterial and fungal microflora grow intensively, which in turn leads to spoilage of the raw material. Most species of storage beetles are small. Their bodies are usually 3–5 mm long, although some are relatively large, because they have 15–25 mm (e.g., millstone meal). The larvae and adult stages of beetles feed on and develop from the egg to the adult outside of the kernels, except for the grain weevil. Hatching eggs, pupae, and young weevil beetles develop from the inside of the corn seeds of the eggs, initially without visible external symptoms. Under favorable conditions, with higher humidity of the raw material and at a higher temperature, the number of generations of storage pest's increases. In most species of storage beetles, lesions are caused by both larvae and adults [15–17]. In butterflies, the harmful stage is larvae, that is, caterpillars, which reside in the top layer of cereal grains. The places where larvae feed, as well as the kernels they feed on, are stuck with threads of sticky yarns [19–21].

Currently, *Sitophilus Zea mays* and *Prostephanus truncates*, the main pests of cereal grains, occur in most parts of Africa and cause huge damage in a very short time [22]. During storage on the farm, there may be more than 30% of the grain loss of maize due to these pests [8–11]. In Ghana, approximately 5%–10% of market value loss is estimated due to invasion by *Sitophilus* spp. and 15%–45% of market value loss because of LGB damage [22–24]. Hussein [25] reported that after 6 months of maize storage, LGB was responsible for more than half (56.7%) of storage losses and then losses due to grain weevil. Patol et al. [5] observed about 25% of *R. dominica* losses in wheat stored for 3 months under laboratory conditions [25].

Pests of Crops

Agricultural pests include 45 types and 55 species of insects from 15 families. Most of them are pests of cereals and millet (3), oilseeds (2), fiber plants (4), vegetables (11), temperate climate (7), tropical and tropical fruits (9), ornamental plants (2), crops field (8), spices (4),

and stored grain (5). Among these families, 12 species (21.8%) predominate in the genus: Curculionidae, followed by Chrysomelidae 10 (18.2%), Cerambycidae 9 (16.3%), Scarabaeridae 6 (10.9%), Coccinellidae 3 (5.4%), Apionidae, Bostrychidae, Brachidae, Buprestidae, and Scolytidae with two species (3.6%). Families such as Dermestidae, Lamiidae, Meloidae, Melolonithidae, and Tenebrionida e contribute to the feeding of one species (1.8%) [5,20,25,26].

Storage Pests

Industrial production forces the need to store both raw materials and finished products for a shorter or longer period. The place that is used to store them and ensures the continuity of production is the warehouse. Among storage pests, several groups of organisms can be mentioned: mites, insects (mainly butterflies and beetles), rodents, and microorganisms (bacteria, fungi) [19,20,27]. The mites spread along with the grain, on the packaging of products or on the clothing of warehouse employees. They can also be transferred on the body of insects, rodents, and birds—other storage pests. Their presence is confirmed during microscopic examination. These pests pollute the grain with faces, broods, and dead birds, give it an unpleasant odor, and affect the temperature and humidity of the grain. Dirt mite contaminated with excrement is unsuitable for human and animal consumption, and if eaten it may cause irritation of the gastrointestinal mucosa. In high temperature and low humidity, the mites die quickly. The most important of the storage organisms for the stored crop are: cereal weevil, mites, bacteria and fungi, and rodents [15,18,20].

Grain weevil

The grain weevil is a beetle with a size of 2.5–4 mm, dark brown with a characteristic head elongated in a snout. It is a pest that feeds mainly on the grain of barley and wheat, biting numerous holes in the grain. The grain settled by larvae of grain weevil is not suitable for sowing. Beetles avoid light, making it difficult to detect them. In the place of concentration of the grain weevil population, the grain mass temperature may be even 10°C higher than in unfertilized places [19,27–30].

Bacteria and fungi

Bacteria and fungi that develop on the surface or inside of the crops are a serious threat. The most dangerous are mold fungi, which can develop on the surface of the seeds, as well as penetrate their depths, causing their damage. Grain infested by mold fungi has a reduced sowing value, its smell, taste, and color change. Developing fungi produce harmful substances—mycotoxins that accumulate in grain. These toxins can cause dangerous poisonings to humans and animals [29,30]. About 25%–40% of cereal grains around the world are contaminated by mycotoxins produced by storage fungi [25,28]. Mycotoxins cause a decrease in the dry matter content of grain, as well as loss of quality and pose a threat in the food supply chain [31,32]. The most common and most important mycotoxins are aflatoxins, fumonisins, deoxynivalenol, and ochratoxin [33–35]. Aflatoxins, produced as a secondary metabolite by two fungal species *Aspergillus flavus* and *A. parasiticus*, are considered the most dangerous group of mycotoxins because they increase the risk of liver cancer and limit the growth of young children [36]. Due to food contamination with mycotoxins, about 4.5 billion people in developing countries are exposed to aflatoxins [22]. High concentrations of aflatoxin can lead to aflatoxicosis, which can cause serious illness and even death [35]. *Penicillium verrucosum* (ochratoxin), the main mycotoxin in mildew is commonly found in humid cool climates (e.g., in northern Europe), while *Aspergillus flavus* is most commonly observed in temperate and tropical climates [24]. During storage, it damages grains and limits germination. Consequently, it causes deterioration of the quality of dishes, change in their taste and smell, and undesirable effects on human health [36]. Seeds of oilseeds with a high oil content require special attention during storage, because high levels of moisture cause the degradation of vegetable oil and produce large amounts of unsaturated fatty acids, which also cause self-heating of seeds [25,36,37]. At the farm level in developing countries, even rodents can damage a large part of the crop, while mushrooms can be the main reason for spoilage when stored at high relative humidity. The use of scientific storage structures and proper handling of grains can reduce storage losses to less than 1% [25,36].

Losses can be minimized by physically avoiding the penetration of insects and rodents and maintaining environmental conditions that prevent the growth of microorganisms. Knowing the control points during collection and drying before storage can help to reduce losses during storage of cereals. The timely prevention of biotic and abiotic factors can be very effective in reducing storage losses [38].

Rodents

Rodents are mostly rats and mice. They are storage pests that are very difficult to eradicate. They get to the warehouses through various gaps. Their activities can destroy a significant amount of grain, and even more contaminate with droppings, scatter, and trample. They can spread infectious diseases such as cholera, typhoid, infectious jaundice, trichinosis, rabies, and others. They are carriers of insects and microorganisms [18,19].

Pests Control Measures

The fight against destructive pests uses mainly the following:

1. Cultivation methods (ecological), such as changing the sowing date or avoiding the same crops being kept constantly in the same place;
2. A chemical fight played a fundamental role in the last century, although its effect is serious ecological problems. In the 1960s (Carson 1962 Silent spring), for the first time, attention was paid to the problems associated with the use of organic compounds for pest control;
3. Biological methods. The biological conservation strategy was proposed by Baker and Snyder in 1965 at the Berkeley Symposium "Ecology of soil-borne plant pathogens: prelude to biological control" [15,18,20].

In the 1970s, the widespread use of pesticides, with a wide range of activities, was replaced by integrated pest control, in other words, chemical warfare was combined with biological combating (using natural enemies—predators and parasites). The term "biocontrol" can be defined as supporting and inciting to naturally occurring "biological wars" [6,16,39,40]. The term 'biological warfare' was used to describe virtually all pest control methods except for the use of nonselective chemical pesticides. Currently, the term has been limited to regulating the number of pests by using their natural enemies. The mechanisms of biocontrol are different and have not been defined until the end. They usually include general and specific inhibition of pathogen development. Another important method is selection of varieties resistant to pests 6, [16,41,42]. There are several methods of biological warfare against pests. These includes the following:

1. The introduction consists of importing a natural enemy from another geographical area, very often from the same from which the pest originally came from. The number of pests is below economic injury level (EIL). This method is often called classic biological struggle. Sometimes it requires the release of the enemy several times where it cannot be maintained throughout the year. Several generations of the pest are thus controlled;
2. The release of the local natural enemy may also be aimed at supplementing the existing population;

therefore, it is carried out repeatedly and usually coincides with the period of rapid growth of the pest population;

3. Colonization—mass release of the natural enemy to destroy the pests present at the time, but without the expectation that it will be a long-term effect. These enemies are sometimes called biological pesticides;
4. Protection—treatments aimed at protecting natural enemies. Recently, more attention has been paid to the fight against insects—pests against insect pathogens as microbiological insecticides;
5. Microbiological insecticides—the use of *Bacillus thuringiensis* bacteria is the main microbiological method used worldwide on an industrial scale in the fight against pests (along with baculovirus and about 100 types of fungi) [15,16,43];
6. Genetic struggle and resistance—numerous methods are known that utilize genetic manipulation to control pests:
   ✔ The self-destructive method, where it uses only pests to increase their own mortality (usually by reducing fertility);
   ✔ For selecting plant varieties resistant to pests (as well as herbicides). The cultivation and use of transgenic plants can be a source of potential environmental benefits. However, social perception and legal aspects related to this technique as well as problems resulting from the evolution of resistance are as difficult as in the case of chemical pesticides [16,42].

Usually three mechanisms are involved in biocontrol:

1. Inhibiting the growth of an organism controlled by antibiotics, enzymes, or biocides (antagonism);
2. Suppressing the growth of the target organism by limiting the availability of the food source or depriving it (competition);
3. Inhibition by direct parasitism or predation on pests or parasites, for example, *Phoma exigua* can also be used for the biocontrol of some weeds—field dandelion (*Sonchus arvensis* L.), field cure (*Cirsium arvense* L. Scop.), due to the production of specific phytotoxins (cytochalasin A and B, p-hydroxybenzaldehyde) [44,45].

In addition, microorganisms can interact with protected plants, increasing their resistance to parasites or pests and to weeds, increasing their sensitivity to damage or death. Bacteria participating in these interactions are representatives of plant growth promoting rhizobacteria. The most vivid example of a naturally occurring biocontrol is competition, where all organisms have developed under the conditions of competitive stress.

Lack of competition is probably responsible for the boomerang effect—many competitors die in sterilized soil, and despite the strong reduction of pathogens, losses of plants grown on it are greater than in natural soil. In the face of environmental problems caused by excessive use of fertilizers and chemical plant protection products, biological control can be an alternative system that can play an important role in achieving the most important agricultural goals [15,18,46].

Integrated pest management

In agriculture, a sustainable food cultivation would be possible if there are reconciled efforts to eliminate crop losses caused by diseases through effective and safe methods, via a practice called integrated pest management (IPM). Integrated pest control is more a philosophy than a specific strategy. It combines physical, ecological, chemical, and biological struggle with the use of resistant varieties. It has ecological foundations and considers factors determining mortality. The IPM strategy is focused on pest control below EIL[1] and is based on data on the number of pests and their natural enemies. Determining the optimal strategy, however, requires time, effort, and money, although to be effective, it should be based on cost effectiveness. The key to an effective IPM program is good field monitoring [45,47−50]. Existing and implemented IPM programs indicate that their use is associated with economic profitability, despite the possibility of a decline in yields.

Biological Plant Protection—Biocontrol

As stated earlier, biological control is the use of one organism to suppress the development of another, such as a parasite or pest in agriculture or an organism that damages the environment. Their antagonism with phytopathogens consists of several mechanisms of action, including competition for living space and nutrients, mycoparasitism, secretion of compounds with antifungal activity, such as antibiotics or killer toxins, induction of immunity in plants, and the production of volatile compounds that may play a large role in the biocontrol phenomenon [51,52]. In terms of origin, they are divided into two main groups, that is, naturally occurring in a given environment and coming from other areas and/or industrially produced, and then introduced or released into a given environment. In a broad sense, biological factors include viruses and pathogenic and/or competitive microorganisms in relation to pests (i.e., bacteria, protozoa, fungi), useful

---

[1]Economic injury level (EIL) is the density of the pest, at which the yield value exceeds the pest control costs to a greatest extent.

macroorganisms (i.e., predatory mites and insects, parasitic insects), substances of vegetable and animal origin (i.e., extracts, active molecules) and so-called semicompounds (pheromones, allomones, kairomones, and other attractants and repellents); and natural products (e.g., plant extracts and their extracts, minerals, and so-called "biologically active molecules" etc.). The second group includes factors that are not registered at the EU level and contain plant protection products for independent registration in the EU. The group includes macroorganisms, that is, beneficial arthropods (insects, mites) and insecticides [48,52,53]. Three basic biological strategies are used, that is, classic, augmenting, and maintenance. Each of them has found a different application and uses various biological factors. The classic biological method involves the introduction into the environment of useful, exotic biological agents to control or reduce the population of pests of local or foreign origin. They are usually obtained from the areas of their natural occurrence and released in new areas, where they expect to be domesticated and effectively keep the pest at an unimaginable level. The greatest hope in biological plant protection is the possibility of using as a protective factor against fungal diseases of bacteria and fungi that would constitute a natural component of the mycosphere population. The area of microsphere is the first line of protection of plant cells against attack of pathogens and harmful microorganisms. For microorganisms with a positive effect on plant growth—PGPR (plant growth—promoting rhizobacteria) and PGPF (plant growth—promoting fungi) to stimulate the development of plants and protect them against pathogens and microorganisms harmful to plant growth—DRMO (deleterious rhizosphere microorganisms), they must quickly settle in the plant root system and be competitive with other rhizosphere organisms. Plant protection against fungal diseases by biological methods is based on the use of specific physiological characteristics of microorganisms. It can be a direct or indirect result microbial influences on pathogen [16,48,49,53—55]. The BCA (biological control agent) mechanisms are as follows:

i. Basic, direct: competition for a niche; competition for nutrients; antibiotics (including antibiotics of the peptaibols type).

ii. Indirect: stimulation of immunity (defense mechanisms) of plants [3,56—61].

To boost plant defense, additional biofertilizer properties can be obtained through growth hormones (auxin: IAA-indolyl acetic acid, ethylene, cytokinin-like compounds: gibberellin) and organic acids (i.e., gluconic, lemon, fumaric) facilitating the uptake of compounds C (mainly glucose) P, Fe, Mn, and Mg [45,50]. Peptaibol is a class of small (15—20 amino acids) linear polypeptides with a strong antimicrobial activity against G+ bacteria and fungi (probably also with the properties of plant resistance elicitors), which belong to the so-called pore forming antibiotics causing the formation of large (0.55 nm) openings in the cell membrane in the presence of sterols, which leads to osmotic autolysis of cells. They are characterized by great microheterogeneity resulting from their postribosomal modification in the process of "thio-template"—the transfer of thiol groups (SH) by specific thioesters [56]. The conditions for effective colonization of the rhizosphere are the biological protection factors and the use of native rhizosphere microorganisms and pre-inoculation, suitable initial inoculum; coinoculation; the use of appropriate carriers; "Soil manipulation"; an appropriate phenotype with antibiotic, siderophore, enzymatic activity; resistance to fungicides; and using root exudates as a source of carbon [46,56]. Supportive features of dissemination: motility, chemotaxis toward root secretions, recognition of plant agglutinin, and application of genetic engineering methods [46,48,49]. For example, the ability to produce compounds complexing Fe (III) among strains isolated from the rye centrosphere is much higher than in soil loosely bound to the roots and in the endosphere. In the centrosphere, the "general synthesis" (number of producers) of both Fe (III) complexing compounds and specific hydroxamic and catechol siderophores is several times higher than in soil loosely bound to roots irrespective of the rye growth period. In the centrosphere, the largest amounts of Fe (III) complexing compounds and hydroxyamide siderophores are produced at the stage of propagation by fungal strains and at the stage of propagation and full maturity by bacterial strains, while catechol siderophores are most intensively generated at the flowering stage. This indicates that the siderophores synthesized by microorganisms colonizing the root zone of cereals can have a significant impact on the amount of available Fe (III) in this zone and on supplying Fe (III). In the conditions of planting the root system by microorganisms, the supply of the Fe plants in both the first and the second strategy is much better. Under conditions of Fe deficiency, strains of plant growth-promoting microorganisms (PGPRs) produce large amounts of siderophores with high affinity to the Fe "scrape" Fe (III) from their environment [56,62,63].

Competitive advantages exists in the organisms that can use, apart from their own, exogenous siderophores (have several receptors). When PGPR siderophore (e.g., pseudobactin) is used, only about 1% of rhizosphere strains synthesize ligands forming complexes from Fe (III), with higher constant durability; they produce more siderophores under given conditions, their transport system is more efficient, and their siderophores are more resistant to degradation (e.g., pseudobactin containing D- and L-amino acids). Siderophores produced by soil microorganisms make Fe (III) available both to microorganisms as well as to plants I and II of the strategy (transport to the plant through diffusion).

They are the factors enabling the biocontrol of phytopathogens. They are also factors that shape soil populations [62,63]. Mechanisms of antagonistic reactions between pathogens and cultivated plants can be observed in Table 1.1.

Indirect mechanisms include stimulation of plant growth and yield, use of microorganisms to create transgenic plants resistant to infection, and induction of plant resistance [39,54,62,68,69]. Preparations used for biological protection and promotion of plant

---

TABLE 1.1
Examples of the Protective Effect of Microorganisms Against Pathogenic Fungi Through the Direct Interaction of Protecting Microorganisms.

Antagonists and Their Mechanism of Action	Pathogen	Protected Plant
ANTIBIOTIC (ACTION THROUGH ANTIBIOTICS)		
Erwinia herbicola	*Fusarium culmorum;*	Wheat, soy, sunflower bean,
Actinoplanes	*Phytophthora megasperma*	poinsettia, white, sugar beet,
Micromonospora sp.	*Macrophomina phaseolina*	Snapdragon, pepper, onion, cotton
Streptomyces sp.	*Pythium:*	Potato
	P. aphanidermatum	
	P. debaryanum	
	P. ultimum	
	Rhizoctonia solani	
Bacillus sp.	*Sclerotium capivorum*	
ANTIBIOTIC, COMPETITION (ACTION THROUGH ANTIBIOTICS, SIDEROPHORES AND HCN)		
Pseudomonas sp.:	*Aphanomyces euteiches*	Peas, cloves, rye, wheat, cucumber,
P. fluorescens	*Fusarium oxysporum*	cotton, corn, sugar beet
P. putida	*Gaeumannomyces graminis*	
	P. aphanidermatum	
	P. ultimum	
PARASITISM/PREDATION (ACTION BY LYTIC ENZYMES)		
Serratia marcescens	*Fusarium oxysporum* ssp. *pisi*	Cotton, cucumber, rice, peas, turf on
Enterobacter cloacae	*P. ultimum*	golf courses, peas, radishes, tomatoes
Trichoderma hamatum	*Sclerotinia homecarpa*	
Trichoderma harzianum	*Pythium ultimum*	
OPERATION BY MYCOPARASITISM AND NUTRIENTS		
Pythium oligandrum	*Pythium ultimum*	Sugar beet
	Aphanomyces cochliodes	
OPERATION BY COMPETITION FOR AN ECOLOGICAL NICHE		
Verticillium biguttatum	*Rhizoctonia solani*	Potato

Sources: CD. Barratt, L.P. Lawson, G.B. Bittencourt-Silva, N. Doggart, T. Morgan-Brown, P. Nagel, & S.P. Loader, New, narrowly distracted and critically endangered species of the throat reed frog (Anura: Hyperoliidae) from the highly endangered coastal forest reserve in Tanzania. Herpetol. J. 17 (1) (2017) 13–24, A. Jamiołkowska, B. Skwaryło-Bednarz, E. Patkowska, Morphological identity and population structure of Hemibiotrophic fungus *Colletotrichum coccodes* colonizing pepper plants. Acta Sci. Pol. Hortorum Cultus, 17(4) (2018), 181–192, ISSN 1644-0692 e-ISSN 2545-1405 doi: 10.24326/asphc.2018.4.16; A. Jamiołkowska. The role of some secondary metabolites in the health status of sweet pepper (*Capsicum annuum* L.) grown in the field. Acta Sci. Pol. Hortorum Cultus, 13(2) (2014), 15–30; A. Jamiołkowska, A.H. Thanoon, Diversity and biotic activity of fungi colonizing pumpkin plants (*Cucurbita pepo* L.) grown in the field. EJPAU, Horticulture, 19, 4 (2016). Available at: http://www.ejpau.media.pl/volume19/issue4/art-11.html, G. Kshirsagar, A.N. Thakre, Plant disease detection in image processing using MATLAB. International Journal on Recent and Innovation Trends in Computing and Communication 6(4) (2018) 113-116; ISSN: 2321-8169, J.M. Whipps, M.P. McQuilken, Aspects of biocontrol of fungal plant pathogenes. In: Exploitation of Microorganisms, Gareth-Jones, ed. Champan & Hall, London (1993) 45–79.

growth and soil "sanitation" (e.g., *Trichoderma*—based measures) have procedural restrictions. In Annex I of the EEC Directive 91/41, there are only four strains: *Ampelomyces quisqualis, Coniothyrium minitans, Paecilomyces fumosoroseus,* and *Pseudomonas chlororaphis.* Despite over 100 registered biological preparations used in plant protection, of which only about 50 are directed to the protection of plants against fungal pathogens. There is still no success in the wide application of these formulations and the reproducibility of the results obtained. The fact that there are mechanisms of induction of plant resistance (largely analogous to the immunization of animals) creates great hopes for the development of an effective and reliable plant vaccine [64].

The augmentative method consists of periodically introducing beneficial micro- and/or macroorganisms from mass breeding into beneficial crops. These organisms are subject to a suitable formulation, used in the form of a biopreparation. In this strategy, the action of a biological agent should be immediate and its establishment in a new area is not expected. This is the most commonly used strategy in commercial crops (growing under cover, mushroom farms, orchards and some field and forest crops). The conservation method uses naturally occurring and specially introduced into agricultural and forest areas of landscape elements, enabling the development of a population of beneficial organisms that naturally occur in these environments. The main goal of these activities is to improve the quality of the environment by diversifying the landscape and, consequently, creating shades and hiding places, appropriate wintering places and securing a diverse food base for natural entomophagy's [70]. The main goal of pest control is to reduce their population to a level where further reduction is no longer beneficial (in some cases, it can be completely eradicated). This is called as the EIL. EIL is the density of the pest, at which the yield value exceeds the pest control costs to the greatest extent. If EIL is greater than zero, complete eradication is not beneficial. If the pest population stays below EIL, pest control is not needed, as the costs would far outweigh the benefits. However, if EIL reaches a level below the natural abundance of the species, it becomes a pest. The local total extermination of the pest by biological warfare is rarely used, as it also causes the loss of the regulatory factor [44,45,47,61]. In practical pest control, it is not so much EIL as the economic threshold called the control action threshold (CAT), that is, the density of the pest, at which actions should be taken to prevent the mass appearance of the pest (there is not one CAT, because it changes over time and depends on the population density of the natural enemies of the

pest). If social or aesthetic benefits are involved, an AET is an aesthetic injury level. In the case of pests carrying diseases, their eradication seems justified (does not consider the economic costs) [51,62,70]. The initiative of the International Organization of Biological Control (IOBC) gathered practitioners and scientists from many fields to define the limitations related to the implementation of biocontrol and to recommend mitigation measures. Limitations in the implementation of BC concern: risk aversion and not the best regulatory processes; bureaucratic barriers to access biological protection measures; lack or insufficient commitment and lack of communication with the society, breeders and politicians in matters related to the benefits of biocontrol; and finally, the fragmentation of biological subdisciplines. There is a need for better information about economic, environmental and social success, as well as the benefits of biological control of pests, weeds and plant diseases, addressed to political and social authorities, breeders, farmers, and other stakeholders. In the future, initiatives that promise biocontrol should be considered [45,49,51,52,55,70]. The ever-growing market for biological agents indicates great popularity of biological methods in plant protection worldwide (Table 1.2). This applies to plant crops whose direct consumption requires the highest safety standards (fruit, vegetables, mushrooms, baby food, etc.) [48,49,52,55].

In recent years, microbial preparations (effective microorganisms—EM) have also become more popular [45,48,49,73,75,77—80]. They are a mixture of naturally occurring microorganisms, mainly lactic acid bacteria (*Lactobacillus casei, Streptococcus lactis*), yeast (*Saccharomyces albus, Candida utilis*), Actinomycetes (*Streptomyces albus, S. griseus*), photosynthetic bacteria (*Rhodopseudomonas palustris, Rhodobacter spae*) and mold fungi (*Aspergillus oryzae, Mucor hiemalis*) [71,72,75,80—82]. Opinions on the impact of EM preparations on plant growth, their appearance and the quality of raw materials obtained from them are generally positive, which contributes to greater plant resistance to diseases and pests and inhibition of pathogen growth in the environment [80,82—85]. Boligłowa et al. [86] and Boligłowa [87] on EM and herbal extracts suggest that they can successfully replace fungicides in crop production and reduce or eliminate the use of plant protection products in the control of pests and plant diseases. However, in recent years, the safety of using certain biological agents has also been questioned, both in relation to human health and the environment [71,72]. Therefore, it is necessary to comply with the relevant provisions defining the principles of

TABLE 1.2
Selected Biological Plant Protection Preparations.

Biocontrol Organisms	The Name of the Biological Plant Protection Preparation
BACTERIA	
Agrobacterium rhizogenes	Galltrol A, Nogall, Diegall, Norbac 84
Agrobacterium tumefaciens	BioJet
Azospirillum	
Bacillus subtilis	Concentrate, HiStick NT, Kodiak, Kodiak HB, Kodiak at, Quantum 4000 HB, Subtilex, system 3
Bacillus subtilis FZB24	Rhizo-Plus, Rhizo-Plus Konz, Serenade
Bacillus subtilis GBO3	Companin, Cocentrate
Bacillus thuringiensis var. *aizawai*	Agree, Design, Math, Xin Tari
Bacillus thuringiensis var. *israelensis*	Aquabac, Bactimos, Bactis, BMP144, Teknar, VectoBac
Bacillus thuringiensis var. *kurstaki*	Agrobac, Bactec, Bactosis K, Bernan, Biobit HP, BMP 123, Condor, Cutlass, Dipel, Foil BFC, Foray, Foray 48B, Foray 68B, Forwabit, Furtura, Javelin WG, Larvo BT, MVP, MVP II, Rapor, Raven, Thuricide, Vault
Bacillus thuringiensis var. *tenebrionis*	Gnatrol, M-Trak, Norodor, Troy-BT
Bacillus popilliae	Doom, Japademic
Burkholderia cepacia	Denny (Blue Circle, Percept)
Heterorhabditis bacteriophora	Cruiser, Lawn Patrol, Otinem
Heterorhabditis spp.	Heterorhabditis Larvanem
Pseudomonas aureofaciens Tx-1	Spotless
Pseudomonas cepacian	Intercept
Pseudomonas fluorescens A506	BlightBan A506, BlightBan A506, Conquer
Pseudomonas fluorescens NCIB	Vivtus
Pseudomonas chlororaphis	Cedomon
Pseudomonas solanacearum	PSSOL
Pseudomonas sp.	BioJet
Pseudomonas syringae ESC-10	Bio-save 100, Bio-save 1000
Pseudomonas syringae ESC-11	Bio-save 110
Streptomyces griseoviridis	Mycostop
Streptomyces lydicus	Acinovate
FUNGI	
Ampelomyces quisqualis	AQ10 Biofungicide
Beauveria bassiana	Botanigard, Mycotrol, Naturalis L, Naturalis O, Naturalis T (turf)
Candida oleophila I-182	Aspire
Coniothyrium minitans	Contans, Koni
Fusarium oxysporum	Biofox C, Fusaclean
Gliocladium catenulatum	Primastop
Gliocladium spp.	Gliomix
Gliocladium	Soil Guard
Gliocladium virens GL-21	Soil Guard (GlioGard)
Metarhizium anisopliae	Bay Bio 1010, Bio Blast
Paecilomyces lilacinus	Paecil (Bioact)
Paecilomyces fumosoroseus	PFR97
Phlebia gigantea	Rotstop, P.G. Suspension

Pythium oligandrum *Phytophthora palmivora*	Polyversum (Polygandron) De vine
Talaromyces flavus	Protus WG
Trichoderma harzianum *Trichoderma polysporum*	Binab T, root Pro, Supresivit, Trichodex, Trichopel, Trichojest, Trichodowels, Trichoseal, T-22G, T-22HB, Planter-Box (Bio-Trek), Binab
Trichoderma spp.	Bio-Fungus (Antifungus), Promot, Trichoderma 2000
Trichoderma viride *Verticillium lecanii*	Eco SOM, Trichopel, Tricjoject, Trichodowles, Trichoseal, Trieco Vertalec
VIRUSES	
Autographa californica (NPV)	VFN80
Granulosis virus	Caapex, Cyd-X
Heliosis zea (NPV) *Mamestra brassicae (NPV)*	Gemstar LC Mamestrin
Spodoptera exigua *Syngrapha falcifera (NPV)*	Otienem-S, Spod-X Celery looper virus, Celery looper virus, Otienem-S
NEMATODES	
Steinernema carpocapsae \ *Steinernema feltiae*	Biovector, Bio-safe, Biovector, Ecomask, Guardian, Scanmask Entonem, Nemasys, Nemasys M

Source: J.M. Whipps, M.P. McQuilken, Aspects of biocontrol of fungal plant pathogenes, in: Exploitation of Microorganisms Gareth-Jones (Ed.), Champan & Hall, London (1993) 45−79, S. Martyniuk. Production of microbial preparations: symbiotic bacteria of legumes as an example. Journal of Research and Applications in Agricultural Engineering 55(4) (2010) 20−23, S. Martyniuk, J., Ksieżniak. Evaluation of pseudo-microbial bio-preparations used in crop production. Polish J. Agronomy 6 (2011) 27−33, M. Jurkowski, M. Błaszczyk. Physiology and biochemistry of lactic acid bacteria. Kosmos, Problems of Biological Sciences, 61(3) (2012) 493−504. (in Polish), J.C. Van Lenteren, Early entomology and the discovery of insect parasitoids. Biol. Control 32 (2005) 2-7. doi:10.1016/j.biocontrol.2004.08.003, I. Paśmionka, K. Kotarba, Possible application of effective microorganisms in environmental protection. Cosmos, Problems of Biological Sciences 64(1) (2015) 173−184. (in Polish), M.R. Finckh, S. Junge H.J. Schmidt, O.D. Weedon. Disease and pest management in organic farming: a case for applied agroecology, in: Improving Organic Crop Cultivation, Prof. U. Köpke (Ed.) Burleigh Dodds Science Publishing (2019) 1−59. (preprint, January 2019).

application, conducting a detailed impact assessment on human health and the environment to which they are released, and registration of biological agents/agents used in plant protection practice. In the European Union, such provisions have been defined for viruses, microbiological agents and substances of natural origin. The rules and decisions regarding the release of beneficial macroorganisms were left to individual member states, only some of whom managed to develop appropriate regulations [45,48,49,52,53].

A special place in international work on biological plant protection products is occupied by macroorganisms, which include parasitic and predatory insects, predatory mites and insecticide nematodes (Table 1.3).

Natural enemies of insect pests, also known as biological control agents, include predators, parasitoids, pathogens, and their competitors. Plant biological control agents are called as antagonists. In turn, the biological control measures of weeds include seed predators, herbivores, and plant pathogens [88,91−93]. These organisms have no impact on the health of consumers and therefore not considered as plant protection products [90]. The use of biological control to manage harmful insects is ahead of the current era of pesticides. The first successes of biological control occurred in the case of exotic pests controlled by natural enemies collected from the origin of the pest (classical control) [88,90−92,94,95]. Bale et al. [90] applied adhesive control against several pests in open spaces and in greenhouses and developed for this purpose biological protection schemes against local predators and parasitoids. They obtained a significant economic effect. It turned out that the cost-benefit ratio for classical biological control was very favorable (1: 250), and for augmentation control it was like that for insecticides (1:2−1:5), at much lower costs. According to van Lenteren [92] and Klapwijk and Koppert [93] many

TABLE 1.3
Biopreparations Based on Living Organisms (Macroorganisms) Available on the European Market for Use in Crops Under Cover.

Name and Characteristics of Organism	The Scope of Application
Steinernema feltiae parasitic nematode	*Diptera* from the family ground worms in vegetable and plant crops decorative under covers and in mushroom farms
Steinernema carpocapsae parasitic nematode	Caterpillars of butterflies in greenhouse cultivation
Phasmarhabditis hermaphrodita parasitic nematode	Snails in crops vegetables and decorative plants under covers
Heterorhabditis bacteriophora parasitic nematode	Beetle larvae (mainly gardeners, mycelium, and cockchafer May) in crops of vegetables
Atheta coriaria predatory ladybird	Ground and pupae thrips in crops of vegetables and decorative plants under covers
Cryptolaemus montrouzieri predatory ladybird	Mealybugs and June w plant crops decorative under covers
Adalia bipunctata predatory ladybird (colon)	Aphids in crops of vegetables and decorative plants under covers
Delphastus pusillus predatory ladybird (whitefly)	Whiteflies in the cultivation of vegetables and decorative plants under covers
Stethorus punctilium predatory ladybug	Spider mites in vegetable crops under covers
Aphelinus abdominalis predatory diphtheria	Aphids in crops of vegetables and decorative plants under covers
Episyrphus balteatus predatory diptera	Aphids in the cultivation of vegetables and ornamental plants under covers
Sphaerophoria rueppellii predatory diptera	Aphids, whiteflies, thrips, and spider mites in vegetable and plant crops decorative under covers
Feltiella acarisuga predatory diptera (minced pimples)	Spider mites in crops of vegetables and decorative plants under covers
Aphidoletes aphidimyza predatory dipter (aphids aphids)	Aphids in crops of vegetables and decorative plants under cover
O. laevigatus, O. majusculus, O. insidiosus, O. strigicollis predatory bugs	Thrips in vegetable and ornamental crops under covers
Macrolophus melanotoma or *pygmeus*, the predatory bug (whitefly perennial)	Whiteflies, spider mites, thrips in crops of vegetables and decorative plants under covers
Nesidiocoris tenuis predatory bugs	Whiteflies and *Tuta absoluta* in the cultivation of greenhouse tomato
Phytoseiulus persimilis predatory mite (benefactor greenhouse)	Spider mites in crops of vegetables and decorative plants under covers
H. miles, H. aculeifer predatory mites	Ground and pupae thrips in crops of vegetables and decorative plants under covers
Amblyseius californicus predatory mite	Spider mites, other mites herbivorous and thrips in crops of vegetables and decorative plants after shields
Neoseiulus fallacis predatory mite	Spider mites, other mites herbivores in crops of vegetables and decorative plants after shields
Amblyseius swirskii predatory mite	Whiteflies, spider mites, and thrips in crops of vegetables and decorative plants under covers
Neoseiulus degenerans predatory mite (benefactor brown)	Thrips in crops of vegetables and decorative plants under covers
Amblyseius cucumeris predatory mite	Thrips in crops of vegetables and decorative plants under covers
Amblyseius barkeri predatory mite	Thrips in crops of vegetables and decorative plants under cover

Amblyseius andersoni predatory mite	Spider mites, other mites herbivores in crops of vegetables and decorative plants under covers
Euseius gallicus predatory mite	Whiteflies and thrips in plant crops decorative under covers
Chrysopa carnea predatory networkers (golden eyed) common)	To combat the thrips, spider mites, and whiteflies
Encarsia Formosa parasitic wasp (greenhouse hygrometer)	Whiteflies in crops of vegetables and decorative plants under covers
Eretmocerus eremicus, *E. ervi* parasitic wasps	Whiteflies in crops of vegetables and decorative plants under covers
Encarsia formosa + *Eretmocerus eremicus* parasitic wasps	Whiteflies in crops of vegetables and decorative plants under covers
Eretmocerus mundus parasitic wasp	Whiteflies in crops of vegetables and decorative plants under covers
Aphidius colemani parasitic wasp (greenhouse aphid)	Aphids in crops of vegetables and decorative plants under covers
Aphidius matricariae parasitic wasp	Aphids in crops of vegetables and decorative plants under covers
Dacnusa sibirica + *Diglyphus* ISAE parasitic wasps	Miners in crops of vegetables and decorative plants under covers
Leptomastix dactylopii parasitic wasp (yellowish reddish)	Mealybugs and June w plant crops decorative under covers

Source: D.I. Shapiro-Ilan R., Gaugler, W.L., Tedders, I., Brown, E.E., Lewis, Optimization of inoculation for in vivo production of entomopathogenic nematodes. J. Nematol. 34(4) (2002) 343–350; M. Kozak-Cieżczyk. Molecular diagnostics in parasitology. Cosmos. Problems of Biological Sciences 54(1) (266) (2005) 49–60 (in polish), M. Pojmańska, M. Parasitism, Parasites and their hosts. Cosmos. Problems of Biological Sciences, 54 (1) (266) (2005) 5–20; S. Bale, J.C., van Lenteren, J.C., Bigler. Biological control and sustainable food production. Phil. Trans. R. Soc. B 363 (2008) 761–776, doi: 10.1098/rstb.2007.2182; J.C. Van Lenteren. The state of commercial augmentative biological control: plenty of natural enemies, but a frustrating lack of uptake. BioControl 57 (2012) 1-20, doi: 10.1007/s10526-011-9395-1; J.C. Van Lenteren, J.S., Bale, F., Bigler, H.M.T., Hokkanen, A.J.M., Loomans, Assessing risks of releasing exotic biological control agents of arthropod pests. Annu. Rev. Entomol. 51 (2006) 609–634, doi: 10.1146/annurev.ento. 51.110104.151374; M.R. Finckh, S. Junge H.J. Schmidt, O.D. Weedon. Disease and pest management in organic farming: a case for applied agroecology, in: Improving Organic Crop Cultivation, Prof. U. Köpke, Ed. Burleigh Dodds Science Publishing (2019) 1–59. (preprint, January 2019) https://www.researchgate.net/publication/330366201.

effective species of natural enemies have already been discovered and 230 are now available commercially. The plant protection industry has developed guidelines for quality control, mass production methods, dispatch, and release as well as appropriate guidelines for farmers. However, augmentative biological control is still used on a small area. However, due to the environmental safety and good of the user and the consumer, a significant number of EU member states have introduced their own regulations governing the import and use of macroorganisms. Freedom in this area means that the same factor and the measure containing it may be subject to completely different regulations in individual countries or be completely excluded from registration. This makes it difficult to market macroorganisms using time and economic barriers, or vice versa, it allows new species to be introduced into Europe, without enough knowledge of

their possible impact on the environment. At present, nine countries (Sweden, Norway, Denmark, Great Britain, Austria, Switzerland, the Czech Republic, Hungary, and Slovenia) have introduced and apply registration regulations for macroorganisms. In several countries (Finland, Germany, Spain, and the Netherlands), such provisions are currently under preparation. The basis for the regulation developed by the International Plant Protection Convention (FAO): there are guidelines for the export, dispatch, import, and issuance of Control Certificates (Agent Biocontrol) of other beneficial organisms—ISPM 3 (2005). These guidelines relate to risk management—the obligations and rights of the contracting parties. The regulation applies to protection organizations at the national level. Potential threats to human health are not considered to be significant, except for some allergies in the production of mites and nematodes

[40]. Nontarget effects include more than 5000 introductions of 2000 exotic arthropod species in 196 countries over the past 120 years, which have rarely resulted in negative environmental risks [90−92]. Despite the large-scale introduction of biological control of aphids and the establishment of key species, no cases have been documented that would show any negative effects except for Harmonia axyridis [93]. The environmental risk assessment (ERA) distinguishes three steps: exotic/native and augmentative/classic control. Key ERA characteristics are establishment, range, species dispersion, direct or indirect effects [90−93]. The registration requirements currently applied to biological control measures in the EU Member States are currently not satisfactory either for the biopesticide industry, scientists, or legislators. Very expensive and very long data collection procedures and final evaluation of already used and new biopreparations make their registration a very severe and disorienting experience for all these parties. This particularly applies to beneficial microorganisms and viruses that are subject to registration requirements like chemical pesticides. Parasitic and predatory arthropods and entomopathogenic nematodes, already widely used in plant protection, are not regulated by EU legislation, and national requirements applied to these macroorganisms differ in individual member states. A European initiative was launched to simplify and harmonize the requirements and procedures for the registration of biological control measures [53,90,93−95]. Therefore, it is worth using the experience of the previous Commission for Registration of Biotechnical and Biological Controls and Transgenic Plants (e.g., central register, participation of the expert group, minimum documentary requirements—larger for exotic species—new, "positive list" and short procedure) [52,93].

CONCLUSION

The management of pests involves the search for suitable control measures. This measure should not interfere with the activities of the natural ecosystems that should at the end assure final, favorable results. This chapter emphasizes on the use of effective biological control agents. It also presents supportive ideas of using BCA, such as contributing to a higher nutritional value of plants and the ability of some antagonists to break down mycotoxins. This applies to the possibility of using microorganisms, such as bacteria, yeasts, and filamentous fungi for plant protection. The next chapter presents comprehensive overview of diseases of agricultural importance and ecofriendly control measures.

REFERENCES

[1] M.P. McQuilken, S.P. Budge, J. Whipps, J, M, Movements of breeding media and environmental factors on conidia germination, pycnid production and extension of hypha Coniothyrium minitans, Mycol. Res. 101 (1997) 11−17.

[2] M.P. McQuilken, J. Gemmell, R.A. Hill, J.M. Whipps, Production of macrosphelide A by mycopar Coniothyrium minitans, FEMS Microbiol. Lett. 219 (2003) 27−31 (21).

[3] S.I.S. Melo, A. Moretini, A.M.R. Cassiolato, J.L. Faull, Development of mutants of Coniothyrium minitans with improved efficiency for control of Sclerotinia sclerotiorum, J. Plant Prot. Res. 51 (2011) 2, https://doi.org/10.2478/v10045-011-0031-y.

[4] A.B. Abass, G. Ndunguru, P. Mamiro, B. Alenkhe, N. Mlingi, M. Bekunda, Post-harvest food losses in a maize-based farming system of semi-arid savannah area of Tanzania, J. Stored Prod. Res. 57 (2014) 49−57, https://doi.org/10.1016/j.jspr.2013.12.004.

[5] S.S. Patol, Review on Beetles (Coleopteran): An Agricultural Major Crop Pests of the World, 2018.

[6] M. Kozłowska, G. Konieczny, Biology of Plant Resistance to Pathogens and Pests, Copyright Agricultural University, Poznań, 2003, p. 213 (in polish).

[7] F.A.O. Food, Agriculture Organization of the United Nations. Toolkit:, Reducing the Food Wastage Footprint, Rome, Italy, 2013.

[8] F.A.O. Food, Agriculture Organization of United Nations, Global Initiative on Food Losses and Waste Reduction, 2014. Rome, Italy.

[9] M. Rutten, M. Verma, N. Mhlanga, C. Bucatariu, FAO. Potential Impacts on Sub-saharan Africa of Reducing Food Loss and Waste in the European Union - A Focus on Food Prices and Price Transmission Effects, 2015. Rome.

[10] FAO, Food and Agriculture Organization of United Nations. Food Loss and Food Waste, 2016. http://www.fao.org/food-loss-and-food-waste/en/.

[11] FAO Save, Food: Global Initiative on Food Loss and Waste Reduction, Key Findings, 2017. http://www.fao.org/save-food/resources/keyfindings/en.

[12] H. Abraha, R. Email, A. Kahsay, Z. Gebreslassie, W. Leake, B.W. Gebremedhin, Assessment of production potential and post-harvest losses of fruits and vegetables in northern region of Ethiopia, Agriculture & Food Security, December 7 (9) (2018). https://link.springer.com/article/10.1186/s40066-018-0181-5.

[13] T. Fox, Global Food: Waste Not, Want Not, Institution of Mechanical Engineers, Westminster, London, UK, 2013.

[14] D. Grover, J. Singh, Post-harvest losses in wheat crop in Punjab: past and present, Agric. Econ. Res. Rev. 26 (2016) 293−297.

[15] K.J. Bradford, P. Dahala, J. Van Asbrouck, K. Kunusoth, P. Belloa, J. Thompson, F. Wu, The dry chain: reducing postharvest losses and improving food safety in humid climates, Trends Food Sci. Technol. 71 (2018) 84−93. https://doi.org/10.1016/j.tifs.2017.1.

[16] G.N. Agrios, Plant Pathology, 5 Edition, Copyright © by Elsevier Academic Press, Amsterdam, Boston, Heidelberg, London, New York, Oxford Paris, San Diego, San Francisco, Singapore, Sydney, Tokyo, 2005, p. 316. ISBN: 0-12-044565-4.

[17] R. Boxall, Damage and loss caused by the larger grain borer Prostephanus truncatus, Integr. Pest Manag. Rev. 7 (2002) 105—121, https://doi.org/10.1023/A:1026397115946.

[18] P.K. Bereś, H. Kucharczyk, D. Górski, Effects of insecticides used against the European corn borer on thrips abundance on maize, Plant Prot. Sci. 53 (1) (2017) 44—49, https://doi.org/10.17221/78/2016-PPS.

[19] P.J. Twardowski, P. Bereś, M. Hurej, Z. Klukowski, R. Warzecha, Effects of maize expressing the insecticidal protein Cry1Ab on non-target ground beetle assemblages (Coleoptera, Carabidae), Rom. Agric. Res. 34 (2017) 351—361.

[20] P.K. Bereś, D. Górski, H. Kucharczyk, Influence of seed treatments and foliar insecticides used against Oscinella frit in maize on the population of thrips, Acta Sci. Pol. Agricultura 16 (1) (2017) 3—15.

[21] J.C. Buzby, H. Farah-Wells, J. Hyman, The Estimated Amount, Value, and Calories of Postharvest Food Losses at the Retail and Consumer Levels in the United States, 2018. https://papers.ssrn.com/sol3/papers.cfm?abstract_id=2501659.

[22] K.E. Ognakossan, A.K. Tounou, Y. Lamboni, K.P. Hell, Post-harvest insect infestation in maize grain stored in woven polypropylene and in hermetic bags, Int. J. Trop. Insect Sci. 33 (2013) 71—81, https://doi.org/10.1017/S1742758412000458.

[23] J. Gliński, J. Horabikk, J. Lipiec, C. Sławiński, Agrophysics, in: J. Gliński, J. Horabik, J. Lipiec, C. Sławiński (Eds.), Processes, Properties, Methods, Institute of Agrophysics name Bohdan Dobrzański, Polish Academy of Sciences in Lublin, 2014, ISBN 978-83-89969-34-7, p. 135.

[24] J. Kaminski, L. Christiansen, Post-harvest loss in sub-Saharan Africa. What do farmers say? Glob, Food Security 3 (2014) 149—158, https://doi.org/10.1016/j.gfs.2014.10.002.

[25] H.C.J. Godfray, J.R. Beddington, I.R. Crete, L. Haddad, D. Lawrence, J.F. Muir, J. Pretty, S. Robinson, S.M. Thomas, C. Toulmin, Food security: the challenge of feeding 9 billion people, Science 327 (2010) 812—818.

[26] E.G. McPhersona, A.M. Berry, N.S. van Doornc, Performance testing to identify climate-ready trees, Urban For. Urban Green. 29 (2018) 28—39. https://doi.org/10.1016/j.ufug.2017.09.003.

[27] L. Kitinoja, S. Saran, S.K. Roy, A.A. Kader, Postharvest technology for developing countries: challenges and opportunities in research, outreach and advocacy, J. Sci. Food Agric. 91 (2011) 597—603, https://doi.org/10.1002/jsfa.4295.

[28] I. Baoua, L. Amadou, L.L. Murdock, Triple cowpea storage bags in rural Niger: questions that farmers ask, J. Stored Prod. Res. 52 (2013) 86—92.

[29] I.B. Baoua, L. Amadou, D. Baributsa, L.L. Murdock, Technology of a hermetic trash bag for conservation of Bambara peanuts after harvest (Vigna subterranea (L.) Verdc.), J. Stored Prod. Res. 58 (2014) 48—52.

[30] I.B. Baoua, L. Amadou, B. Ousmane, D. Baributsa, L.L. Murdock, PICS storage bags after harvesting maize grain in West Africa, J. Stored Prod. Res. 58 (2014) 20—28.

[31] D. Baributs, K. Jibo, J. Lowenberg-DeBoer, M. Bokar, I. Baoua, The fate of three-layer plastic bags used to store cowpea, J. Stored Prod. Res. 58 (2014) 97—102.

[32] K. Hell, C. Mutegi, P. Fandohan, Strategies for the control and prevention of aflatoxins in maize for Sub-Saharan Africa, in: 10th International Working Conference on Stored Product Protection, Nr. 425, 2010, https://doi.org/10.5073/jka.2010.425.388.

[33] L. Amadou, I.B. Baoua, D. Baributsa, S.B. Williams, L.L. Murdock, Triple hermetic technology for fighting bruch (Spermophagus sp.) (Coleoptera, Chrysomelidae) in stored Hibiscus sabdariffa, J. Stored Prod. Res. 69 (2016) 22—25, https://doi.org/10.1016/j.jspr.2016.05.004.

[34] D. Garcia, A.J. Ramos, V. Sanchis, S. Marín, Prediction of mycotoxins in food: a review, Microbiol. Food 26 (2009) 757—769.

[35] M. Singh, S.K. Biswas, D. Nagar, K. Lal, J. Singh, Impact of biofertilizer on growth parameters and yield of potato (2), Int. J. Curr. Microbiol. Appl. Sci. 6 (5) (2017) 1717—1724. ISSN: 2319-7706.

[36] J. Gustavsson, C. Cederberg, U. Sonesson. Global food Loss and food waste. SIK — Swedish Institute of Food and Biotechnology Food Congress, Düsseldorf May 16, (2011). Available from: https://www.researchgate.net/publication/267919405_Global_Food_Losses_and_Food_Waste [accessed Apr 19 2019].

[37] J. Aulakh, A. Regmi, J.R. Fulton, C. Alexander. Estimating post-harvest food losses: developing a consistent global estimation framework. Proceedings of the Agricultural & Applied Economics Association's AAEA & CAES Joint Annual Meeting; Washington, DC, USA, 4—6 August, (2013).

[38] E. Kannan, P. Kumar, K. Vishnu, H. Abraham, Assessment of Pre and Post Harvest Losses of Rice and Red Gram in Karnataka, Agricultural Development and Rural Transformation Centre, Institute for Social and Economic Change, Banglore, India, 2013.

[39] S. Grzesiuk, I. Koczowska, R. Górecki, The Physiological Foundations of Plant Resistance to Disease, Agricultural University of Technology, Olsztyn, 1999, p. 132.

[40] K. Dettner, W. Peters, Lehrbuch der Entomologie, Spektrum Akademischer Verlag, 2010.

[41] Z.Z. Htike, S.L. Win, Classification of eukaryotic splice-junction genetic sequences using averaged one-dependence estimators with subsumption resolution", Procedia Computer Science 23 (2013b) 36—43. ISSN: 1877-0509.

[42] M. Grzegorczyk, M.M. Grzegorczyk, A. Szalewicz, B. Żarowska, M. Wojtatowicz, Microorganisms in biological plant protection against muscle diseases, Acta Sci. Pol. Biotechnol. 14 (2) (2015) 19–42 (in polish).

[43] H.L. Yang, W.K.J. Xing, X. Qiao, L. Wang, Z.S. Gao, Research on insect identification based on pattern recognition technology, in: 6th International Conference on Natural Computation, 2010, 08.09.

[44] A. Cimmino, A. Andolfi, A. Berestetskiy, A. Evidente, Production of phytotoxins by *Phoma exigua* var. exigua, a potential mycoherbicide against perennial thistles, J. Agric. Food Chem. 56 (2008) 6304–6309.

[45] B. Sawicka, P. Pszczółkowski, A.H. Noaema, Nanotechnology in agriculture and food processing, in: D. Łuczycka (Ed.), Agriculture of the 21st Century – Problems and Challenges. Processing of Food Raw Materials, Publisher: Idea Knowledge Future, Wroclaw, 2018a, ISBN 978-83-945311-9-5, pp. 582–599 (in polish).

[46] A. Sharma, V.D. Diwevidi, S. Singh, K.K. Pawar, M. Jerman, L.B. Singh, S. Singh, D. Srivastawa, Biological control and its important in agriculture, Int. J. Biotechnol. Bioeng. Res. 4 (3) (2013) 175–180. ISSN: 2231-1238, http://www.ripublication.com/ijbbr.html.

[47] B. Sawicka, P. Barbaś, The rate of spread of Alternaria spp. on the potato plant in an ecological and integrated system of cultivation, Post. w Ochronie Roślin/Prog. Plant Prot. 51 (3) (2011) 1–5.

[48] P. Pszczółkowski, B. Sawicka, The effect of using fungicides, microbiological preparations and herbal extracts on shaping of potato yield, Fragmenta Agronomica 35 (1) (2018b) 81–93 (in Polish).

[49] P. Pszczółkowski, B. Sawicka, The effect of application of biopreparations and fungicides on the yield and selected parameters of seed value of seed potatoes, Acta Agrophysica 25 (2) (2018c) 239–255, https://doi.org/10.31545/aagr.

[50] T.S. Hameed, B. Sawicka. Level of practical use of recommendations to protect against potato blight (Phytopthora infestans Mont. de Bary) in the region of south-eastern Poland. International Conference: Bioeconomy in Agriculture. Puławy, 21–22 June (2016), IUNG-PIB Puławy, ISBN: 978-83-7562-219-5, 33-34.

[51] G. Lazarovits, M.S. Goettel and C. Vincent, (Eds.), 2007. Adventures in Biocontrol. In Biological control: a global perspective. Case stories from around the world. CABI Publishing, Wallingford, United Kingdom, 440, 1–6

[52] A. Pacholczak, P. Petelewicz, K. Jagiełło-Kubiec, A. Ilczuk, The effect of two biopreparations on rhizogenesis in stem cuttings of *Cotinus coggygria* Scop, Eur. J. Hortic. Sci. 80 (4) (2015) 183–189. ISSN: 1611-4426, https://doi.org/10.17660/eJHS.2015/80.4.6.

[53] M. Tomalak, Registration of biological control agents of plant protection in Europe – new perspectives, Prog. Plant Prot./Postepy w Ochronie Roślin 47 (4) (2007) 233–240.

[54] S. Kryczyński, Z. Weber, Phytopathology Vol. I, II, Państwowe Wydawnictwo Rolnicze i Leśne, Poznań, 2011, ISBN 978-83-09-01077-7 (in Polish).

[55] B. Sawicka, P., Sawicki, P., Pszczółkowski, Attempt to predict outbreak of potato late blight (Phytophthora infestans) basing on meteorological data. (in:) Abstracts of Papers & Posters. 16th Triennial Conf. Of the EAPR, 17–22 July (2005), Bilbao, Basque Country. (ed. E. Ritter – A. Carrascal) Vitoria-Gasteiz, Part II: 828-833.

[56] A. Borcean, I. Imbrea, Stem rot on wild peppermint species on South-western part of Romania, Res. J. Agric. Sci. 49 (1) (2017) 75–80.

[57] B. Sawicka, Studies on the Variability of Selected Traits and Degeneration of Potato Varieties in the Biala-Podlaska Region, in: Series Publishing - Scientific Dissertations, vol. 141, Publisher of the Agricultural University of Lublin, 1991, p. 76. ISSN: 0860-43-55 (in: polish).

[58] B. Sawicka, Infection of Streptomyces scabies tubers with 37 potato varieties in varying conditions of acidity and chemical composition of the soil, Zesz. Prob. Post. Nauk. Roln. 456 (1998) 591–597 (in polish).

[59] B. Sawicka, Effects of growth regulators, Mival and Poteitin, application in potato cultivation. Part II. The influence of growth regulators on incidence of Rhizoctonia solani sclerotia bearing tubers, Ann. Agric. Sci., Ser. E 28 (1/2) (1999) 55–65.

[60] B. Sawicka. Dates of potato late blight appearance and spread (Phytophthora infestans (Mont.) De Bary) in changing of atmospheric and soloic conditions. Conf. EAPR, Hamburg, 14–19.07 (2002) 103.

[61] B. Sawicka, J. Kapsa J, Effect of varietal resistance and chemical protection on the potato late blight (Phytophthora infestans (Mont.) de Bary) development (in:) Report of the meeting of the pathology section of the EAPR, 10–14 July 2001, Poznan (Poland), Potato Res. 44 (3) (2001) 287–307, https://doi.org/10.1007/BF02357906 (2001).

[62] J.M. Whipps, M.P. McQuilken, Aspects of biocontrol of fungal plant pathogenes, in: Gareth-Jones (Ed.), Exploitation of Microorganisms, Champan & Hall, London, 1993, pp. 45–79.

[63] G. Kshirsagar, A.N. Thakre, Plant disease detection in image processing using MATLAB, Int. J. Recent Innovation Trends Comput. Commun. 6 (4) (2018) 113–116. ISSN: 2321-8169.

[64] C.D. Barratt, L.P. Lawson, G.B. Bittencourt-Silva, N. Doggart, T. Morgan-Brown, P. Nagel, S.P. Loader, New, narrowly distracted and critically endangered species of the throat reen frog (Anura: hyperoliidae) from the highly endangered coastal forest reserve in Tanzania, Herpetol. J. 17 (1) (2017) 13–24.

[65] A. Jamiołkowska, B. Skwaryło-Bednarz, E. Patkowska, Morphological identity and population structure of Hembiotrophic fungus *Colletotrichum coccodes* colonizing pepper plants, Acta Sci. Pol. Hortorum Cultus 17 (4) (2018) 181–192, https://doi.org/10.24326/asphc.2018.4.16. ISSN 1644-0692 e-ISSN 2545-1405.

[66] A. Jamiołkowska, The role of some secondary metabolites in the health status of sweet pepper (*Capsicum annuum* L.) grown in the field, Acta Sci. Pol. Hortorum Cultus 13 (2) (2014) 15–30.

[67] A. Jamiołkowska, A.H. Thanoon, Diversity and biotic activity of fungi colonizing pumpkin plants (*Cucurbita pepo* L.) grown in the field, EJPAU Horticulture 19 (2016) 4. Available at: http://www.ejpau.media.pl/volume19/issue4/art-11.html.

[68] M. Kozak-Cieżczyk, Molecular diagnostics in parasitology (266), Cosmos Problems of Biological Sciences 54 (1) (2005) 49–60 (in polish).

[69] T. Rumpf, A.K. Mahlein, U. Steiner, E.C. Oerke, H.W. Dehne, L. Plumer, Early detection and classification of plant diseases with Support Vector Machines based on hyperspectral reflectance, Comput. Electron. Agric. 74 (1) (2010) 91–99, https://doi.org/10.1016/j.compag.2010.06.009. ISSN 0168-1699.

[70] M. Grzegorczyk, A. Szalewicz, B. Żarowska, X. Połomska, W. Wątorek, M. Wojtatowicz, Drobnoustroje w biologicznej ochronie roślin przed chorobami grzybowymi, Acta Sci. Pol., Biotechnologia 14 (2) (2015) 19–42.

[71] S. Martyniuk, Production of microbial preparations: symbiotic bacteria of legumes as an example, J. Res. Appl. Agric. Eng. 55 (4) (2010) 20–23.

[72] S. Martyniuk, J. Ksieżniak, Evaluation of pseudo-microbial biopreparations used in crop production, Polish J. Agronomy 6 (2011) 27–33.

[73] M. Jurkowski, M. Błaszczyk, Physiology and biochemistry of lactic acid bacteria, Kosmos, Problems Biol. Sci. 61 (3) (2012) 493–504 (in Polish).

[74] J.C. Van Lenteren, Early entomology and the discovery of insect parasitoids, Biol. Control 32 (2005) 2–7, https://doi.org/10.1016/j.biocontrol.2004.08.003.

[75] I. Paśmionka, K. Kotarba, Possible application of effective microorganisms in environmental protection, Cosmos, Problems of Biol. Sci. 64 (1) (2015) 173–184 (in Polish).

[76] M.R. Finckh, S. Junge, H.J. Schmidt, O.D. Weedon, Disease and pest management in organic farming: a case for applied agroecology, in: U. Köpke (Ed.), Improving Organic Crop Cultivation, Burleigh Dodds Science Publishing, 2019, pp. 1–59 (preprint, January 2019), https://www.researchgate.net/publication/330366201.

[77] M. Kołodziejczyk, Effectiveness of nitrogen fertilization and application of microbial preparations in potato cultivation, Turk. J. Agric. For. 38 (2014a) 299–310, https://doi.org/10.3906/tar-1305-105.

[78] M. Kołodziejczyk, Effect of nitrogen fertilization and microbial preparations on potato yielding, Plant Soil Environ. 60 (8) (2014b) 379–386, https://doi.org/10.17221/7565-PSE.

[79] Z. Zydlik, P. Zydlik, The effect of microbiological products on soil properties in the conditions of replant disease, Zemdirbyste-Agricul. 100 (1) (2013) 19–24, https://doi.org/10.13080/z-a.2013.100.003.

[80] T. Higa, Effective Microorganisms, concept and recent advances in technology, in: Proceedings of the Conference on Effective Microorganisms for a Sustainable Agriculture and Environment. 4th International Conference on Kyusei Nature Farming, Bellingham-Washington USA, 1998, pp. 247–248.

[81] P.J. Valarini, M.C.D. Alvarez, J.M. Gasco, F. Guerrero, H. Tokeshi, Assessment of soil properties by organic matter and EM – microorganism's incorporation, R. Bras. Ci. Solo 27 (2003) 519–525, https://doi.org/10.1590/S0100-06832003000300013.

[82] D.P.C. Stewart, M.J. Daly, Influence of "effective microorganisms" (EM) on vegetable production and carbon mineralization – a preliminary investigation, J. Sustain. Agric. 14 (2) (1999) 3–15.

[83] Y. Hoshino, N. Satou, T. Higa, Remediation and preservation of natural ecosystems through application of effective microorganisms (EM), in: Proceedings of 7th International Conference on Kyusei Natural Farming. New Zealand, 2002, pp. 129–137.

[84] Z. Kaczmarek, A. Wolna-Murawka, M. Jakubs, Changes of the number of selected microorganism groups and enzymatic activity in the soil inoculated with effective microorganism (EM), J. Res. Appl. Agric. Eng/ 53 (3) (2008) 122–127.

[85] R. Janas, Possibilities of using effective microorganisms in ecological crop production systems, Problems of Agricultural Engineering 3 (2009) 111–119.

[86] E. Boligłowa, Protecting potato against diseases and pests using Effective Microorganisms (EM) with herbs, in: Z. Zbytek (Ed.), Selected Ecological Issues in Modern Agriculture, PIMR Publishing House, Poznań, 2005, pp. 165–170 (in Polish).

[87] E. Boligłowa, K. Gleń, P. Pisulewski, The effect of herbicides on the yield and some quality characteristics of potato tubers, Zesz. Probl. Postepow Nauk. Rol. 500 (2004) 391–397 (in Polish).

[88] D.I. Shapiro-Ilan, R. Gaugler, W.L. Tedders, I. Brown, E.E. Lewis, Optimization of inoculation for in vivo production of entomopathogenic nematodes, J. Nematol. 34 (4) (2002) 343–350.

[89] M. Pojmańska, M. Parasitism, Parasites and their hosts, Cosmos Problems Biol. Sci. 54 (1) (2005) 5–20 (266).

[90] S. Bale, J.C. van Lenteren, J.C. Bigler, Biological control and sustainable food production, Phil. Trans. R. Soc. B 363 (2008) 761–776, https://doi.org/10.1098/rstb.2007.2182.

[91] J.C. Van Lenteren, The state of commercial augmentative biological control: plenty of natural enemies, but a frustrating lack of uptake, BioControl 57 (2012) 1–20, https://doi.org/10.1007/s10526-011-9395-1.

[92] J.C. Van Lenteren, J.S. Bale, F. Bigler, H.M.T. Hokkanen, A.J.M. Loomans, Assessing risks of releasing exotic biological control agents of arthropod pests, Annu. Rev. Entomol. 51 (2006) 609–634, https://doi.org/10.1146/annurev.ento. 51.110104.151129.

[93] J. Klapwijk, B.V. Koppert, International Regulation of Invertebrate Biological Control Agents, International Biocontrol Manufacturers Association (IBMA). Biocontrol Workshop, Cairo, January, 2018, p. 16.

[94] EPPO, Safe use of biological control: import and release of biological control agents. EPPO Standard PM6/2(1), EPPO Bull. 31 (2001) 33–35.

[95] EPPO, List of biological control agents widely used in the EPPO region. EPPO Standard PM6/3(2), EPPO Bull. 32 (2002) 447–461, https://doi.org/10.1046/j.1365-2338.2002.00600.x.

FURTHER READING

[1] S. Hussein, Molecular Plant Phatology, in: A. Abbas (Ed.), 2008. https://www.researchgate.net/publication/308642520.

[2] C. Szewczuk, D. Sugier, S. Baran, E.J. Bielińska, M. Gruszczyk, The impact of fertilizing agents and different doses of fertilizers on selected soil chemical properties as well as the yield and quality traits of potato tubers, Ann. UMCS, Agric. 71 (2) (2016) 65–79 (in Polish).

CHAPTER 2

Plant Diseases, Pathogens and Diagnosis

BARBARA SAWICKA, PHD • CHUKWUEBUKA EGBUNA, BSC, MSC •
ASHISH KUMAR NAYAK, PHD • SMRITI KALA, PHD

INTRODUCTION

Plant diseases and pests attacks are key issues in agriculture, because they can cause a significant reduction in both the quantity and quality of agricultural products [1−6]. Quantitative losses are more common in developing countries [5,7] than in developed countries. At the global level, the amount of lost and wasted agricultural produce in high income regions is higher in the further phases of the food chain, and vice versa as in low-income regions where more food is lost and wasted in the start-up phase [5]. In turn, the quality losses also include those that affect the composition of nutrition, caloric, digestibility, or the acceptability of agricultural produce. These losses are generally more common in developed countries [8,9]. It was estimated that in cereals, root crops, fruits, and vegetables, these losses account for respectively: approx. 19%; 20%, and 44% of losses [5,9,10]. Root plants and vegetables are particularly exposed to these losses [6,11−13]. It is estimated that in Georgia (USA), there are over 2046 plant diseases which causes losses, amounting to approximately USD $1,040 million in loss. Of this amount, about US $ 185 million was spent on combating leaf diseases, and the rest is the damage caused by other diseases [6]. Due to the calorific value of products, cereals (53%), such as wheat, rice, and maize, have the largest share in losses [6,14]. According to the World Bank's report, only Sub-Saharan Africa is losing grains worth more than USD 4 billion a year. In India, losses due to crop damage by diseases and pests, inadequate storage of yield reach 47% [14−17]. These losses play a key role for small agricultural producers [14,18,19]. Therefore, the purpose of this chapter is to attempt to classify diseases, pests, and pathogens, and assess their harmfulness and impact on yield losses of the most important crop species of high economic importance for feeding the population, as well as an attempt to diagnose them earlier.

Plant Diseases

All manifestations of plant life, for example, growth, maturation, and fruiting, are a result of physiological processes taking place inside the plant in a harmonious manner, in a specific order, and in closely consecutive time sequences. These processes therefore represent the internal environment of the plant, which is influenced by external factors. If the plant's organism can adapt its processes to influences from the external environment, it remains healthy. However, for example, failure to supply the plant with water for 3−5 days results in the loss of turgor and disturbance of physiological balance. After this time, plant water supply usually leads to its rapid return to normal. Physiological imbalance due to temporary loss of water is not a disease, but if you leave the plant without access to water for the next few days, changes in plant physiology will become irreversible. Therefore, considering the time and intensity of the external factor's impact on the plant, the disease can be defined as follows: the disease is the effect of strong disturbance of physiological processes of the plant [20−22].

Classification scheme for plant diseases

Plant diseases can be classified in many ways (Table 2.1). For instance, they can be divided into noninfectious and infectious diseases. Plant infectious diseases are caused by bacteria, viruses, and viroids, while plant noninfectious diseases are mainly caused by abiotic factors.

As described in Table 2.1, there are several criteria that may be used as basis for plant disease classification. For instance, plant diseases are classified based on the following:

Symptoms they cause:
  i. Root rots.
  ii. Cankers.

TABLE: 2.1
Classification of Plant Diseases.

S/No	Class	Category	Diseases/Organisms/Hosts
1.	Type of infection	Localized diseases Systematic diseases	Leaf spot Downy mildew
2.	Part of host affected	Foliage diseases Stem diseases Root diseases	Blight of rice Stem rot Root rot
3.	Kind of symptoms produced	Root rots Fruit rots Powdery mildew Downy mildew Leaf spot Rust Leaf blight Mosaic Wilt Damping off Smut Canker	Bean Tomato Grapes, cucumber Rose, llettuce Corn Wheat Rice Tobacco Cucurbit, potato Tobacco, tomato Wheat, bbarley Citrus
4.	Multiplication of inoculum	Simple interest disease Compound interest disease	
5.	Type of perpetuation and spread	Soilborne disease Seedborne disease Airborne disease	Root rot, wilt Blight, rust Damping off
6.	Extent of occurrence and geographic distribution	Endemic diseases Epidemic diseases Sporadic diseases Pandemic diseases	Wart disease of potato Rust, mildews Leaf blight, wilt Late blight of potato
7.	Based on causal agent	Biotic Abiotic Mesobiotic	Fungi, protozoa, algae, infects, nematodes, phytoplasma, spiroplasma Deficiency/excess of nutrients Low/extreme light Unfavorable oxygen supply and soil moisture Atmospheric impurities Air pollutants Viruses Viroids
8.	Based on pathogen generation	Monocyclic disease Polycyclic disease	Loose smut of wheat Late blight of potato

iiiv. Wilts'
 iv. Leaf spots.
  v. Scabs.
 vi. Blights.
vii. Anthracnoses.
viii. Rusts.
 ix. Smuts.

Plant organ affected:

i. **Root diseases:** Root rots (*Verticillium, Colletotrichum, Macrophomina*), club root disease of crucifer, vascular wilts (*Fusarium* spp., *Verticillium*), and other root diseases caused by *Gaeumannomyces graminis, Phymatotrichum omnivorum*, and *Armillaria mellea*.

ii. **Stem diseases:** Stem rot of jute, Black stem rust of wheat.

iii. **Foliage diseases:** Leaf spot of turmeric, leaf blight of wheat, leaf curl of peach.

iv. **Fruit diseases:** Crown gall of stone fruits, citrus canker.

v. **Seedling diseases:** Seedling blight (*Pythium, Phytophthora, Fusarium, Corticium*).

Extent to which plant disease is associated with plant:

(i) **Localized disease:** Affects only a part of the plant.

(ii) **Systemic disease:** Affects the entire plant.

Host plants affected:

i. Vegetable diseases.

ii. Cereal diseases.

iii. Ornamental diseases.

iv. Forest diseases.

v. Fruit tree diseases.

Mode of natural perpetuation and mode of infection:

(i) **Airborne:** The microorganisms are spread through the air and attack the plants causing diseases, for example, blight, rust, powdery mildew.

(ii) **Soilborne:** Inoculums of the diseases causing pathogen remains in the soil and penetrate the plant resulting in diseased condition, for example, root rot, wilt.

(iii) **Seedborne:** The microorganisms are carried along with seeds and cause diseases when the congenial condition occurs, for example, damping off.

Occurrence and distribution of plant disease geographically:

(i) **Endemic:** When a disease is more or less constantly prevalent from year to year in a moderate-to-severe form in a particular country, for example, Wart disease of potato is endemic to Darjeeling.

(ii) **Epidemic or epiphytotic:** A disease occurring periodically but in a severe form involving a major area of the crop. It may be constantly present in the locality but assume severe form occasionally, for example, rust, late blight, mildews.

(iii) **Sporadic:** Diseases that occur at a very irregular interval and location in a moderate-to-severe form, for example, leaf blights, wilt.

(iv) **Pandemic:** Diseases occurring throughout the continent or subcontinent resulting in mass mortality, for example, late blight of potato.

However, the most commonly used criterion in plant disease classification is the type of pathogen that causes the disease. On the basis of causal factors, diseases are classified into two groups:

1. **Nonparasitic disease:** The causal factors of these are mainly physiological or environmental like freezing injury caused by low temperature, high temperature, unfavorable oxygen or soil moisture, mineral deficiency or excess mineral, etc. Example: Red leaf of cotton and Kharia disease of rice is caused due to mineral deficiency, Black heart of potato is caused due to high temperature, Bark necrosis of red delicious apple is caused due to excess mineral. In summary, diseases can be classified based on the following factors:

   (a) Nutrient deficiencies.

   (b) Mineral toxicities.

   (c) Lack or excess of soil moisture.

   (d) Too low or too high temperature.

   (e) Air pollution.

   (f) Lack of oxygen.

   (g) Lack or excess of light.

   (h) Soil acidity or alkalinity (pH).

2. **Parasitic disease:** The causal factors of the disease are parasitic micro- or macroorganisms that need a host plant to survive or to complete the life cycle. Various fungi, bacteria, viruses, mycoplasma, algae, and animal parasites such as nematodes parasitize and cause disease in host plants. Example: Club root of crucifer caused by mycoplasma, Bacterial blight of paddy, smut, and rusts caused by fungi, Tobacco mosaic caused by a virus, ergot etc. On this basis, plant diseases are classified as follows:

   (a) Diseases caused by fungi

   (b) Diseases caused by bacteria

   (c) Diseases caused by parasitic higher plants

   (d) Diseases caused by viruses

   (e) Diseases caused by nematodes

Pathogenic Agents

A pathogen is a parasite that adversely affects its host to such an extent that it causes its illness or death. Not all parasites are pathogens and not all pathogens are parasites (an example of an antiparasitic pathogen is *Clostridium botulinum* that produces exotoxin—neurotoxin—inhibits the release of acetylcholine and, consequently, botulism—muscle paralysis—which can cause death) [23,24].

Among pathogens, the following can be distinguished: noninfectious and infectious. In the case of noninfectious agents, the disease does not spread; therefore, there are no new plants with clear symptoms. Noninfectious agents include too low air humidity, low air temperature and high temperature, lack or insufficiency of light, environmental pollution, soil reaction, genetic defects, plant protection chemicals, acid rain, hurricane, wind, hail, and tsunamis. Noninfectious plant diseases, however, most often cause lack or excess of nutrients [25–27].

Infectious factors are those that infect after physical contact with the plant. Then, they grow, grow, and reproduce on it or in it. Infectious agents are called **pathogens**. Infection factors include viruses and viroids, phytoplasmas, bacteria, fungi, parasitic seed plants, chromista, and protozoa [28].

Classification of pathogens

In terms of parasitism, pathogens can be divided into the following:

1. Obligatory parasites (absolute, strict)—they develop and reproduce only in living cells of plants, for example, viroids or fungi from the order *Erysiphales*.
2. Facultative parasites (relative, facultative), which differ from the absolute parasites, that in certain circumstances they can develop on dead organic matter, for example, Ascomycota (*Ascomycota*) mushrooms [27].
3. Occasional parasites (relative, optional), that is, pathogens of weakness. They attack plants that are weakened or damaged [29].

Concepts of parasitism and pathogenicity

Parasitism and pathogenicity are not unambiguous concepts. A parasite is an organism that lives on a living plant and receives nutrients from it, without providing anything to the plant in exchange. Dependencies between the parasite and its host are called parasitism. Parasitism is a kind of intimidation. Among the parasites, there are pathogens, nonpathogenic parasites, symbiotes [29,30]. The pathogen is the body, often a microorganism that causes disease. However, the main goal of the pathogen is to get food. It causes disease in plants to weaken their defenses. For most diseases, the amount of damage to plants caused by the pathogen is usually greater than the losses resulting from the absorption of nutrients by them. Additional losses result from, for example, introducing toxins to plants, damaging cells and tissues that interfere with transpiration, causing wilting, abnormal cell division, etc. The parasite's pathogenicity is usually not proportional to its nutrient needs. For example, local damage to the base of the stems usually causes the plants to freeze. Nonpathogenic parasites use the nutrients of the living plant, but they do not cause disease. The nonpathogenic parasite is, for example, *Phialophora graminicola*, a fungus inhabiting the roots of many grass species [28].

Symbiotes, on the one hand, use the plant's food; on the other hand, they either directly supply it with nutrients or facilitate the plant's collection [28,31].

Parasitic affinity, aggressiveness, pathogenicity:

Parasitic affinity is the mutual tendency of the parasite and the host. The pathogenic agent then has a tropism against the host and enzymatic adaptation that allows tissue infection and colonization. Lack of affinity prevents infection or colonization. Pathogenicity is a genetically determined ability of a parasite to interfere with one or more of the physiological functions of a plant. Parasitism plays a significant but often not very significant role in this process [29,30].

Aggressiveness is the ability to infect, colonize, and use plant nutrients. The degree of aggressiveness depends on the characteristics of the pathogen and host and on external factors. The measure of aggressiveness is (a) the infection threshold, that is, the smallest number of infectious agents of the pathogen (e.g., fungal spores): needed to infect the plant; (b) duration of infection; (c) the length of the incubation period of the disease, that is, the time from the infection to the first symptoms manifested. Pathogenicity is the parasite's ability to cause disease. Virulence is the measure of pathogenicity [29,30]. Virulence is a constant, genetic ability of a pathogen to cause disease. However, the vortex can sometimes change, for example, after prolonged breeding on artificial substrates [28].

Mandatory parasites, facultative parasites, facultative saprophytes, half-parasites. The obligatory parasites are represented, for example, by the fungi of the order *Uredinales*, fungi of the order *Erysiphales*, most of the representatives of the *Peronosporaceae* family, and fungal organisms from the order *Plasmodiophorales* (e.g., *Plasmodiophora brassicae*, the perpetrator of *Cruciferous syphilis*). Parasitic parasites develop only on living plants. They infect plants through undamaged skin, natural holes, or less often through wounds. Most parasites from this group cannot be grown on artificial substrates. They usually infect plants in a good condition, because they are the source of more nutrients of higher quality [29].

Optional parasites live commonly as saprotrophs, but under certain conditions they eat parasitically. This group includes most fungi and fungus-causing organisms that cause diseases, among others: *Botrytis cinerea* and *Pythium* spp. Optional parasites infect plants mainly through wounds, sometimes through undamaged peels. Symptoms often are accompanied by various necrosis. These parasites are easier to paralyze weakened plants. Organisms from this group grow well on artificial substrates [29].

Hosts. There are several categories among the hosts. There are two main categories associated with parasitic ontogenesis: final hosts and intermediate hosts. The final host is the organism in which the sexual parasite propagates. In the case of Metazoan, this is the host of the adult figure, and in the case of Protozoa—the host, in which the parasite produces male and female gametes, joining the zygote. However, in the case of digenetic flaws, where in ontogenesis occurs at least in two generations of parthenogenetic and one sexual generation, the final host is the last-generation host of this generation, also known as the adult figure. The intermediate host is the organism in which Metazoan larvae parasitize, or those protozoan generations that reproduce asexually. The exception are larvae of few tapeworms (including the genus *Echinococcus* and some *Hymenolepidiidae*), whose larvae produce in the intermediate host numerous so-called. "Heads" tapeworms, and each of the "heads" may in the final host develop into an adult individual. In ontogenesis of digenetic digestion, intermediate hosts are mollusks in which the parasite produces at least two generations of parthenogenetic, including a larva, from which the adult will grow in the final host. Many parasitic trematodes produce two larval forms in the developmental cycle, and the development of each of them is associated with a change in the host. In this case, the first and second intermediate hosts are distinguished [29,30].

Half-parasites are photosynthetic-able organisms, but they take up water and mineral salts from their host plant. A parasite is, for example, mistletoe (*Viscum album*), which is particularly often parasitized on poplar [28].

The scope of parasitism, parasitic specialization. Depending on the number of hosts, we divide pathogens into monophagous, polyphagia, omni phages (omnivorous). Monophagous exist among all pathogen groups. Monofagic are, for example, numerous perpetrators of rust, powdery mildew, and *Venturia inaequalis*. Polyphagias are, for example, *Botrytis cinerea*, *Pythium* spp., *Fusarium* spp., and *Verticillium* spp. and *Rhizoctonia solani* [32]. Most monophagous are obligatory parasites. Among the polyphagous, occasional parasites dominate. The group of omni phages is represented by, for example, *Botrytis cinerea*, the perpetrator of gray molds of various plant species. Specialization may also apply to the organ of the plant, for example, roots, leaves, and shoots. The proper form of specialization is the occurrence of special forms and physiological races within the same species. The special form is the result of the adaptation of the pathogen to the host plant species, for example, *Fusarium oxysporum* f. sp. *Dianthi*

parasitizes on *Dianthus caryophyllus* (garden carnation). The breed is a form of adaptation of the pathogen to the variety or species of the plant. Breeds are marked with numbers, for example, *Fusarium oxysporum* f. sp. *Dianthi* rasa 1—it can infect only, for example, the A variety of carnation, and *Fusarium oxysporum* f. sp. *Dianthi* rasa 2—only variety B of this species. Special forms usually do not differ in morphological features or these differences are very small. The physiological race does not show such differences; breed is differentiated based on pathogenicity [29,33,34].

Life cycle of the pathogen versus disease cycle. The life cycle is the successive stages of parasitic or parasitic and saprophytic development interrupted by periods of rest, for example, due to winter. During both types of development, reproduction (reproduction) and dissemination often occur. All parts, fragments, or structures of the organism from which a new individual may develop, for example, a fungal or bacterial colony, are propagation units [29,30,33,34]. In summary, a **disease cycle** is a sequence of events during which the disease begins, develops, and maintains. The disease cycle sometimes coincides with the pathogen life cycle, but not always. The disease cycle is distinguished by infection, inoculation, and specific disease (symptomizing). The infection is preceded by inoculation [28,31].

Infection. An infection is a process in which a pathogen contacts a sensitive cell or tissue and uses nutrients from it. During the infection, the pathogen grows and multiplies, penetrates, and colonizes the plant to a greater or lesser extent. Thus, the penetration of a pathogen into plant tissues and its growth as well as reproduction in or on infected tissues are components of disease development as part of the infection stage [24,28].

The factors determining the occurrence of infection are as follows: contact of the pathogen with the plant, for example, inoculation; presence of a sensitive plant variety on a specific race of pathogens (the pathogen must be virulent toward the plant); the presence of the plant in a pathogen-sensitive development phase; presence of pathogen in the pathogenic stage; temperature and humidity conducive to the growth, reproduction, and spread of the pathogen. An effective infection ends in establishing a parasitic contact and after the incubation period leads to the formation of discolorations or necrosis on the surface of the plant, which are called disease symptoms. Sometimes the infection does not lead to the occurrence of disease symptoms. It is then that the disease is latent or asymptomatic. Its manifestation often depends on the environmental conditions or the age of the plant [29,30].

Depending on the extent of the infection, the following are distinguished:

1. A local infection that involves a single cell, several cells, or a small part of the plant; this infection is caused by the plant anthracnose offenders,
2. Systemic infection—this includes all or most of the plant; this infection is caused by viruses, viroids, mycoplasmas, and some fungi, for example, *Graphium ulmi*, the metamorphic stage of *Ceratocystis ulmi*, the author of the Dutch elm disease [28].

Spread of pathogens

Pathogens are spread directly and indirectly. The direct spread of pathogens occurs through seeds, organs, or other vegetative parts:

- **Seeds**: Infected seeds are the primary source of infection of plants by some viruses, bacteria, and many fungus-like organisms and fungi. The seeds carry, for example, alfalfa mosaic virus, transferring a mosaic of lucerne, *Colletotrichum lindemuthianum*, the perpetrator of bean anthracnose, *Pseudomonas syringae* var. *lachrymans*, causing bacterial squared cucumber, *Pseudomonas syringae* var. *phaseolicola*, causing peripheral bacteriologist of the beans, *Septoria apiicola*, cause of celery septoriosis, *Stagonospora nodorum*, causing wheat chaff disease, *Ustilago tritici*, cause of wheat dusting head [27,35].
- **Vegetative Organ**: Organs of vegetative reproduction originating from plants infected systemically, mainly by viruses and bacteria. In this way, all viruses found in the organs of plants used for their breeding because of grafting, budding, planting in the form of seedlings, bulbs, bulbs, and rhizomes are transmitted;

Factors that facilitate spread of pathogens

The indirect transfer of the perpetrators of diseases is usually facilitated by a number of factors:

1. **Air currents:** Pathogens transmitted by air currents are usually the most serious causes of an epidemic. With air currents move, among others urediniospora of fungi from the *Pucciniaceae* family, causing rust of plants. Urediniospores can be carried over several thousand kilometers;
2. **Flowing water and raindrops:** With flowing water are transferred, among others spores of fungus-like organisms from the family *Pythiaceae*, fungi of the genus *Fusarium*, bacteria cells *Erwinia carotovora* subsp. *carotovora*, causes potato rot wet rot, spores of *Plasmodiophora brassicae*, fungal-like organism causing syphilis cabbage and swimming spores of representatives of the genus *Phytophthora* causing root diseases and ground organs of many plant species. Raindrops are spreading with raindrops mitospores *Spilocea pomi* (mitomorphic stage *Venturia inaequalis*), fungus causing scab of apple trees [27,28].

3. **Animals:** For example, the thrush spreads mistletoe seeds. Flies and mosquitoes spread spores of *Sphacelia segetum* (mitomorphic stage *Claviceps purpurea*). Bees and flies carry along with pollen *Erwinia amylovora*, the perpetrator of the fire blight of many plant species from the Rosaceae family. Insect, the green fodder (*Tortixviridana*) participates in the spread of *Chalara quercina* (mitomorphic stage *Ceratocystis fagacearum*), causes of oak dieback. Many species of aphids and jumpers with piercing and sucking mouth transmits viruses [28,34].
4. **Human:** Humans spreads pathogens because of the acquisition and sale of agricultural products and carrying out care work. For example, during potato peeling, it can carry a bacterium that causes the black leg of a potato (*Erwinia carotovora* subsp. *atroseptica*), and various viruses that transmit in a mechanical manner (EPPO 1994, Rataj-Guranowska et al., 2017).
5. **Infection, pollution (contamination):** When a dead material contains or carries infectious units of a pathogen, it is contaminated. Pollution often also means contamination with toxic substances. Contamination is the loss of cleanliness. Dead material cannot be infected. Live material can be contaminated, contaminated and infected. Infected material carries a pathogen [27,28].
6. **Monocyclic and polycyclic pathogens:** Pathogens that close their disease cycle only in one or in part of 1 year are monocyclic pathogens. Monocyclic pathogens are numerous species of fungi from the *Ustilaginaceae* family, the perpetrators of plant heads, because they produce spores at the end of the growing season; in the following year, these spores are the primary and only source of infection. Other monocyclic pathogens are different perpetrators of tree rust requiring two separate hosts and at least 1 year for the end of the cycle. In monocyclic pathogens, the primary inoculum is the only source of infection throughout the season, because the pathogen does not cause secondary infections and therefore does not form a secondary inoculum. The amount of inoculum produced at the end of the growing season is usually greater than the amount presents at the beginning of the. In the case of pathogens that propagate with propagating material, in both terms the amount of inoculum may be similar [28].

7. **Polycyclic pathogens:** Polycyclic pathogens are pathogens that can create more than one generation of infectious units in one growing season. The number of generations in this group of pathogens can range from 2 to 30 during the year. The amount of inoculum produced thus increases with each cycle and is eventually increased many times. These pathogens are the main contributors to the fastest-growing epidemics, for example epidemics caused by perpetrators of pseudo and mildew mildews and contagious potato and tomato plague. In polycyclic fungi, the primary inoculum is generally represented by spores of the monomorphic stage. In fungi rarely forming or not forming this stage, the primary inoculum is spores or mycelium developing from the wintering organs. Pathogens terminating the disease cycle over many years are called poly-bicyclic pathogens. Such pathogens are, for example, the perpetrators of odorous diseases, mycoplasmas and viral trees. The amount of inoculum at the end of the growing season is usually not significantly different from the amount of inoculum present at the beginning of the season. In many years, the amount of inoculum may gradually increase, and the pathogen may cause an onerous epidemic [27,28].

Resistance, Tolerance, and Disease Outbreak

During the growing season, plants are attacked by a huge number of infectious units of a specific pathogen. Resistance to disease is a feature of the host plant that does not allow disease to be caused by a pathogen adapted to it. The types of plant resistance to diseases are real resistance and apparent immunity. Disease resistance, which is genetically controlled by the presence of one, several, or many plant resistance genes, is called real resistance. In the real resistance, the host and pathogen are mutually incompatible, because of the lack of the following:
a. Chemical recognition between host and pathogen.
b. The plant's ability to be protected against a pathogen by means of various protective mechanisms already present in the plant or activated after infection of the pathogen [28,31].

Types of real resistance
There are two types of real resistance: horizontal and vertical resistance.

Horizontal resistance. The distinguishing features of horizontal resistance are as follows:
1. Controlling immunity by many genes (polygene or multigene resistance). Each of the genes is rather inefficient to the pathogen or plays a minor role in creating this resistance.
2. The influence of genes on many physiological processes conditioning the emergence of plant defense mechanisms. However, noticeable immune responses in, for example, different varieties are too little differentiated to be able to divide these varieties into resistance groups using horizontal resistance.
3. Variability depending on environmental conditions.
4. Lack of protective capacity against infection of the plant.
5. Horizontal immunity only delays the infection and the spread of the disease [27,28,31]. Horizontal immunity is sometimes called as nonspecific, general, quantitative, or permanent immunity.

Vertical resistance. Distinguishing features of vertical resistance are as follows:
i. No variability depending on environmental conditions.
ii. Inhibiting the development of an epidemic by limiting the size of the initial inoculum.
iii. Controlling immunity by one or several genes (monogenic or oligogenic resistance). These genes are likely to control the main phase of interaction of the pathogen with the host.
iv. Maladjustment of the pathogen to the host. One of the ways of this maladaptation is the hypersensitivity reaction.
v. Suppression of initiation of the pathogen association with the host [27,28].

The vertical resistance guarantees the total resistance of the plant to certain races of the pathogen. However, in relation to his other breeds, the same plant is sensitive. Vertical resistance is also called specific or qualitative resistance. In turn, apparent immunity is expressed by avoidance of disease or tolerance. Avoidance of the disease occurs when the sensitive plant is not infected, because three factors are necessary for the onset of the disease (sensitive variety, virulent pathogen, and favorable environment) and do not cooperate at the right time or long enough. Plants can avoid the disease because of the following:
1. Their seeds germinate faster, or seedlings strengthen before the advent of a pathogen-friendly temperature.
2. Some plants are sensitive to the pathogen only in certain stages of growth; in the absence of a pathogen or it's a virulence, the plant remains healthy. For example, young plants are much more sensitive to *Pythium* spp., perpetrators of powdery mildews, and most bacteria and viruses. However, fungi from

the genera *Alternaria* and *Botrytis* are easier to infect mature or dying parts of plants [22,28,34].

3. In field conditions, sensitive plants are separated with resistant plants. The amount of inoculum reaching sensitive plants is too small to initiate an infection [31,36].

4. Environmental factors can reduce the production of spores, survival, and infectiveness of the pathogen (unfavorable climatic factors, the presence of ecto-parasites or antagonists). For example, most potato plants "escape" the disease caused by *Phytophthora infestans*, with too little air humidity. At low temperatures, *Fusarium* and *Rhizoctonia* spp. usually do not affect sensitive varieties of cereals [21,27,32,37].

5. Lack of enough humidity generally limits the occurrence of all diseases, especially fungal diseases. However, in the case of *Streptomyces scabies*, the perpetrators of ordinary potato scab, plants are more likely to "escape" the disease in irrigation or moist soils, because plants are better protected themselves in the absence of water stress or the pathogen is suppressed by organisms requiring high humidity. Higher pH limits plant infection by *Plasmodiophora brassicae*, and lower pH suppresses *Streptomyces scabies* [36,38].

Tolerance to diseases and the concepts of symbiosis

Tolerance to disease is the ability of a plant to yield a good crop, even if it is infested with a pathogen. Tolerance results from the specific features of the host plant, which allow the development of the pathogen and its reproduction, but because of the possibility of compensating for the adverse effect of the pathogen, the plant is still able to yield well. Tolerant plants are sensitive to the pathogen, but they are not killed by them, but only relatively slightly damaged [21,31,39].

Symbiosis versus disease. The etymology of the word "symbiosis" comes from the words: bios—live and sym.—together. Symbiosis is the coexistence of two living organisms that differ in their taxonomy, in which each of the partners of this relationship benefits. A classic example of symbiosis is the coexistence of bacteria that bind nitrogen (bacteria of the genus *Rhizobium*) with legumes. After entering the root, the bacterium lives only based on plant nutrients. The plant is the host, and the bacterium is a parasite. The plant uses nitrogen formed because of bacterial activity, so it is also a parasite [27,28].

Although the presence of warts with nitrogen-binding bacteria is common in legumes, they are not natural, innate structures of these plants. Warts are caused by the irritation of root cells by bacteria. Thus, they are on the one hand an expression of the disease, and on the other hand, a manifestation of the connection between the bacteria and the plant. This thesis is confirmed by the results of observations that (a) certain strains of papillary bacteria do not bind nitrogen in certain host plant varieties, although these bacteria initiate the formation of warts and (b) some warts are effective in plants growing in the sun, whereas they lose the ability to bind nitrogen when plants grow under full shade conditions. Furthermore, in other types of symbiotic compounds, for example, in the case of mycorrhizae, that is, the intercourse of a plant with a fungus, the effects of this compound in certain phases of plant development can be detrimental to the host and even cause the disease symptoms. Therefore, symbiosis and disease are not mutually exclusive but reflect the result of the interaction of co-occurring organisms [27,28,31].

Disease outbreak

Epidemic is the occurrence and continuous increase of disease in a significant part of the population of fragile plants. An important part of the plant population may already be 10% of infected plants. The rate of disease growth is not a factor that defines an epidemic, because many epidemics are developing rapidly and quickly, but some epidemics are slow, even within a few years. The word "epidemic" means "among people" although it is a term commonly used about any disease. Factors influencing the development of epidemic are as follows:

1. Pathogen's ecology. Most fungus-like organisms and fungi and parasitic seed plants produce an inoculum on the surface of above-ground parts of plants. Hence, the infection units of this inoculum are easily spread over long distances. Other organisms, for example, viruses and mycoplasmas, reproduce inside the plant. In this case, the spread of the pathogen occurs mainly with the participation of vectors. Soil pathogens or underground plant organs usually spread slowly and rarely lead to epidemics.

2. Introduction of new pathogens. International travel favors the transfer of seeds, bulbs, and cuttings. Along with them, pathogens are imported, against which the plants did not create resistance genes or did not introduce them to the grower. Such pathogens usually lead to the appearance of dangerous epidemics [27,30].

3. The level of genetic resistance or host susceptibility. Plants with vertical resistance are completely protected against infection and epidemic development, unless a new breed of pathogen is produced or

introduced. Plants with low level of resistance will probably be infected, but the extent of disease and epidemic development will depend on the level of immunity and environmental factors. Sensitive varieties are the most favorable for the occurrence of the disease and its transformation into an epidemic [24,28].

4. Level of virulence of the pathogen. The virulent pathogen establishes contact with the plant more quickly and creates a larger amount of inoculum [24,31].

5. Type of soil and its preparation. Shallow, poorly permeable, windproof soils, especially located near other soils that maintain many potential pathogens, are the most favorable sites for epidemics [21,35,39].

6. Type of crop. Cultivation of plants in monocultures, application of high doses of nitrogen fertilizers, excessive irrigation, too intensive use of herbicides increases the possibility of epidemics (Sawicka 1999, Kozłowska & Konieczny 2003, Rataj-Guranowska et al., 2017). In annual crops, epidemics develop much faster than in long-term crops, for example in fruit and forest trees. Longer-lasting contact of the pathogen with the plant allows it to develop the necessary forms of immunity. In addition, annual plants contain more parenchymal tissues that are more easily overcome by pathogens than woody tissues [22,35].

7. The way the pathogen spreads. The spores of most pathogenic fungi are spread by the wind, often at very long distances. These include mainly the perpetrators of rust, powdery mildew, and fungi that cause leaf spots, for example, the anthracnose culprit. They are the ones that most often cause epidemics. In terms of the ability to cause sudden and widespread outbreaks, pathogens transmitted by vectors such as insects, nematodes, fungal organisms, and fungi are in second place. Vectors carry viruses, viroids, mycoplasmas, and some fungal organisms and fungi, for example, *Graphium ulmi*, the perpetrator of the Dutch elm disease and bacteria, for example, *Diabrotica vitata*, carry *Pseudomonas syringae* var. *lachrymans*, the culprit of the bacterial angular spotted cucumber. Responsible for the usual local epidemics are pathogens carried in a raindrop, mainly fungi that cause anthracnoses, scab of apple and pear. Pathogens transmitted with seeds or other vegetative organs rarely cause epidemics, this depends on the intensity of propagation of propagating material. After all, soil pathogens are rarely caused by sudden and wide epidemics, but they are the causes of slowly expanding, but very harmful diseases [24,28].

8. The degree of genetic uniformity of the host plants. Genetically uniform plants, especially in possessing resistance genes, grown on large areas, are conducive to the creation of new breeds of pathogen that overcome previous immunity. Epidemics occur more frequently and develop faster in vegetative propagated crops and self-pollinating crops. This partly explains why in natural plant populations epidemics are slower [21,31].

Plant Disease Diagnosis

Diagnosing factors that cause disease symptoms of plants (pathogens: fungi, bacteria, viruses and pests: nematodes, mites, insects) is difficult and takes place using both traditional methods (symptomatology, etiological symptoms) and the latest techniques based on hybridization, use molecular markers and specific antibodies [27,40–44]. Symptoms of diseases visible to the naked eye are called the macroscopic, noticed only under the microscope—the microscopic and are subject to pathological anatomy. In the case of the reaction of living cells to pathogens, physiological and biochemical symptoms are reported [20,31]. The total of macroscopic symptoms in a specific stage of the disease of a given species, plant varieties constitutes the so-called picture of the disease, or syndrome, or a set of disease symptoms. In contrast, a series of consecutive images of the disease on the same plant is only the pathogenesis or course of the disease. Therefore, it is possible to speak about significant, or irrelevant, primary or secondary disease symptoms, specific or nonspecific [20,27]. Significant symptoms are still associated with the occurrence of a given plant disease, while insignificant ones do not show permanence. Primary symptom is a symptom that occurs first on a diseased plant due to the action of a pathogen. The secondary symptom and its occurrence are a consequence of the conditions created by the primary symptom. In turn, a specific (specific) symptom is a significant symptom associated with only one disease, whereas a symptom significant in more than one disease is a nonspecific (nonspecific) symptom. A distinction is made between two groups of disease symptoms. The first one covers all the symptoms resulting from the host—plant reaction to the effect exerted on it by the disease agent—these are the specific symptoms. The second group of symptoms refers to the presence on the diseased plant of fructification and vegetative pathogens, for example, fruiting fungi, spores or vegetative mycelium, bacterial slime, mistletoe shoots, etc. These symptoms are described as inappropriate or an etiological [20]. Very often, the etiological symptoms are caused by fungi. These are clusters of spores, sporocarps, and spores of the fungus. Bacteria cause leakage with pathogen cells. For example, *Erwinia amylovora*, causing fire blight, causes yellowish spills on the apple tree shoots, and *Pseudomonas syringae*, causing bacterial angular spotted cucumber, causes leaks on the leaves [27]. Aetiological signs are a very important diagnostic feature when identifying a pathogen. They are characteristic of individual species or their groups, for example, mealy, white coatings on the surface of plants testify to the presence of powdery mildew (*Erysiphe*), brown-brown and dusty clusters of spores with the presence of the *Ustilago* [20,27,29].

Marker systems significantly influence the progress in breeding and plant genetics. The current state of knowledge on the structure of plant genomes has enabled the progress of work related to the mapping of plant genomes, analysis of quantitative and qualitative characteristics loci, while the analysis of sequences coupled with important economic features has become an important element of breeding work. Intensive work on improving quantitative trait loci mapping methods with the use of complex computer analyzes enables detailed elaboration of existing plant gene maps. Molecular marker technologies have become an inseparable part of not only breeding programs, finding application in the transfer and modification of genes between distant plant species and breeding of species and varieties with high resistance to pathogens, but also the possibility of diagnosing them. When conducting research based on molecular techniques, one should consider not only their great advantages, but also be aware of their limitations, which may influence the obtained test results. The choice of a marker system should depend on the research objective, the required level of polymorphism, the scale of analyzes, and laboratory equipment [40,42,44,45]. Parasitologists now have a wide range of molecular biology methods that can identify parasites with very high accuracy and sensitivity. The main limitation of the use of these methods, however, remains their high cost, while the availability is virtually unlimited due to the huge number of companies offering equipment and reagents for molecular biology.

Relying solely on blind eye observation or marker assessment for the detection and classification of plant diseases may be too expensive, especially in developing countries [12,46–48]. Pathogen threat assessment based on image processing for the automatic detection and classification of plant leaf diseases, which has been used recently, indicates that it can have a great realistic meaning for agricultural practice. The methodology of this solution is based on image processing and consists of four main phases:

1. Creating a color transformation structure for the RGB leaf image for the transformation of the color space;
2. Dividing images into segments using the K-means grouping technique;
3. Calculation of texture features for segmented infected objects;
4. Separation of features transmitted through the neural network.

The results of automatic disease detection indicate that the proposed approach can significantly support accurate and automatic detection and recognition of leaf diseases. The developed neural network classifier based on statistical classification works well in all types of leaf diseases tested and can successfully detect and classify diseases with an accuracy of about 93%, and proposed neural networks based on detection models are very effective in diagnosing leaf diseases [12]. By entering the diagnosis of diseases, one should fully appreciate the advantages and limitations of these methods to receive reliable results. One should also be aware that the use of such sensitive methods is not always necessary, because often conventional methods can fully answer the questions that bother us. According to Refs. [28,41,45,48–51] molecular, statistical, and traditional methods, in diagnosing pathogens, should complement each other, and a modern bacteriologist, virologist, phytopathologists, or parasitologist should know the morphology of their research object as well as its genetics.

CONCLUSION

Diseases are the result of some disturbance in the normal life process of the plant. As there are vast diversity of plants, there are a large number of different kinds of diseases that affect them. On average, a 100 or more plant diseases can affect each kind of plant. To simplify the study of plant diseases, these have been grouped in some generalized categories. It also becomes essential to emphasis on disease diagnosis and identification and subsequent control measures.

REFERENCES

[1] T. Rumpf, A.K. Mahlein, U. Steiner, E.C. Oerke, H.W. Dehne, L. Plumer, Early detection and classification of plant diseases with support vector machines based on hyperspectral reflectance, Comput. Electron. Agric.

74 (1) (2010) 91–99. https://doi.org/10.1016/j.compag.2010.06.009. ISSN 0168-1699.

[2] FAO, Food and Agriculture Organization of the United Nations, Toolkit: Reducing the Food Wastage Footprint, Italy, Rome, 2013.

[3] FAO. Food and Agriculture Organization of United Nations, Global Initiative on Food Losses and Waste Reduction, Italy, Rome, 2014.

[4] FAO, M. Rutten, M. Verma, N. Mhlanga, C. Bucatariu, LEI. Potential Impacts on Sub-saharan Africa of Reducing Food Loss and Waste in the European Union - A Focus on Food Prices and Price Transmission Effects, 2015. Rome.

[5] FAO. Food and Agriculture Organization of United Nations, Food Loss and Food Waste, 2016. http://www.fao.org/food-loss-and-food-waste/en/.

[6] FAO, Save Food: Global Initiative on Food Loss and Waste Reduction, Key Findings, 2017. http://www.fao.org/save-food/resources/keyfindings/en.

[7] L. Kitinoja, S. Saran, S.K. Roy, A.A. Kader, Postharvest technology for developing countries: challenges and opportunities in research, outreach and advocacy, J. Sci. Food Agric. 91 (2011) 597–603. https://doi.org/10.1002/jsfa.4295.

[8] A. Obiedzińska, Impact of food losses and waste on food security, Scientific Journals of the Warsaw University of Life Sciences in Warsaw. Problems of World Agriculture 17 (1) (2017) 125–141. https://doi.org/10.22630/PRS.2017.17.1.12, 1.

[9] J.C. Buzby, H. Farah-Wells, J. Hyman, The Estimated Amount, Value, and Calories of Postharvest Food Losses at the Retail and Consumer Levels in the United States, 2018. Available online: https://papers.ssrn.com/sol3/papers.cfm?abstract_id=2501659.

[10] J. Aulakh, A. Regmi, J.R. Fulton, C. Alexander. Estimating post-harvest food losses: developing a consistent global estimation framework. Proceedings of the Agricultural & Applied Economics Association's AAEA & CAES Joint Annual Meeting; Washington, DC, USA, 4–6 August, 2013.

[11] T.D. Banjaw, Review of post-harvest loss of horticulture crops in Ethiopia, its causes and mitigation strategies, J. Plant. Sci. Agric. Res. 2 (1) (2017) 1–4. http://www.imedpub.com/plant-sciences-and-agricultural-research.

[12] H. Abraha, R. Email, A. Kahsay, Z. Gebreslassie, W. Leake, B.W. Gebremedhin, Assessment of production potential and post-harvest losses of fruits and vegetables in northern region of Ethiopia, Agric. Food Secur. 7 (2018) 9. https://link.springer.com/article/10.1186/s40066-018-0181-5.

[13] B. Zhang, J. Zhou, Y. Meng, N. Zhang, B. Gu, Z. Yan, S. Idris, Comparative study of mechanical damage caused by a two-finger tomato gripper with different robotic grasping patterns for harvesting robots, Biosyst. Eng. 171 (2018). https://doi.org/10.1016/j.biosystemseng.2018.05.003.

[14] E. Kannan, P. Kumar, K. Vishnu, H. Abraham, Assessment of pre and post harvest losses of rice and red gram in Karnataka, in: Agricultural Development and Rural Transformation Centre, Institute for Social and Economic Change, Bangalore, India, 2013.

[15] D. Grover, J. Singh, Post-harvest losses in wheat crop in Punjab: past and present, Agric. Econ. Res. Rev. 26 (2013) 293–297.

[16] J. Kaminski, L. Christiansen, Post-harvest loss in sub-Saharan Africa. What do farmers say? Glob, Food Secur. 3 (2014) 149–158. https://doi.org/10.1016/j.gfs.2014.10.002.

[17] M. Singh, S.K. Biswas, D. Nagar, K. Lal, J. Singh, Impact of biofertilizer on growth parameters and yield of potato (2), Int. J. Curr. Microbiol. Appl. Sci. 6 (5) (2017) 1717–1724. ISSN, 2319-7706.

[18] S. Zorya, N. Morgan, L. Diaz Rios, R. Hodges, B. Bennett, T. Stathers, P. Mwebaze, L.J., Missing Food: The Case of Postharvest Grain Losses in Sub-saharan Africa, The International Bank for Reconstruction and Development/The World Bank, Washington, DC, USA, 2011.

[19] A.B. Abass, G. Ndunguru, P. Mamiro, B. Alenkhe, N. Mlingi, M. Bekunda, Post-harvest food losses in a maize-based farming system of semi-arid savannah area of Tanzania, J. Stored Prod. Res. 57 (2014) 49–57. https://doi.org/10.1016/j.jspr.2013.12.004, 2014.

[20] G.N. Agrios, Plant Pathology, 5 Edition, Copyright © by Elsevier Academic Press, Amsterdam, Boston, Heidelberg, London, New York, Oxford Paris, San Diego, San Francisco, Singapore, Sydney, Tokyo, 2005, ISBN 0-12-044565-4, p. 316.

[21] B. Sawicka, J. Kapsa, Effect of varietal resistance and chemical protection on the potato late blight (*Phytophthora infestans* (Mont.) de Bary) development (in:) Report of the Meeting of the Pathology Section of the EAPR, 10–14 July 2001, Poznan (Poland), Potato Res. 44 (3) (2001) 303–304, 287–307. https://doi.org/10.1007/BF02357906.

[22] B. Sawicka, P. Barbaś, The rate of spread of *Alternaria* spp. on the potato plant in an ecological and integrated system of cultivation, Post. Ochron Roślin/Prog. Plant Prot. 51 (3) (2011) 1–5.

[23] D.I. Shapiro-Ilan, R. Gaugler, W.L. Tedders, I. Brown, E.E. Lewis, Optimization of inoculation for in vivo production of entomopathogenic nematodes, J. Nematol. 34 (4) (2002) 343–350.

[24] M. Kozłowska, G. Konieczny, Biology of Plant Resistance to Pathogens and Pests, Copyright Agricultural University, Poznań, 2003, p. 213 (in polish).

[25] W. Jun, S. Wang, Image thresholding using weighted parzen-window estimation, J. Appl. Sci. 8 (2008) 772–779. https://doi.org/10.3923/jas.2008.772.779. URL: http://scialert.net/abstract/?doi=jas.2008.772.779.

[26] P.M.S. Babu, S.B. Rao, Leaves Recognition Using Back-Propagation Neural Network - Advice for Pest and Disease Control on Crops, Department of Computer Science & Systems Engineering, Andhra University, India, 2010. Technical report. http://www.indiakisan.net.

[27] S. Kryczyński, Z. Weber, Phytopathology vols. I, II, Państwowe Wydawnictwo Rolnicze i Leśne, Poznań, 2011, ISBN 978-83-09-01077-7, p. 312 (in polish).

[28] M. Rataj-Guranowska, S. Stepniewska-Jarosz, K. Sadowska, N. Łukaszewska-Skrzypniak, J. Wojczyńska, Compendium of symptoms of plant diseases and morphology of their perpetrators, in: M. Rataj-Guranowska, K. Sadowska (Eds.), Copyright © by Institute of Plant Protection — National Research Institute, Bank of Plant Pathogens and Their Biodiversity Research, Poznań, 2017, ISBN 978-83-7986-169-9.

[29] M. Pojmańska, Parasitism, parasites and their hosts, Cosmos Probl. Biol. Sci. 54 (1) (2005) 5–20, 266.

[30] H. Yamasaki, J.C. Allan, O. Sato, M. Nakao, Y. Sako, K. Nakaya, D. Qiu, W. Mamuti, P.S. Craig, A. Ito, DNA differential diagnosis of taeniasis and cysticercosis by multiplex PCR, J. Clin. Microbiol. 42 (2004) 548–553.

[31] S. Grzesiuk, I. Koczowska, R. Górecki, The Physiological Foundations of Plant Resistance to Disease, Publisher: Agricultural University of Technology, Olsztyn, 1999, p. 132.

[32] B. Sawicka, Effects of growth regulators, Mival and Poteitin, application in potato cultivation. Part II. The influence of growth regulators on incidence of *Rhizoctonia solani* sclerotia bearing tubers, Ann. Agric. Sci., Ser. E 28 (1/2) (1999) 55–65.

[33] OEPP/EPPO, EPPO Standard PP 2/1(1) Guideline on good plant protection practice: principles of good plant protection practice, Bulletin OEPP/EPPO Bulletin 24 (1994) 233–240.

[34] OEPP/EPPO, Phytosanitary Procedures. *Phoma exigua* var. foveata — Inspection and Test Methods. EPPO Standards PM 3/23(1), 1998, p. 248.

[35] B. Sawicka, Studies on the variability of selected traits and degeneration of potato varieties in the Biala-Podlaska region, in: Series Publishing — Scientific Dissertations, vol. 141, Publisher of the Agricultural University of Lublin, 1991, p. 76. ISSN: 0860-43-55 (in: polish).

[36] A. Borcean, I. Imbrea, Stem rot on wild peppermint species on south-western part of Romania, Res. J. Agric. Sci. 49 (1) (2017) 75–80.

[37] B. Sawicka. Dates of potato late blight appearance and spread (*Phytophthora infestans* (Mont.) De Bary) in changing of atmospheric and soloic conditions. Conf. EAPR, Hamburg, 14-19.07 (2002) 103.

[38] B. Sawicka, Infection of *Streptomyces scabies* tubers with 37 potato varieties in varying conditions of acidity and chemical composition of the soil, Probl. J. Agric. Prog. 456 (1998) 591–597 (in polish).

[39] T.S. Hameed, B. Sawicka. Level of practical use of recommendations to protect against potato blight (*Phytopthora infestans* Mont. de Bary) in the region of south-eastern Poland. International Conference: Bioeconomy in Agriculture. Puławy, 21–22 June (2016) 33-34. ISBN: 978-83-7562-219-5.

[40] J. Sztuba-Solińska, Molecular markers systems and their application in plant breeding, Cosmos Probl. Biol. Sci. 54 (2005) 227–239, 2-3(267-268).

[41] M. Kozak-Cieżczyk, Molecular diagnostics in parasitology, 266, Cosmos Probl. Biol. Sci. 54 (1) (2005) 49–60 (in polish).

[42] S. Hussein, Molecular Plant Phatology, in: A. Abbas (Ed.), 2008. https://www.researchgate.net/publication/308642520.

[43] H.L. Yang, W.K.J. Xing, X. Qiao, L. Wang, Z.S. Gao. Research on insect identification based on pattern recognition technology". 6th International Conference on Natural Computation, 08.09. (2010).

[44] S.L. Win, Gene expression mining for predicting survivability of patients in early stages of lung cancer, Digital Object Identifier 4 (1) (2014) 32–39. ISSN: 1839-9614.

[45] G. Kshirsagar, A.N. Thakre, Plant disease detection in image processing using MATLAB, International Journal on Recent and Innovation Trends in Computing and Communication 6 (4) (2018) 113–116. ISSN: 2321-8169.

[46] L.Q. Zhu, Z. Zhang, Auto-classification of Insect Images Based on Color Histogram and GLCM, in: 7th International Conference on Fuzzy Systems and Knowledge Discovery August 10–12, 2010. Date of Addition to IEEE Xplore: September 9, 2010. Location of the Conference: Yantai, China, Publisher: IEEE, 2010. https://doi.org/10.1109/FSKD.2010.5569848. ISBN information: INSPEC access number: 11526422.

[47] D. Al-Bashish, M. Braik, S. Bani-Ahmad, Detection and classification of leaf diseases using K-means based se mentation and neural-networks-based classification, Inf. Technol. J. 10 (2011) 267–275. https://doi.org/10.3923/itj.2011.267.275.

[48] N.S. Bharti, R. Mulajkar, Detection and classification of plant diseases, Int. Res. J. Eng. Technol. 2 (02) (2015) 2267–2272, e-ISSN: 2395-0056. www.irjet.net. p-ISSN: 2395-0072.

[49] Z.Z. Htike. Can the future really be predicted? Signal Processing: Algorithms, Architectures, Arrangements and Applications 26–28 September (2013). SPA: 360-365.

[50] Z.Z. Htike, S.L. Win, Recognition of promoters in DNA sequences using weightily averaged one dependence estimators, Procedia Computer Sci. 23 (2013a) 60–67. ISSN 1877-0509.

[51] Z.Z. Htike, S.L. Win, Classification of eukaryotic splice-junction genetic sequences using averaged one-dependence estimators with sub sumption resolution", Procedia Computer Sci. 23 (2013b) 36–43. ISSN 1877-0509.

CHAPTER 3

Pesticides, History, and Classification

YUSUF ABUBAKAR • HABIBU TIJJANI, PHD • CHUKWUEBUKA EGBUNA, BSC, MSC •
CHARLES OLUWASEUN ADETUNJI, PHD • SMRITI KALA, PHD • TOSKË L. KRYEZIU •
JONATHAN C. IFEMEJE, PHD • KINGSLEY C. PATRICK-IWUANYANWU, PHD

INTRODUCTION

Pesticide is any substance or mixture of substances used to prevent, destroy, or control pest including insects, fungus, rodents or, unwanted species of plants causing harm during production and storage of crops [1]. The word "pesticide" is a broad term that includes insecticides, herbicides, fungicides, and rodenticides that may be used to kill some specific pests. Pesticides are classified according to the sources of origin as either being a chemical pesticide or biopesticides. Biological pesticides are host specific. They are highly specific in the sense that it acts on the target pest and strongly related organisms, whereas chemical pesticides are nonspecific with wide range of activities on a large group of nontarget organisms. Biopesticides are environmentally friendly because they are less toxic, decomposed easily, and required in small quantities. Chemical pesticides cause several environmental pollution because they are quite toxic and may not be biodegradable. More so, biopesticides are of important advantage being less susceptible to genetic modification in plant populations. This confirms the little chance of pesticide resistance in pests, which is hardly encountered in case of chemical pesticides. Chemical pesticides are further subdivided into organochlorine, organophosphate, carbamate, and pyrethroids. Biopesticides are group of pesticides derived from natural materials such as animal, plant, and microorganism (bacteria, viruses, fungi, and nematodes). They include microbial pesticides, plant incorporated protectants, and biochemical pesticides.

Pesticides act through several mechanisms. Some are termed growth regulators as they either stimulate or retard the growth of pest, while repellents are known to repel pests, and attractants attract pests or chemosterilants, which sterilize pests. Pesticides with a wide range of activities and used to control more than one class of pests are difficult to classify [2]. Examples are aldicarb, which is used in Florida citrus production and may be considered an acaricide, insecticide, or nematicide for the reason that it controls mites, insects, and nematodes, respectively.

Aside from established chemicals employed as insecticides, other traditional means are also employed to decrease the growth of insect or limit their activities. Some of these compounds are generally meant for other purposes but are being used as insecticides. They include alcohols and oils. Alcohols are used as surface compounds for the control of mosquito larvae. It is the active ingredient in Agnique, a known larvicidal product. The aim for the uses of alcohol is to reduce surface tension of water, thus reducing the adherence of mosquito larvae or pupae leading to their drowning and death. Oils similarly are useful for surface applications. However, they prevent supply of oxygen for the survival of the larvae or pupae. The use of this water surface control has several advantages because their actions are more physical than conventional insecticides that are more biochemical in actions. These ensure that the organisms do not develop pesticide resistance.

HISTORY

Ever since ancient times, human civilizations have tried to apply the most effective and less time-consuming approaches for cultivating and preserving their food resources. An actual illustration of this is how they cultivated venomous and nutritious vegetation in the same place due to the shielding effect of toxic plants for insect elimination. Correspondingly, throughout this period, elemental sulfur has also been used. These would be the initial methods for removing pests for several millennia.

Later on came the Ebers papyrus, one of the oldest still existing documents that contain some of the preparation techniques for the removal of insects from foods. In parallel, traditional Chinese medicine also uses primitive sulfides. It is likewise interesting to note that Homer's epic work "Odysseus," written around the same time, describes the use of substances to remove insects.

Around 1500's, the early stages of the use of the "para-pesticides," namely mercury and arsenic, emerged. These substances were used until the start of synthetic pesticide era (1940 and beyond), initially for the destruction of food reserves during the World War II and later on as precious tools for cultivating processes of foods consumed daily. It is crucial to note that through this time, several scientists have highlighted the adverse effects of pesticides on human health when used for a long time. For instance, the drastic increase in the number of lymphoma patients is a topic that would be discussed until today [1−3].

Present Era

A vital event in the history of pesticides represents the discovery of the initial modern pesticide: dichloro-diphenyl-trichloroethane (DDT) by Paul Muller in 1939. This revelation granted him years later the Nobel Prize in Medicine, primarily because of the decrease of the damages caused by pesticides in agriculture and likewise health-related problems such as malaria or typhus.

Despite its sensation, the use of DDT would not last long. In 1962, the book "Silent Spring" by author Rachel Carson, elaborated the harmful effects of DDT. Immediately after that, numerous states banned the use of DDT in favor of lower risk organophosphates and carbamates in the forthcoming years.

Presently, there are still public concerns about the consequences of using pesticides in health (especially in old people and children). In contrast, because of the exponential growth of the population in the last decades, global production of pesticides is growing at the same trend [2]. In the meantime, according to the new EU regulations, producers are required to minimize the manufacture of pesticides to decrease the number of serious illnesses in the population. However, high attention is now being dedicated to the arrival of pesticide-resistant herbs, a factor that will definitely have a high impact in the future [1−3].

GENERAL MODE OF ENTRY OF PESTICIDES

The mode of entry of pesticides can either be systemic or nonsystemic.

Systemic Contact Pesticides

Pesticides with systemic contact are absorbed by animals or plants and transferred to tissues that are not treated. Systemic herbicides pass through the plant and can reach in the areas of leaves, stems, or roots that are untreated. They are effective in killing of weeds even with partial spray coverage of the pesticide. They effectively penetrate the plant tissues and pass via plant vascular system to kill target pests. In addition, few pesticides are considered locally systemic and pass only to a distance that is not far from the point of contact. Such insecticides include 2,4-dichlorophenoxyacetic acid (2, 4-D) and glyphosate [3].

Nonsystemic Contact Pesticides

The nonsystemic pesticides are also known as contact pesticides, as they produce the desired effect when they come in contact with the pest. Nonsystemic pesticides must come into physical contact with the pest for it to be active. Pesticide enters the body of pests via their epidermis while in contact which result to death by poisoning. These pesticides must not necessarily penetrate the tissues of the plant and consequently not transported via the vascular system of the plant. Contact pesticides include paraquat, diquat, and dibromide.

Stomach poisoning pesticide enters the pest's body via their mouth and to the digestive system resulting to death by poisoning. These insecticides kill the vector by destroying the midgut (or stomach) of the larvae, for example, malathion.

Fumigants are pesticides that act or kill the target pests through production of vapor. These pesticides form poisonous gases after application. The vapor formed enter the body of pests through their respiratory system (tracheal) via spiracles, which result, to death by poisoning. Most of their active ingredients are in liquid form when packaged under high pressure and change to gases when they are released. Fumigants are used for the removal of pest from stored product such as fruits, vegetables, and grains. It is also useful in controlling pests in soil.

Repellents also belong to this class of pesticides. Their uniqueness is the ability to repel the pest and not exactly kill, but they are distasteful enough to keep pests away from treated areas. They also interfere with the ability of pests to locate crop. Repellants are useful in public health applications to prevent mosquitoes and bloodsucking insects including ticks and black flies from biting pets, livestock, or humans.

USEFULNESS OF PESTICIDES IN COMMERCIALIZED AGRICULTURE

The use of pesticides has gained the attention of several stakeholders, farmers, scientists, governments, and agroindustries. The usefulness of pesticides were summarized:

1. Pesticides have been highlighted as an effective weapon of man for the management of agricultural pests affecting the increase in agricultural production. The process of applying these pesticides has led to a drastic reduction of insect pests, diseases, and weeds that can strikingly decrease the amount of harvestable agricultural yield and economic margin.

2. The utilization of pesticides has resulted in enhanced agricultural production of quality foods. For example, the consumption of diet having fruit and vegetables prevail over impending risks from consumption of low residues of pesticides available in some agricultural crops after application of pesticides [4].

3. The application of pesticides helps in the preservation of wood from destruction by termites and other wood-boring insects as well as in controlling the spread of malaria vector, the *Anopheles* mosquitoes [5].

4. The reception of pesticides for the management of agricultural pests by farmers has boosted and encouraged several scientists and researchers to divert most of their energy into more research and growth in the pesticide subsector. This has led to the discovery of more types of pesticides with different modes of action.

PROBLEMS ASSOCIATED WITH THE USE OF PESTICIDES

Some of the demerits of pesticide use include the following:

1. The consequence of environmental pollution that has led to uncontrolled outward influence on the quality of the environment makes the utilization of pesticides counterproductive. This might cause a long-term or short-term effect that entails contamination of groundwater or drinking water. Some pesticides also have adverse effect on nontarget plant which results to the bioaccumulation of the pesticide's residues [6].

2. Many pesticides have been shown to exhibit a nontarget effect on beneficial microorganism, soil quality soil enzymes as well as other beneficial aquatic microorganisms and algae [7].

3. The continuous use of pesticides has led to the development of resistance by target organism as well as cross-resistance to other vigorous constituents from some pesticides. This later on might lead to the development of more enhanced vicious populaces of pests or target organisms.

4. Severe usage of pesticides has made the farmers to depend and ignore other available alternatives such as organic agricultural practices and biopesticides [8].

5. The high cost of production and the socioeconomic aspect of the pesticides have discouraged so many investors most especially taking into consideration the standard benefit:cost ratio [9]. The blend of secondary pest outbursts, insect resistance, government rules, and legal clashes over well-being and the surroundings, consistently have repercussions for the economics of unadventurous pesticide utilization.

6. The registration of most pesticides has been observed to be very tedious, time consuming, and highly demanding with a lot of requirements. This has resulted to a lot of economic challenges due to the high costs implication involved in research and development necessary for utilization that has led to less preference of pesticides for control of agronomic pests.

7. The health challenges and food safety issue have become paramount most especially during postharvest storage of agricultural crops whenever pesticides are used to prevent the incident of agricultural pest, that is, Gamalin 20 and malathion. This includes acute poisoning, chronic poisoning, neurobehavioral effects, developmental and reproductive effects, carcinogenic effects, and immunological effects. Most farmers in developing country do not allow the pesticides to decay before they are being send to market and most consumers do not wash these agricultural produces before consumption that has led to the accumulation of ingestion of pesticide residues. This has led to the issue of food poisons and increase in death of many families [10]. Another typical challenge is the issue of increase in human infertility and birth defects [11].

8. The continual utilization of pesticides has increased the issue of environmental pollution whenever pesticides are used for the management of pests. This has resulted to pesticide applications that contribute significantly to pollution of the environment. This happens whenever pesticides are introduced into the environment that normally leads to run-off into water body, leakage of pollutants and irrigation water into groundwater, and drift of pesticides through wind and soil erosion. All these have led to systemic introduction of pesticides into the ecosystem that has increased the number of killing among the wildlife while some grieve impairment to vital roles such as reproductive failure.

9. The continual utilization of pesticides has led to high level of soil impoverishment by affecting the water retention, soil structure, physicochemical properties, and porosity that normally results into less water permeation and more run-offs thereby, leading to flooding down streams.

10. The incessant application of pesticides has resulted into pollution of water bodies. This has led to the continual washing of pesticides into the water bodies by irrigation water and rainwater. The non-biodegradability of pesticides that easily build-up in the bodies of plant and animal causes a great instability along the food chain and high level of imbalance in the ecosystem, and causes biodiversity collapse. A typical example is the percolation or leakages of pesticide residues into water bodies, which consequently led to the killing of aquatic invertebrates and fishes whenever pesticides are applied, that is, application of DDT for the killing of fishes.

11. The continual usage of pesticides has increased the rate of attenuation of soil fertility. Most especially it also affects the activity of soil microorganism that perform decomposition of organic matters, soil aeration, and most especially these microorganisms that are involved in the biodegradation of soil nutrient (nitrogen cycles, carbon cycle, phosphorus cycle), that is, free-living nitrogen-fixing bacteria, nitrogen-fixing blue-green algae, blue-green algae, and arbuscular mycorrhizal fungus.

12. The erroneous dosage of pesticide has been identified as one of the major problems affecting agricultural produce. This might be linked to the large financial losses experienced by primitive farmers most especially in the developing countries. They normally experience high level of phytotoxicity, which will normally reduce yield of agricultural produce because the rate of photosynthesis has been affected. In addition, there is a large tendency for the pests to cultivate resistance against pesticides, which normally have devastating effect on the large-scale production of agricultural produce [12].

13. The problem of low government intervention has greatly affected the pesticide regulatory role of the government most especially in most developing country. This might be linked to the inadequate of staffs in pesticide regulatory events, absence of appropriate legislative expert, shortage of infrastructure, equipment and materials, transportation, reduced budget provision on operating money,

absence of formulation control, and pesticide residue investigative amenities and competencies [13].

14. Inadequate safety measures has led to several issues of contaminations and poisoning in the field. For example, most farmers in the rural area are ignorant about some safety measures that are necessary whenever they apply pesticides on their farm, that is, protective clothing and gloves as well as other protecting gears wear in the course of spraying of pesticides. Other preventive actions are barely observed by these farmers as the farmers have been several times without number to be eating, or drinking in-between smoking spraying actions. Mostly, these farmers utilize all or some of these pesticide containers for domestic purposes that make some pesticide residues to cause a lot of hazards.

15. The challenges of poor extension services from staffs involved in the extension services have shown considerable effect on rural farmers that dwells most especially in the rural areas. This might be linked to reduced staffing and mobility. Extensionists in general are deprived of necessary support, poorly educated in the area of pesticide administration, lack incentive, and are scarcely any supplement arrangement to educate farmers most especially from developing country on safety protections as well as overall understanding on pesticide utilization. Moreover, lack of close intimacy between the farmers and the extension officer has increased the reliability and overreliance of the extension officer on pesticide vendors. Mismanagement and maintenance of equipment used for the application of pesticides due to high level of illiteracy of the farmers henceforth indulge in severe misuses of pesticide usage such as incorrect timing of application and dearth of awareness on the time required for complete degradation of pesticides, inappropriate utilization of use of nozzles, incompetent to differentiate one pest from the other, use of incorrect doses and formulations, and mingling together of various types of pesticides [14].

CLASSIFICATION OF PESTICIDES

A pesticide varies in identity (physical and chemical properties). According to Drum [15], pesticides can be broadly classified in different ways depending on their origin (Table 3.1) and target pest species (Table 3.2). They can also be classified based on function (Table 3.3).

TABLE 3.1
Classification of Pesticides Based on Origin.

Based on Origin	Sources and Examples
Organic sources	Natural—plant phytochemical (essential oil, plant extracts, leftover oilseed cakes) Synthetic—produced by chemical synthesis e.g., Pyrethroids, organophosphates, carbamates, organochlorine
Inorganic sources	Inorganic—mixture of inorganic salts [1] Bordeaux mixture Cu $(OH)_2.CaSO_4$ Malachite Cu $(HO)_2.CuCO_3$ and sulfur
Biological	Biological: microbial pesticide (bacteria, virus, and fungi)

TABLE 3.2
Classifications of Pesticides Based on Target Pest Species.

Pesticide Class	Target Pests	Example
Acaricides	Mites	Bifonazole
Algaecides	Algae	Copper sulfate
Avicides	Birds	Avitrol
Bactericides	Bacteria	Copper complexes
Fungicides	Fungi	Azoxystrobin
Herbicides	Weeds	Atrazine
Insecticides	Insects	Aldicarb
Larvicides	Larvae	Methoprene
Molluscicides	Snail	Metaldehyde
Nematicides	Nematodes	Aldicarb
Ovicides	Egg- prevents hatching of egg in insects and mites	Benzoxazine
Piscicides	Fishes	Rotenone
Repellents	Insects	Methiocarb
Rodenticides	Rodents	Warfarin
Termiticides	Kills termites	Fipronil
Viricides	Viruses	Scytovirin

TABLE 3.3
Classification of Pesticides Based on Function.

Action	Function	Examples
Feeding deterrents	Prevent an insect or other pest from feeding	(*Azadirachta Indica* A. Juss)
Ovipositor deterrent	Prevent egg laying by gravid female	*Azadirachta indica*
Repellents	Deters pest from approaching toward crops	Plant essential oil
Attractants	A chemical that lures pests	Gossyplure
Fumigants	Kills the target pests by producing vapor	Phosphine
Insect growth regulator	A substance that works by disrupting the growth or development of an insect	Diflubenzuron
Synergist	A chemical that enhances the toxicity of a pesticide but not by itself toxic to pest	Piperonyl butoxide

Natural Organic Pesticides

Natural organic pesticides include class of pesticides such as the phytochemicals present in plants; alkaloids, terpenes, and phenolic compounds, which have proven pesticidal potential [16]. Plant-derived essential oil, plant extracts, and leftover oilseed cakes are found to be active against varieties of pests [17]. These are particularly attractive on the grounds of low mammalian toxicity, short environmental persistence, and complex chemistry that do not develop resistance in pest [18].

Inorganic Pesticides

Inorganic pesticides include inorganic salts such as copper sulfate, ferrous sulfate and lime, and sulfur [19]. The chemicals in inorganic pesticides tend to be simpler and are soluble in water than those of organic pesticides. Many of them work by causing stomach poisoning in insects [20].

Synthetic Pesticides

The major classes of synthetic pesticides are organochlorines, organophosphates, carbamates, and pyrethroides.

Organochlorines

Organochlorine pesticides (also called chlorinated hydrocarbons) are organic compounds attached to five or more than five chlorine atoms. They represent one of the first categories of pesticides ever synthesized and are used in agriculture. Most of them are usually used as insecticides for the control of a broad range of insects, and have a long-term residual effect in the environment. These insecticides may alter the proper function of the nervous system of the insects leading to disorders such as convulsions and paralysis followed by eventual death [1]. Common examples of these pesticides include DDT, lindane, endosulfan, aldrin, dieldrin, heptachlor, toxaphene, and chlordane (Fig. 3.1). Although, the production and usage of DDT has been banned in most developed countries including the United States many years back, it is still produced and being used in most tropical developing countries for the control of vector.

Organophosphorus

Organophosphorus are phosphoric acid-derived pesticides (Fig. 3.2), considered to be one of the wide spectrum pesticides consisting of a heterogeneous group of chemicals [21], which control broad range of pests, weeds, or plant diseases because of their multiple functions. They are acetylcholine cholinesterase inhibitors, disturbing neurotransmitter across a synapse [22]. As a result, nervous impulses fail to move across the synapse causing a rapid twitching of voluntary muscles, hence, leading to paralysis, which is associated to death. Some of the widely used organophosphorus insecticides include parathion, malathion, dichlorvos, diazinon, and glyphosate (Fig. 3.2). The compounds in this group are characterized by a covalent binding of carbon to phosphate (C–P) bond replaced by one of its four carbon to oxygen to phosphorus bonds of phosphate ester [23]. The C–P direct linkage is described to be chemically and thermally inert, making organophosphonate compounds resistant to some drastic conditions such as chemical hydrolysis, photolysis compared and thermal decomposition with analogous compounds characterized with more reactive N–P, S–P, or O–P linkages.

Organophosphate pesticides have contributed drastically to improved agricultural productivity as well as effective crop yields [24]. Developing countries are increasingly adapting to its use. For example, Iran

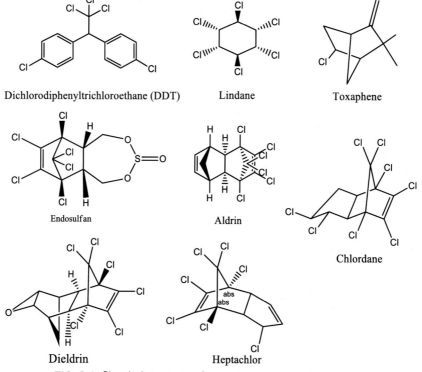

Dichlorodiphenyltrichloroethane (DDT)   Lindane   Toxaphene

Endosulfan   Aldrin   Chlordane

Dieldrin   Heptachlor

FIG. 3.1 Chemical structures of some organochlorine compounds.

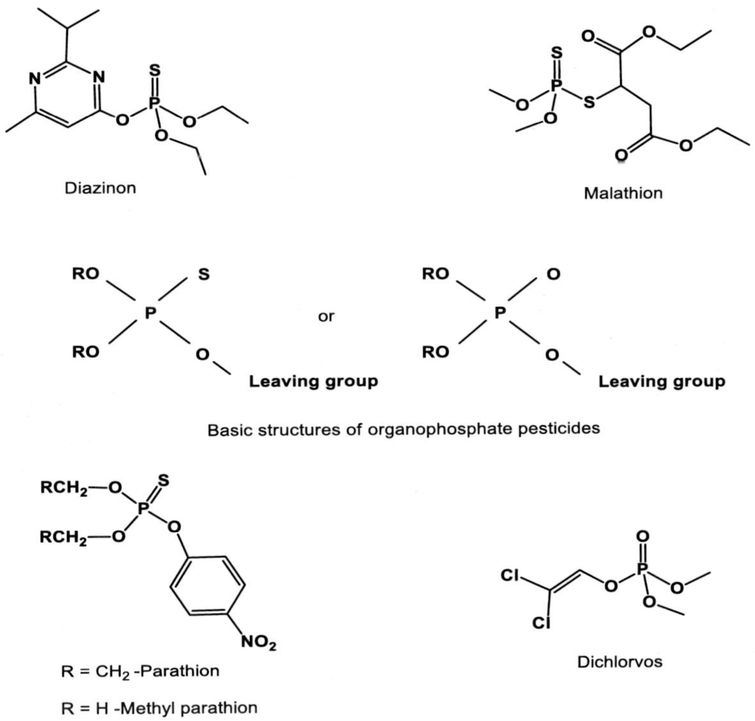

Diazinon

Malathion

Basic structures of organophosphate pesticides

R = CH₂ -Parathion

R = H -Methyl parathion

Dichlorvos

FIG. 3.2 Chemical structures of some organophosphate compounds.

agriculturists use pistachios for orchards to pest control being the largest producer [25]. These pesticides are also biodegradable, also cause minimum environmental pollution, and may be classified as slow pest resistance compounds.

Carbamates

Carbamates are organic pesticides derived from carbamic acid. These include carbaryl, carbofuran, and aminocarb [1]. They are similar in structure to organophosphates. However, they differ from their origin. Organophosphates are derivatives of phosphoric acid, while carbamates are derived from carbamic acid. The principle behind the use of carbamate pesticides is similar to that of organophosphate pesticides by affecting the transmission of nerve signals resulting in the death of the pest through poisoning [26]. Sometimes, they are also used as stomach and contact poisons as well as fumigant. This pesticide can be easily be degraded under natural environment with minimum environmental pollution.

Pyrethroides

Pyrethroids are organic compounds isolated from the naturally occurring flowers of pyrethrums

(*Chrysanthemum Coccineum* and *Chrysanthemum cinerariaefolium*) [1]. The insecticidal properties of pyrethrins are due to pyrethroic acids. Pyrethroides affect the sodium channels and lead to paralysis of the organism. The most widely used synthetic pyrethroids includes permethrin, cypermethrin, deltamethrin, lethrin, furethrin, fenevelerate, and alphcyperamethrin. The synthetic pyrethroids also belong to this group of pesticide, which can be synthesized by duplicating the structure of natural occurring pyrethrins. Relatively, they are more stable and effective than natural pyrethrins. The major active components are pyrethrin I and pyrethrin II plus smaller amounts of the related cinerins and jasmolins. Synthetic-pyrethroid pesticides are highly neurotoxic to insects and fish but less toxic to mammals and birds. Most of synthetic insecticides are nonpersistent, and can be broken easily on exposure to light. They are considered the safest insecticides for use in food. Modern pyrethroids slightly resemble their pyrethrin I (progenitor) and exert different mode of action from the original natural products. The commonly used synthetic-pyrethroid pesticides are cypermethrin, fenvalerate, fluvalinate, deltamethrin, lethrin, furethrin, and permethrin (Fig. 3.3).

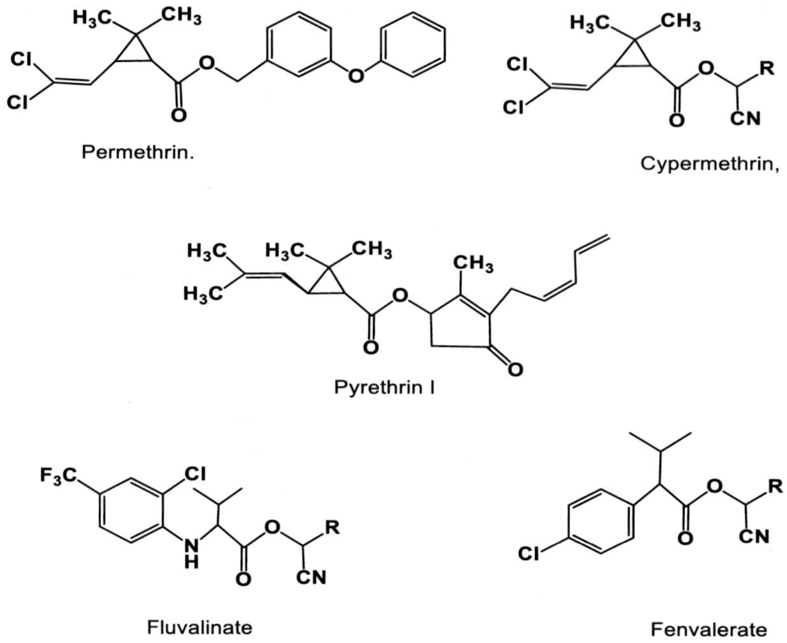

FIG. 3.3 Chemical structures of some synthetic pyrethroids.

Biorationals

Biorational pesticides (biopesticides) represent a specific group of compounds that are considered relatively safe to the environment and nontoxic to humans. Biorational pesticides are derived from natural materials including plants, animals, bacteria, and certain mineral elements. They can be subdivided into two major groups, the biochemically derived ones which include hormones, enzymes, pheromones and the microbial derived which are viruses, bacteria, fungi, protozoa, and nematodes. Biochemical pesticides are less selective compared to microbial pesticides. In arthropods, the mechanism of action is based on the interruption of their natural growth processes. Diflubenzuron, methoprene, and *Bacillus thuringiensis* serotype israelensis (Bti) are specific examples in biorational class of pesticides. Another notable microbial pesticide is those derived from fungus, *Trichoderma* [27]. In addition, some insect viruses found in nature such as those of *Baculovirus* family (*Baculoviridae*) are considered as biopesticides [27,28].

Pesticides of Plant Origin

Natural plant defense systems are found useful in the prevention of insects and other arthropods. Plant chemical constituents are subdivided into two categories namely secondary and primary metabolites derived from distinct biochemical pathways and are responsible for diverse biological activities [29]. The secondary metabolites are not directly involved in reproduction, growth, or development but are useful as sources of vital compounds with diverse biochemical activities [29–31]. Natural pesticides are found in terpenoids, phenolics, and alkaloids. Research evidences have shown that over hundreds of isolated compounds from plants have demonstrated bioactivities. However, rather than killing the insects, more of the compounds are active as feed deterrents or growth inhibitors [32]. Some of the identified natural pesticides were reviewed below.

Terpenoids

Terpenoids are secondary metabolites which have hydrocarbon isoprene (C_5H_8) as their simplest unit [29,33]. Volatile isoprene produced in large quantities during the process of photosynthesis protects plants' cell membrane from damages caused by light radiations and high temperature [33]. The protective roles of terpenoids are attributed to their bitter taste, preventing plants containing them from being eaten by animals [34]. *Chrysanthemum* plants have monoterpene ester (Pyrethroids) in their leaves and flowers, which are strong insecticides and protect the plant from insects. Monoterpenoids are neurotoxic to insects and to mites after significant contact with them.

Primary components of essential oils include monoterpenoids and sesquiterpenoids, which are high volatile compounds that contribute to plant fragrances. These essential oils can act as insect toxins and protect plants against the attack of bacterial and fungal. Examples of plants with terpenoids insecticidal properties include *Cinnamomum* spp., *Laurus* spp., *Mentha* spp., *Ocimum* spp., *Origanum* spp., *Piper* spp., *Rosmarinus* spp., *Salvia* spp., *Satureja* spp. and *Thymus* spp. among others.

Diterpenoids (C_{20}) synthesized from sesquiterpenoids (C_{15}) [29] similarly exhibit insecticide activities. Diterpenoids from *Gossypium hirsutum* (cotton) are strong antibacterial and antifungal compounds.

Triterpenoids (C_{30}) are secondary metabolites whose biosynthesis is a complex one [29]. They have similar structural identification with animal steroids and other steroid hormones. Because of this structural relationship, phytoecdysones, which mimic insect molting hormones, can disrupt larvicidal development and increase mortality of insects. Phytoecdysones are found in plants such as *Spinacia oleracea*. Limonoids are also examples of triterpenoids, which are found in *Azadirachta indica* (neem trees) seed as azadirachtin [35,36]. Azadirachtin are strong limonoid with insect repelling properties at low concentrations (Fig. 3.4). The mechanism by which azadirachtin exact its insecticidal effects is through the interference of insect neuroendocrine system, disrupting moulting, metamorphosis, and reproduction process in the insects. *Cymbopogon citratus* (lemon grass) is another example of plant with high limonoid levels.

Melia volkensii is a source of insecticidal compounds that can be obtained from its large fruit extract [37]. The active principles are identified to be limonoids, predominantly volkensin triterpene that is closely related to nimbilin from *A. indica* [32].

Rocaglamides are Aglaia compounds obtained from the family of *Meliaceae* [38,39] among interested biological compounds such as lignans [40] and triterpenes [41]. Rocaglamides occurs in nature with over 50 identified compounds [39]. They are aromatic with cyclopentatetrahydrobenzofuran skeleton, and are strong insecticidal compounds with broad-spectrum activities against neonate larvae of *Spodoptera littoralis*, *Ostrinia* sp. and *Helicoverpa armigera* [36,42,43]. The mode of actions of rocaglamides is derived from their ability to inhibit protein synthesis, which also account for their potential anticancer activity [44].

Isobutylamides from *Piper nigrum* (black pepper) such as its pipericide have been known with acute insecticidal properties. The Piperaceae from which pipericide are rich with numerous potentials of natural insecticides [45].

Other insecticidal compounds from the family include conocarpan (*P. decurrens*) and dillapiol (*P. aduncum*) [32].

Essential oils obtained from plants through various methods from aromatic medicinal plants have shown great potentials as insecticidal compounds or as insect repellants. They have been commercialized as natural oils for their use as insect repellants or insecticidal purposes [32]. The activities vary with structure, amount of saturation, and functional groups of the oils [46]. Thymol a typical ovicidal compound from this class of compounds is toxic to housefly [46]. Thymol was more active compared with its derivatives that were evaluated. The enhanced fumicidal and ovicidal activities are attributed to increase volatility and lipophilicity of the new acylated derivative [46].

Saponins

Saponins belong to the group of nonnitrogenous plant secondary metabolites commonly found in plants [29]. They are glycosylated triterpenoids with detergent properties, and they disrupt the cell membranes of invading microbial organism. Avenacins (triterpenoid saponins) contained in wheat prevent the invasion of *Gaeumannomyces graminis*, the pathogen for wheat.

Trilactone terpenes extracted from *Ginkgo biloba* foliages are potent insecticides on brown planthopper (*Nilaparvata lugens*) rice pest. The array of promising compounds in Ginkgo has paved way for their cultivation and use for the preparation of several pharmaceuticals and natural insecticidal compounds [32]. Grayanoid diterpenes from Rhododendron molle are strong grayanotoxins, which are strong neurotoxic compounds in invertebrates and mammals [32].

Phenolics

Phenolics are a class of secondary metabolites with hydroxyl group [29]. They are synthesized in plants as part of their host defense mechanism against pathogens. Phenolics have large and complex chemical structure [47]. Flavonoids are the largest class of phenolics. Phytoalexins are isoflavonoids that present antibiotic and antifungal properties in response to invading pathogen attacks. They are toxic molecules that disrupt pathogen biochemical process or the cellular structure. Specific examples of phytoalexin include rishitin obtained from tomatoes and potatoes (Solanaceae), camalexin from *Arabidopsis thaliana* and medicarpin obtained from *Medicago sativa* (Fig. 3.5).

Furanocoumarin is another example of phenolics synthesized by a wide variety of plants as response to herbivore attack or invading pathogen. Furanocoumarins are highly toxic to some vertebrate and invertebrate

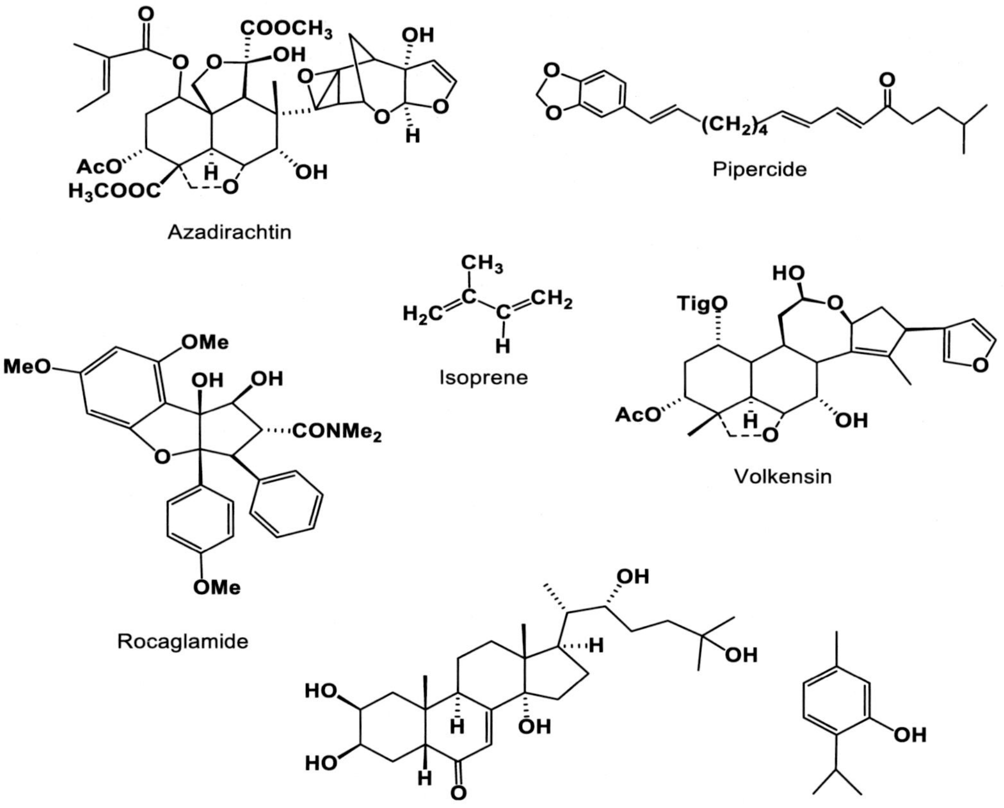

FIG. 3.4 Chemical structures of some terpenoids derived pesticides.

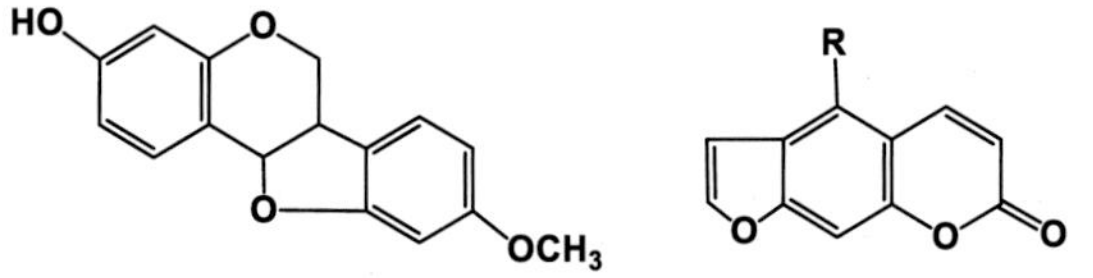

Medicarpin (Phytoalexin) Psoralen (Furanocoumarin)

FIG. 3.5 Chemical structures of some phenolic pesticides.

herbivores. They are usually inactive until exposed to ultraviolet light, and act by their integration into vertebrate DNA leading to the rapid death of cells. Furanocoumarins are found in smaller quantities in grapefruits.

Tannins

Tannins are a class of flavonoid polymers, which are water-soluble. They are stored in vacuoles after production by plants. Tannins binds to insect salivary proteins and digestive enzymes (trypsin and chymotrypsin) leading to the inactivation of the proteins. This action makes tannins toxic to insects. Tannins similarly prevent weight gains in insect herbivores who inject high amount of tannins leading to their premature death.

Lignins are strong components of plant cell walls, which confer them with rigidity, insolubility, and indigestibility making them excellent barriers against the attack of various pathogen attacks.

Alkaloids

Alkaloids are a class of secondary metabolite with characteristic bitter-tasting nitrogenous compound. They are synthesized from aspartic acid, tyrosine, lysine, or tryptophan [29]. Alkaloids can be found in several vascular plants: major classes are morphine, caffeine, nicotine, and cocaine. Caffeine are found in *Camellia sinensis* (tea), *Coffea arabica* (coffee), and *Theobroma cacao* (cocoa). Caffeine is toxic to insects and fungi as beneficial to the host plant (Fig. 3.6). Atropine are products of *Atropa belladonna* also known as "deadly nightshade plant." They are neurotoxic and cardiac stimulant compounds.

Plant-derived compounds have been used in the field and farms as traditional means for the control of pest and vermifuges [48]. Specifically, acetogenins isolated from *Asimina triloba, A. longifolia,* and *Annona muricata* have been evaluated for their pesticidal activity

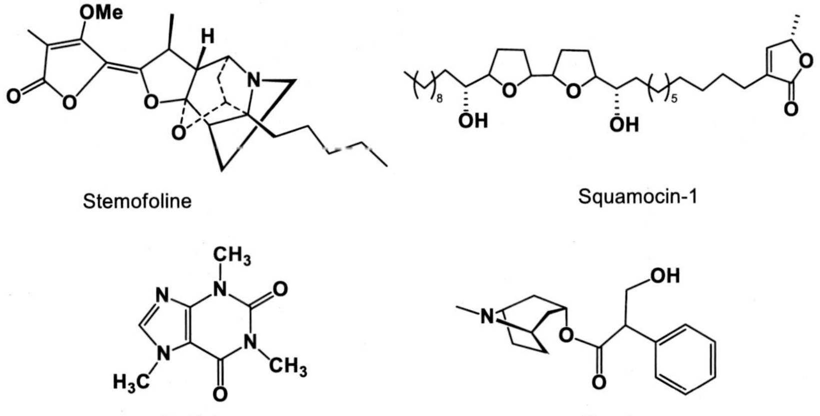

FIG. 3.6 Chemical structures of some alkaloid pesticides.

[49]. Squamocin, a complex alkaloid, was among the compound studied (Fig. 3.6). Rotenone is also from this group of chemicals. They are commercial botanical insecticide derived from the plant *Derris elliptica*. Acetogenins and rotenone exhibit similar mechanism of action, which is the inhibition of mitochondrial NADH: ubiquinone oxidoreductase enzyme [50].

Alkaloids from the Stemonaceae family contain useful biological compounds with insecticidal activities. They are the only source of stemona alkaloids [51], that are structurally characterized due to the presence of pyrrolo(1,2-α)azepine (5,7 bicyclic AB-ring system) nucleus [52]. Japanese and Chinese traditional medicine system for centuries have explored the use of plant extract from Croomia and Stemona for the management of respiratory diseases, enteric helminths as well as ectoparasites of humans and cattle [53,54]. The insecticidal activities of the plants have also been explored [55,56].The alkaloid stemofoline from crude extract of *Stemona cochinchinensis* and *S. curtisii* roots are attributed to the strong insecticidal activities against *Spodoptera littoralis* [56]. Similarly, the extract from *S. collinsae* possesses insecticidal activities against *S. littorallis*. This has been attributed to the presence in large quantities of didehydrostemofoline and smaller stemofoline. The insect-toxic potencies of these two alkaloids are higher than that of standard pyrethrum extract. The mechanism of action of the stemofoline is by agonism to insect nicotinic acetylcholine receptor [57].

Napthoquinone
Napthoquinone compounds (BTG 504 and BTG 505) from *Calceolaria andina*, a Chilean plant from Scrophulariaceae are important pesticides. The compounds have

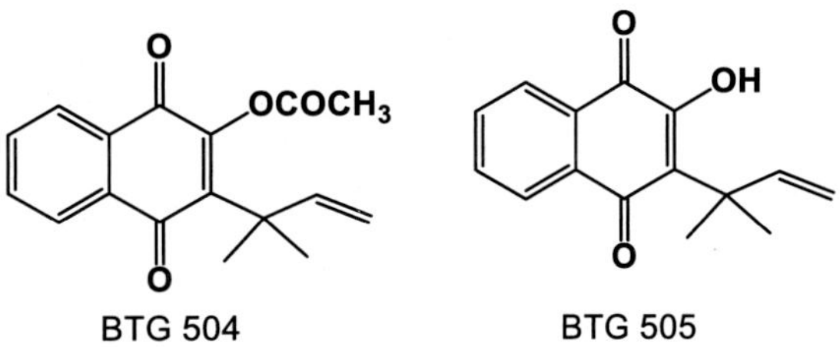

FIG. 3.7 Chemical structures of some napthoquinone pesticides.

been characterized as 2-acetoxy-3-(1,1-dimethylprop-2-enyl)-1,4-naphthoquinone and 2-(1,1-dimethylprop-2-enyl)- 3-hydroxy-1,4-naphthoquinone, respectively (Fig. 3.7). The compounds are effective against aphids, spider mite, and tobacco whitefly [58,59]. BTG 504 and BTG 505 are good insecticidal and fungicidal when compared to dunnione. The compounds are potent against *Musca domestica* (house fly), *B. tabaci* (whitefly), *Phaedon cochleariae* (beetle), and *T. urticae* (spider mite) unlike dunnione [60]. The mechanism of action for the compounds was found to be different. Although BTG 505 acts through the inhibition of mitochondrial complex III, dunnione acts primarily by the initiation of redox cycling [60].

Defense proteins
Plant proteins can play protective roles against pest and pathogens in some plants producing them. These proteins, which may be present in plants, or plant seeds specifically can inhibit pathogens and pest enzymes through the formation of enzyme complexes, which form complexes that block the active sites and may alter the conformation of enzymes, in turn reducing the

functions of enzymes. They serve as defensins, lectins, amylase, proteinase inhibitors, inhibiting wide range of pathogens including nematodes, bacteria, fungi, and insect herbivores. The proteins are usually small peptide chains, which are rich in cysteine. Defensive proteins are generally produced in significant quantity by plants after an attack by pathogens or pests through mechanisms that involve more resources and energy than that in biosynthesis of simple defense phytochemicals (terpenoids, saponins, phenolics, tannins, and alkaloids) discussed earlier.

FUTURE PROSPECTS

Pesticides are tremendously helpful in the control of pest because they provide means for the control of loss of crops during production or storages. The use of pesticide in developing countries is increasing especially as it pertains organophosphate pesticides among others [25]. These pesticides in some cases find their way to human through animal that feed on contaminated feeds. The contaminated feeds exact soome kind effects on animals and thus reduce their performances [25].

CONCLUSION

The application of chemical pesticides has been given a lot of attention and support for the management of agricultural pests but there are lots of limitations to their application. This might be linked to the safety issue of a serious hazard from the various active ingredients. Therefore, there is a need to make sustainable, green, eco-friendly biopesticides.

REFERENCES

[1] Z.J. Tano, Identity physical and chemical properties of pesticides, in: Pesticides in the Modern World: Trends in Pesticides Analysis, InTechopen, UK, 2011, pp. 1—18. Available: ISBN: 978-953-307-437-5, http://www.intechopen.com/books/show/title/pesticides-in-the-modern-world-trends-in-pesticidesanalysis.

[2] F.M. Fishel, J.A. Ferrell, Managing Pesticide Drift. Agronomy Department. PI232, University of Florida, Gainesville, FL, USA, 2013. Available: https://edis.ifas.ufl.edu/pi232.

[3] K.H. Buchel, Chemistry of Pesticides, John Wiley & Sons, Inc., New York, USA, 1983.

[4] reportPesticides Residue Committee Report 2004. Available: http://www.fao.org/fileadmin/templates/agphome/documents/Pests_Pesticides/JMPR/Reports_1991-2006/report2004jmpr.pdf).

[5] G. Ross, Risks and benefits of DDT, Lancet 366 (9499) (2005) 1771—1772.

[6] P.C. Struik, F. Bonciarelli, Resource use at the cropping system level, Eur. J. Agron. 7 (1997) 133—143.

[7] C.O. Adetunji, J.K. Oloke, O.O. Osemwegie, Environmental fate and effects of granular pesta formulation from strains of Pseudomonas aeruginosa C1501 and Lasiodiplodia pseudotheobromae C1136 on soil activity and weeds, Chemosphere 195 (2018) 98—107. https://doi.org/10.1016/j.chemosphere.2017.12.056.

[8] J.C. Zadoks, H. Waibel, From pesticides to genetically modified plants: history, economics and politics, Neth. J. Agric. Sci. 48 (2000) 125—149.

[9] A. Rola, P. Pingali, Pesticides, Rice Productivity, and Farmers' Health. An Economic Assessment. IRRI, International Rice Research Institute, Los Baños. US, 1993, pp. 25—30.

[10] R.I. Kola, S.L. Lawal, Degradation of aquatic environment by agro-chemicals in the middle −belt of Nigeria, in: A. Osuntokun (Ed.), Environmental Problems of Nigeria; Ibadan, Davidson Press, 1999, pp. 78—88.

[11] R.L. Smith, T.M. Smith, Elements of Ecology, fourth ed., Addison Wesley Longman Inc., California, 1998, pp. 235—240, 9, 10, 144.

[12] G.V. Meijden, Pesticide Application Techniques in West Africa. A Study by the Agricultural Engineering Branch of FAO through the FAO Regional Office for Africa, 1998, p. 17.

[13] A. Youdeowei, Provisional Report on Pesticide Management in Anglophone West Africa, Prepared for FAO Rome, 1989.

[14] E.U. Asogwa. Crop protection practices for cashew and cocoa production in Nigeria. Paper Presented at the National Training Workshop for Farmers Development Union (FADU) Officials on Cocoa and Cashew Production in Nigeria. Ibadan, May 8—9, 2008. 43 slides.

[15] C. Drum, Soil Chemistry of Pesticides, PPG Industries, Inc, USA, 1980.

[16] A.H. Paul, M.H. Robert, New Applications for Phytochemical Pest-Control Agents, Phytochemicals for pest control, 1997, pp. 1—12, https://doi.org/10.1021/bk-1997-065 (Chapter 1).

[17] P. Roman, History, presence and perspective of using plant extracts as commercial botanical insecticides and farm products for protection against insects − a review, Plant Protect. Sci. 52 (4) (2016) 229—241.

[18] D. George, R. Finn, K. Graham, O. Sparagano, Present and future potential of plant-derived products to control arthropods of veterinary and medical significance, Parasit Vectors 7 (2014) 28.

[19] D. Gunnell, M. Eddleston, M.R. Phillips, F. Konradsen, The global distribution of fatal pesticide self-poisoning: systematic review, BMC Public Health 7 (2007) 357—371.

[20] K. Ki-Hyun, K. Ehsanul, A.J. Shamin, Exposure to pesticides and the associated human health effects, Sci. Total Environ. 575 (2017) 525—535.

[21] L. Gámiz-Gracia, A.M. García-Campaña, J.J. Soto-Chinchilla, J.F. Huertas-Pérez JF, A. González-Casado, Analysis of pesticides by chemiluminescence detection

in the liquid phase, Trac. Trends Anal. Chem. 24 (11) (2005) 927–942.

[22] B. Dipsikha, K. Bulbuli, G. Hiren, Plant based pesticides: green environment with special reference to silk worms, Pestic.: Adv. Chem. Bot. Pestic. (2012) 171–206 (Chapter 8).

[23] B.L. Wanner, W.W. Metcalf, Molecular genetic studies of a 10.9-kb operon in *E. coli* for phosphonate uptake and biodegradation, FEMS Microbiol. Lett. 100 (1992) 133–140.

[24] C. Bolognesi, Genotoxicity of pesticides: a review of human biomonitoring studies, Mutat. Res. 543 (2003) 251–272.

[25] M. Kazemi, A.M. Tahmasbi, R. Valizadeh, A.A. Naserian, A. Soni, Organophosphate pesticides: a general review, Agric. Sci. Res. J. 2 (9) (2012) 512–522.

[26] I.C. Yadav, N.L. Devi, J.H. Syed, Z. Cheng, J. Li, G. Zhang, K.C. Jones, Current status of persistent organic pesticides residues in air, water, and soil, and their possible effect on neighboring countries: a comprehensive review of India, Sci. Total Environ. 511 (2015) 123–137.

[27] D. Kachhawa, Microorganisms as a biopesticides, J. Entomol. Zool. Studies 5 (3) (2017) 468–473.

[28] P. Bhandari, M. Pant, P.K. Patanjli, S.K. Raza, Advances in bio-botanicals formulations with incorporation of nanotechnology in intensive crop management, in: R. Prasad (Ed.), Advances and Applications through Fungal Nanobiotechnology. Fungal Biology, Springer, Cham, 2016, pp. 291–305.

[29] H. Tijjani, C. Egbuna, C.D. Luka, Biosynthesis of phytochemicals, in: Phytochemistry, Chapter 2, Fundamentals, Modern Techniques, and Applications, vol. 1, Apple Academic Press Inc, Canada, 2018, pp. 37–78.

[30] A.K. Pandey, S. Kumar, Perspective on plant products as antimicrobials agents: a review, Pharmacologia 4 (7) (2013) 469–480.

[31] K.L. Compean, R.A. Ynalvez, Antimicrobial activity of plant secondary metabolites: a review, Res. J. Med. Plants 8 (5) (2014) 204–213.

[32] I.B. Murray, A. Yasmin, in: I. Ishaaya, R. Nauen, A. Rami Horowitz (Eds.), Plant Natural Products as a Source for Developing Environmentally Acceptable Insecticides, Insecticides Design Using Advanced Technologies, Springer-Verlag Berlin Heidelberg, 2007, pp. 235–248.

[33] B.C. Freeman, G.A. Beattie, An overview of plant defenses against pathogens and herbivores, Plant Health Instr. 94 (2008), https://doi.org/10.1094/PHI-I-2008-0226-01.

[34] J. Degenhardt, J. Gershenzon, I.T. Baldwin, A. Kessler, Attracting friends to feast on foes: engineering terpene emission to make crop plants more attractive to herbivore enemies, Current Opinion Biotech. 14 (2003) 169–176.

[35] G. Schmutterer (Ed.), The Neem Tree, Mumbai Neem Foundation, 2002, 892 pp.

[36] O. Koul, J.S. Multani, G. Singh, S. Wahab, Bioefficacy and mode of action of rocaglamide from Aglaia elaeagnoidea against gram pod borer, Helicoverpa armigera, J. Appl. Entomol. 128 (2004) 177–181.

[37] H. Rembold, R.W. Mwangi, in: H. Schmutterer (Ed.), Melta Volkensu Gurke, M the Neem Tree, VCH, Wemheim, Germany, 1995, pp. 647–652.

[38] F. Ishibashi, C. Satasook, M.B. Isman, GHN. Towers, insecticidal 1H-cyclopentatetrahydro[b]benzofurans from Aglaia odorata (Lour.) (Meliaceae), Phytochemistry 32 (1993) 307–310.

[39] P. Proksch, R. Edrada, R. Ebel, I.F. Bohnenstengel, W.B. Nugroho, Chemistry and biological activity of rocaglamide derivatives and related compounds in Aglaia species (Meliaceae), Curr. Org. Chem. 5 (2001) 923–938.

[40] B. Wang, H. Peng, H. Huang, X. Li, G. Eck, X. Gong, P. Proksch, Rocaglamide, aglain, and other related derivatives from *Aglaia testicularis* (Meliaceae), Biochem. Syst. Ecol. 32 (2002) 1223–1226.

[41] S. Weber, J. Puripattanavong, V. Brecht, A.W. Frahm, Phytochemical investigation of Aglaia rubiginosa, J. Nat. Prod. 63 (2000) 636–642.

[42] B.W. Nugroho, B. Gussregen, V. Wray, L. Witte, G. Bringmann, M. Gehling, P. Proksch, Insecticidal rocaglamide derivatives from *Aglaia elliptica* and A. harmsiana, Phytochemistry 45 (1997) 1579–1585.

[43] B.W. Nugroho, R.A. Edrada, V. Wray, L. Witte, G. Bringmann, P. Proksch, An insecticidal rocaglamide derivative and related compounds from Aglaia odorata, Phytochemistry 51 (1999) 367–376.

[44] G. Satasook, M.B. Isman, P. Wiriyachita, Activity of rocaglamide, an insecticidal natural product, against the variegated cut worm, Peridroma saucia, (Lepidoptera: Noctuidae), Pestic. Sci. 36 (1993) 53–58.

[45] S. MacKinnon, D. Chauret, M. Wang, R. Mata, R. Pereda-Miranda, A. Jiminez, C.B. Bernard, H.G. Krishnamurty, L.J. Poveda, P.E. Sanchez-Vindas, J.T. Arnason, T. Durst, Botanicals from the Piperaceae and Meliaceae of the American Neotropics: phytochemistry, in: P.A. Hedin, R.M. Hollingworth, E.P. Masler, J. Miyamoto, D.G. Thompson (Eds.), Phytochemicals Forest Control American Chemical Society, Washington, DC, 1997, pp. 49–57.

[46] P.J. Rice, J.R. Coats, Insecticidal properties of several monoterpenoids to the house fly (Diptera: Muscidae), red flour beetle (Coleoptera: Tenebrionidae), and southern corn rootworm (Coleoptera: Chrysomelidae), J. Econ. Entomol. 87 (1994) 1172–1179.

[47] N.J. Walton, M.J. Mayer, A. Narbad, Molecules of interest: Vanillin, Phytochemistry 63 (2003) 505–515.

[48] J.L. McLaughlin, L. Zeng, N.H. Oberlies, D. Alfonso, H.A. Johnson, B.A. Cummings, Annonaceous acetogenins as new natural pesticides: recent progress, in: P. Hedin, R. Hollingworth, J. Mujamoto, E. Masler, D. Thompson (Eds.), Hytochemical Pest Control Agents, Am Chem Soc, Washington, DC, 1997, pp. 117–133.

[49] K. He, L. Zeng, Q. Ye, G. Shi, N.H. Oberlies, G.X. Zha, C.J. Njoku, J.L. McLaughlin, Comparative SAR evaluations of Annonaceous acetogenins for pesticidal activity, Pestic. Sci. 49 (1997) 372–378.

[50] M. Londershausen, W. Leight, F. Lieb, S. Moeschler, Molecular mode of action of annonins, Pestic. Sci. 33 (1991) 427–438.

[51] R.A. Pilli, M.C. Ferreira de Oliviera, Recent progress in the chemistry of the Stemona alkaloids, Nat. Prod. Rep. 17 (2000) 117–127.

[52] B. Brem, C. Seger, T. Pacher, O. Hofer, S. Vajrodaya, H. Gregger, Feeding deterrence and contact toxicity of Stemona alkaloids—a source of potent natural insecticides, J. Agric. Food Chem. 50 (2002) 6383–6388.

[53] R.S. Xu, Y.J. Lu, J.H. Chu, T. Iwashita, H. Naoki, Y. Naya, K. Nakanishi, Studies on some new stemona alkaloids, Tetrahedron 38 (1982) 2667–2670.

[54] K. Sakata, K. Aoki, C.F. Chang, A. Sakurai, S. Tamura, S. Murakoshi, Stemospironine, a new insecticidal alkaloid of Stemona japonica Miq. Isolation, structural determination and activity, Agric. Biol. Chem. 42 (1978) 457–463.

[55] R.S. Xu, Some bioactive natural products from Chinese medicinal plants, in: Atta-urRahman (Ed.), Studies in Natural Products Chemistry, vol. 21, Elsevier Science Publishers, Amsterdam, 2000, pp. 729–772.

[56] E. Kaltenegger, B. Brem, K. Mereiter, H. Kalchhauser, H. Kahlig, O. Hofer, S.H. Vajrodaya, Insecticidal pyrido [1,2-a]azepine alkaloids and related derivatives from Stemona species, Phytochemistry 63 (2003) 803–816.

[57] C. Godfrey, J. Benner, M. Clough, S. Dunbar, F. Earley, A. Russel, C. Urch, A. Ware, in: 10th IUPAC International Congress on the Chemistry of Crop Protection, 2002, p. 236.

[58] B.P.S. Khambay, D. Batty, H.M. Niemeyer, Naphthoquinone Derivatives, U.K. Patent Appl. 2289463A, PCT Appl, 1995. WO 95/32176.

[59] B.P.S. Khambay, D. Batty, M. Cahill, I. Denholm I, Isolation, characterization, and biological activity of naphthoquinones from *Calceolaria andina* L, J. Agric. Food Chem. 47 (1999) 770–775.

[60] B.P.S. Khambay, D. Batty, P.J. Jewess, G.L. Bateman, D.W. Hollomon, Mode of action and pesticidal activity of the natural product dunnione and of some analogues, Pest Manage Sci. 59 (2003) 174–182.

FURTHER READING

[1] A.S. Perry, I. Yammamoto, I. Ishaaya, R.Y. Perry, Insecticides in Agriculture and Environment: Retrospects and Prospects, Springer, Berlin Heidelberg New York, 1998, p. 261.

[2] G. Brader, S. Vajrodays, H. Greger, M. Bacher, H. Kalchhauser, O. Hofer, Bisamides, lignans, triterpenes, and insecticidal cyclopenta[b]benzofurans from Aglaia species, J. Nat. Prod. 61 (1998) 1482–1490.

[3] B.W. Nugroho, R.A. Edrada, B. Gussregen, V. Wray, L. Witte, G. Bringmann, P. Proksch, Insecticidal rocaglamide derivatives from *Aglaia duppereana*, Phytochemistry 44 (1997) 1455–1461.

Biopesticides, Safety Issues and Market Trends

CHUKWUEBUKA EGBUNA, BSC, MSC • BARBARA SAWICKA, PHD •
HABIBU TIJJANI, PHD • TOSKË L. KRYEZIU • JONATHAN C. IFEMEJE, PHD •
DOMINIKA SKIBA, PHD • COLIN B. LUKONG, PHD

INTRODUCTION

As described in Chapters 1—3, the losses in the cultivation of agricultural and horticultural plants caused by diseases and pests are the main production limitation in all countries of the world and were estimated at 30%—40% of world production, and in developing countries at 60%—80% [1]. Therefore, synthetic plant protection agents have been used throughout the world in conventional agriculture as the mid-20th century. Their widespread use as well as considerable durability has made them ubiquitous in the natural environment. Some of them break down for a very long time, even those that were withdrawn from production and use even 30—40 years ago, for example, DDT and its decomposition products, are still detected in agricultural crops. Due to the persistence of pesticides, as well as possible threats to them, the number of scientific studies assessing the impact of pesticides on the environment in the last 30 years has not increased [2].

In recent years, consumers have been paying increasing attention to the potential health impact of synthetic chemicals in food production [3—7]. The interest in food safety issues has caused considerable pressure in Europe and globally, not only by consumers, but also by various EU committees and organizations, to reduce pesticide residue levels in foods from farms where plant protection products are used. Since 2009, new rules have been introduced, exacerbating the requirements for chemical compounds used as pesticides. The introduction of new provisions, especially in the field of food safety, led to the withdrawal from the market of many synthetic active substances, in the light of their unacceptable potential or real detrimental effects on human and animal health. Annex II to Regulation (EC) No 1107/2009 of the European Parliament [8] and of the Council of 21 October 2009 [9] concerning the placing of plant protection products on the market introduces the so-called "Cut-off criteria" that directly prohibit the use of many substances as pesticides. This applies, among others, carcinogenic, mutagenic, toxic (reproduction) substances, especially endocrine disrupters and extremely persistent. Substances meeting these criteria have been called "substitute candidates" Compounds receiving a negative evaluation, in accordance with the provisions of the previously mentioned regulation, should be replaced by other compounds with more safe properties. The list of candidate substances for substitution should be published by the European Commission in the shortest possible time, in accordance with art. 80 paragraph 7 of Regulation (EC) 1107/2009 [8] of the European Parliament and the Council of Europe. Pesticides replacing forbidden chemicals should undergo a comparative assessment procedure. Preparation of such a procedure is currently under way. In addition, the element that influences the search for new pest control tools is the growing evolution of pest population immunity to currently used pesticides. Another factor supporting the development of the biopesticide market is the increase in demand for so-called ecoproducts [10,11]. There are also various alternatives to withdrawn, synthetic pesticides in the form of natural products, that is, of biopesticides [11—13].

Biopesticides—Characteristics and Benefits

The use of biopesticides in plant protection can lead to many beneficial changes, such as reduced residues of pesticides in food and reduced risk for consumers. Biopesticides are typical for harmful and low-risk organisms other than target organisms. Usually, they break down quite quickly, and some, semiochemical are used in very small doses. Biopesticides are a special group of active substances for the protection of plants that occur naturally or are synthetic substances identical to natural ones. They also include many living organisms (so-called biological protection organisms) [8,13,14]. They can be divided into several groups, as the mechanisms of action of pesticides and their impact on human health are better understood. There is a statistically significant relationship between the exposure to plant protection products and the increased risk of developmental disorders, neurological or immunological diseases and some types of cancer [15–18]. Finding out if contact with a pesticide caused a disease to develop or another disorder is not easy. Detecting the molecular basis of resistance to agents against new generations of chemical and antimicrobial agents is an important task, so for this purpose, tools of molecular genetics, molecular and microbiological physiology, and others are used, Although this is still a serious challenge for science. The exposure of human populations to pesticide residues may also include many substances that come from many sources, such as the environment, food, and drink. The adverse effects of such combinations, also known as the "cocktail effect" may appear in different, unpredictable ways, leading to a change in the assessment of the risk of human health, compared to exposure to individual substances [17,19]. In the light of recent scientific results and legal regulations at European and international level, the combined impact of mixtures of pesticide residues used in plant protection products (PPP) aims to map various actors involved in the regulation of PPP marketing, conduct risk assessments and set maximum residue limits for pesticides and planning and strict enforcement of control [3–7,17–22].

The implementation of integrated pest management in agriculture on a large scale has led to an increase in interest in nonchemical methods. Ecological and biodynamic agriculture are also lacking effective Colorado potato beetle control methods [19,23–27]. The application of Cow Horn Manure (BD 500 biodynamic preparation), applied by Levickienė [27], increased, among others, biomass of soil microbes and biomass conversion and dehydrogenase activity. In the soil sprayed with BD 500, the amount of microbial biomass was by as much as 11% higher in the entire cultivation period, in comparison to the control object. Significant increases in bacterial biomass in soil were recorded after 14 and 115 days after application, respectively, by 19% and 18%. The activity of dehydrogenase in the soil was on average 21% higher than in soil not treated with this preparation; moreover, the content of available phosphorus, potassium, and nitrogen in the soil increased, which significantly influenced the resistance of plants

to soil pathogens. According to Keidan [23], weed management using nonchemical methods, that is, thermal and mechanical methods, and using biological preparations, has reduced the number of rape plants in autumn, compared to the use of self-regulating weed control methods; however, the plants showed better postwinter survival and were characterized by higher productivity. The use of nonchemical methods of weed control (thermal, mechanical) combined with the application of biological preparations improves the soil structure, upsurges the number of earthworms in the soil, and reduces CO_2 emissions from the soil; however, these practices reduce the soil enzymatic activity [16,23,27]. The outcomes of research on biological preparations will allow to select innovative weed management practices more effectively and combine them with biological preparations in winter crops (winter oilseed rape, winter triticale, winter wheat, alfalfa, red clover) in a temperate climate and provide appropriate biological agents to improve the survival of wintering plants and increasing soil fertility [16]. The application of biodynamic preparations in the cultivation of agricultural and horticultural plants will form the opinion on the feasibility of practical use of biodynamic preparations to improve agrochemical and biological properties of soil, increase crop resistance to pathogens, and enable targeted selection of desired biologically active compounds.

Research on the usage of entomopathogenic mushrooms to combat potato beetles has been conducted in Europe for several years [8–13,27–29]. The results of many studies indicate the average or good effectiveness of *Beauveria bassiana* in potato protection against Colorado beetle beetles [12,13,15,30–33]. Despite the positive results obtained, no preparation based on entomopathogenic fungi has yet been registered. However, natural substances are still more and more important. Most often, they are plant compounds that are products of plant metabolism. These substances include, but are not limited to saponins, farnesyls, and NEEM extract [34]. Entomopathogenic fungi are considered promising agents for protection against many crop plant pests not only in Europe, but also around the world [13,20,35,36]. However, the practical use of entomopathogenic mushrooms in plant protection is generally limited. Some argue that entomopathogenic mushrooms, such as *B. bassiana* or *Isaria fumosoroseus*, are effective in controlling pests not only found in the soil, but also those that damage the above-ground parts of plants [30]. The results of the Ropek and Kołodziejczyk studies [13] showed that the efficiency of natural preparations in controlling *Leptinotarsa decemlineata* larvae, however, showed to be much inferior than the chemical insecticides [37,38]. Chemical agents more effectively protected potatoes against *L. decemlineata* larvae, as the reduction of this pest was about 90%, 10 days after one application in the form of imidacloprid or lambda-cyhalothrin spray. The most effective protection against this pest was obtained by treating tubers with thiamethoxam. *L. decemlineata* larvae did not appear on potato plants protected with this insecticide. The lower effectiveness of *B. bassiana* or *I. fumosoroseus* can be explained by their slower action than by the action of chemical insecticides [37,38].

Natural insecticides are more dependent on weather than chemical insecticides. In previous studies, the foliar application of *B. bassiana* was ineffective or of low or moderate efficacy against *L. decemlineata* [13,37,39]. The term foils application of entomopathogenic fungi is also important. Osman [40], in turn, showed the relative toxicity of some biopreparations and insecticides compared to the younger stages of potato beetle larvae and beetles. The first application of *B. bassiana* made by Ropek & Kołodziejczyk [13] was when most *L. decemlineata* larvae were at an early stage of development (L1-L2). Thus, although field tests are promising, further research is needed to improve the control strategies of *L. decemlineata* larvae with entomopathogenic fungi. Products with pheromones or other semiochemicals and their active substances are chemical compounds excreted by animals or plants (or their synthetic counterparts) to transmit information or influence their environment in some way, usually for defensive purposes, such as alerts, territory marking, or informing partners about sexual readiness. They usually act interspecifically (e.g., pheromones, auto-inhibitory regulators, autotoxins, necromania's) or interspecific (such as allomones, kairomones, and depressors). An example of this group of pesticides is the straight chain Lepidoptera pheromones, which are used in insecticide traps [15,37].

Microbiological pesticides include fungi, bacteria, and viruses. *Beauveria bassiana* strain GHA is used as an insecticide to control sucking insects, feeding on greenhouse vegetables and decorative plants. The use of *B. bassiana* is an example of a fungal plant protection product. *Bacillus thuringiensis* subsp. *tenebrionis* strain NB-176, destroying the *Leptinotarsa decemlineata* larvae, is an example of a bacterial insecticide [19,30,41,42]. In turn, *Cydia pomonella* ssp. *granulovirus*, used in the protection of fruit trees against moth larvae, which is a dangerous pest feeding on fruit, is an example of a virus as a pesticide [43]. The list of pesticides containing microorganisms is very long. Only for one type of *Bacillus* several bioactive components of pesticides are registered in Europe [26,27]. These are *Bacillus amyloliquefaciens* ssp. strain FZB24; *B. amyloliquefaciens* subsp. D747; *B. amyloliquefaciens* MBI 600; *B. firmus* I-1582; *B. subtilis* strain QST 713; *B. subtilis*—strain IBT 711; *B. pumilus* QST 2808; *B. sphaericus*; *B. thuringiensis* subsp. *israeliensis* strain AM 6552; *B. thuringiensis* ssp. *aizawai* ABTS-1857; *B. thuringiensis* ssp. *aizawai* GC 91 strain; *B. thuringiensis* ssp. strain *kurstaki* ABTS 351; *B. thuringiensis* ssp. *kurstaki* SA11_SA12_EG2348; *B. thuringiensis tenebrionis* NB-176; and *B. thuringiensis* ssp. *kurstaki* strain PB 54 [13,22,27].

Products containing living organisms—invertebrates (e.g., predatory insects) and nematodes of the genus *Heterorhabditis* and *Steinernema*. In the US regulatory system, they are exempted from registration of biopesticides [44,45].

Preparations based on extract and vegetable oil. This group of pesticides is diverse, both in chemical and functional terms. Pesticides in this group often contain complex mixtures difficult to classify (e.g., citronella or orange oil, garlic extract, tea tree extract) [22,46]. Many of them have been safely used for years. For example, in the United States, biopesticides include genetically modified crops that have transgenes, which in turn encode natural plant protection agents (e.g., Bt toxin). The biopesticides of this group are called as protected by plants (PIP) by the USEPA [11,44]. Pesticides potentially reduce the risk for the consumer and in addition to these, include inorganic salts and substances such as fatty acids. The groups of pesticides perform various functions in plant protection and can be used as:

Biodynamic preparations, for example, BD 500 and BD 501 in Swiss production, which increase the content of phenolic compounds and total flavonoids in leaves, and have strong antiradical activity, which increases the resistance to viral and fungal diseases [23,27,47];

Fungicides: *B. subtilis* (specific strain) is used in the protection of fruits and lettuce in fungal infections; *Coniothyrium minitans* (a specific strain) has a strong parasitic character against *Sclerotinia sclerotiorum*; laminarin from *Laminaria digitata*, is used in cereal protection to induce resistance to plant pathogens [44] and other natural preparations and bioregulators are used to protect potato from *Ph. infestans* [48]. One of the most devastating diseases in almost all countries where cocoa is cultivated. So-called the "black pod" of cocoa, caused by various species of *Phytophthora*, is difficult to eradicate. Until now, there are no effective chemicals or biological or cultivation measures to reduce this disease. Due to the lack of resistant varieties and ineffective control strategies, the development of a genetically improved variety on *Phytophthora* is the only possible option. Expression of target genes is enhanced when cocoa plants are infected by *Phytophthora* fungus, with the orientation of highly expressed genes, suggesting that they may play a role in modulating the response to cocoa fungal infections and may lead to resistance to *Phytophthora* infection [49];

Gibberellins: these are the hormones that occur naturally in the parts of the apical plant and in some fungi. Gibberellic products are classified as growth regulators. Presently, nearly 100 diverse gibberellic compounds are identified, but only a limited number of them are biologically active, including GA3, GA4, and GA7. Gibberellic products containing GA4 + 7 have been used for many years around the world to improve the yield and quality of fruit, giving very good results. GA4 + 7 gibberellic products used in the horticulture contain a mixture of gibberellin GA4 and gibberellin GA7. GA4 is a component that can be attributed to beneficial effects. In contrast, GA7 is a ballast gibberellin, which has a negative effect on the formation of flower buds. Separation of both compounds would be too expensive, which is why industrial production usually occurs together. It is important that the farmer chooses the preparation containing the least amount of GA7. The purest product available on the European market is Novagib 010 SL (contains over 90% GA4). Similar products contain only 60%–70% gibberellin GA4 and the remainder are noxious impurities in GA7 form [44,46,48];

Herbicides: citronella oil, vinegar, fatty acids. Vinegar, as a herbicide for a long time is known and used by English gardeners to control weeds [15,28,50];

Insecticides: *B. thuringiensis* var. *kurstaki*; *C. pomella granuloma*; *Verticillium lecanii*; *spinosad*, a biologically active substance isolated from the bacteria *Saccharopolyspora spionsa* is used in the protection of fruits and vegetables and registered in the European Union (in the United States is registered as a pesticide with reduced risk, not as a biopesticide); fatty acids [13,51]. In India, insecticide soap (soap spray insecticide) is used as an insecticide against mites, aphids, skins, skydivers, spider mites, whiteflies, beetles, and other small insects. Another method used against insects is the Nephrite insecticide (Neem), which controls aphids, trips, and whiteflies. Neem oil can disrupt the life cycle of insects at all stages of development (adult insects, larvae, and eggs). Neem oil acts as a hormone—a disintegrator and as a "preventive" for insects feeding on leaves and other parts of plants [49]. Diatomaceous earth is also used as a natural pesticide in reducing the number of snails and other creeping insects. This material works not by poisoning or throttling insects, but because of its abrasive properties and its affinity to absorb lipids from the exoskeleton of insects, which then dehydrates them. Diatomaceous earth must be reapplied after each rain. As insecticides, you can treat some plants, such as powdered chili fruits or tomato leaves, as a natural insecticide that controls the number of aphids [52];

Yellow zucchini mosaic virus: strain, registered to stimulate the defense mechanisms of vegetable plants; *Verticillium albo-atrum* strain WCS850, used as injections as a preventive agent for trees, causing systemic acquired resistance; seaweed extract, plant growth regulator [13,28];

Others: strain of *Candida oleophila*, a biological control agent directed against other fungal species, for example, on gray mold (*Botrytis cinerea*) and blue mold, (*Penicillium expansum; Peniophora gigantea*), used to protect tree trunks against fungal infections; Pepper dust extraction residues (*Piper nigrum*) are used as a repellent for cats and dogs [28].

Biologically active ingredients classified as biopesticides are very diverse. Some biopesticides have several uses. For example, garlic extract may be used as an insecticide, nematicide, repellent for birds and mammals, or as a fungicide [11]. It also has molluscicide properties [3,4]. Applying the same criteria for assessing pesticide residues at the same time can be qualified as basic substances [53]. Examples of such substances are talcum, cinnamon, ferric citrate, potassium chloride, calcium dichloride, carbon dioxide, lemon oil, and urea [46].

Annex IV of the European Parliament Regulation No 396/2005 [14] contains a list of active substances for which maximum residue limits (MRLs) are not required due to risks for the consumer [14]. Guideline SANCO [54] specifies the criteria that must be met by active substances to enable their inclusion in Annex IV. Substances with a low degree of toxicological risk [within the meaning of point 5 of Annex II to Regulation (EC) No 1107/2009] [21], do not require the establishment of an acceptable daily intake and an acute reference dose. In laboratory animals, these substances show no adverse effects. This group may include nontoxin and noninfectious microorganisms as well as substances for which the normal daily food intake is higher than from food after application of a pesticide containing the same chemical substance (e.g., sulfur, iron phosphate, or some plant extracts). For substances that have a nutritional value or are food ingredients, residue levels after application are not a problem. This applies to garlic, some marine algae extract, laminar oil, mint, and many other biopesticides from this group [20,23].

Annex IV may also include substances that leave no residue in protected foodstuffs. These include, for example, carbon dioxide used as an insecticide and acaricide or pheromones used in plant protection [with a simple *Lepidoptera* chain (SCLP)]. Pheromones used as insecticides contain as many as 30 chemical compounds [55]. Currently, other substances from this group are analyzed [56]. To approve the new substance tested in Annex IV, answer such questions as follows:

- Can the assessed substance be qualified as food?
- Does it cause toxicological concerns?
- Is the substance naturally found in the human environment or is it for use in other food-related areas?
- If yes, is the exposure to this substance after use of the plant protection product and is it lower than in other sources naturally present in the environment or from other food applications? [57].

For naturally occurring substances that are not food ingredients, it is necessary to know whether consumers can be exposed to pesticide residues because of their use as active substances. Examples of biologically active ingredients with multifaceted exposure are carbon dioxide or gibberellins [22]. Because carbon dioxide occurs naturally in the environment, being a product of biological and chemical processes, there is no need to show its residues after application as a bioactive pesticide [5]. In the case of gibberellins, which are natural plant hormones, it is not possible to distinguish between exogenous and endogenous sources [3,4]. Most biopesticides are included in the basic or low risk of exposure to substances with a minimum level of risk for which it is not necessary to set an MRL. Many of them are used at concentrations like their natural occurrence [6,7].

Health Issues and Legal Requirements

On a daily basis, people are exposed to many pesticides found in food. Plant protection products get into the air, contaminate the soil and water reservoirs, and are sometimes also absorbed by organisms that were not sprayed with plant protection products [3,4]. Group of people particularly exposed or sensitive to pesticides are first farmers and persons carrying out plant protection treatments, and especially spraying in greenhouses, are exposed to high concentrations of chemicals during work. Unborn and young children are particularly vulnerable. If women during pregnancy get in contact with pesticides, these substances could reach the unborn child directly. The fetus during an individual development is particularly exposed to the effects of contact with harmful substances. Young children are more exposed to adverse effects of pesticides, in addition children have much smaller body sizes than adults, and the metabolism of toxins in their bodies is still inefficient [19,20,22].

Diagnosed health effects in children exposed to pesticides during fetal life primarily cause delayed cognitive development, behavioral problems, and congenital malformations. It has been proven that there is a strong relationship between pesticide exposure and childhood leukemia. There was also a relationship between the exposure to plant protection products and the increased incidence of prostate, lung, and neurodegenerative diseases such as Parkinson's disease and Alzheimer's disease [15,21,58−60]. Pesticides also impair the various functions of the endocrine glands and the immune system of humans and animals. In some cases, the functions of enzymes and signaling mechanisms at the cellular level are disturbed. Some chemicals also impair gene expression by influencing the mechanisms of epigenetic inheritance on future generations that have not yet been exposed to direct contact with pesticides. Therefore, the negative effects of pesticides may be long term and occur many years after their withdrawal from production and marketing [17,19]. The current legal requirements for assessing the risk of combined exposure to chemicals through multiple routes of exposure have also been reviewed, focusing on human health. The main objective is to identify the regulatory needs and current approaches to this type of risk assessment, as well as the challenges associated with the implementation of appropriate, harmonized guidelines at the international level. The review of current legal requirements in the European Union, in the United States and Canada should give an insight into how important the subject matter is in research [61]. Differences in legal requirements have been identified to assess the risks associated with combined exposure to many chemicals and their implementation between the EU and non-EU countries and in many regulatory and regulatory sectors. The methods used to assess the risks associated with combined exposure to many chemicals were assessed. To avoid significant discrepancies between regulations in sectors or countries, the approach to assessing risks associated with total pesticide exposure should be based on similar rules for all types of chemicals. OECD and EFSA have identified the development of harmonized methods for combined exposure to many chemicals as the main priority area [18,20,21,50,61−65]. The Horizon 2020 project "Euro-Mix" is expected to contribute to the further development of an internationally harmonized approach and risk assessment approach by developing an integrated test strategy using in vitro and in silico tests previously verified for chemical mixtures based on relevant data on potential combined effects [61]. These approaches and testing strategies should be integrated with a scientific approach to address the complexity of the topic and improve the risk assessment.

Trends and Market Demand

At the end of the 20th century, a very dynamic development of the biotechnology industry related to the production of biological plant protection products was predicted. OECD member countries have registered over 250 biological plant protection products based on microorganisms. Over 1000 different biopesticides are produced in the world. These comprise not only preparations based on living organisms, but also products containing natural chemical substances (plant extracts), plant growth regulators, or pheromone traps. A large group of microbiological biopreparations indicates that the research was successful, and thus the implementation of preparations containing very different microorganisms, viruses, bacteria (*Bacillus, Pseudomonas*), fungi (*Beauveria, Metarhizium, Pythium, Trichoderma*) and microscopic nematodes (*Heterorhabditis, Steinernema*), which in various ways limit (fight) pathogens and plant pests. Estimates of sales for the second decade of the 21st century exceed $ 1 billion. In the last decade, the sales of products containing macroorganisms (parasitic and predatory insects, predatory mites), that is, biopreparations used in more controlled conditions (greenhouses, tunnels), increased. Their share currently amounts to 55%−60% of total sales of biological plant protection products, and microbiological preparations to approximately 28%. The registration of biological plant protection products based on microorganisms is carried out on similar principles as chemical protection measures, that is, it is a process that is not only rigorous but also very costly. For example, the Swedish company for registration in the EU of two biopreparations containing the bacterium *Pseudomonas chlororaphis* spent over 4.0 million euros. Difficult and expensive registration procedure for biopesticides causes that some manufacturers give up

registration and even manufacture of these preparations, which, unfortunately, reduces the range and limits the scope of application. The number of recorded biopreparations in individual countries is very diverse. It is leading in the United States. In our country, there are only about 40, of which three-fourths are used by gardeners in greenhouse crops and mushroom farms.

Question: why are biological plant protection products used so little in agricultural practice? This is due to quite a few reasons, as no effective biopreparations have been developed so far to combat or reduce the most important diseases (powdery mildew, rust, septicemia, fusariosis) of commodity plants, especially cereals, the effectiveness of most biopesticides is less than that of chemicals. This is especially true for biopreparations based on microorganisms (bacteria and fungi), which are applied to the soil in field conditions, that is to the environment characterized by a very high complexity of interactions between soil microorganisms and other soil inhabitants. Soil is also an environment with a high variability of abiotic factors (humidity, temperature, pH), which have a very significant impact on the development, and thus the effectiveness of organisms introduced into the soil.

Much more effective are those formulations that can be applied directly to the plant's protected organ, for example, by dipping the roots of the seedlings into suspensions of various vaccines. However, they are more expensive, and their application is more troublesome than chemicals, especially in large-scale field crops. Other factors affecting the use of biopesticides may in practice be mentioned that they are usually more expensive, and their application is much more difficult and more troublesome than chemicals, especially in large field crops.

It seems, however, that the future of biological plant protection depends largely on scientific progress in solving the most important problems, such as increasing the reliability and effectiveness of biopreparations in field conditions and developing appropriate formulations of biopesticides based on living organisms adapted to large-area cultivation technologies of the most important agricultural plants.

Not so long ago, it seemed that biotechnology, and especially the technique of genetic modification, would be very helpful in increasing efficiency and reliability. This method would allow a relatively rapid increase in the ability of microorganisms to settle various plant organs, especially roots, and to produce antibiotic substances that inhibit the development of pathogens. Currently, due to the high opposition of various scientific bodies and social organizations and the release of GMOs into the environment, one cannot count on rapid progress in this area. As for the formulations of biopreparations that protect the roots, the optimal solution would be presowing application to seeds, especially cereals, oilseed rape, or bean plants. However, in the case of preparations containing microorganisms, this method of application is very difficult.

Drying of seeds after treatment with biopreparations causes the death of most microorganisms. You can also store stored seeds for longer, as is the case with chemical mortars.

The combined use of chemical and biological mortars is likely to be excluded because the former is usually harmful to living organisms contained in biopesticides. Therefore, the search for microorganisms with features such as resistance to drying, the ability to intensively settle the roots, or a wide range of antifungal activity is difficult. However, biological plant protection products containing microorganisms are successfully registered in OECD countries.

Perspectives and Conclusion

Low-risk substances decompose very quickly (except for basic substances), and therefore they leave little or no residue in the environment or in foods that can affect living organisms. Microorganisms approved for use in the European Union meet the requirements for qualification as biologically active low-risk factors. Nevertheless, during the assessment of microorganisms, a thorough check should be made to determine if they interact with other organisms living in the environment and pay attention to the secondary metabolites they produce, which may have a negative effect on humans and animals. Other problems that arise in the assessment of microorganisms are, to date, insufficiently researched issues of interspecific transfer of genetic material and its expression. Such doubts were expressed, for example, during the assessment of *B. thuringiensis* subsp. *tenebrionis*, strain NB-176.26 [47]. In the assessment of plant extracts, problems may arise during the identification of a mixture of many active substances [53,54,66–68]. However, the clear majority of biopesticides meet the criteria for low-risk active substances or they belong to basic substances. According to Annex VI to Regulations (EC) No 1095/2007 and No 283/2013 [64,69,70], they can be characterized as not constituting harmful effects on human health, either for animals or for groundwater or have other unacceptable effects on the environment. They are also included in Annexes 2229/2004 and IV of Regulation (EC) 396/2005 of the European Parliament and the Council of Europe [14,62] as active substances for which no MRLs are required due to the risk of danger. It should be emphasized that a very important branch of plant protection [47,62] actively participates in the promotion of biopesticides (Table 4.1).

TABLE 4.1
Natural bioactive compounds to control of bacteria and fungi pathogenic to plants.

Natural Bioactive Compounds	CONTROL FROM			
	Bacteria	Fungi *sensu lato*	Plants/product form	References
Allicin	*Agrobacterium* spp., *Erwinia* sp., *Pseudomonas* spp., *Xanthomonas* spp.	*Fusarium oxysporum, Sclerotinia sclerotiorum*	*Allium* sp./garlic extracts and homogenates in vitro	[23]
		Alternaria alternata, Botrytis cinerea, Colletotrichum coccodes, Rhizoctonia solani	Preparation bioczos płynny and bioczos standard (garlic pulp) in vitro and in vivo	[26,27,31,33]
		F. oxysporum f. sp. *callistephi, Fusarium moniliforme,*	Garlic as forecrop/garlic extract in vitro	[24,33]
		Pythium aphanidermatum, Pythium ultimum, Phytophthora cinnamomi	*Allium* sp./garlic extract in vitro	[25]
		Cercospora arachidicola		[33]

Continued

TABLE 4.1
Natural bioactive compounds to control of bacteria and fungi pathogenic to plants.—cont'd

Natural Bioactive Compounds	CONTROL FROM		Plants/product form	References
	Bacteria	Fungi *sensu lato*		
Naringin		*Botrytis cinerea, Alternaria alternate*	Biosept 33 SL (grapefruit extract) in vitro and in vivo	[27,34–36]
		Phytophthora cryptogea, P. cinnamomi, Fusarium oxysporum f. sp. *cyclaminis, Phomopsis sojae, Fusarium* spp., *Sclerotinia sclerotiorum, Phoma exigua*	Biosept 33 SL (grapefruit extract) in vivo	[38,39]
		Cercospora beticola	Biosept 33 SL (grapefruit extract) in vivo	[29]
		Fusarium culmorum, F. oxysporum, F. solani, Rhizoctonia solani, Sclerotinia sclerotiorum	Biosept 33 SL (grapefruit extract) in vitro	[34]
Terpenes		*Botrytis cinerea, Aspergillus fumigatus, Chaetomium globosum, Penicillium chrysogenum, Fusarium graminearum, F. culmorum, Pyrenophora graminea, Ascochyta rabiei, Colletotrichum lindemuthianum, Drechslera avenae, Alternaria Radecina, A. dauci*	*Melaleuca alternifolia* L. Timorex gold 24 EC	[40–42]
		Bremia lactucae	Timorex gold 24 EC in vitro and in vivo	[43]
		Erysiphe cichoracearum	Timorex gold 24 EC in vivo	[46]
Thymol		*Cladosporium sphaerospermum, Rhizopus* sp., *Ascochyta rabiei, Colletotrichum lindemuthianum, Fusarium graminearum, F. culmorum*	*Thymus* sp./thyme herb	[47]
		Alternaria alternata, Penicillium cyclopium, Trichothecium roseum	*Thymus* sp./oil, hydrogel, dried and powdered thyme herb	[48]
		Aspergillus flavus, Fusarium oxysporum, Cladosporium herbarum, Aspergillus niger, Alternaria alternata, Botryodiplodia theobromae, Aspergillus fumigatus, Curvularia lunata	*Thymus* sp./thyme oil in vitro	[51]
		Fusarium oxysporum, F. solani	Thyme oil in vitro	[26]
Chitosan		*Botrytis cinerea*	Preparation biochikol 020 PC	[34,47]
		Fusarium spp., *Pythium* spp., *Botrytis cinerea, Rhizoctonia solani*	Chitosan	[57]

REFERENCES

[1] FAOSTAT, Faostat Data Base, 2017. http://www.fao.org/faostat/en/.

[2] B. Sawicka, The possibilities of uses of biomarkers to the estimation of exposure of potato to herbicides, in: 9th International Conference on Biomarkers Expert Speakers. Evidence Based. Health Care Professionals. Osaka, Japan, October 26-28, 2017. http://www.omicsgroup.com/conferences/ACS/conference/pdfs/molecular-cancer-biomarkers2017_workshop-program.pdf.

[3] European Food Safety Authority, Conclusion on the peer review of the pesticide risk assessment of the active substance garlic extract, EFSA J. 10 (2) (2012a) 2520−2559.

[4] European Food Safety. Authority, Conclusion on the peer review of the pesticide risk assessment of the active substance gibberellins, EFSA J. 10 (1) (2012b) 2502−2551.

[5] European Food Safety Authority, Conclusion on the peer review of the pesticide risk assessment of the active substance *Bacillus thuringiensis* ssp. *tenebrionis* strain NB-176, EFSA J. 11 (1) (2013a) 3024−3059.

[6] European Food Safety Authority, Conclusion on the peer review of the pesticide risk assessment of the active substance carbon dioxide, EFSA J. 11 (5) (2013b) 3053−3153.

[7] European Food Safety Authority, Conclusion on the peer review of the pesticide risk assessment of the active substance orange oil, EFSA J. 11 (2) (2013c) 3090−3144.

[8] Regulation (EC). No. 1107/2009, OJ L 309 of 24.11, 2009, pp. 1−50.

[9] European Commission Art. 80 Paragraph 7 of Regulation (EC) 1107/2009 of the European Parliament and the Council of Europe, 2009.

[10] Commission Implementing Regulation (EU) No. 540/2011, OJ L 153 of 11.06, 2011, pp. 1−186.

[11] J.J. Villaverde, B. Sevilla-Morán, P. Sandín-España, C. López-Goti, J.L. Alonso-Prados, Biopesticides in the framework of the European. Pesticide regulation (EC) No. 1107/2009, Pest Manag. Sci. 70 (2013) 2−5.

[12] M. Kołodziejczyk, A. Szmigiel, D. Ropek, Effectiveness of potato protection production using selected insecticides to combat potato beetles (*Leptinotarsa decemlineata* Say), Acta Sci. Pol. Agric. 8 (4) (2009) 5−14.

[13] D. Ropek, M. Kołodziejczyk, Effectiveness of selected insecticides and natural preparations against *L. decemlineata*, Potato Res. 62 (85) (2019). https://doi.org/10.1007/s11540-018-9398-8.

[14] Regulation (EC) No. 396/2005, OJ L 70 of 16.03, 2005, pp. 1−16.

[15] N. Defarge, E. Takács, V.L. Lozano, R. Mesnage, J. Spiroux de Vendômois, G.E. Séralini, A. Székács, A Co-formulants in glyphosate-based herbicides disrupt aromatase activity in human cells below toxic levels, Int. J. Environ. Res. Public Health 13 (2016) 264−280.

[16] NFUP, Nationales Fremd Stoffuntersuchungs Programm, Jahresbericht, Bern, Schweiz, 2017.

[17] N. Roth, M.F. Wilks, Combination ("cocktail") Effects of Pesticide Residues in Food SCAHT Report for FSVO, Combination effects of pesticide residues in food Technical Report, 11.2018, Final Version, 06 November 20, 2018, https://doi.org/10.13140/RG.2.2.30899.07208, file:///C:/Users/Barbara/Downloads/CombinationEffectsofPesticideResiduesinFoodSCAHTReportforFSVO2018.pdf.

[18] S. Rotter, A. Beronius, A.R. Boobis, A. Hanberg, J. van Klaveren, M. Luijten, K. Machera, D. Nikolopoulou, H. van der Voet, J. Zilliacus, R. Solecki, Overview on legislation and scientific approaches for risk assessment of combined exposure to multiple chemicals: the potential EuroMix contribution, Crit. Rev. Toxicol. (2019) 796−814, https://doi.org/10.1080/10408444.2018.1541964. https://doi.org/10.1080/10408444.2018.1541964.

[19] J.I. Barčić, R. Bažok, S. Bezjak, T.G. Čuljak, J. Barčić, Combinations of several insecticides used for integrated control of Colorado potato beetle (*Leptinotarsa decemlineata*, Say., Coleoptera: *chrysomelidae*), J. Pest. Sci. 79 (2006) 223−232. https://doi.org/10.1007/s10340-006-0138-5.

[20] S. Barlow, ILSI (International Life Sciences Institute) Europe Concise Monographs Series 2005, 2018, pp. 1−31. Brussels, Belgium, http://ilsi.org/publication/threshold-of-toxicological-concern-ttc/BfR (Bundesinstitut für Risikobewertung.

[21] A. Boobis, R. Budinsky, S. Collie, K. Crofton, M. Embry, S. Felter, R. Hertzberg, D. Kopp, G. Mihlan, M. Mumtaz, P. Price, K. Solomon, L. Teuschler, R. Yang, R. Zaleski, Critical analysis of literature on low-dose synergy for use in screening chemical mixtures for risk assessment, Crit. Rev. Toxicol. 41 (2011) 369−383.

[22] K. Czaja, K. Góralczyk, P. Strucinski, A. Hernik, W. Korcz, M. Minorczyk, M. Łyczewska, J.K. Ludwicki, Biopesticides − towards increased consumer safety in the European Union, Pest Manag. Sci. 71 (2015) 3−6.

[23] M. Keidan, Optimization of Winter Oilseed Rape Technological Parameters in the Organic Farming System (Ph.D. thesis), Aleksandras Stulginskis University, Kaunas, 2018, p. 221.

[24] B. Kulig, A. Lepiarczyk, A. Oleksy, M. Kołodziejczyk, The effect of tillage system and fore crop on yield and values of LAI and SPAD indicators of spring wheat, Eur. J. Agron. 33 (2010) 43−51.

[25] A. Lepiarczyk, B. Kulig, K. Stepnik, The effect of simplified soil cultivation and fore crops on LAI development of selected winter wheat varieties for crop rotation, Fragmenta Agronomica 2 (86) (2005) 98−105.

[26] D. Levickienė, E. Jarienė, M. Gajewski, H. Danilčenko, N. Vaitkevičienė, J.L. Przybył, M. Sitarek, Influence of harvest time on biologically active compounds and antioxidant activity in mulberry leaves in Lithuania//Notulae Botanicae, Horti Agrobotanici Cluj-Napoca. ISSN: 0255-965X 45 (2) (2017) 431−436.

[27] D. Levickienė, The Influence of the Biodynamic Preparations on the Soil Properties and Accumulation of Bioactive Compounds in the Leaves of White Mulberry (*Morus alba* L.) (Ph.D. thesis), Aleksandras Stulginskis University, Kaunas, 2018, p. 212.

[28] F.E. Dayan, C.L. Cantrell, S.O. Duke, Natural product in crop protection, Bioorg. Med. Chem. 1 (2009) 4022–4034.

[29] J.J. Lipa, D. Sosnowska, S. Pruszyński, Advances in biological control of *Leptinotarsa decemlineata* in Poland, Bull. OEPP/EPPO 28 (1998) 463–469.

[30] C. Daniel, E. Wyss, Field application of Beauveria bassiana to control the European fruit fly Rhagoletis cerasi, J. Appl. Entomol. 134 (2010) 675–681. https://doi.org/10.1111/j.1439-0418.2009.01486.

[31] J. Fargues, J.P. Cugier, P. van de Weghe, Experimentation in Parcells du champignon Beauveria bassiana (Hyphomycete) contra *Leptinotarsa decemlineata* (Col. *Chrysomelidae*), Acta Ecol. 1 (1980) 49–61.

[32] A.E. Hajek, R.S. Soper, D.W. Roberts, T.E. Anderson, K.D. Biever, D.N. Ferro, D.N. LeBrun, R.H. Storch, Foliar applications Beauveria bassiana (Balsamo) Vuillemin to control potato beetle, *Leptinotarsa decemlineata* (Say) (Coleoptera: *chrysomelidae*). An overview of the results of pilot tests from the northern United States, Can. Entomol. 119 (1987) 959–974.

[33] T.J. Poprawski, R.I. Carruthers, J.I. Speese, D.C. Vacek, L.E. Wendel, The early application of *Beauveria bassiana* mushrooms and the introduction of the predator hemipteran *Perillus bioculatus* to control potato beetle, Biol. Control 10 (1997) 48–57.

[34] D. Waligóra-Rosada, Saponins as the agents influencing the fecundity of Colorado potato beetle (*Leptinotarsa decemlineata* Say), Prog. Plant Prot. 50 (2010) 394–397.

[35] M.I. Rudeen, S.T. Jaronski, J.L. Petzold-Maxwell, A.J. Gassmann, Entomopathogenic fungi in cornfields and their potential to fight the larvae of western corn rootworm *Diabrotica virgifera* var. *virgifera*, J. Invertebr. Pathol. 114 (2013) 329–332.

[36] C. Wang, M.G. Feng, Advances in fundamental and applied studies in China of fungal biocontrol agents for use against arthropod pests, Biol. Control 68 (2014) 129–135.

[37] Ž. Laznik, T. Tóth, T. Lakatos, M. Vidrih, S. Trdan, Inspection of Colorado potato beetle (*L. decemlineata* [Say]) on potato under field conditions: comparison of the effectiveness of the two strains *Steinernema feltiae* (Filipjev) and spraying with thiamethoxam, J. Plant Dis. Prot. 117 (3) (2010) 129–135.

[38] D.W. Long, F.A. Drummond, E. Groden, Sensitivity of Colorado potato beetle (*Leptinotarsa decemlineata*) to *Beauveria bassiana*, J. Invertebr. Pathol. 71 (1998) 182–183. https://doi.org/10.1006/jipa.1997.4714.

[39] S.P. Wraight, M.E. Ramos, Application parameters affecting field efficacy of Beauveria bassiana foliar treatments against Colorado potato beetle *Leptinotarsa decemlineata*, Biol. Control 23 (2) (2002) 164–178.

[40] M.A.M. Osman, Biological efficacy of some insecticides taking part in research and conventional in control of various stages of Colorado potato beetle, *Leptinotarsa decemlineata* (Say) (Coleoptera: chrysomelidae), Plant Prot. Sci. 46 (3) (2010) 123–134.

[41] G.C. Cutler, C.D. Scott-Dupree, J.H. Tolman, C.R. Harris, The field efficacy of novaluron to control potato beetle (Coleoptera: *chrysomelidae*) on potato, Crop Protect. 26 (2007) 760–767. https://doi.org/10.1016/j.cropro.2006.07.002.

[42] J.E. Dripps, Z. Smilowitz, Analysis of potato plant growth damaged by potato beetle (Coleoptera: *chrysomelidae*) at various stages of plant growth, Environ. Entomol. 18 (1989) 854–867.

[43] G. Boiteau, Time of application of insecticides to control potato beetle, *Leptinotarsa decemlineata* (Say), on potatoes in New Brunswick, Can. Entomol. 120 (1988) 587–591.

[44] D. Chandler, A.S. Bailey, G.M. Tatchell, G. Davidson, J. Greaves, W.P. Grant, The development, regulation and use of biopesticides for integrated pest management, Phil. Trans. R Soc. B 366 (2011) 1987–1998.

[45] G.,V.P. Reddy, S. Tangtrakulwanich, S. Wua, J.H. Miller, V.L. Ophus, J. Prewett, S.T. Jaronski, Evaluation of the effectiveness of entomopaths in the control of wireworms (Coleoptera: *Elateridae*) on spring wheat, J. Invertebr. Pathol. 120 (2014) 43–49. https://doi.org/10.1016/j.jip.2014.05.005.

[46] R.S. Mann, P.E. Kaufman, Natural product pesticides: their development, delivery and use against insect vectors, Mini-Reviews Org. Chem. 9 (2012) 185–202.

[47] European Union Pesticides Database, 2019. Available: http://ec.europa.eu/food/plant/protection/evaluation/database_act_subs_en.html.

[48] B. Sawicka, Rate of spread of fungal diseases on potato plants as affected by application of a biostimulator and foliar fertilizer, Published by the Editorial House: Wieś Jutra, Limited, in: Z. Dąbrowski (Ed.), Biostimulators in Modern Agriculture. Solanaceous Crops, Plantpress, Warsaw, 2008, ISBN 83-89503-55-7, pp. 68–76.

[49] S.M.N. Khalid, Natural and Homemade Pesticides, 2018, ISBN 978-953-307-531-0, p. 458. Margarita Stoytcheva p. cm. free online editions of InTech Books and Journals can be found at: www.intechopen.com. Mexicali, Baja California Mex.

[50] M.R. Rapagnani, M. Magliuolo, M. Picciolo, L. Nencini, T. Galassi, F. Mazzini, Future Availability of Pesticides in the Integrated Pest Management Agricultural Programme in Italy in Accordance with the Application of the New European Regulation No. 1107/2009 Concerning the Placing of Plant Protection Products on the Market: Impact of the Application of Cut-Off Criteria and Selection Criteria for Substances that Are Candidates for Substitution, 2011. RT/2011/8/ENEA. Available: http://www.csa.it/centri/ra/docs/allegatoENEA.pdf.

[51] S.O. Duke, C.L. Cantrell, K.M. Meepagala, D.E. Wedge, N. Tabanca, K.K. Schrader, Natural toxins for use in pest management, Toxins 2 (2010) 1943–1962.

[52] S.M.N. Khalid, Natural and Homemade Pesticides, 2018. http://file:///C:/Users/Barbara/Downloads/Natural_and_homemade_pesticides.pdf.

[53] SANCO/10472/2003 –rev.5 6.7.2004 Draft working document concerning the data requirements for active

substances of plant protection products made from plants or plant extracts (http://ec.europa.eu/food/plant/protection/evaluation/chem_subst.pdf).

[54] SANCO/2609/08. 2008, rev. 1.

[55] SANCO/2633/2009. 2009.

[56] SANCO/5272/2009. 2010, rev. 3.

[57] K. Góralczyk, G. Kostka, J.K. Ludwicki, P. Strucinski, Guide to terminology: toxicology, food safety, public health and risk assessment (in polish), in: J.K. Ludwicki (Ed.), National Institute of Public Health — National Institute of Hygiene, Warsaw, Poland, 2013.

[58] BLV (Bundesamt für Lebensmittelsicherheit und Veterinärwesen), Erläuterungen zur Verordnung des EDI über die Höchstgehalte für Pestizid-rückstände in oder auf Erzeugnissen pflanzlicher und tierischer Herkunft (Verordnung über Pestizidrückstände, VPRH, Bern, Schweiz, 2017a. Mai.

[59] BLV (Bundesamt für Lebensmittelsicherheit und Veterinärwesen), 2017.

[60] BLV (Bundesamt für Lebensmittelsicherheit und Veterinärwesen), Jahresbericht 2017 zu den Kontrollprogrammen an der Grenze. Überwachung von pflanzlichen Lebensmitteln und Gebrauchsgegenständen, 2018. Bern, Schweiz.

[61] Anonymous, What Does the Future Hold for Harmonized Human Health Risk Assessment of Plant Protection Products? Workshop Summary Report, BfR, Berlin, Germany, November 2017, pp. 23—24. https://www.bfr.bund.de/cm/349/workshop-what-does-the-future-hold-for-harmonised-human-health-riskassessment-of-plant-protection-products-conference-report.pdf Bliss CI.

[62] Commission Regulation (EC), No, 2229/2004, OJ L 379 of 24.12.2004., 2004, pp. 13—63.

[63] Commission Regulation (EC), No. 1095/2007, OJ L 246 of 21.09., 2007, pp. 19—28.

[64] Commission Regulation (EU) (2013a) No. 283/2013, OJ L 93 of 3.4.2013. 2013. 1—84.

[65] Commission Regulation (EU) (2013b) No. 284/2013, OJ L 93 of 3.4.2013. 2013. 85—152.

[66] SANCO/2621/08. 2008, rev. 2.

[67] SANCO/10363/2012. 2012a, rev. 7.

[68] SANCO/11188/2013. 2013a. rev. 2.

[69] Commission Regulation (EC), No. 1095/2007 (2007). OJ L 246 of 21.9.2007., 2007, pp. 19—28.

[70] European Commission Neonicotinoids. 2018. https://ec.europa.eu/food/plant/pesticides/approval_active_substances/approval_renewal/neonicotinoids_en.

CHAPTER 5

Natural Compounds Against Plant Pests and Pathogens

AGNIESZKA JAMIOŁKOWSKA, PHD • MAREK KOPACKI, PHD

INTRODUCTION

Every year, the damage done to crops by pests and diseases (according to data of the United Nations Food and Agricultural Organization—FAO) constitutes approximately 20%–25% of the potential world yield of crops. Crop plants are continuously under attack by pathogens and pests, both during pre- and postharvest stages, often causing economically important yield losses. Therefore, plant protection play an extremely important role in increasing the production of agricultural crops and in protecting them. Plant protection is a branch of agricultural science that devises ways and means of controlling diseases, pests, and weeds of crops and trees, as well as a set of measures used in agriculture and forestry to prevent and eliminate the damage done to plants by harmful organisms. The goal of plant protection is not only to destroy harmful organisms or limit their activity but also to forecast the time they appear and the possible extent to which they might spread as well as to prevent especially harmful organisms moving from some countries and regions to others.

Since ancient times, diseases and pests have been known to cause harm to plants (especially rusts, smuts, and desert locusts). In the latter half of the 19th century, scientists made new discoveries among which are species of phytopathogenic fungi and harmful insects through the study of their morphology and developmental characteristics. The tremendous damage done to the economy of many countries in the latter half of the 19th century by pests and diseases (including *phylloxera*, locusts, and potato blight) made it necessary to centralize research efforts and the ways of devising control measures. By the early 20th century, thousands of new species of phytopathogenic fungi, bacteria, viruses, nematodes, and insects had been discovered, and the species composition of the principal pests and their biology and physiology were being studied. Research in plant pathology and entomology was based on the principles and methods of ecology and biotechnology. The concept of "crop protection" as a discipline in its own rights dates from the 19th century. At the turn of the 20th century, several authors looked back on crop protection in the general or on its contributing disciplines' entomology, phytopathology, and virology [1]. Methods of controlling harmful organisms were improved.

Agrotechnical, biological, chemical, biophysical, and other control methods, including those involving direct extermination of harmful organisms as well as indirect actions through environmental factors, plant hosts, or a complex of other organisms associated with the development of pests or other pathogens, were devised. Scientists were looked for different methods of plant protection and, above all, preventive methods, which generally are the most effective. After II World War, agricultural intensity with using chemical pesticides has changed considerably with strong intensification especially in developed countries [2]. From the beginning, pesticides met with criticism and environmental concern and alternatives were actively sought for. Official services stimulated the use of pesticides, often for their own purposes, and credit for pesticide purchases was always available. Often, farmers were forced into package deals including seed of high-yielding varieties, fertilizers, and pesticides. In 1968, responding to the outcry following the publication of "Silent Spring" [3], the Food and Agriculture Organization of the United Nations (FAO) created a Panel on Integrated Pest Control, which functioned from 1968 to 1994. With the support from the Panel, FAO introduced IPM in South East Asian rice cultivation [1]. This concept was determined at the Entomological Congress in Tokyo in 1976.

The main objective of integrated pest management (IPM) of plants is to obtain optimal yields of high quality without destroying the natural environment

Natural Remedies for Pest, Disease and Weed Control. https://doi.org/10.1016/B978-0-12-819304-4.00005-1

and human health. This system is based on the sustainable use of pesticides, mainly on the application of nonchemical methods of plant protection against diseases, pests, and weeds. The integration of methods in plant protection was established in the 20th century, in the course of functioning of crop production systems. In the process of IPM, they are used mainly biological and physiological mechanisms of plants supported by a rational use of conventional, natural, and biological preparations [4]. Furthermore, there is an increasing market, for organic produce, for which most pesticides and inorganic fertilizers are unacceptable. For most countries, organic food comprises a small but rapidly growing part of the food industry. The fact that people are becoming more interested on how their food is sourced and where it derives from has however led to an increase in demand for organic food among the consumers, especially in the last few years [5]. There is therefore a pressing need to develop more effective, sustainable, and environmentally friendly tools for pests' control. It is connected with theory of sustainable development in which the needs of the present generation can be met without diminishing the opportunities of future generations to satisfy them. The use of beneficial microorganisms and natural preparations for the control of pests is very attractive, and the availability of novel application and molecular techniques open unexplored avenues for plant protection approaches.

Extracts from Asteraceae Family Plants Against Pests

Numerous species of plants from the Asteraceae family have an impact on pests. The most important of them include plants of the genus *Achillea, Artemisia, Ambrosia, Chrysanthemum, Matricaria, Tagetes, Tanacetum,* and *Taraxacum* [6−8]. The substances contained in these plants act on pests as stimulants, antifeedants, repellents, or attractants. Because secondary metabolites in leaves and flowers of some chrysanthemum cultivars (unsaturated fatty acid isobutyl amides) have influence on host-plant resistance against the western flower trips (WFT) *Frankliniella occidentalis* [9]. Aqueous extracts of marigold were used against root-knot nematode in tomato [10] and insects [6].

The most important of natural materials is pyrethrin. Pyrethrins are insecticidal second metabolites of *Chrysanthemum cinerariaefolium* which accumulate in high concentration especially in the flowers [11,12]. In ancient time, natural pyrethrins, as well as dried flower powders (in China 1000 BC and "Persian Powder" in Iran), were employed to control field and household

pests, but have been replaced by synthetic pyrethroids—the human-made form of pyrethrins [13]. Pyrethrin from plants was produced in Europe after 1820, and early commercial production occurred in Dalmatia and later in Japan in 1881. Introduced into eastern Africa in the 1920s, *Chrysanthemum cinerariaefolium* thrived, increasing in quality and items of beauty and utility [14].

Chemical composition, mode of action, and industrial production

Pyrethrins comprise a group of six closely related monoterpene esters. Pyrethroids include ester compounds derived from chrysanthemum and quinin acid and alcohols: pyrerol, cinerol, and jasmolin. The ester of chrysanthemum acid is pyrethrin I, cinerin I, and jasmolin I. Ferric acid esters are pyrethrin II, cinerin II, and jasmolin II [11]. The pyrethrum extract according to the world standard is included: jasmolin I—8%, pyrethrin I—46%, pyrethrin II—24%, cinerin I—11%, cinerin II—7%, jasmolin II—4% [15]. Pyrethroids act specifically on invertebrates, disrupting neurotransmission in the nervous system. Pyrethroids inhibit the activity of pyrophosphatase and calcium and magnesium-dependent ATPases, adenylyl cyclases, and phosphodiesterase. They act on pests quickly, initially causing movement disorders, convulsions, and nerve conduction and ultimately death. Some have a strong and irritating smell, and others are pleasant to use and do not cause irritation to the respiratory system [6].

The industrial production is based on their extraction from *Chrysanthemum cinerariaefolium*. The world production of natural pyrethrins still falls short of global market demand stimulating the research in vitro production as an alternative to conventional cultivation methods. The most popular method is extraction from petals [15,16]. The different biotechnological alternatives such as callus cultures, shoot and root cultures, plant cell suspension cultures, and bioconversion of precursors by means of enzymatic synthesis or genetically engineered microorganisms, as well as the progress achieved in methods for the identification and quantitation of insecticidal compounds have been reviewed [7,11]. Although technology for plant cell culture exists, industrial applications have, to date, been limited due to both the low economic viability and technological feasibility at large scale. Bioconversion of readily available precursors looks more attractive, but more research is needed before this technology is used for the industrial production of pyrethrins [6]. In herbal terms, the Dalmatian stonecrop is provided by the flower—Flos Pyrethrin. For medicinal and industrial purposes, this plant is cultivated in Kenya, Tasmania, Tanzania, Rwanda, and Ecuador. Smaller crops are in Brazil, Croatia, and India. Similar insecticidal properties have species (subspecies) of *Chrysanthemum coccinea* and *Chrysanthemum marshalii*. Raw material originating in Kenya contains not less than 1.3% of pyrethrin; raw material of Japanese origin 0.9%—1% of pyrethrin, and from Dalmatia 0.7%—0.8% of pyrethrin. Insecticidal and antiparasitic retrains are also found in the Roman Bertram root—*Anacyclus pyrethrum* DC. = *Anacyclus pseudopyrethrum* Asch. = *Anthemis pyrethrum* Desf./provides Radix Pyrethrin Romani—*Anacycli pyrethri* radix, also rich in inulin 30%—55%. According to the literature, in the root of the medical nursery—*Anacyclus officinarum* Hayne (raw material Radix *Pyrethri germanici*, contains approx. 50% inulin) [17].

Advantages and disadvantages

Pyrethrins are included in many preparations for combating flies, mosquitoes, beetles, aphids, etc., in the form of powder and suspensions (for sprinkling, preparation of spray solutions). The flower extract showed active biological effect against beetle flour *Tribolium castanum* (*Coleoptera*) and reached 100% at the concentration 40% [16]. High mortality and repellent effect were noted. Padin et al. [8] showed that plant extracts from *Ambrosia tenuifolia*, *Matricaria chamomilla*, and *Tagetes minuta* were used against adult beetles of Tribolium *castaneum* in stored grains. Pyrethrins of plant origin are approved for pest management in organic farming by European Crop Protection Association [18]. They are biodegradable in the environment, and they do not accumulate in the system [6]. Pyrethroids exhibit very low toxicity to humans and other mammals. Compounds found in flowers of chrysanthemums may inhibit the process of tumor formation [19]. Unfortunately, pyrethrins can irritate and sensitize and affect the severity of asthma symptoms [20,21].

Natural Compounds Against Plant Pathogens

Natural pesticides are substances of natural origin (plant or animal) that limit the development of pathogens and insects. Natural bioactive compounds that have an impact on pathogens are called natural fungicides. The action of these organic compounds is not specific, and their effect on pathogens is comprehensive (Table 5.1). Natural bioactive compounds used in plant protection kills fungi (fungicidal effect) or limit their development (fungistatic effect), as well as involve plant defense reactions (elicitors) [22]. Direct action of these compounds is based on inhibition of fungal sporulation, germination of spores, and reduction of hyphae growth [23]. Natural secondary metabolites produced by plants under the influence of elicitors also have a protective role of plants in relation to pathogens. As a result of the indirect action of plant extracts in plant cells, there is the synthesis of phytoalexins, the accumulation of callose and the lignification of cell walls, which protect plants against pathogens [23].

Plant origin compounds

Allicin is an organosulfur compound obtained from garlic, which has antibacterial and antifungal activity. It occurs in the tissues of garlic and onions (*Allium* L.), as well as in other plants of Amaryllidaceae family. It arises from the transformation of alliin under the influence of the alliinase enzyme. Antifungal properties of allicin (contained in garlic) against various pathogenic fungi are described by many authors (Table 5.1). Saniewska [24] showed that garlic extracts, and homogenates inhibit the development of *Fusarium oxysporum* and *Sclerotinia sclerotiorum*. The garlic juice used in vitro limits the growth of bacteria of the genus *Agrobacterium*, *Erwinia*, *Pseudomonas*, and *Xanthomonas*. The use of garlic as fore crop resulted in a strong inhibition of the development of fusariosis of the Astra vessels grown in the medium containing *F. oxysporum* f. sp. *callistephi* [25]. The influence of garlic extracts on soil pathogens (*Pythium aphanidermatum*, *P. ultimum*, *Phytophthora cinnamomi*) was also demonstrated by Sealy and coauthors [26]. Antifungal properties of garlic pulp contained in the preparation of Bioczos Liquid (Himal—Poland) have been confirmed by Jamiołkowska and Wagner [27]. The preparation in vitro inhibited linear growth of *Alternaria alternata*, *Botrytis cinerea*, *Colletotrichum coccodes*, and *Rhizoctonia solani*. The garlic pulp contained in preparation of Bioczos Liquid, which is used in pepper protection against pathogenic fungi, proved to be more effective than azoxystrobin (Amistar 250 SC fungicide) [28]. Many authors write about the high efficiency of allicin [29—31]. Horoszkiewicz-Janka et al. [32] showed a low degree of diseases index (DI) of leguminous plants whose seeds were treated with garlic pulp, and the effectiveness of the treatment was similar to that of a chemical protection based on carboxin and thiuram (Vitavax 200 FS). Garlic pulp additionally influenced the seed germination energy improvement. Marjańska-

TABLE 5.1
Natural Bioactive Compounds to Control of Bacteria and Fungi Pathogenic to Plants.

Natural Bioactive Compounds	CONTROL FROM		Plants/Product Form	References
	Bacteria	Fungi *sensu lato*		
Allicin	*Agrobacterium* spp., *Erwinia* sp., *Pseudomonas* spp., *Xanthomonas* spp.	*Fusarium oxysporum, Sclerotinia sclerotiorum*	*Allium* sp./garlic extracts and homogenates in vitro	[24]
		Alternaria alternata, Botrytis cinerea, Colletotrichum coccodes, Rhizoctonia solani	Preparation bioczos płynny and bioczos standard (garlic pulp) in vitro and in vivo	[27,28,31,33]
		F. oxysporum f. sp. *callistephi, Fusarium moniliforme,*	Garlic as forecrop/ garlic extract in vitro	[25,33]
		Pythium aphanidermatum, Pythium ultimum, Phytophthora cinnamomi	*Allium* sp./garlic extract in vitro	[26]
		Cercospora arachidicola		[33]
Naringin		*Botrytis cinerea, Alternaria alternata*	Biosept 33 SL (grapefruit extract) in vitro and in vivo	[28,34,35,39]
		Phytophthora cryptogea, P. cinnamomi, Fusarium oxysporum f. sp. *cyclaminis, Phomopsis sojae, Fusarium* spp., *Sclerotinia sclerotiorum, Phoma exigua*	Biosept 33 SL (grapefruit extract) in vivo	[37,38]
		Cercospora beticola	Biosept 33 SL (grapefruit extract) in vivo	[29]
		Fusarium culmorum, F. oxysporum, F. solani, Rhizoctonia solani, Sclerotinia sclerotiorum	Biosept 33 SL (grapefruit extract) in vitro	[39]
Terpenes		*Botrytis cinerea, Aspergillus fumigatus, Chaetomium globosum, Penicillium chrysogenum, Fusarium graminearum, F. culmorum, Pyrenophora graminea, Ascochyta rabiei, Colletotrichum lindemuthianum, Drechslera avenae, Alternaria Radecina, A. dauci*	*Melaleuca alternifolia* L./Timorex gold 24 EC	[40–42]
		Bremia lactucae	Timorex gold 24 EC in vitro and in vivo	[43]
		Erysiphe cichoracearum	Timorex gold 24 EC in vivo	[44]

Thymol	*Cladosporium sphaerospermum, Rhizopus* sp., *Ascochyta rabiei, Colletotrichum lindemuthianum, Fusarium graminearum, F. culmorum*	*Thymus* sp./thyme herb	[46]
	Alternaria alternata, Penicillium cyclopium, Trichothecium roseum	*Thymus* sp./oil, hydrogel, dried, and powdered thyme herb	[48]
	Aspergillus flavus, Fusarium oxysporum, Cladosporium herbarum, Aspergillus niger, Alternaria alternata, Botryodiplodia theobromae, Aspergillus fumigatus, Curvularia lunata	*Thymus* sp./thyme oil in vitro	[50]
	Fusarium oxysporum, F. solani	Thyme oil in vitro	[27]
Chitosan	*Botrytis cinerea*	Preparation biochikol 020 PC	[39,59]
	Fusarium spp., *Pythium* spp., *Botrytis cinerea, Rhizoctonia solani*	Chitosan	[56]

Cichoń and Sapieha-Waszkiewicz [31] pointed out the effectiveness of allicin (in Bioczos Standard) in the protection of strawberries against gray mold. The effectiveness of allicin (containing in biopreparation—Bioczos Liquid) also depends on the weather conditions. In the vegetative season, abundant in rainfall, their effectiveness may be unsatisfactory. Antifungal properties of allicin obtained from garlic are described by Abdulrahman and Alkhail [33]. The extracts of garlic were very effective in reducing the growth of *A. alternata, F. moniliforme, Cercospora arachidicola*, and *R. solani* in vitro cultures. The effectiveness of allicin (preparations based on garlic) depends not only on weather conditions, but also on the type of pathogenic fungi. Sadowski and coauthors [30] showed lack of effectiveness of Bioczos BR in the protection of onion bulbs against downy mildew of onion. The scientific research also shows the effect of Bioczos Liquid on the increase of fungi biodiversity occurring in the soil communities of cultivated plants, which minimizes the risk of dominance of one species that can cause disease [28].

Naringin is a flavonoid glycoside found in the grapefruit pulp and seeds, as well as in the epidermis of lemon and orange (*Citrus* L.) and it is responsible for their bitter taste. In the human body, it is metabolized to naringenin (naringin aglucon). Naringin has antimicrobial properties and is used against bacterial and fungal diseases (Table 5.1). This organic compound contained in the preparation Biosept 33SL (Poland) is also used to control pathogenic fungi on cultivated plants. Biosept 33SL contains grapefruit pulp and seeds extract (33%). Many authors have demonstrated its effectiveness against gray mold, fusariosis, and alternariosis in vegetable and ornamental plants [34—36]. The preparation had a significant efficacy in the control of some soil pathogens in ornamental plants (*Phytophthora cryptogea, P. cinnamomi, Fusarium oxysporum* f. sp. *cyclaminis*) and in soybean crops (*Phomopsis phaseoli, Fusarium* spp., *S. sclerotiorum, Phoma exigua*) [37,38]. Its effectiveness in protection of potato and onion against fungal diseases was confirmed by Sadowski and coauthors [30]. The preparation limited the development of potato alternariosis grown in the ecological system. A few treatments also limited the development of beetroot (*Cercospora beticola*) on organic beet plantations [29]. Jamiołkowska [36] reports that the preparation strongly limited the growth and development of

Alternaria alternata in vitro, and its effectiveness was higher than the tested fungicide based on azoxystrobin (Amistar 250 SC). The preparation showed a fungistatic effect (after 24 days) by inhibiting the growth of *A. alternata* colonies in 70% compared to the control. In laboratory tests, the preparation not only inhibited the growth of mycelium hyphae, but also the development of the spores. Studies carried out by the author [28,35] indicate the high efficiency of grapefruit extract to protect the sweet pepper fruits grown in the field, against early blight (*A. alternata*). Its antifungal activity is related to the presence of many biologically active compounds contained in the pulp and seeds of grapefruit. The compounds included in the preparation are not only endogenous flavonoids and glycosides (mainly naringin), but also terpenes, coumarins, and furanocoumarins. The compounds show strong antifungal properties by inhibiting germination of spores, growth of infectious hyphae and development of mycelium [39]. The positive effect of Biosept 33 SL on soil fungi communities is also described by Patkowska [38]. The author describes the high efficiency of the grapefruit extract in the protection of soybean seedlings against pathogenic soil fungi. Extract from grapefruit seeds and pulp (Biosept 33 SL) influenced the increase in the number of saprotrophic fungi (*Trichoderma* spp. and *Penicillium* spp.) in the soil community. Patkowska [37] reports that the biopreparation increased the number of colonies forming units of antagonistic bacteria of the genus *Bacillus* and *Pseudomonas* in the rhizosphere soil of soybean. In vitro studies [39] indicate that the grapefruit extract effectively inhibited the growth of fungal hyphae, sporulation, and the emergence of spores of fungi such as *A. alternata*, *B. cinerea*, *F. culmorum*, *F. oxysporum*, *F. solani*, *R. solani*, and *S. sclerotiorum*.

Terpenes (terpinen-4-ol, gamma-terpinene, 1,8-cineole) are organic chemical compounds contained in tea tree oil (*Malaleuca alternifolia* L.). The oil is obtained from leaves and small branches of trees, growing in the natural state on the territory of Australia. The oil, obtained by distillation, is the main ingredient of Timorex Gold 24 EC (23.8%) [40]. Timorex Gold shows a strong antiseptic effect and is used in the control of phytopathogenic fungi such as *Botrytis cinerea*, *Aspergillus fumigatus*, *Chaetomium globosum*, *Penicillium chrysogenum*, *Fusarium graminearum*, *F. culmorum*, *Pyrenophora graminea*, *Ascochyta rabiei*, *Colletotrichum lindemuthianum*, *Drechslera avenae*, *Alternaria radicina*, and *A. dauci* (Table 5.1) [40–42]. Laboratory and field studies conducted by the Institute of Horticulture in Skierniewice (Poland) show the high efficiency of Timorex Gold 24 EC in limiting *Bremia lactucae* on lettuce

and high efficiency in protection of this plant against downy mildew [43]. Tea tree oil is also recommended for the protection of cucumber in crops under cover against powdery mildew (*Erysiphe cichoracearum*). The fungistatic and fungicidal influence of the oil on *E. cichoracearum* is manifested by the degeneration of conidial spores and deformation of the mycelium of the pathogen [44].

Thymol is a monoterpene phenol and component of the essential oils present in thyme (*Thymus vulgaris* L.), oregano (*Origanum vulgare* L.), and satureja (*Satureja* L.) [45]. It has antibacterial and antifungal properties [46]. Due to its properties, it was already used in ancient Egypt. Scientific research indicates that thyme herb has an inhibitory effect on the growth of many filamentous fungi (Table 5.1), including *Cladosporium* (*C. sphaerospermum*), *Trichoderma* sp., *Rhizopus* sp., *Ascochyta rabiei*, *Colletotrichum lindemuthianum*, or *Fusarium* (*F. graminearum*, *F. culmorum*). The antifungal properties of thyme depend on the type of herb, the concentration of the extract, as well as the sensitivity of fungi species [42,47]. In the studies in vitro, the authors compared the activity of oil, hydrogel, and dried, powdered thyme herb to pathogenic fungi such as *Alternaria alternata*, *Penicillium cyclopium*, and *Trichothecium roseum* [48]. The results show that the greatest inhibiting effect on the growth of the tested fungi was oil, while the weakest was the hydrogel. Both oil and dried herb were the most effective in inhibiting *T. roseum* growth and the weakest to limiting *P. cyclopium* development. The strong antifungal activity of thyme (especially thyme oil) has also been demonstrated by other authors [49]. Thyme oil used at the low concentration $(0.6-0.7 \, \mu L \, mL^{-1})$, completely inhibited the growth of *Aspergillus flavus* and production of dangerous aflatoxin B1. It also effectively limited the development of *Fusarium oxysporum*, *Cladosporium herbarum*, *Aspergillus niger*, *Alternaria alternata*, *Botryodiplodia theobromae*, *Aspergillus fumigatus*, and *Curvularia lunata*. In the case of *Penicillium digitatum* (green mold of citrus fruits), a thyme oil concentration of $600 \, \mu L \, mL^{-1}$ was the most effective [50]. In the current studies on the effect of thyme oil, its minimal concentration inhibiting the growth of individual fungi species (MIC—minimal inhibition concentration) was determined. It was shown that for *Aspergillus sulphureus* and *A. versicolor*, which can produce dangerous toxins (Ochratoxin A), the MIC was in the range of $3.2-10.88 \, \mu L \, mL^{-1}$, and for *Penicillium chrysogenum* and *P. brevicompactum*, often present in closed air, MIC ranged from 18.95 to $19,6 \, \mu L \, mL^{-1}$. It was also reported that thymol, extracted from thyme oil, strongly limited the growth of filamentous fungi rather than the thyme oil [51]. The antifungal activity of thyme oil was also demonstrated under field conditions. Jamiołkowska and Wagner [52] used thyme oil to protect pepper plants against pathogens. The oil had a significant impact on the increase in the number of saprotrophic fungi on the underground parts of peppers and it reduced the number of pathogenic fungi colonies on the leaves and stems of cultivated plants. It was shown that compared to the control, thyme oil was particularly effective in the protection of pepper plants against *F. oxysporum* and *F. solani* but was less effective against *A. alternata*. Orzeszko-Rywka and co-authors [53] evaluated the usefulness of thyme oil for protection of seeds (parsley, carrots, and radishes). Thyme oil used at concentrations at 3% and 9% for seeds protection had similar effect as tebuconazole (fungicide Funaben T).

Animal origin compound

Chitosan is a polysaccharide present in the natural state as an ingredient in the shellfish of marine crustaceans. The organic compound is obtained by chitin distillation using sodium hydroxide at elevated temperature or with the help of enzymes [54]. Scientific research shows the activity of this organic substance against viruses, bacteria, fungi, and other pathogens (Table 5.1) [55,56]. Chitosan is also known as the plant immune elicitor, stimulating the formation of phytoalexins, PR proteins, lignin synthesis, and callosis [57]. In vitro studies conducted by Patkowska and Pieta [58] confirmed the efficacy of chitosan in limiting the development of *Botrytis cinerea*. Pastucha [56] observed good results after using an aqueous solution of chitosan in the form of a microgel (0.1%) for soybean seeds treatment as prophylactic protection. The author has observed on the soybean plants a smaller number of pathogenic fungi such as *Fusarium* spp., *Pythium* spp., *Botrytis cinerea*, and *Rhizoctonia solani* in comparison to the control. Mazur [59] showed the inhibitory effect of preparation Biochikol 020 PC (chitosan) on the development of *Botrytis cinerea* on strawberry fruits. Beta-Chikol is also a preparation containing chitosan. The product is used for the protection of agricultural, vegetable, and ornamental plants against diseases. Compared to typical fungicides, this product has not only directed limiting effect on pathogens and stimulates the plant's immunity mechanism [60].

Conclusions

Plants produce a wide variety of secondary metabolites that serve as plant defense mechanisms against pests and pathogens. These compounds named natural pesticides are found in plants growing in various climate zones. Bioactive compounds can also be obtained from certain animal species. Their extracts have a direct impact on pathogens by limiting their growth and development (inhibition of sporulation, mycelium deformation) and indirect as elicitors of defensive reactions. The scientific research shows the effect of many natural compounds on the increase of microorganism's biodiversity occurring in the soil and plant communities of cultivated plants, which minimizes the risk of dominance of one species that can cause disease. Currently, many bioactive substances have the basis for the production of biotechnical preparations, which are an alternative to synthetic chemicals used in agriculture [28,30,31,34–42,59,60].

REFERENCES

[1] J.C. Zadoks, Fifty years of crop protection 1950 − 2000, Neth. J. Agric. Sci. 50 (2) (2002) 181−193.

[2] J.C. Zadoks, H. Waibel, From pesticides to genetically modified plants: history, economics and politics, Neth. J. Agric. Sci. 48 (2000) 125−149.

[3] R. Carson, Silent Spring, Fawcett Crest, New York, 1962, ISBN 978-0-618-24906-0, 368 pp.

[4] R.L. Metcalf, Changing role of insecticides in crops protection, Annu. Rev. Entomol. 25 (1980) 219−256.

[5] K.T. Caliyurt, D. Crowther, No accounting for a Silent Spring: the discouragement of organic agriculture, Soc. Responsib. J. 1 (3/4) (2005) 179−189. https://doi.org/10.1108/ebo45808.

[6] A. Hitmi, A. Coudret, C. Barthomeuf, The production of pyrethrins by plant cell and tissue cultures of *Chrysanthemum cinerariaefolium* and *Tagetes* species, Crit. Rev. Biochem. Mol. Biol. 35 (5) (2000) 317−337.

[7] O. Boussaada, M. Ben Halima Kamel, S. Ammar, D. Haouas, S. Gannoun, A.N. Helal, Insecticidal activity of some Asteraceae plant extracts against *Tribolium confusum*, Bull. Insectol. 61 (2) (2008) 283−289.

[8] S.B. Padin, C. Fuse, M.I. Urrutia, G.M. Dal Bello, Toxicity and repellency of nine medicinal plants against *Tribolium castaneum* in stored wheat, Bull. Insectol. 66 (1) (2013) 45−49.

[9] R. Tsao, A.B. Attygalle, F.C. Schroeder, C.H. Marvin, B.D. Mc Garvey, Isobutylamides of unsaturated fatty acids from *Chrysanthemum morifolium* associated with host-plant resistance against western flower trips, J. Nat. Prod. 66 (2003) 1229−1231.

[10] H. Tibugari, D. Mombeshora, R. Mandumbu, C. Karavina, C. Parwanda, A comparison of the aqueous extracts of garlic, castor beans and marigold in the biocontrol of root-knot nematode in tomato, J. Agric. Technol. 8 (2) (2012) 479−492.

[11] K. Matsuda, Y. Kikuta, A. Haba, K. Nakayama, Y. Katsuda, A. Hatanaka, K. Komai, Biosynthesis of pyrethrin I in seedlings of *Chrysanthemum cinerariaefolium*, Phytochemistry 66

(13) (2005) 1529−1535, https://doi.org/10.1016/j.phytochem.2005.05.005.

[12] Y. Kikuta, H. Ueda, K. Nakayama, Y. Katsuda, R. Ozawa, J. Takabayashi, A. Hatanaka, K. Matsuda, Specific regulation of pyrethrin biosynthesis in *Chrysanthemum cinerariaefolium* by a blend of volatiles emitted from artificially damaged conspecific plants, Plant Cell Physiol. 52 (3) (2011) 588−596. https://doi.org/10.1093/pcp/pcr017.

[13] J.J. Schleier III, R.K.D. Paterson, Pyrethrins and pyrethroid insecticide, in: O. Lopez, J. Fernandez-Bolanos (Eds.), Green Trends in Insect Control, 2011, pp. 94−131. https://doi.org/10.1039/9781849732901-00094.

[14] G. Prance, M. Nesbitt (Eds.), The Cultural History of Plants, Routledge, 2012, 460pp.

[15] K. Essig, Z.J. Zhao, Preparation and characterization of a pyrethrum extract standard, LCGC 19 (7) (2001) 722−730.

[16] M.S. Shawkat, A.Q. Khazaal, M.R. Majeed, Extraction of pyrethrins from *Chrysanthemum cinerariaefolium* petals and study its activity against beetle flour *Tribolium castanum*, Iraqi J. Sci. 52 (4) (2011) 456−463.

[17] European Food Safety Authority, Compendium of botanicals reported to contain naturally occurring substances of possible concern for human health when used in food and food supplements, EFSA Journal 10 (5) (2012) 2663, https://doi.org/10.2903/j.efsa.2012.2663.

[18] R.J. Hillocks, Farming with fewer pesticides: EU pesticide review and resulting challenges for UK agriculture, Crop Protect. 31 (2012) 85−93, https://doi.org/10.1016/j.croppro.2011.08.008.

[19] M. Ukiya, T. Akihisa, H. Tokuda, Constituents of *Composite* plants III. Antitumor promotion effects and cytotoxic activity against human cancer cell lines of triterpene, diols and trials from edible chrysanthemum flowers, Cancer Lett. 177 (2002) 7−12.

[20] L. Wagner Sheldon, Fatal asthma in a child after use of an animal shampoo containing pyrethrin, West. J. Med. 173 (2) (2000) 86−87.

[21] C.A. Sousa, Fleas, flea allergy and flea control, a review, Dermatol. Online J. 3 (2) (2001) 7.

[22] A.V. Babosha, Changes in lecithin activity in plants treated with resistance inducers, Biol. Bull. Rus. Acad. Sci. 31 (1) (2004) 51−55.

[23] M. Kozłowska, G. Konieczny, Biology of Plant Resistance to Pathogens and Pests, first ed., Publishing House of Agricultural Academy in Poznań, Poland, 2003. ISBN 87-7160-320-7 173 pp., (in Polish).

[24] A. Saniewska, Antifungal activity of grapefruit (*Citrus paradisi*) endogenic flavonoids, Probl. J. Agric. Sci. 496 (2004) 609−617.

[25] A. Saniewska, I. Żuradzka, Comparison of antifungal activity of four cultivars of garlic (*Allium sativum* L.) for several pathogenic fungi, Folia Hortic. 13/1A (2001) 405−412.

[26] R. Sealy, M.R. Evans, C. Rothrock, The effect of garlic extracts and root substrate on soilborne fungal pathogens, HortTechnology 17 (2) (2007) 169−173.

[27] A. Jamiołkowska, A. Wagner, Effect of garlic pulp (Bioczos Plynny) on some fungi pathogenic to vegetables, in: Fourth International Conference on Non-chemical Crop Protection Methods, AFPP, Lille, France, 2011, pp. 213−220.

[28] A. Jamiołkowska, Biotechnical and biological preparations in the protection of sweet peppers (*Capsicum annuum* L.) against pathogenic fungi and in the induction of plant defense reactions, Sci. Diss. Univ. Life Sci. 376 (2013). ISSN 1899-2347 117pp., (in Polish).

[29] C. Sadowski, L. Lens, W. Koala, Investigations on the possibility of protection of organically grown red beet against fungal diseases, J. Res. Appl. Agric. Eng. 52 (4) (2007) 38−44.

[30] C. Sadowski, L. Lenc, A. Łukanowski, Pytopathological aspect of onion seed production in organic farm, J.Res.Appl.Agric. Eng. 54 (4) (2009) 80−84.

[31] B. Marjańska−Cichoń, A. Sapiecha-Waszkiewicz, Effectiveness of biotechnical preparations based on garlic extract against gray mold on strawberry, Prog. Plant Prot. 50 (1) (2010) 378−382 (in Polish).

[32] J. Horoszkiewicz-Janka, E. Jajor, M. Korbas, Usage of biopreparations as seed dressing in legume cultivation, J. Res.Appl.Agric. Eng. 57 (3) (2012) 162−166.

[33] A. Abdulrahman, A. Alkhail, Antifungal activity of some extract against some plant pathogenic fungi, Pak. J. Biol. Sci. 8 (3) (2005) 413−417.

[34] A. Saniewska, A. Jarecka, Influence of endogenous grapefruit flavonoids (*Citrus paradisi* Macf.) on the growth and of development of two special forms of *Fusarium oxysporum* Schlecht, Prog. Plant Prot. 46 (2) (2006) 517−520 (in Polish).

[35] A. Jamiołkowska, The influence of bio-preparation Biosept 33 SL on fungi colonizing of sweet pepper plants (*Capsicum annuum* L.) cultivated in the field, EJPAU 12 (2009) 3. http//www.ejpau.media.pl/volume12/issue3/art-13.html.

[36] A. Jamiołkowska, Laboratory effect of azoxystrobin (Amistar 250 SC) and grapefruit extract (Biosept 33 SL) on growth of fungi colonizing zucchini plants, Acta Sci. Pol. Hortorum Cultus 10 (2) (2011) 245−257.

[37] E. Patkowska, The effect of biopreparations on the healthiness of soybean cultivated in a growth chamber experiment, EJPAU 8 (2005) 4. http://www.ejpau.media.pl/volume8/issue4/art-08.html.

[38] E. Patkowska, Effectiveness of grapefruit extract and *Pythium oligandrum* in the control of bean and peas pathogens, J. Plant Prot. Res. 46 (1) (2006) 18−27.

[39] E. Patkowska, D. Pieta, Introductory studies on the use of bioprearations and organic compounds for seed dressing of runner bean (*Phaseolus coccineus* L.), Folia Universitatis Agriculturae Stetinensis, Agricultura 239 (95) (2004) 295−300.

[40] V. Terzi, C. Morcia, P. Faccioli, G. Valé, G. Tacconi, M. Malnati, *In vitro* antifungal activity of the tea tree (*Mealeuca alternifolia*). Essentials oil and its major components against plant pathogens, Lett. Appl. Microbiol. 44 (6) (2007) 613−618.

[41] P. Angelini, R. Pagiotti, A. Menghini, B. Vianello, Antimicrobial activities of various essential oils against foodborne pathogenic or spoilage moulds, Ann. Microbiol. 56 (1) (2006) 65–69.

[42] L. Roccioni, L. Orzali, Activity of tea tree (*Melaleuca alternifolia*, Cheel) and thyme (*Thymus vulgaris*, Linnaeus) Essential oils against some pathogenic seed borne fungi, J. Essent. Oil Res. 23 (6) (2011) 43–47.

[43] A. Włodarek, J. Robak, Possibilities of using the natural preparations in the protection of lettuce cultivated in the field and under cover against diseases, Sci. Pap. Hortic. Inst. 21 (2013) 117–126 (in Polish).

[44] A. Włodarek, B. Dyki, New possibilities of cucumber protection cultivated under cover against powdery mildew (*Erysiphe cichoracearum*) using natural preparations, Scientific Papers of Horticulture Institute 22 (2014) 147–155 (in Polish).

[45] O. Borugă, C. Jianu, C. Mişcă, I. Golet, A.T. Gruia, F.G. Horhat, *Thymus vulgaris* essential oil: chemical composition and antimicrobial activity, J. Med. Life 7 (3) (2014) 56–60.

[46] B. Wójcik-Stopczyńska, D. Kądziołka, Thyme – natural fungicide, Panacea 3 (48) (2014) 28–29 (in Polish).

[47] B. Kedzia, E. Hołderna-Kedzia, Study of the influence of essential oils against bacteria, fungi and human dermathophytes, Adv. Phytother. 2 (2007) 71–77 (in Polish).

[48] Wójcik-Stopczyńska, M. Wójcikowska, D. Szaferska, Influence of various derivatives of thyme (*Thymus vulgaris* L.) on growth of some filamentous fungi, in: Materials of 14th Herbal Symposium Herbs – Medicines, Foods, Cosmetics, 2012, p. 72. Żerków, Poland.

[49] A. Kumar, R. Shukla, P. Singh, C.S. Prasad, N.K. Dubey, Assessment of *Thymus vulgaris* L. essential oil as a safe botanical preservative against postharvest fungal infestation of food commodities, Innovative Food Sci. Technol. 9 (2008) 575–580.

[50] M. Yahyazadch, R. Omidbaigi, R. Zare, H. Taheri, Effect of some essential oils on mycelial growth of *Penicillium digitatum* Sacc, World J. Microbiol. Biotechnol. 24 (2008) 1445–1450.

[51] M.Š. Klarić, I. Kosalec, J. Mastelić, E. Piecková, S. Pepeljnak, Antifungal activity of thyme (*Thymus vulgaris* L.) essential oil and thymol against moulds from damp dwellings, Lett. Appl. Microbiol. 44 (2007) 36–42.

[52] A. Jamiołkowska, A. Wagner, Attempts to the use of thyme oil in protection of sweet pepper cultivated in the field against pathogenic fungi, Prog. Plant Prot. 47 (4) (2007) 149–152 (in Polish).

[53] A. Orzeszko-Rywka, M. Rochalska, M. Chamczyńska, Evaluation of usefulness of vegetable oils for seed treatment of selected crop plants, J. Res. Appl. Agric. Eng. 55 (4) (2010) 36–41 (in Polish).

[54] M. Placek, A. Dobrowolska, K. Wraga, K. Zawadzińska, P. Żurawik, The use of chitosan in cultivation, storage and protection of horticultural plants, Prog. Plant Prot. (3–4) (2009) 101–110.

[55] D. Pieta, The use of Biosept 33 SL, Biochicol 020 PC and Polyversum to control soybean (*Glycine max* (L.) Merrill) diseases against pathogens. Part I. Healthiness and yielding of soybean after using biopreparations, Acta Sci. Pol. Hortorum Cultus 5 (2) (2006) 35–41.

[56] A. Pastucha, Chitosan as compound inhibiting the occurrence of soybean diseases, Acta Sci. Pol. Hortorum Cultus 7 (3) (2008) 41–55.

[57] A. El Hadrami, L.R. Adam, L. El Hadrami, F. Daayf, Chitosan in plant protection, Mar. Drugs 8 (4) (2010) 968–987.

[58] D. Pieta, E. Patkowska, A. Pastucha, The impact of biopreparations on the growth and development of some fungi pathogenic for legumes, Acta Sci. Pol. Hortorum Cultus 3 (2) (2004) 117–177 (in Polish).

[59] S. Mazur, The impact of strawberry protection with natural products on fruits and leaves infection with some pathogenic fungi, Prog. Plant Prot. 49 (1) (2009) 379–382 (in Polish).

[60] A. Jamiołkowska, B. Hetman, Mode of action the biological preparations used in plant protection against pathogens, Ann. UMCS, E. Agricultura 71 (1) (2016) 13–29 (in Polish).

CHAPTER 6

Phytochemicals of Plant-Derived Essential Oils: A Novel Green Approach Against Pests

ARVIND SAROJ, PHD • OLUMAYOWA VINCENT ORIYOMI, MSC •
ASHISH KUMAR NAYAK, PHD • S. ZAFAR HAIDER, PHD

INTRODUCTION

Agriculture is the foremost necessity of man to fulfill food requirements since ancient time. During ancient time, people correlated plant diseases that were caused by nonvisible organisms as curse of God. Similarly, Romans created two Gods, Robigus (male) and Robigo (female) to get rid of dreaded rust disease. Great famine in human history changed many civilizations and affected human race drastically like Irish potato famine (1845), grapevine famine of France (1863), and paddy famine of Bengal (1943). The introduction of first chemical fungicide was introduced after grapevine famine and since then paradigm shift was noticed in agriculture sector. Similarly, many chemicals were introduced to curtail losses due to weeds, insects, and microorganisms. Estimated economic losses due to the plant diseases are more than 12% globally [1]. Pests not only affect production of crops but also deplete the shelf life and market value. Mycotoxins present in infected plant food commodities may cause serious human health problems. *Fusarium* and *Aspergillus* spp. are the most common among mycotoxin producing phytopathogens causing serious health problems after consumption [2]. The management of pre- and postharvest diseases is mainly performed by conventional methods using synthetic pesticides (herbicides/plant growth regulators, insecticides, and fungicides).

As stated in Chapter 3, pesticide is broadly defined as any compound or combination of compounds proposed for preventing, killing, or deterring any pest (insects, fungus, bacteria) or planned for use as a plant regulator or defoliant (weeds). According to Zhang [3], herbicides or weed killers accounted for a large proportion of total pesticide consumption (25.1%) followed by fungicides and bactericides (12.06%), insecticides (7.50%), and plant growth regulators (1.24%), while other pesticides including nematicides, algaecides, miticides, and acaricides contribute 53.84% of total pesticide consumption. It is estimated that about 2.5 million tonnes of pesticides are used on crops every year and due to them the damage caused by pesticides in the globe reaches $100 billion annually [4]. Global use of synthetic pesticides has many disadvantages including cost, hazardous effects, pesticide residues, resistant phytopathogen strains, and threats to human health [5]. To counter resistant fungal strains, higher concentration of pesticides was used increasing the level of resistant and high level toxic residues further increasing human health concerns many folds. The consumption of pesticides increases day by day to meet up the current demand of agricultural-based food commodities. Therefore, in current scenario, the consequences of uncontrolled use of synthetic pesticides (insecticides, fungicides, etc.) seek urgent attention to find out its alternative. Essential oils (EOs) are one of the most studied and potential alternatives to the synthetic pesticides. The EOs extracted from plant families Lamiaceae, *Brassicaceae, Apiaceae, Poaceae, Myrtaceae, Lauraceae* are thoroughly investigated for their antiphytopathogenic activities.

Plant-derived phytochemicals, especially essential oil–based formulations are the best alternatives of synthetic chemicals and being utilized as "biopesticides" or "green pesticides." EOs are considered as risk-free and environment friendly pesticides, usually having strong activities against all sorts of pests (insects as well as plants and foodborne pathogens). Many EOs and their bioactive compounds have strong activities against pests due to several properties ranging from antibacterial, antifungal, antioxidant, insecticidal, antifeedant, fumigant, repellent, miticides, oviposition deterrent, nematicidal, growth regulatory, and antivector [6]. The presence of low molecular weight terpenes and phenolic constituents in the essential oils plays a major role in plant chemical defense against insects, fungal pathogens, and nematodes [7]. Therefore, since last few decades, much effort has been put to focus on essential oils and their constituents as potential sources of commercial products for the control of pre- and postharvest agriculture diseases. In particular, flavor and fragrance industries have made possible the commercialization of essential oil–based pesticides with the widespread availability of essential oils. This chapter describes novel green approach of various biopesticidal properties of essential oils and their phytochemicals against pests.

Antiinsect Properties of Essential Oils and Their Constituents

Broadly, plants are rich reservoirs of active volatile compounds (essential oils) capable of preventing insect attacks on plants. Antiinsect compounds are mainly classified as insecticidal, repellents, fumigant, antifeedant, oviposition deterrents, miticides, ovipositional deterrent, and growth regulators. Plant essential oils are isolated from different species of plant having mixed complex functional groups called as active constituents that are mainly phenol-derived aromatic components, aliphatic components, alkaloids, tannins, quinines, coumarins, saponins, polypeptides, flavonoids, phenols, terpenoids, lectins, and flavones, applied either as fumigant or contact agents against different strains of microbes, insect pests, protozoans, helminths, mites, etc. In this section, secondary metabolites/constituents responsible for the antiinsect properties of essential oils are discussed with a view to establish constituent-effect relationships of the essential oils as antiinsect compounds.

Insecticidal property of essential oils and their constituents

The essential oil from *Cymbopogon winterianus*, rich in citronellal and citronellol has shown strong insecticidal properties and repellency against *Spodoptera frugiperda* larvae [8]. Furthermore, 1,8-cineole present in the essential oil of *Ocimum kenyense* (Ayobangira) is a very active insecticidal compound against stored-product beetles [9]. Terpenes and many oxygenated

constituents such as terpineol, farnesol, cineole, citral, capric acid, lauric acid, and carvone in *Acorus calamus* oil have demonstrated great insecticidal activity toward *Heteropsylla cubana* [10]. Eugenol and methyl salicylate isolated from the leaf buds of *Eugenia caryophyllata* are also well reported for significant toxicity against *Pediculus capitis* [11]. Cinnamaldehyde isolated from *Cinnamomum osmophloeum* has also shown larvicidal activity against *Aedes aegypti*. The essential oil (Cinnamaldehyde) together with cinnamyl acetate has been reported as excellent inhibitor of *A. aegypti* [12]. Similarly, marjoram, pennyroyal, and rosemary essential oils have demonstrated interesting insecticidal activity when tested against *Pediculus humanus* [13]. The volatile oils of *Calendula micrantha* produced insecticidal effects by inhibiting reproductive potential of Mediterranean fruit fly, *Ceratitis capitata* [14]. Insecticidal properties of rosemary oil and eugenol, thymol, and citronellal against instar larvae of *Agriostes obscurus* were also reported [15]. In the bioassay, citronellal caused highest insect mortality ($LC_{50} = 6.3 \, \mu g \, cm^{-3}$) than rosemary oil ($15.9 \, \mu g \, cm^{-3}$) and showed a higher toxicity against *A. obscurus* than thymol ($17.1 \, \mu g \, cm^{-3}$) and eugenol ($20.9 \, \mu g \, cm^{-3}$). It is noteworthy that these phytocompounds have been exempted from Environmental Protection Agency (EPA) regulations of many countries due to their low persistence in the environment with little or no risk to humans and the environment. Excellent modulatory effect of γ-aminobutyric acid (major inhibitor of neurotransmitter in the insect peripheral and central nervous systems) receptor by essential oils has also been used in pest management [16]. They reported that monoterpenoids such as thymol, carvacrol, and pulegone have positive allosteric modulatory effect against *Drosophila melanogaster* GABA receptor. The essential oils significantly increase the binding of GABA receptor to [³H]-TBOB. Furthermore, it was observed that allosteric binding of thymol, carvacrol, and pulegone to the insect's GABA receptors increased Clinflux into the neurons causing hyperpolarization of the chlorine channel and hyperexcitation of the nervous system leading to death of the *D. melanogaster*. Ebadollahi and Ashouri [17] reported that *Heracleum persicum*, *Achillea millefolium* and *Artemisia dracunculus* essential oils showed 100% mortality against adults *Plodia interpunctella* when exposed to 50%, 65%, and 80% concentrations, respectively. Furthermore, the LC_{50} values of essential oils significantly decreased with increasing exposure times. Other reports on insecticidal properties of essential oils stated that monoterpenoids such as 1,8-cineole, α-pinene, carvone, linalool, and phenolic compounds exhibit great toxicity toward insects [18]. Investigation of insecticidal activity on *Tamarindus indica, Azadirachta indica, Cucumis sativus, Eucalyptus* species, *Swietenia mahagoni,* and *Psidium guajava* extracts against *Tribolium castaneum* using film residue method revealed that their essential oils showed moderate (50%) to high (80%) toxicity against *T. castaneum* [19]. According to Chu et al. [20], the principal constituents identified in *Ostericum grosseserratum* that were responsible for its insecticidal activity against maize weevil, *Sitophilus zeamais,* were (D)-limonene (16.2%), 4-terpineol (13.5%), myristicin (11.3%), γ-terpineol (8.3%), β-pinene (5.1%), β-caryophyllene (4.6%), and linalool (4.1%). The compounds have also been identified in other plants that exhibit insecticidal effects. Sendi and Ebadollahi [21] found that citrus peels essential oil, which contains ∼92% D-limonene has strong insecticidal effect and showed 68% mortality against *Coptotermes formosanus* within 5 days of application. In another study, Kasrati et al. [22] reported that menthone, pulegone, and isomenthone found in the essential oil of mint timija (a popular medicinal plant in Morocco) implicated in contact toxicity to several species of insects.

Repellence property of essential oils and their constituents

Mwangi et al. [23] reported that monoterpene hydrocarbons are more active as repellent than oxygenated constituents of plant essential oils. Maize weevils are repelled by essential oils of *Lippia ukambensis.* Other promising repellents of maize weevils are essential oils from *L. javanica, L. dauensis, L. somalensis,* and *L. grandifolia.* Essential oils isolated from *Antrum sowa* and *Artemisia annuua* have been reported for repellent, toxic, and developmental inhibitory activities against *Tribolium castaeneum* [24]. Tapondjou et al. [25] reported that essential oils consisting of 1,8-cineole, terpineol, and α-pinene as major compounds demonstrate high repellent properties. Throughout the world, essential oils obtained from Lamiaceae (mint family), Pinaceae (cedar and pine family), and *Poaceae* (aromatic grasses) are mainly used as insect repellents. Traditionally, crude extracts from various species in these families are naturally used for driving away biting insects like mosquitoes [26]. Furthermore, plants belonging to *Myrtaceae, Verbenaceae, Meliaceae, Asteraceae, Caesalpiniaceae,* Fabaceae, Rutaceae, and Zingiberaceae families have demonstrated huge repellent activity against crop pests. Commercial repellents

are commonly formulated with essential oils such as clove, peppermint, patchouli, geraniol, lemongrass, pennyroyal, pine oil, thyme oil, cedar oil, and patchouli. The essential oil of *Laurus nobilis* was reported for repellence activity against *Rhyzopertha dominica* and *Tribolium castaneum.* The oil exhibited repellence at 0.04 mL cm^{-2} against adult *R. dominica* with a percentage repellence of 87.5% [27]. Assessment of repellence and antioviposition activities of essential oils from *Thymus vulgaris, Achillea millefolium, Foeniculum vulgare, Cuminum cyminum,* and *Citrus sinensis* was carried out against *Trialeurodes vaporariorum* (greenhouse whitefly) by Dehghani and Ahmadi [28]. It was reported that *A. millefolium, F. vulgare,* and *C. sinensis* demonstrated significant repellence and antioviposition activities against *T. vaporariorum.* The results underline the significance of essential oils as veritable tools in pest management. The repellence activities of essential oils from *Eucalyptus dundasii, Eucalyptus floribunda,* and *Eucalyptus kruseana* against *Rhyzopertha dominica* were found to be high at 70, 140, and 280 μL l^{-1} air, respectively, with *E. kruseana* showing highest repellence effect at low concentrations of all the oils tested [29]. *Melaleuca leucadendron* essential oil having (E)-nerolidol, ledol, and 1,8-cineole as major compounds was reported for high repellence, antiinflammatory, and antiseptic properties [30]. In this case, the repellence activity of the essential oils was measured by observing their behavior against female beetles exposed to treated and untreated beans in a linear olfactometer. The essential oils of *Ceiba pentandra* and *Pongamia pinnata* have been used to repel the notorious *Helicoverpa armigera* (common pest of field crops), and it is reported that oils from *C. pentandra* and *P. pinnata* demonstrated larvicidal activity at LC$_{50}$ as 228.01 ppm [31]. *Tanacetum tomentosum* and *T. dolichophyllum* essential oils have effective fumigant and repellent activities against *T. castaneum, T. tomentosum* oil, having β-bisabolene (50.0%) as the major compound, emerged as more potent with LC$_{50}$ values of 6.85 and 4.32 μL.0.25 L^{-1} air after 24 and 48 h of exposure, respectively. Similarly, the oil also exhibited repellent activity in the range of 38.70%–82.35% [32]. Examples of chemical constituents having repellent activities are presented in Table 6.1.

Fumigant property of essential oils and their constituents

Generally essential oils rich in monoterpenes and ketones are used as insect fumigants [4]. Compounds that showed fumigant effect in *Aphyllocladus decussatus, Aloysia polystachya, Minthostachys verticillata,* and *Tagetes minuta* against stored product pest like *Sitophilus zeamais* were identified

TABLE 6.1
Repellent Activity and Chemical Constituents of Common Plant Families.

S. No.	Family	Species	PhytoConstituents Having Repellent Activities
1	Myrtaceae	Corymbia citriodora	Citronellal, p-menthane-3,8-diol, limonene, geraniol, ısopulegol, pinene
		Eucalyptus spp.	1,8-Cineole, citronellal, citral, α-pinene
		Syzygium aromaticum	Eugenol, carvacrol, thymol, cinnamaldehyde
2	Verbenaceae	Lippia spp.	Myrcene, linalool, α-pinene, eucalyptol
		Lippia javanica	Alloparinol, camphor, limonene, α-terpeneol, verbenone
		Lippia cheraliera	Eucalyptol, caryophyllene, ipsdienone, p-cymene
		Lantana camara	Caryophylene
3	Lamiaceae	Ocimum spp.	Linalool, eugenol, p-cymene, eucalyptol, camphor, citral, thujone, limonene, ocimene
		Hyptis spp.	Myrcene
		Thymus spp.	α-Terpinene, carvacrol, thymol, p-cymene, linalool, geraniol
4	Poaceae	Cymbopogon nardus	Citronellal
		C. martini	Geraniol
		C. citratus	Citral
5	Meliaceae	Azadirachta indica	Azadirachtin, saponins
6	Asteraceae	A. vulgaris, Artemisia spp.	Camphor, linalool, terpenen-4-ol, α- and β-thujone, β-pinene
		Artemisia monosperma	Myrcene, limonene, cineole
7	Caesalpiniaceae	Daniellia oliveri	Not yet determined
8	Fabaceae	Glycine max	Not yet determined
9	Rutaceae	Zanthoxylum limonella	Not yet determined
		Citrus hystrix	Not yet determined
10	Zingiberaceae	Curcuma longa	Not yet determined

as α-thujone, R-carvone, S-carvone, (−) menthone, R (+) pulegone, and E-Z- ocimenone that are mainly rich in ketones. The reported fumigant toxicities of the plants are 11.8 $\mu L\,l^{-1}$ air (pulegone), 17.5 $\mu L\,l^{-1}$ air (R-carvone), 28.1 $\mu L\,l^{-1}$ air (S-carvone), and 42.3 $\mu L\,l^{-1}$ air (E-Z-ocimenone) [33]. Some compounds such as thymol, trans-anethole, carvacrol, terpineol, linalool, and 1,8-cineole act as strong fumigants against T. castaneum [4]. The essential oils of anise (Pimpinella anisum) and peppermint (Mentha piperita) have been found to have fumigant toxicity against stored product pests T. castaneum, R. dominica, Oryzaephilus surinamensis, and Sitophilus oryzae [34]. Thymol, a constituent of thyme essential oil binds GABA receptor of human and house fly to act as modulator [35]. Essential oil isolated from Ipomoea cairica also shows larvicidal effect against Aedes aegypti, Anopheles stephensi, Culex tritaeniorhynchus, and Culex quinquefasciatus [36]. Volatile compound (diallyl disulphide), present in Azadirachta indica show potent toxic, fumigant and feeding deterrent activities against stored grain pests S. oryzae and T. castaneum [37]. Lee et al. [38] indicated that 1,8-cineole induced fumigant toxicity against major stored grain insects. Pulegone, linalool and limonene are known fumigants effective against rice weevil, S. oryzae. Essential oils from some African plants have shown fumigant toxicity against Anopheles gambiae [39]. Amizadeh et al. [40] showed that Eucalyptus microtheca, Heracleum persicum, Satureja sahendica and Foeniculum vulgare essential oils exhibited fumigant toxicity against egg and adult female Tetranychus urticae, a major pest of ornamental crops. Fumigant toxicity of garlic (Allium sativum) was investigated against adults and larvae of beetle, T. castaneum and showed that the essential oil caused mortality of T. castaneum with increasing concentration and time at LC_{50} and toxicity values (267.37 and 145.8 $\mu L\,l^{-1}$

air) for larvae and (90.8 and 127.90 µL l^{-1} air) for adult at 24 and 48 h, respectively [41].

Ovipositional deterrence property of essential oils and their constituents

Essential oils isolated from *Salvodora oleoides* and *Cedrus deodara* have shown significant oviposition deterrence against *Phthorimaea overcalls* [24]. Cardamom oil has been used to reduce hatching of *S. zeamais* and *T. castaneum* eggs as well as the survival rate of the larvae. It was discovered that the oil drastically reduced emergence of adult *T. castaneum* when applied on the eggs [34]. *Artemisia annua* oil also affects viability of *Callosobruchus maculatus* eggs [42,43]. Essential oils of *Piper guineense* and *Xylopia aethiopica* was also reported for ovicidal activity and greatly reduced emergence of against *T. castaneum* [44]. Ovipositional deterrence of α-pinene and β-caryophylene tested by fumigation against *T. castaneum* was high. The ovicidal activity of the oils increased when combined together and applied as a mixture against *T. castaneum*. The synergistic effect shown by the oils was responsible for inability of *T. castaneum* to lay more eggs [45]. Ovipositional effects of *Eucalyptus citriodora*, *E. globulus* and *E. staigerana* essential oils were tested against *Zabrotes subfasciatus* and *Callosobruchus maculatus* at different concentration and reported for reduction in egg viability and emergence of adult insects. Similarly, eugenol, citronellal, thymol, pulegone, rosemary oil and cymene have been reported for high oviposition deterrence activities different from borneol, camphor and β-pinene that increased egg laying ability of the insect [21]. Essential oils of *Pongamia pinnata* and *Ceiba pentandra* deterred the development of *Helicoverpa armigera* egg by 96.2% and 98.8%, respectively [31]. For long, essential oils from these plants have been useful in the management of important agricultural pests.

Miticidal property of essential oils and their constituents

Chang et al. [46] reported α-cadinol, muurolol, ferruginol and cadinol as the dominant compounds responsible for the antimite activity of *Taiwania cryptomerioides* heartwood against *Dermatophagoides pteronyssinus* and *D. farinae*. The study revealed a miticidal activity of 67.0% and 36.7% against *D. pteronyssinus* and *D. farinae*, respectively at a dosage of 12.6 mg cm^{-2} after 48 h. The highest miticidal activity was found in α-cadinol when compared with other essential oils derived from *T. cryptomerioides* heartwood. Miticidal activities of thyme (bioactive compounds as p-cymene and thymol), eucalyptus (1,8-cineole, α-pinene,

α-terpineol), cinnamon (cinnamaldehyde, cinnamic aldehyde), chenopodium (p-cymene, pinene-2-ol, ascaridole), clove (eugenol, β-caryophyllene), fennel (anethole, pinene, 1,8-cineole), garlic (pentdecane, hexadecane), geranium (citronellol, citronellol formate), lemongrass (limonene, terpinene), rosemary (1,8-cineole, camphor), peppermint (menthol, p-menthone), rose (citronellol, geraniol) and caraway (L-carvone, limonene) were tested on house dust mite, *D. pteronyssinus* [47]. The effect was observed to be time-dependent as LC$_{50}$ values of the essential oil decreased as exposure time prolonged. They also reported that chlorothymol was the most effective miticidal essential oil against *Tetranychus urticae* followed by thymol, carvacrol and cinnamaldehyde. It was suggested that criteria for choosing most suitable oil should be based on persistence of acaricidal activity in relation to volatility, temperature, human acceptability and adverse effects on fabrics [47]. Miticidal activities of α-pinene, β-pinene, aromadendrene, limonene, eucalyptol (1,8-cineole), p-cymene, pinocarvone, trans-pinocarveol, α-terpineol, terpinyl acetate and viridiflorol from *Eucalyptus globulus* essential oils were analyzed on varroa destructor (*Paenibacillus larvae*). Mortality of the mite against tested essential oil increased as concentration increased [48].

Antifeedant property of essential oils and their constituents

Essential oils that are able to suppress, deter or repel insects without making direct contact with them are best used as antifeedant. Antifeedant activity of cardamom (*Elletaria cardamomum*) was investigated by topical application on *Sitophilus zeamais* and *Tribolium castaneum*. The oil expressed contact toxicity level (LD$_{50}$) of 56 µg mg^{-1} against *S. zeamais* and 52 µg mg^{-1} against *T. castaneum* [34]. Antifeedant activity of essential oils from *Syzygium aromaticum*, *Cinnamomum zeylanicum*, *Lavendula latifolia*, *L. angustifolia*, *Mentha crispa*, *M. arvensis* and *M. piperita* were tested against cabbage looper. *C. zeylanicum* caused highest mortality followed by *S. aromaticum*, *M. crispa*, *M. arvensis*, *M. piperita* and *L. angustifolia* oils [49]. The efficacy of essential oil of *Zanthoxylum alatum* was tested by the leaf dip method against the polyphagous pest *Spodoptera litura* and reduced up to 90.0% of larval growth [50].

Efficacy of Phytochemicals Against Nematodes

Nematodes are the most devastating group plant pathogens worldwide and their control is extremely challenging [51]. These are like tiny worms and parasites to the plants. As they are parasites to the host plant, many of them can play an important role in the predisposition of the host plant to the invasion by secondary pathogens [52]. The parasitic natures of the nematodes destroy a large variety of crops worldwide, resulted billions of dollars losses annually [53]. Different species of root-knot nematode *Meloidogyne* spp. are cosmopolitan and responsible for considerable crop losses. Another notorious species reported to cause heavy damage to the pine wood, *Bursaphelenchus xylophilus* is the causal agent of pine wilt disease. The continuous and indiscriminate use of synthetic or chemical nematicides has also induced undesirable effects including toxicity to non-target organisms and environment related consequences. In the last decades, on the basis of environmental and human health concerns, the use and availability of efficient commercial nematicides have steadily reduced [54]. Hence, strong interest generated to find out nematicides of natural origin [55].

Nematicides developed from the plant phytochemicals have a great potential in nematode control or can serve as model compounds for the development of derivatives with enhanced activity [56]. Essential oils including many plant constituents and metabolites have been investigated for activity against plant-parasitic nematodes. The presence of volatile monoterpenes provides an important defense strategy to the plant against nematodes and pathogenic organisms. These monoterpenes also play a role in plant parasitic interactions, acting as signaling molecules [57]. A large number of essential oils from different botanical families have been analyzed*in vitro* for their nematicidal activities, mainly against root-knot nematodes and the pinewood nematode *Bursaphelenchus xylophilus* [58]. In particular, high toxicity to root-knot nematodes has been reported for the essential oils from *Mentha* spp., *Eucalyptus* spp., *Cymbopogon* spp., *Pelargonium graveolens* and *Ocimumbasilicum* [59]. Some plants and main components of essential oils with nematicidal activity are reported in Table 6.2.

Chemical Groups of Nematicidal Plant Compounds
Cyanogenic glycosides
Cyanide releasing cyanogenic glucosides is amino acid derived glycosides, involved in the defense of more than 2500 plant species against nematodes and predators. Cyanide releasing cyanogenic glucosides present in different plants of *Poaceae* family like green manure of sudan grass, *Sorghum sudanense* (Piper), are widely reported for its suppressiveness on root-knot nematodes and due to the soil fumigating effect released by the hydrolysis of dhurrin [62].

Glucosinolates
Glucosinolates (GLSs) are thioglucosides secondary metabolites, mainly present in the Brassicaceae and *Capparidaceae* families, which coexist in vivo with the myrosinase enzyme. Myrosinase-catalyzed hydrolysis of glucosinolates, upon tissue damage by harvesting, processing or mastication, results in the release of a variety of isothiocyanate derivatives with nematotoxic action.

Saponins
Saponins are a large group of glycosidic secondary metabolites produced by many plant species, including major food crops. It is belonging to three major chemical classes: steroid alkaloid glycosides, steroid glycosides, and triterpene glycosides. Due to their physical, chemical and physiological characteristics, naturally occurring saponins display a broad spectrum of pharmacological and biological effects, including nematicidal, activities [63].

Limonoid triterpenes
Limonoids are a group of metabolically altered triterpenes occurring in species belonging to Rutaceae and *Meliaceae* families. Presence of limonoids in Neem tree (*Azadirachta indica*) is the most widely investigated for their biological activities [64]. More than 100 limonoid compounds present in Neem, including azadirachtin, salannin and nimbin, which are mainly working as nematodes growth inhibitors, feeding deterrents and repellents [65]. Azadirachtin is most preventive against growth of phytoparasitic nematodes and insects [66].

Polythienyls
Polythienyls substances present in different species of the Asteraceae family (mainly in the genus *Tagetes*) having both nematicidal and insecticidal properties [56]. A large number of studies reported about the nematicidal effect of *Tagetes* species used either as a source of nematode-antagonistic formulations or as cover, green manure or rotation crops [67].

Alkaloids
Alkaloids are nitrogen-containing natural secondary metabolites present in several botanical families, such as: Solanaceae, Fabaceae, Liliaceae, Apocynaceae, and Papaveraceae. Pyrrolizidine alkaloidswere reported as the highest activity against phytonematodes found from different species of Fabaceae, Liliaceae, Apocynaceae and Papaveraceae families [68]. A nematicidal activity has been documented also for steroidal

TABLE 6.2
Plants and Main Components of Essential Oils With Nematicidal Activity.

S. No.	Common Name	Botanical Name	Extraction Part	Main Components	References
1	Wormwood	*Artemisia judaica*	Foliage	Artemisia ketone (34%)	[60]
2	Caraway	*Carum carvi*	Umbels	(+)-carvone (50%), limonene (48%)	[60]
3	Wild thyme	*Coridothymus capitatus*	Foliage	Carvacrol (70%)	[60]
4	Lemongrass	*Cymbopogon citratus*	Foliage	Geranial (47%), neral (31%)	[60]
5	Fennel	*Foeniculum vulgare*	Foliage	*t*-Anethole (43%), limonene (6%)	[60]
6	Applemint	*Mentha rotundifolia*	Foliage	Isomers of 1,2-epoxymenthyl acetate (74%), piperitone (13%)	[60]
7	Spearmint	*Mentha spicata*	Foliage	Carvone (58%), limonene (19%)	[60]
8	White micromeria	*Micromeria fruticosa*	Foliage	Pulegone (75%)	[60]
9	Syrian oregano	*Origanum syriacum* (C-type)	Foliage	Carcavol (76%)	[60]
10	Syrian oregano	*Origanum syriacum* (CT-type)	Foliage	Carvacrol (68%), thymol (2%)	[60]
11	Oregano	*Origanum vulgare* (C-type)	Foliage	Carvacrol (80%)	[60]
12	Oregano	*Origanum vulgare* (T-type)	Foliage	Thymol (60%)	[60]
13	Thyme	*Thymus vulgaris*	Foliage	Thymol (40%), carvacrol (7%)	[60]
14	Cinnamon	*Cinnamomum zeylanicum*	Bark	Methyl chavicol (74%), 1,8 cineole (2.1%), linalool (2.6%), caryophyllene (3.1%), eugenol (2.5%)	[61]
15	Lemongrass	*Andropogon nardus*	Leaves	Citral (75.2%), myrcene (18.5%), geraniol (16.9%), limonene (7.7%)	[61]
16	Sweet orange	*Citrus sinensis*	Leaves	Limonene (40%), linalyl acetate (37%), minolool (16%), bergaptene (4%)	[61]
17	Eucalyptus	*Eucalyptus* spp.	Leaves	1,8 cineol (30%), camphor (18%), α-pinene (19%), borneol (17%), borneol acetate (5%)	[61]
18	Fennel	*Foeniculum vulgare*	Flower	Limonene (63.6%), anethole (25.5%), fenchyl acetate (2.6%), α-pinene (0.97%), myrcene (0.98%), estragole (1.1%)	[61]
19	Laurel	*Laurus nobilis*	Leaves	1,8 cineole (44%), α-pinene (4.5%), β-pinene (2.2%), sabinene (8.9%), methyl eugenol (1.9%)	[61]

Continued

TABLE 6.2
Plants and Main Components of Essential Oils With Nematicidal Activity.—cont'd

S. No.	Common Name	Botanical Name	Extraction Part	Main Components	References
20	Lavender	*Lavandula stoechas*	Flowers	Fenchone (40.1%), 1,8 cineole (11.7%), bornyl acetate (5.8%), myrtenyl acetate (2.8%), myrtenol (2.1%), α-pinene (2%), viridiflorol (1.9%)	[61]
21	Chamomile	*Matricaria discoidea*	Leaves, flowers	Piperitenone (54.2%), pulegone (10.7%), piperitenone oxide (11.3%), menthone (3.3%), 1,8 cineole (2.8%)	[61]
22	Myrtle	*Myrtus communis*	Leaves	1,8 cineole (40%), α-pinene (17%), linalool (9.9%), α-terpineol (7.9%), geranyl acetate (4.5%), myrtenyl acetate (7%), α-terpinyl acetate (2%)	[61]
23	Oregano	*Origanum syriacum*	Leaves	Carvacrol (61%), thymol (21.8%), δ-terpinene (4%), p-cymene (5.5%), myrcene (1.2%), α-terpinene (1.3%)	[61]
24	Geranium	*Pelargonium graveolens*	Leaves	Citronellol (21.7%), linalool (17.3%), geraniol (14%), menthone (1.1%)	[61]
25	Anise	*Pimpinella anisum*	Seeds	Transanethole (85%), estragole (0.57%), linalool (1.5%), α-terpineol (1.5%)	[61]
26	Pine	*Pinus pinea*	Leaves	Limonene (74.6%) 1,8 cineole (4.3%), α-pinene (3.8%), myrcene (2.7%), β-caryophyllene (3.4%)	[61]
27	Sage	*Salvia officinalis*	Leaves, stem, flower	α-thujene (26.8%), limonene (22.9%), linalyl acetate (17.4), ocimene (10.7%), linalool (5.7%), myrcene (3.6%) α-pinene (2.5%)	[61]
28	Marigold	*Tagetes patula*	Leaves, stem, flower	Linalool (26.8), limonene (22%, 9%), linalyl acetate (17.4%), ocimene (10.7%)	[61]

alkaloids, such as α-tomatine and α-chaconine and solanine [56].

Phenolics, flavonoids, and tannins
Phenolics are known as toxic to nematodes, insects, weeds, fungi and also bacteria [69]. Flavonoids are a large group of secondary metabolites reported as consistent nematicidal activity and play a key role in plant defense against insects, fungal and bacterial pathogens [70]. Polyphenolic compounds are widely present in many plant species and documented for their activity on phytoparasitic nematodes [71]. The root-knot infection on squash (*Cucurbita pepo*) by *Meloidogyne arenaria* was found to be effectively control by soil treatments with tannic acid [72].

Fungicide and Bactericide Activity of Essential Oils
Estimated economic losses due to the plant diseases are more than 12% globally [1]. The management of pre

and post-harvest diseases are mainly performed by conventional methods successfully using synthetic pesticides (herbicides, insecticides, fungicides, plant growth regulators). Although, frequent malpractice of synthetic pesticides also leads to the development of resistant insect and fungal verities. To counter resistant fungal strains, plant derived pest control agents have long been utilized as alternatives to synthetic pesticides for integrated pest management. Furthermore, bioactivity of plant-based compounds is well documented in the literatures [73].

Essential oils are very complex combination of many monoterpenes (C10), sesquiterpenes (C15) and some contains phenols (phenylpropenes and cinnamates). Some of these compounds have been reported as strong fungicides and bactericides alone like menthol (*Mentha arvensis*), eugenol (*Syzygium aromaticum*), thymol (*Trachyspermum ammi*), methyl chavicol (*Ocimum basilicum*), piperitenone oxide (*Mentha spicata*), camphor (*Cinnamomum camphora*), ocimene (*Syzygium cumini*), carvacrol (*Origanum vulgare*). However, other compounds of essential oils especially monoterpenes (limonene, pinene etc.) can enhance the activity of these compounds. Many essential oils showed antimicrobial activities due to the presence of synergistic effect among compounds. Recently, many EOs are reported as antifungal, listed in Table 6.3.

TABLE 6.3
Anti-fungal Activity of Essential Oils.

Source of Essential Oil	Tested Phytopathogens	Applied Technique	Minimum Inhibitory Concentration (MIC)	References
BRASSICACEAE				
Brassica nigra	*A. niger* *A. ochraceous* *Penicillium citrinum*	Vapor phase and poisoned food technique (PFT)	4700 and 400 ppm respectively in both methods	[74]
APIACEAE				
Cicuta virosa	*A. ochraceous* *A. niger* *A. flavus* *A. alternata*	PFT	300 ppm	[75]
Cuminum cyminum	*A. alternata* *Penicillium citrinum*	PFT	600 ppm	[76]
Trachyspermum ammi	*A. glucans* *A. flavus* *A. alternata*	PFT	800 ppm	[77]
POACEAE				
Cymbopogon citratus	*A. niger* *Botrytis cinerea*	PFT	500 ppm	[78]
Cymbopogon martini	*A. fumigates* *A. flavus* *A. niger* *Fusarium* spp. *Penicillium* spp.	PFT	500 ppm	[79]
LAMIACEAE				
Lippia rugosa	*A. flavus*	Agar medium Assay	1000 ppm	[80]
Mentha spicata	*A. niger* *A. glutans* *A. nidulans*	PFT	1000 ppm	[76]

Continued

TABLE 6.3
Anti-fungal Activity of Essential Oils.—cont'd

Source of Essential Oil	Tested Phytopathogens	Applied Technique	Minimum Inhibitory Concentration (MIC)	References
Mentha spicata var. *viridis*	*R. solani* *C. lunatus* *B. australiensis*	PFT	450, 400 and 600 ppm, respectively	[81]
Mentha arvensis	*A. niger*	Volatile phase	600 ppm	[82]
Ocimum basilicum	*R. solani* *C. cucurbitarum*	PFT	1200 ppm	[83]
Ocimum sanctum	Choanephora cucurbitarum Rhizoctonia solani	PFT	900 and 1200 ppm, respectively	[84]
Ocimum sanctum	*C. cucurbitarum* *R. solani*	PFT	730 and 1200 ppm, respectively	[85]
Satureja hortensis	*A.flavus*	Agar dilution method	500 ppm	[86]
Thymus pulegioides	*A. flavus* *A.fumigatus* *A. niger*	PFT	320 ppm	[87]
MYRTACEAE				
Syzygium cumini	*R. solani* *C. cucurbitarum*	PFT and volatile phase	1200 ppm	[88]
Syzygium aromaticum	*A. flavus* *A. fumigatus* *A. niger*	PFT	640 ppm	[89]
LAURACEAE				
Cinnamomum camphora	*C. cucurbitarum*	PFT	1200 ppm	[85]

Antifungal and antibacterial mechanism

Considering complex nature of essential oils, antimicrobial (antifungal and antibacterial) activity comprises more than one specific mechanism and involves many cell targets [88,90,91]. Compounds having hydroxyl-(OH) groups showed broad antimicrobial activities against many phytopathogens. However, low molecular weight (mw) alcohol (α-terpineol, mw 154) showed good antimicrobial activity in volatile phase-in contrast to high molecular weight alcohol (7-hydroxycalamenene, mw 218), because less carbon number contributes to increase the volatility of the essential oil compounds giving more exposure to the target phytopathogens [88].

More than one mechanism involves against fungal cell including cell wall disruption [92], degradation to cytoplasmic membrane [93–95], leakage of cell materials [96–99], cytoplasm coagulation [85,97], lipid protein disruption [100] and reduction in the proton motive force [101]. Likewise, many mechanisms involve simultaneously against plant pathogenic bacteria like disruption in cell membrane permeability [99], chelation of cations [98], dissolution of phospholipid bilayer [102], disruption in passive cell permeability [103], disturbing enzyme action [104], reduction in ATP synthesis/ATP hydrolysis, disintegration of cytoplasmic membrane.

Biopesticidal Products and Uses

After many decades rigorous research on pesticidal properties of essential oils and their constituents throughout the world resulted few commercially available pesticides in recent years. Essential oil-based pesticides are attractive to both consumers and manufacturers, because these are generally regarded as safe (GRAS) and less expensive to market products. These products are also exempt from the normal registration requirements of U.S. Environmental Protection

Agency [US EPA 2014] under FIFRA Section 25b [105]. Cinnamaldehyde rich cinnamon essential oil based insecticide and herbicide products named as Weed Zap and Repellex are traded by the companies [106−108]. US based Mycotech Corporation introduced cinnamaldehyde rich fungicide for horticultural crops. Insecticide containing eugenol aimed crawling and flying insects, named as EcoPCOR were introduced by EcoSMART Technologies commercially. GreenMatch EX, a natural insecticide and herbicide for organic crop production is based on citral rich lemongrass essential oil [106,107]. Rosemary oil based formulation of insecticide, acaricide and fungicide has been introduced under the name EcoTrol and Sporan for horticultural crops [106,107,109], while clove oil based insecticide and herbicide formulations introduced as Matran EC and Burnout II for weed control companies [106−109]. Several neem oil (*Azadirachta indica*) based products (Ecozin, Azatrol EC, Agroneem, Trilogy) are commercially available as insecticides, acaricides, and fungicides [108]. Menthol and thymol based insecticides (Apilife VARTAM) were introduced for the management of diseases caused by mites in beehives. A pesticide product 'Bed Bug Bully', manufactured by Optimal Chemical LLC, Tamarac, FL is having mint, clove, citronella and rosemary oils as active ingredients [105].

Conclusion and Future Prospects

The increasing concerns of environmental pollution, food poisoning and persistence of residues in the environment resulting from synthetic pesticide application have caused a shift of interest to essential oils which are promising botanical pesticides that are cheap, environment friendly and capable of enhancing sustainable crop protection. Essential oil based formulations can be used as fumigants, emulsion, granular or direct spray. Generally, the effectiveness of essential oil falls short in contrast to synthetic pesticides due to high volatility of its active compounds. However, recent patent WO2017091096 [110] on the essential oils encapsulation led to commercialize essential oils as pesticide. Synergistic effects among compounds and essential oil must be investigated to capitalize on the antimicrobial activity and reduce the MIC to achieve the desired antimicrobial activity. Different combination of essential oil and its compounds may guide to activate multiple mechanisms to initiate antimicrobial activities, resulted to resolve the problem of pathogens getting resistance against synthetic pesticides [111]. If the current intensity of campaign against the use of synthetic compounds in pest management is sustained,

the future of essential oils looks bright as they have presented themselves as veritable candidates in pest management with only small dosage needed to cause insect toxicity through inhalation. As plants are a rich natural reservoir of essential oils, farmers, industrialists, perfumers, technologists, phytochemists, environmentalists, agronomists and researchers should look into the future with keen interest to explore the constituents of essential oils in the management of insects and pests.

REFERENCES

[1] G.N. Agrios, Plant Pathology, fifth ed., Elsevier Academic Press, London, 2005. ISBN: 9780120445653.

[2] H.J. Williams, T.D. Philliips, E.P. Jolly, K.J. Stiles, M.C. Jolly, D. Aggrawal, Human aflatoxicosis in developing countries: a review of toxicology, exposure, potential health consequences, and interventions, Am. J. Clin. Nutr. 80 (2004) 1106−1122.

[3] W. Zhang, Global pesticide use: profile, trend, cost/benefit and more, Proc. Int. Acad. Ecol. Environ. Sci. 8 (1) (2018) 1−27.

[4] O. Koul, S. Walia, G.S. Dhaliwal, Essential oils as green pesticides: potential and constraints, Biopestic. Int. 4 (1) (2008) 63−84.

[5] N. Paster, L.B. Bullerman, Mould spoilage and mycotoxin formation in grains as controlled by physical means, Int. J. Food Microbiol. 7 (1988) 257−265.

[6] M. Mohan, S.Z. Haider, H.C. Andola, V.K. Purohit, Essential oils as green pesticides: for sustainable agriculture, Res. J. Pharm. Biol. Chem. Sci. 2 (4) (2011) 100−106.

[7] F. Bakkali, S. Averbeck, D. Averbeck, M. Idaomar, Biological effects of essential oils - a review, Food Chem. Toxicol. 46 (2008) 446−475.

[8] A.M. Labinas, W.B. Crocomo, Effect of Java grass (*Cymbopogon winterianus* Jowitt) essential oil on fall armyworm *Spodoptera frugiperda* (J. E. Smith, 1797) (Lepidoptera, Noctuidae), Scientiarum 24 (2002) 1401−1405.

[9] D. Obeng-Ofori, C. Reichmuth, J. Bekele, A. Hassanali, Biological activity of 1,8-cineole, a major component of essential oil of *Ocimum kenyense* (Ayobangira) against stored product beetles, J. Appl. Entomol. 121 (1997) 237−243.

[10] T. Nakatsu, A.T. Lupo, J.W. Chinn, R.K.L. Kang, Biological activity of essential oils and their constituents, Stud. Nat. Prod. Chem. 21 (2000) 571−630.

[11] Y.C. Yang, S.H. Lee, W.J. Lee, D.H. Choi, Y.J. Ahn, Ovicidal and adulticidal effects of *Eugenia caryophyllata* bud and leaf oil compounds on *Pediculus capitis*, J. Agric. Food Chem. 51 (2003) 4884−4888.

[12] S.S. Cheng, J.Y. Liu, K.H. Tsai, W.J. Chen, S.T. Chang, Chemical composition and mosquito larvicidal activity of essential oils from leaves of different *Cinnamomum osmophloeum* provenances, J. Agric. Food Chem. 52 (14) (2004) 4395−4400.

[13] Y.C. Yang, H.S. Lee, J.M. Clark, Y.J. Ahn, Insecticidal activity of plant essential oils against *Pediculus humanus capitis* (Anoplura: pediculidae), J. Med. Entomol. 41 (4) (2004) 699–704.

[14] K.T. Hussein, Suppressive effects of *Calendula micrantha* oil and gibberelic acid (PGR) on reproductive potential of the mediterranean fruit fly *Ceratitis capitata* Wied. (Diptera: tephritidae), J. Egypt. Soc. Parasitol. 35 (2) (2005) 365–377.

[15] R. Waliwitiya, M.B. Isman, R.S. Vernon, A. Riseman, Insecticidal activity of selected monoterpenoids and rosemary oil to *Agriotes obscurus* (Coleoptera: elateridae), J. Econ. Entomol. 98 (5) (2005) 1560–1565.

[16] F. Tong, J.R. Coats, Effects of monoterpenoid insecticides on [^{3}H]-TBOB binding in house fly GABA receptor and ^{36}Cl uptake in American cockroach ventral nerve cord, Pestic. Biochem. Physiol. 98 (2010) 317–324.

[17] A. Ebadollahi, S. Ashouri, Toxicity of essential oils isolated from *Achillea millefolium* L., *Artemisia dracunculus* L. and *Heracleum persicum* Desf. against adults of *Plodia interpunctella* (Hubner) (Lepidoptera: Pyralidae) in Islamic Republic of Iran, Ecol. Balk. 3 (2) (2011) 41–48.

[18] P.U. Rani, Fumigant and contact toxic potential of essential oils from plant extracts against stored product pests, J. Biopestic. 5 (2) (2012) 120–128.

[19] M. Mostafa, H. Hossain, M.A. Hossain, P.K. Biswas, M.Z. Haque, Insecticidal activity of plant extracts against *Tribolium castaneum* Herbst, J. Adv. Res. 3 (3) (2012) 80–84.

[20] S.S. Chu, Q.Z. Liu, S.S. Du, Z.L. Liu, Chemical composition and insecticidal activity of the essential oil of the aerial parts of *Ostericum grosseserratum* (Maxim) Kitag (Umbelliferae), Trop. J. Pharm. Res. 12 (1) (2013) 99–103.

[21] J.J. Sendi, A. Ebadollahi, Biological activities of essential oils on insects, in: J.N. Govil, S. Bhattacharya (Eds.), Recent Progress in Medicinal Plants (RPMP): Essential Oils—II, vol. 37, Studium Press LLC, 2013, pp. 129–150.

[22] A. Kasrati, C.A. Jamali, R. Spooner-Hart, L. Legendre, D. David Leach, A. Abbad, Chemical characterization and biological activities of essential oil obtained from mint timija cultivated under mineral and biological fertilizers, J. Anal. Methods Chem. (2017) 1–7.

[23] J.W. Mwangi, I. Addae-Mensah, G. Muriuki, R. Munavu, W. Lwande, A. Hassanali, Essential oils of *Lippia* species in Kenya. IV: maize Weevil (*Sitophilus zeamais*) repellancy and larvicidal activity, Int. J. Pharmacogn. 30 (1992) 9–16.

[24] V. Tare, Bioactivity of some medicinal plants against chosen insect pests/vectors, J. Med. Aromat. Plant Sci. 22 (2000) 35–40.

[25] A.L. Tapondjou, C. Adler, D.A. Fontem, H. Bouda, C. Reichmuth, Bioactivities of cymol and essential oils of *Cupressus sempervirens* and *Eucalyptus saligna* against *Sitophilus zeamais* motschulsky and *Tribolium confusum* du val, J. Stored Prod. Res. 41 (2005) 91–102.

[26] M.F. Maia, S.J. Moore, Plant-based insect repellents: a review of their efficacy, development and testing, Malar. J. 10 (11) (2011) 1–14.

[27] J.M. Ben Jemâa, N. Tersim, K.T. Toudert, M.L. Khouja, Insecticidal activities of essential oils from leaves of *Laurus nobilis* L. from Tunisia, Algeria and Morocco, and comparative chemical composition, J. Stored Prod. Res. 48 (2012) 97–104.

[28] M. Dehghani, K. Ahmadi, Anti-oviposition and repellence activities of essential oils and aqueous extracts from five aromatic plants against greenhouse whitefly *Trialeurodes vaporariorum* Westwood (Homoptera: aleyrodidae), Bulg. J.Agric. Sci. 19 (2013) 691–696.

[29] S.P. Aref, O. Valizadegan, Fumigant toxicity and repellent effect of three Iranian *Eucalyptus* species against the lesser grain beetle, *Rhyzopertha dominica* (F.) (Col.: bostrichidae), J. Entomol. Zool. Stud. 3 (2) (2015) 198–202.

[30] E. Adjalian, P. Sessou, B. Yehouenou, F.T.D. Bothon, J. Noudogbessi, D. Kossou, et al., Anti-oviposition and repellent activity of essential oil from *Melaleuca leucadendron* leaf acclimated in Benin against the angoumois grain moth, Int. J. Biol. Pharm. Allied Sci. 4 (2) (2015) 797–806.

[31] S. Lakshmanan, S. Thushimenan, V. Tamizhazhagan, Antifeedant, larvicidal and oviposition deterrent activity of *Pongamia pinnata* and *Ceiba pentandra* against pod borer larvae of *Helicoverpa armigera* (Noctuidae: Lepidoptera), Indo Am. J. Pharma. Sci. 4 (1) (2017) 180–185.

[32] S.Z. Haider, M. Mohan, A.K. Pandey, P. Singh, Use of *Tanacetum tomentosum* and *Ta. dolichophyllum* essential oils as botanical repellents and insecticidal agents against storage pest *Tribolium castaneum* (Coleoptera: tenebrionidae), Entomol. Res. 47 (2017) 318–327.

[33] J.M. Herrera, M.P. Zunino, Y. Massuh, R.P. Pizzollito, J.S. Dambolena, N.A. Ganan, J.A. Zygadlo, Fumigant toxicity of five essential oils rich in ketones against *Sitophilus zeamais* (Motschulsky), AgriScientia 31 (1) (2014) 35–41.

[34] Y. Huang, S.L. Lam, S.H. Ho, Bioactivities of essential oil from *Elletaria cardamomum* (L.) Maton. To *Sitophilus zeamais* motschulsky and *Tribolium castaneum* (Herbst), J. Stored Prod. Res. 36 (2000) 107–117.

[35] C.M. Priestley, E.M. Williamson, K.A. Wafford, D.B. Sattelle, Thymol, a constituent of thyme essential oil, is a positive allosteric modulator of human GABA(A) receptors and a homo-oligomeric GABA receptor from *Drosophila melanogaster*, Br. J. Pharmacol. 140 (8) (2003) 1363–1372.

[36] T.G. Thomas, S. Rao, S. Lal, Mosquito larvicidal properties of essential oil of an indigenous plant, Ipomoea cairica Linn, Jap. J. Infect. Dis. 57 (4) (2004) 176–177.

[37] O. Koul, Biological activity of volatile di-n-propyl disulfide from seeds of neem, *Azadirachta indica* (Meliaceae), to two species of stored grain pests, *Sitophilus oryzae* (L.) and *Tribolium castaneum* (Herbst), J. Econ. Entomol. 97 (3) (2004) 1142–1147.

[38] B.H. Lee, P.C. Annis, F. Tumaalii, S.E. Lee, Fumigant toxicity of *Eucalyptus blakelyi* and *Melaleuca fulgens* essential oils and 1,8-cineole against different development stages of the rice weevil *Sitophilus oryzae*, Phytoparasitica 32 (5) (2004) 498−506.

[39] M.O. Omolo, D. Okinyo, I.O. Ndiege, W. Lwande, A. Hassanali, Fumigant toxicity of the essential oils of some African plants against *Anopheles gambiae* Sensustricto, Phytomedicine 12 (3) (2005) 241−246.

[40] M. Amizadeh, M.J. Hejazi, G.A. Saryazdi, Fumigant toxicity of some essential oils on *Tetranychus urticae* (Acari: tetranychidae), Int. J. Acarol 39 (4) (2013) 285−289.

[41] M. Mobkia, S.A. Safavia, M.H. Safaralizadeha, O. Panahib, Toxicity and repellency of garlic (*Allium sativum* L.) extract grown in Iran against *Tribolium castaneum* (Herbst) larvae and adults, Arch. Phytopathol. Plant Prot. 47 (1) (2014) 59−68.

[42] A.K. Tripathi, V. Prajapati, K.K. Agarwal, S.P.S. Khanuja, S. Kumar, Toxicity towards *Tribolium castenum* in the fractions of essential oil of *Anethum sowa* seeds, J. Entomol. 98 (5) (2000a) 1560−1565.

[43] A.K. Tripathi, V. Prajapati, K.K. Aggrawal, S.P.S. Khanuja, S. Kumar, Repellency and toxicity of oil from *Artemisia annua* to certain stored product beetles, J. Econ. Entomol. 93 (2000b) 43−47.

[44] F.A. Ajayi, A. Olonisakin, Bio-activity of three essential oils extracted from edible seeds on the rust-red flour beetle, *Tribolium castaneum* (Herbst.) infesting stored pearl millet, Trakia J. Sci. 9 (1) (2011) 28−36.

[45] M.K. Chaubey, Acute, lethal and synergistic effects of some terpenes against *Tribolium castaneum* Herbst (Coleoptera: *tenebrionidae*), Ecol. Balk. 4 (1) (2012) 53−62.

[46] S. Chang, P. Chen, S. Wang, H. Wu, Antimite activity of essential oils and their constituents from *Taiwania cryptomerioides*, J. Med. Entomol. 38 (3) (2001) 455−457.

[47] Z. Saad, R. Hussien, F. Saher, Z. Ahmed, Acaricidal activity of some essential oils and their monoterpenoidal constituents against the house dust mite, *Dermatophagoides pteronyssinus* (Acari: pyroglyphidae), J. Zhejiang Univ. Sci. B 7 (12) (2006) 957−962.

[48] L. Gende, M. Maggi, C. van Baren, A. di Leo, A. Bandoni, R. Fritz, M. Eguaras, Antimicrobial and miticide activities of *Eucalyptus globulus* essential oils obtained from different argentine regions, Span. J. Agric. Res. 8 (3) (2010) 642−650.

[49] Akhtar, E. Pages, A. Stevens, R. Bradbury, A.G. da Camara, M.B. Isman, Effect of chemical complexity of essential oils on feeding deterrence in larvae of the cabbage looper, Physiol. Entomol. 37 (2012) 81−91.

[50] A. Kumar, N. Negi, S.Z. Haider, D.S. Negi, Composition and efficacy of *Zanthoxylum alatum* essential oils and extracts against *Spodoptera litura*, Chem. Nat. Comp. 50 (5) (2014) 920−923.

[51] D.M. Bird, V.M. Williamson, P. Abad, J. McCarter, E.G. Danchin, P. Castagnone-Sereno, C.H. Opperman, The genome of root-knot nematodes, Annu. Rev. Phytopathol. 47 (2009) 333−351.

[52] U.L.B. Jayasinghe, B.M.M. Kumarihamy, A.G.D. Bandara, E.A. Vasquez, W. Kraus, Nematicidal activity of some Sri Lankan plants, Nat. Prod. Res. 17 (2003) 259−262.

[53] T. Bleve-Zacheo, M.T. Melillo, P. Castagnone-Sereno, The contribution of biotechnology to root-knot nematode control in tomato plants, Pest Technol. 1 (2007) 1−16.

[54] J. Sorribas, C. Ornat, Estragias de control integrado de nematodos fitopara'sitos, in: M.F. Andre's, S. Verdejo (Eds.), Enfermedades causadas por nematodos fitopara'sitos en España. Phytoma-SEF. Valencia, 2011, pp. 115−127.

[55] R. Ghorbani, S. Wilcockson, A. Koochek, C. Leifert, Soil management for sustainable crop disease control: a review, Environ. Chem. Lett. 6 (2008) 149−162.

[56] D.J. Chitwood, Phytochemical based strategies for nematode control, Annu. Rev. Phytopathol. 40 (2002) 221−249.

[57] D.R. Batish, H.P. Singh, R.K. Kohli, S. Kaur, Eucalyptus essential oil as a natural pesticide, For. Ecol. Manag. 256 (2008) 2166−2174.

[58] M.F. Andres, A. Gonzalez-Coloma, J. Sanz, J. Burillo, P. Sainz, Nematicidal activity of essential oils: a review, Phytochem. Rev. 11 (2012) 371−390.

[59] N.K. Sangwan, B.S. Verma, K.K. Verma, K.S. Dhindsa, Nematicidal activity of some essential plant oils, Pestic. Sci. 28 (1990) 331−335.

[60] Y. Oka, S. Nacar, E. Putievsky, U. Ravid, Z. Yaniv, Y. Spiegel, Nematicidal activity of essential oils and their components against the root-knot nematode, Phytopathology 90 (2000) 710−715.

[61] S.K. Ibrahim, A.F. Traboulsi, S. El-Haj, Effect of essential oils and plant extracts on hatching, migration and mortality of *Meloidogyne incognita*, Phytopathol. Mediterr. 45 (2006) 238−246.

[62] T.L. Widmer, G.S. Abawi, Mechanism of suppression of *Meloidogyne hapla* and its damage by a green manure of Sudan grass, Plant Dis. 84 (2000) 562−568.

[63] A. Tava, P. Avato, Chemical and biological activity of triterpene saponins from Medicago species, Nat. Prod. Commun. 1 (2006) 1159−1180.

[64] M. Akhtar, Nematicidal potential of the neem tree *Azadirachta indica* (A. Juss), Integr. Pest Manag. Rev. 5 (2000) 57−66.

[65] H. Schmutterer, Properties and potential of natural pesticides from the neem tree, *Azadirachta indica*, Annu. Rev. Entomol. 35 (1990) 271−297.

[66] Y. Oka, N. Tkachi, S. Shuker, U. Yerumiyahu, Enhanced nematicidal activity of organic and inorganic ammonia-releasing amendments by *Azadirachta indica* extracts, J. Nematol. 39 (2007) 9−16.

[67] K.H. Wang, C. Hooks, A. Ploeg, Protecting Crops from Nematode Pests: Using Marigold as an Alternative to Chemical Nematicides, Cooperative Extension Service,

2007. CTAHR, PD-35. Available from: http://www.ctahr.hawaii.edu/oc/freepubs/pdf/PD-35.pdf.

[68] T.C. Thoden, M. Boppre, J. Hallmann, Effects of pyrrolizidine alkaloids on the performance of plant-parasitic and free-living nematodes, Pest Manag. Sci. 65 (2009) 823–830.

[69] P. Ohri, S.K. Pannu, Effect of phenolic compounds on nematodes-A review, J. Appl. Nat. Sci. 2 (2010) 344–350.

[70] N.G. Ntalli, P. Caboni, Botanical nematicides: a review, J. Agric. Food Chem. 60 (2012) 9929–9940.

[71] T.E. Hewlett, E.M. Hewlett, D.W. Dickson, Response of Meloidogyne spp., Heterodera glycines and Radopholus similis to tannic acid, J. Nematol. 29 (1999) 737–741.

[72] I.H. Mian, R. Rodriguez-Kabana, Organic amendments with high tannin and phenolic contents for control of Meloidogyne arenaria in infested soil, Nematropica 12 (1982) 221–234.

[73] O. Koul, Phytochemicals and insect control: an antifeedant approach critical reviews in plant sciences 27 (2008) 1–24.

[74] B. Mejia-Garibay, E. Palou, A. Lopez-Malo, Composition, diffusion, and antifungal activity of black mustard (Brassica nigra) essential oil when applied by direct addition or vapor phase contact, J. Food Prot. 4 (2015) 843–848.

[75] J. Tian, X. Ban, H. Zeng, J. He, B. Huang, Y. Wang, Chemical composition and antifungal activity of essential oil from Cicuta virosa L. var. Latisecta celak, Int. J. Food Microbiol. 145 (2011) 464–470.

[76] A. Kedia, B. Prakash, P.K. Mishra, C.S. Chanotiya, N.K. Dubey, Antifungal, antiaflatoxigenic and insecticidal efficacy of spearmint (Mentha spicata L.) essential oil, Int. Biodeterior. Biodegrad. 89 (2014) 29–36.

[77] A. Kedia, B. Prakash, P.K. Mishra, A.K. Dwivedy, N.K. Dubey, Trachyspermum ammi L. essential oil as plant based preservative in food system, Ind. Crops Prod. 69 (2015) 104–109.

[78] N.G. Tzortzakis, C.D. Economakis, Antifungal activity of lemongrass (Cympopogon citratus L.) essential oil against key postharvest pathogens, Innov. Food Sci. Emerg. Technol. 8 (2) (2007) 253–258.

[79] P.K. Mishra, A. Kedia, N.K. Dubey, Chemically characterized Cymbopogon martinii (Roxb.) Wats. essential oil for shelf life enhancer of herbal raw materials based on antifungal, antiaflatoxigenic, antioxidant activity, and favorable safety profile, Plant Biosyst. 50 (2015) 1313–1322.

[80] N.L. Tatsadjieu, P.J. Dongmo, M.B. Ngassoum, F.X. Etoa, C.M.F. Mbofung, Investigations on the essential oil of Lippia rugosa from Cameroon for its potential use as antifungal agent against Aspergillus flavus Link ex. Fries, Food Control 20 (2009) 161–166.

[81] P.V. Shanmugam, A. Saroj, R. Maurya, A. Yadav, N. Gupta, A. Samad, C.S. Chanotiya, Enantioselective GC analysis of C3−oxygenated p-menthane type Indian Mentha spicata var. viridis 'Ganga' essential oil, Nat. Prod. Commun. 12 (2017) 427–430.

[82] J. Varma, N.K. Dubey, Efficacy of essential oils of Caesulia axillaris and Mentha arvensis against some storage pests causing biodeterioration of food commodities, Int. J. Food Microbiol. 68 (2001) 207–710.

[83] R.C. Padalia, R.S. Verma, A. Chauhan, P. Goswami, C.S. Chanotiya, A. Saroj, A. Samad, A. Khaliq, Compositional variability and antifungal potentials of Ocimum basilicum, O. tenuiflorum, O. gratissimum and O. kilimandscharicum essential oils against Rhizoctonia solani and Choanephora cucurbitarum, Nat. Prod. Commun. 9 (2014) 1507–1510.

[84] A. Kumar, R. Shukla, P. Singh, N.K. Dubey, Chemical composition, antifungal and anti-aflatoxigenic activities of Ocimum sanctum L. essential oil and its safety assessment as plant based antimicrobial, Food Chem. Toxicol. 48 (2010) 539–543.

[85] V.S. Pragadheesh, A. Saroj, A. Yadav, C.S. Chanotiya, M. Alam, A. Samad, Chemical characterization and antifungal activity of Cinnamomum camphora essential oil, Ind. Crops Prod. 49 (2013) 628–633.

[86] M. Omidbeygi, M. Barzegar, Z. Hamidi, H. Naghdibadi, Antifungal activity of thyme, summer savory and clove essential oils against Aspergillus flavus in liquid medium and tomato paste, Food Control 18 (2007) 1518–1523.

[87] E. Pinto, C. Pina-Vaz, L. Salgueiro, M.J. Goncalves, S. Costa-de-Oliveira, C. Cavaleiro, A. Palmeira, A. Rodrigues, J. Martinez-de-Oliveira, Antifungal activity of the essential oil of Thymus pulegioides on Candida, Aspergillus and dermatophyte species, J. Med. Microbiol. 55 (2006) 1367–1373.

[88] A. Saroj, V.S. Pragadheesh, Palanivelu, A. Yadav, S.C. Singh, A. Samad, et al., Anti-phytopathogenic activity of Syzygium cumini essential oil, hydrocarbon fractions and its novel constituents, Ind. Crops Prod. 74 (2015), 337–335.

[89] E. Pinto, L. Vale-Silva, C. Cavaleiro, L. Salgueiro, Antifungal activity of the clove essential oil from Syzygium aromaticum on Candida, Aspergillus and dermatophyte species, J. Med. Microbiol. 58 (2009) 1454–1462.

[90] C.F. Carson, B.J. Mee, T.V. Riley, Mechanism of action of Melaleuca alternifolia (tea tree) oil on Staphylococcus aureus determined by time-kill, lysis, leakage and salt tolerance assays and electron microscopy, Antimicrob. Agents Chemother. 46 (6) (2002) 1914–1920.

[91] P.N. Skandamis, G.J.E. Nychas, Effect of oregano essential oil on microbiological and physico-chemical attributes of minced meat stored in air and modified atmospheres, J. Appl. Microbiol. 91 (6) (2001) 1011–1022.

[92] J. Thoroski, G. Blank, C. Biliaderis, Eugenol induced inhibition of extracellular enzyme production by Bacillus cereus, J. Food Prot. 52 (6) (1989) 399–403.

[93] K. Knobloch, A. Pauli, B. Iberl, H. Weigand, N. Weis, Antibacterial and antifungal properties of essential oil components, J. Essent. Oil Res. 1 (1989) 119–128.

[94] K. Oosterhaven, B. Poolman, E.J. Smid, S-carvone as a natural potato sprout inhibiting, fungistatic and bacteristatic compound, Ind. Crops Prod. 4 (1995) 23–31.

[95] J. Sikkema, J.A.M. De Bont, B. Poolman, Interactions of cyclic hydrocarbons with biological membranes, J. Biol. Chem. 269 (11) (1994) 8022–8028.

[96] S.D. Cox, C.M. Mann, J.L. Markham, H.C. Bell, J.E. Gustafson, J.R. Warmington, S.G. Wyllie, The mode of antimicrobial action of essential oil of *Melaleuca alternifola* (tea tree oil), J. Appl. Microbiol. 88 (2000) 170–175.

[97] J.E. Gustafson, Y.C. Liew, S. Chew, J.L. Markham, H.C. Bell, S.G. Wyllie, J.R. Warmington, Effects of tea tree oil on *Escherichia coli*, Lett. Appl. Microbiol. 26 (1998) 194–198.

[98] I.M. Helander, H.L. Alakomi, K. Latva-Kala, T. Mattila-Sandholm, I. Pol, E.J. Smid, L.G.M. Gorris, A. Von Wright, Characterization of the action of selected essential oil components on Gram-negative bacteria, J. Agric. Food Chem. 46 (1998) 3590–3595.

[99] R.J.W. Lambert, P.N. Skandamis, P.J. Coote, G.J.E. Nychas, A study of the minimum inhibitory concentration and mode of action of oregano essential oil, thymol and carvacrol, J. Appl. Microbiol. 91 (2001) 453–462.

[100] A. Ultee, E.P.W. Kets, E.J. Smid, Mechanisms of action of carvacrol on the food-borne pathogen *Bacillus cereus*, Appl. Environ. Microbiol. 65 (10) (1999) 4606–4610.

[101] A. Ultee, E.J. Smid, Influence of carvacrol on growth and toxin production by *Bacillus cereus*, Int. J. Food Microbiol. 64 (2001) 373–378.

[102] A. Ultee, E.P.W. Kets, M. Alberda, F.A. Hoekstra, E.J. Smid, Adaptation of the food-borne pathogen *Bacillus cereus* to carvacrol, Arch. Microbiol. 174 (4) (2000) 233–238.

[103] A. Ultee, M.H.J. Bennink, R. Moezelaar, The phenolic hydroxyl group of carvacrol is essential for action against the food-borne pathogen *Bacillus cereus* (Ph. D. thesis),

ISBN 90-5808-219-9, Appl. Environ. Microbiol. 68 (4) (2002) 1561–1568.

[104] C.N. Wendakoon, M. Sakaguchi, Inhibition of amino acid decarboxylase activity of *Enterobacter aerogenes* by active components in spices, J. Food Prot. 58 (3) (1995) 280–283.

[105] N. Singh, C. Wang, R. Cooper, Potential of essential oil-based pesticides and detergents for bed bug control, J. Econ. Entomol. 107 (6) (2014) 2163–2170.

[106] F.E. Dayan, C.L. Cantrell, S.O. Duke, Natural products in crop protection, Bioorg. Med. Chem. 17 (2009) 4022–4034.

[107] D. Fischer, C. Imholt, H.J. Pelz, M. Wink, A. Prokopc, J. Jacoba, The repelling effect of plant secondary metabolites on water voles, Arvicola amphibious, Pest Manag. Sci. 69 (2013) 437–443.

[108] O. Pino, Y. Sanchez, M.M. Rojas, Plant secondary metabolites as an alternative in pest management. I: background, research approaches and trends, Rev. Prot. Veg. 28 (2) (2013) 81–94.

[109] M.B. Isman, C.M. Machial, Pesticides based on plant essential oils: from traditional practice to commercialization, in: Rai, Carpinella (Eds.), Naturally Occurring Bioactive Compounds, Elsevier, 2006, pp. 29–44.

[110] C.L. Nistor, C. Ianchis, F. Oancea, M.L. Jecu, L. Raut, D. Donescu, Process for Essential Oils Encapsulation into Mesoporous Silica Systems and for Their Application as Plant Biostimulants, Patent No. WO/2017/091096; PCT/RO2016/000025, 2017.

[111] A. Saroj, A.K. Srivastava, A. Nayak, C.S. Chanotiya, A. Samad, Essential oil in pest control and disaese management, in: C. Egbuna, I.J. Chineye, S. Kumar, N. Sherif (Eds.), Phytochemistry: Marine Phytochemistry, Industerial Applications and Phytochemical Pesticides, vol. 3, Apple Academic Press, 2019, pp. 341–368.

CHAPTER 7

Semiochemicals: A Green Approach to Pest and Disease Control

SHAHIRA M. EZZAT, PHD • JAISON JEEVANANDAM, PHD •
CHUKWUEBUKA EGBUNA, BSC, MSC • RANA M. MERGHANY •
MUHAMMAD AKRAM, PHD • MUHAMMAD DANIYAL • JAWERIA NISAR •
AAMIR SHARIF

INTRODUCTION

Chemical signaling among insects play a key role in their survival and facing up different environmental conditions, by modifying their physiology or behavior. Insects use semiochemicals (a group of organic compounds) to transmit these chemical messages over long distances that help them in surviving, through locating a host, mate, and food and beating a host defense system or a natural enemy [1]. Semiochemicals, a term derived from the Greek word (semeion) meaning a signal, differs in their molecular weights according to their carbon chain [2].

Insects live in an environment surrounded by a lot of semiochemicals produced from insects itself or the host plant. Insects interact with host plant semiochemicals in a way where these volatile chemicals relate with each other and lastly modify the behavior of the pest species. Some insects use host plant chemicals as sex pheromone for mating [3]. Many insects like butterflies and moths use pyrrolizidine alkaloids from their host plants as food to survive from their natural enemies [4]. The African palm weevil *Rhynchophorus phoenicis* use a mixture of volatile esters from the host plant oil palm *Elaeis guineensis*, where the ethyl acetate makes male weevils to liberate the aggregation pheromone rhyncophorol to attract both sexes for mating [5]. Males of orchid bees use terpenoids mixture from orchids as pheromone to establish certain sites where males compete for females to mate [6].

Semiochemical characterization can either be by extraction (when the organ that produces the semiochemical is known) or headspace collection where a charcoal-purified air is passed along the antenna of the insect in an isolated chamber and the air is then retrieved by vacuum to be analyzed [7]. Then identification of semiochemicals can be accomplished using different equipment's which include gas chromatography–mass spectrometry, solid-phase microextraction, nuclear magnetic resonance, and gas chromatography–electroantennography. The elucidation of behavioral response of the insect to the active product is also an essential step in the identification of semiochemicals [8,9]. Semiochemicals have been used for pest control for more than 100 years ago. The carob moth *Ectomyelois ceratoniae*, tomato leaf miner *Tuta absoluta*, and the armyworm *Spodoptera frugiperda* are different pests that have been effectively controlled using semiochemicals [2].

Many factors can affect the success of using pheromones in pest control programs like the biological differences in the behavior of different species for mate-finding. Some of these factors are the diverse chemical composition of pheromones, different economic and political regulations of pheromones' use in different countries and the use of controlled-release dispenser, proper trap design and density [4].

Semiochemical usage would be effective in pest control through these strategies: recognizing the insidious species, selecting the suitable timing for the insecticide usage, assessing a post-application evaluation, improving of traditional methods used in pest control, increasing the rates of predators and parasitoids, putting into consideration at least one of the techniques developed that are used in pest control like attract and kill, mating disruption, mass trapping, or repellency techniques [10].

So far, many techniques has been developed for the use of semiochemicals in pest control like attract and

Natural Remedies for Pest, Disease and Weed Control. https://doi.org/10.1016/B978-0-12-819304-4.00007-5

kill, mating disruption, mass trapping, and repellency techniques.

Attract and kill is a technique that uses a semiochemical to attract an insect to a certain location having a killing agent (pathogen or insecticide) to reduce its fitness or killing it by causing certain disease [6].

Mating disruption is a technique that affects an insect sexual behavior to reduce population by saturating the surrounding environment by synthetic semiochemicals especially sex pheromones where the ability of males to recognize the natural sex pheromone produced by females are disrupted so delaying the time for females to mate and so will affect their ability to locate the appropriate sites for oviposition. Mating disruption can be explained by three mechanisms:

1. Competitive attraction: This involves a situation where males respond to synthetic sex pheromones rather than the natural sex pheromones produced by the female. This is a density-dependent mechanism and its efficiency decreases when population increases [6].

2. Camouflage: This is also a density-independent mechanism that needs complete saturation of the surrounding environment with the synthetic sex pheromone.

3. Desensitization: where the male central nervous system adapts to the synthetic sex pheromone due to overexposure.

Mass trapping is a technique where females are trapped so reducing egg laying in case of males acting as an attractant through producing the natural sex pheromone such as red palm weevil [11].

Repellency is a technique where insects are deterred from locating food on an attractive host [6]. This technique is limited because of the availability of other cheap alternatives and lack of suitable formulations for delivery [12]. These repellents may be used in combination with some attractants as a method of a push–pull approach [13].

Classification of Semiochemicals

Depending on their function, semiochemicals can be classified. Serious consideration must be put in place since similar particle could be active as a kairomone or allomone for one insect species and as pheromone for others. Semiochemicals are isolated into two general groups: pheromones that intercede connections among population of similar species (intraspecific reactions) and allele chemicals that intervene communications among population of various species (interspecific collaborations). Pheromones are more subdivided as indicated by the behavioral response into (i) releaser pheromones that have immediate or short-term behavioral response and (ii) primer pheromones that produces long-term physiological changes, which further include aggregation attachment pheromones, sex pheromones, and a primer pheromone. Allelochemicals are divided into (i) kairomones that intercede the collaborations supporting the beneficiary whereas (ii) allomones support producer only and (iii) synomones support both the producer and the beneficiary, and (iv) pneumones are mainly produced by nonliving material that educes behavioral response that is supportive for beneficiary organism but unsafe for the second organism living on that dead material. The semiochemicals classification is presented in schematic diagram as shown in Fig. 7.1.

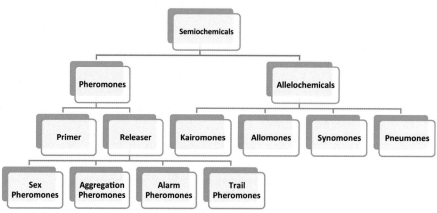

FIG. 7.1 Classification of semiochemicals.

Insect parasitoid kairomones

In host—parasitoid relationship, semiochemicals has a significant role which was categorized into different stages such as habitat location, host location, host acceptance, and oviposition.

Parasitoids produced two types of kairomones to find their hosts:

(a) **External to the host**: These are esterified cholesterols or proteins, which are long-chain hydrocarbons; ketones of fatty acids found in host glue are used to attach eggs to the substrate [14].

(b) **Internal kairomones:** These are amino acids and salts represented in hemolymph, which detect the ovipositors and act as indicators to provide the appropriateness of parasitoid offspring's to the host.

Pheromone

Chemical substance secreted externally into the environment influences the physiology or behavior of other members of the same species. Pheromones are a type of semiochemical: chemical ecology, alarm pheromones, and sex pheromones. The term "pheromone" is obtained from two Greek words, pherein, denoting to carry, and horman, denoting to excite. Many insects secrete pheromones, a special class of semiochemicals, to attract mates. Insects use chemical cues as signals to find mates, food, oviposition, and hibernation sites. Molecular genetic methods are now being employed to study different characteristics of pheromone response behavior. The genes encoding proteins responsible for pheromone synthesis, the processing of the signals, and the perception of semiochemicals are being cloned and characterized. The biosynthesis of pheromone seems to use one or more enzymes that change the metabolites of normal primary metabolism into products that operate as pheromones. For example, pheromones arise from biosynthesis of isoprenoid, or by the transformation of fatty acids or amino acids. Sometimes both sexes produce a pheromone that functions in some aspect of mating. Alternatively, only the female or the male may produce an attractant pheromone, commonly called a sex pheromone. The sex pheromone is usually composed of several chemical components secreted in a particular ratio or blend. The blend or ratio of components allows more information to be encoded, and allows greater discrimination among closely related species. The multifunctional compounds secreted by insects are termed semiochemicals and primarily include pheromones (communication chemicals) and alimonies (defensive chemicals). Pheromones have important role to intervene many actions of social insects, including recruitment to food, colony defense, individual's recognition and nest mates, and caste development regulation. Various pheromones are produced by the exocrine glands that scattered throughout the body, and leafcutter ants exemplified it [15].

Use of pheromones by insects and humans.

Chemical communication via pheromones has tremendous benefit to insect species that may not have extensive adaptations for visual or acoustic communication. Thus, pheromone production is extremely common and widespread among the class Insect [15]:
- Social insect pheromones
- Alarm pheromones
- Sex attractant pheromones

Insect pheromones. First, the term pheromone is proposed in previous study to depict chemical signals that intervene intraspecific connections [16]. First pheromone was *Bombyx mori* (silkworm moth's sex pheromone) that was chemically recognized in 1959. In pest management, it is considered as the most important utilized semiochemical. Different pheromones consist of aggregation pheromones that are released by males that attract both genders of nonspecific individuals. The sex pheromone produced by moths is one of the usually studied and broadly utilized pheromones than other pheromones in insect pest management [3]. Moths have specialized posture and behavior that is sponsored by pheromones to produce attraction for mating [5,17].

The parsimonious activities of eclectic alarm pheromones. Alarm pheromones, which are used by both eusocial and solitary species, are generally produced during traumatic interactions that result in dispersion or defense of the alarm pheromone producers. These are similar in many aspects to allomones and produced in large quantities than other pheromones. Some eusocial species have adapted their alarm pheromones to regulate important social activities in a variety of specific contexts. Honey bee workers, *Apis mellifera*, suppress visitations of workers to nectar-depleted flowers by marking these flowers with 2-heptanone, an alarm pheromone produced in the capacious mandibular glands. Some species of beetles, cockroaches, and mutillid wasps, as well as noninsect arthropods that have frequent encounters with formicides have the basis for avoiding attacking ants when identified with the use of cryptic alarm pheromones. Displacement behavior by some species of ants results when workers undertake digging after exposure to high concentrations of their alarm pheromones. Buried worker bees during digging use alarm pheromones as signals. Several alarm pheromones of ants have been demonstrated to be potently fungistatic, that is, (E)-2-hexenal and (E)-2-decenal of *N. virudula*. Workers of ants and eusocial species of bees and wasps may mark assailants with alarm pheromones that function to attract additional workers to the labeled individual [18].

Defensive allomones as alarm pheromones

If the defensive allomones of solitary insects are remarkably eclectic, it is also rather significant that many of these compounds are structurally congruent with the alarm pheromones of eusocial insects. Alarm pheromones are generally secreted in considerably greater quantities than other classes of pheromones, further emphasizing their suitability as defensive allomones. Indeed, defensive allomones and alarm pheromones share many common properties, a fact consistent with the suggestion that the latter were derived from the former [19].

Sex pheromones as parsimonious agents

A variety of insect species have adapted secretary compounds with clearly established defensive functions that also serve as sex pheromones. In addition to defensive compounds, in some hymenopterous species, these sexual pheromones constitute versatile releasers and primers of social behavior that are essential to colonial organization [20].

Defensive secretions as sexual pheromones

A variety of insects have adapted defensive exudates (defensive allomones) to function as either female or male sex attractants (sex pheromones), for example, the glandular secretion is fortified with 1,4-benzoquinones, hydrocarbons, and aliphatic aldehydes, a formidable mixture of defensive allomones. Three compounds—(Z)-4-tridecene, dodecanal, and (Z)-5-tetradecenal—are powerful male attractants and release strong copulatory behavior in these insects [21].

When pheromones become kairomones

It has been reported that predators and parasitoids have the ability to manipulate and utilize their prey pheromones. For instance, a cleric scarabs group utilizes the bark beetles pheromonal signal to find and assault the prey's passages in the bark and conifers cambium layers. In the same way, species of pentatonic bug often go through raise parasitism from parasitic flies from numerous families, or from authority wasp predators.

It has been clearly established that the bugs' pheromones as kairomone signals are utilized by these parasitoids to find host. The flies and wasps become attentive to their host's pheromone blend components. The attraction to their prey pheromones can be so vigorous to traps fly with the pheromones of prey actually grab greater amount of the predators or parasitoids than the target species, for the predatory clerids offensive to beetles of the bark and also for the fly and wasp species assaulted pentatonic bugs. In any case, unlawful utilization of the prey pheromones can go well further than basically intrude on pheromonal signals. In an intriguing case of coevolution, bolas creepy crawlies in the *Mastophora* genus (a few other genera) create the pheromonal cues of their prey to bait the prey close enough to be gotten by sticky string of silk and swinging. The preys are male noctuid moths reacting to duplicates of the female pheromone. More uncommon, there is proof to recommend that in an hour or somewhere in the vicinity, the creepy crawlies can change the pheromone baits they create, to maximally utilize the diverse period of prey species fight that react to distinct pheromone mixes. An assortment of creepy crawlies that live inside their homes and once in a while social bugs parasitize groups, for example, ants and termites likewise have built up the capacity to forcefully imitate their hosts' pheromonal signals. For instance, a few staphylinid species of insect exist within termite homes, where they get all their nourishment from their hosts, and are prepped and thought about by their hosts just as they were termite brood. The concoction signals that both keep the termites from perceiving the inquilines and prompt the grooming and nourishing behaviors firmly impersonate the genuine pheromones utilized by the termites for these works. In a much progressively forceful precedent, the hatchlings of some hover flies are committed predators on their ant host's progeny. The fly hatchlings create a mixture of cuticular hydrocarbons that intently coordinates the host's progeny hydrocarbon profile. This camouflage mechanism is fine to the point that if the home is assaulted, the laborer ants will convey the fly hatchlings to security as if they were insect brood [22].

Pros and Cons of the Use of Semiochemicals

Semiochemicals are very specific to insect species, so they will not affect nontarget organisms. They are nontoxic to the environment due to their natural origin, so nowadays they are of great interest to be used in pest control rather than using traditional agents [23]. However, semiochemical formulations are physically unstable and volatile, and the active ingredient may be sensitive to temperature and light. Again, there is high cost compared to traditional methods that employ manual application and irregularity that comes with the formulation to achieve a certain release rate.

Depending on the technique used like mating disruption technique in producing the formulation, a huge amount of semiochemical that entails more cost is needed. However, this problem is now solved by ISCA technologies that developed certain semiochemical formulation for agricultural pests. The technique used is named "specialized pheromone and lure application technology" (SPLAT). It is a flowable and controlled-release formulation, where its composition can be adjusted by small changes through application. The formulation is flexible in application due to its fluid that enters into a liquid state by agitation but rapidly solidifies when agitation is stopped [6].

Mechanism of Action of Semiochemicals

Semiochemicals have been reported to possess enhanced pesticidal activity via various mechanisms of actions. The knowledge on the mechanisms of semiochemicals is highly significant in formulating distinct strategies for efficient pest management systems. Inhibition by the attraction (attract and kill), mating disruption, repellents, and mass trapping are some of the mechanisms followed by semiochemicals to control pests.

Attract (lure) and kill

Inhibition of pest growth by attraction, also called attract and kill, is the most common mechanism that is reported to be followed by semiochemicals. Pheromones [2], kairomones [24], and attractants [25] are the semiochemicals that are used to attract pests and inhibit their growth. The factors that decide the benefits of semiochemicals in attraction and killing-based pest management are deployed of semiochemicals to be effectively perceived by all the male and female pest

population in the treated area, their ability to attract pest populations is greater than natural and common odor sources, they attract adult pests toward the insecticide that inhibits their growth and the complete inhibition of pest population ranging from instars to adults [26]. Previously, semiochemicals are used as standalone pesticide that can attract and kill pests. However, drawbacks such as expensive and the inability to be used in large farm fields lead to a combination of semiochemicals with insecticides for pest management [27]. Most common commercial formulations of semiochemicals for pest control are droplets of paste or gel such as Sirene, SPLAT, and GF-120. In these formulations, combinations of insecticide and semiochemicals are incorporated in a wax, gel, or paste, which is applied as droplets per hectare on the farm. The paste or gel is usually slowly degradable in contact with the atmosphere and releases semiochemicals to attract pests and insecticides that help to inhibit their growth [28]. In addition, degradable paste or gels are replaced by micro or nanoencapsulations using emulsifiers [29]. Semiochemicals and insecticides are encapsulated in a plastic capsule as microencapsulation and are uniformly distributed in the farm via conventional spray technique [30]. The toxicity of plastic capsule and their inefficient degradability leads to novel nanoencapsulation-based semiochemicals formulation. Currently, nanoencapsulations with biodegradable polymers are under extensive research to replace microencapsulation and other conventional methods [31,32].

Fig. 7.2 shows the mechanism of effective pest control using a combinatorial formulation of semiochemicals and insecticides.

Mating disruption

Certain semiochemicals acts on pest populations by disrupting their mating processes. A novel polyvinyl chloride polymer dispenser of semiochemicals such as codlemone and pear ester of *Cydia pomonella* (codling moth) were recently evaluated with commercial

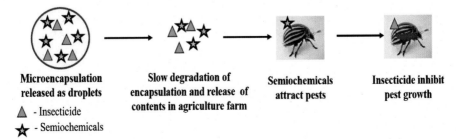

Microencapsulation released as droplets

Slow degradation of encapsulation and release of contents in agriculture farm

Semiochemicals attract pests

Insecticide inhibit pest growth

▲ - Insecticide

★ - Semiochemicals

FIG. 7.2 Mechanism of pest control by combinatorial semiochemical and insecticide formulation.

codlemone loaded Isomate and CheckMate dispensers. The result revealed that the novel semiochemical dispensers showed enhanced attraction and captured male moth to disrupt their mating processes, compared to commercial semiochemical dispensers [33]. Semiochemicals such as sex pheromones are useful to attract specific gender of pests (mostly males) and inhibit their growth, similar to attract (lure) and kill method as mentioned in the previous section. Disruption of mating processes in pests that belong to the Lepidoptera order of insects and olive fly (*Bactrocera oleae*) are observed, when synthetic sex pheromone semiochemicals are used on the farm to control pests [34]. The initial mechanism of mating disruption in pests via semiochemicals involves releasing high quantities of synthetic female sex pheromones to alter the mating behavior of males. Later, the attracted male can be inhibited via mass trapping approach using other semiochemicals [35]. In recent times, 4-Play formulation of semiochemicals was used for the mating disruption of codling moth as a replacement of conventional synthetic sex pheromone semiochemical formulations [36]. Furthermore, nanoparticle-based aerosol formulation of sex pheromone semiochemicals showed enhanced efficiency in disrupting the mating process of *Lobesia botrana* [37]. Furthermore, novel polymeric nanofiber-based, controlled pheromone release dispensers were prepared in recent times for accurate attraction and mating disruption of *Grapholita molesta* in the laboratory, which is considered as a futuristic method in pest management [38]. Fig. 7.3 is the schematic representation of mating disruption via sex pheromone-based semiochemicals.

Repellents

Apart from attract and kill, and mating disruption-mediated pest control ability, semiochemicals also possess repellent properties against pests in integrated pest management systems. Repellents are components, either natural or synthetic, that help to repel pests of economically significant plants and agricultural products. Semiochemicals such as nonhost volatiles are employed as repellents of striped ambrosia beetles (*Trypodendron lineatum*). The result demonstrated that these nonhost volatiles extracted from angiosperm tree barks possess ability to disrupt the visual, gustatory, and tactile stimuli of beetles with aggregated lineatin pheromone. Thus, these volatile-based semiochemicals are used as repellents against ambrosia beetles that drastically affected commercial timber production in British Columbia with its pesticide activity [39]. Similarly, 19 classified semiochemicals including aliphatic alcohols, ketones, terpenoids, and carboxylic esters are reported to be toxic toward housefly (*Musca domestica*) and the stable fly (*Stomoxys calcitrans*). Furthermore, the study demonstrated that Rosalva, citronellol, geranyl acetone, and permethrin possess ability to inhibit the susceptible strains of house and stable fly. Thus, these semiochemicals are proposed to be beneficial as potential insecticide and insect repellent in fly control programs [40]. However, pests started to develop resistance against conventional synthetic semiochemical-based pest repellents, which was evident from several recent reports. Thus, plant semiochemicals with the ability to alter the behavior of pests via natural volatile signaling processes are employed as efficient pest repellents. Tobacco whitefly (*Bemisia tabaci*) and greenhouse whitefly (*Trialeurodes vaporariorum*) that are considered as pests in agriculture are effectively controlled by these putative plant semiochemicals as repellents [41]. Moreover, plant-derived semiochemical repellents have been demonstrated recently to disrupt host seeking behavior in salmon louse (*Lepeophtheirus salmonis*) [42] and blow fly (*Chrysomya megacephala*) [43].

Mass trapping

Mass trapping is one of the unique mechanisms for effective pest management. Semiochemicals were initially used to control mosquitoes via mass trapping technology. These mass trapping techniques are effective only against pests with low population densities; however, several pests possess enormous reproductive potential [44]. Odor-baited traps are one of the conventional mass trapping methods that utilizes semiochemicals to attract masses (large number) of pests and instantly inhibit their growth [45]. These semiochemical mass traps are used to control potential pests that disturb agricultural economy, such as biting flies [46],

Sex pheromone semiochemicals → Semiochemicals attracting male pests → Disrupts mating efficiency → Mating with female pests → Absence of offspring

FIG. 7.3 Schematic representation of mating disruption via sex pheromone-based semiochemicals.

European buprestid beetles in oak forests [47] green bottle fly [48], and adult plum curculio [49]. Similarly, mass trapping efficiency of semiochemical-baited traps via Mark-recapture approach is proven to effectively control *Monochamus galloprovincialis*, also called as pine sawyer, which is a well-known vector of pine wood nematode in Europe that causes pine wilt disease. These baited traps attract pests, reduce their population density, and alter their flight behavior that eventually leads to the mass trap of the pests and inhibit their growth [50]. Certain studies also demonstrated that the mass trapping approach is an extended version of standalone lure and kill potential of semiochemicals [51]. Usually, semiochemicals are used to mass trap several pests for integrated pest management systems and not a specific pest. In recent times, semiochemical mass traps are used to reduce the population of mirid in strawberry [52], raspberry crops [53], *Frankliniella occidentalis* in roses [54], and phytoplasma vectors [55]. The drawbacks of mass trap techniques such as nonspecificity of pests, inefficiency in controlling large-scale pest population, and ineffectiveness in attracting pest masses lead to nanoformulated semiochemicals [56]. These nanosized semiochemical formulations help in overcoming the challenges of conventional mass trapping semiochemicals with enhanced efficiency in controlling the pest populations.

Other mechanisms

It is noteworthy that the current semiochemical formulations are fabricated to exhibit all the previously mentioned mechanisms for effective pest management systems. Push–pull strategy and biological control are some of the other mechanisms followed by semiochemicals to inhibit the growth of pests [55]. Push–pull approach is relatively a new concept in pest management, which utilizes formulation of semiochemicals that contains both attractive and repulsive stimuli to alter the pest behavior [57]. This mechanism is followed by semiochemicals to control the population of pests such as aphids [58], mealworm (*Alphitobius diaperinus*) [59], and ambrosia beetle [60]. The drawbacks of synthetic semiochemicals such as toxicity to the environment and resistance in pests lead to biological control mechanisms. These biological pest control agents belong to botanical semiochemicals, which are less toxic to the environment and also help in plant growth [61]. In future, nanoformulated botanical semiochemicals are proposed to be beneficial in effective and sustainable pest management.

Future Prospects and Conclusion

In many insects, semiochemicals are a predominant type of correspondence, and the messages passed on helps in its physiological and behavioral abilities. Current microanalytical systems allow recognition of many of these messages, despite the fact that they happen in little amounts. Research studies are required to increase our understanding to characterize these chemicals produced by insects and their effect on behavior, particularly about the complex communication system of social insects. Recent frontiers of inspection in pheromones of insects incorporate setting up how biosynthesis is organized by genes, how in responder receptor cells these chemical components are converted into electrical cues, and how these cues are handled in mind, stimulating a social yield. A few number of preliminary pheromones structures have been clarified and hence a lot stays to be found out about their method of activity.

REFERENCES

[1] A.M. El-Sayed. The pherobase: database of insect pheromones and semiochemicals. HortResearch, LiA. ncoln, New Zealand, (baza danych), Available: http://www.pherobase.com/. Retrieved 17/02/2019.

[2] V.A. Soroker, V. Harari, J.R. Faleiro, The role of semiochemicals in date pest management, in: Sustainable Pest Management in Date Palm: Current Status and Emerging Challenges, Springer, 2015, pp. 315–346.

[3] G.V. Reddy, A. Guerrero, Interactions of insect pheromones and plant semiochemicals, Trends Plant Sci. 9 (5) (2004) 253–261.

[4] R. Nishida, Sequestration of defensive substances from plants by Lepidoptera, Annu. Rev. Entomol. 47 (1) (2002) 57–92.

[5] A. Mafra-Neto, R.T. Cardé, Fine-scale structure of pheromone plumes modulates upwind orientation of flying moths, Nature 369 (6476) (1994) 142.

[6] K. Jaffé, et al., Chemical ecology of the palm weevilRhynchophorus palmarum (L.)(Coleoptera: Curculionidae): attraction to host plants and to a male-produced aggregation pheromone, J. Chem. Ecol. 19 (8) (1993) 1703–1720.

[7] A.M. Fraser, W.L. Mechaber, J.G. Hildebrand, Electroantennographic and behavioral responses of the sphinx moth Manduca sexta to host plant headspace volatiles, J. Chem. Ecol. 29 (8) (2003) 1813–1833.

[8] K.C. Park, et al., Odor discrimination using insect electroantennogram responses from an insect antennal array, Chem. Senses 27 (4) (2002) 343–352.

[9] J. Faleiro, et al., Screening date palm cultivars for resistance to red palm weevil, Rhynchophorus ferrugineus (Coleoptera: Curculionidae), Fla. Entomol. 97 (4) (2014) 1529–1536.

[10] W. Lewis, et al., Kairomones and their use for management of entomophagous insects: I. Evaluation for increasing rates of parasitization byTrichogramma spp. in the field, J. Chem. Ecol. 1 (3) (1975) 343−347.

[11] A.C. Oehlschlager, et al., Control of red ring disease by mass trapping of Rhynchophorus palmarum (Coleoptera: Curculionidae), Fla. Entomol. 85 (3) (2002) 507−513.

[12] M.B. Isman, Botanical insecticides, deterrents, and repellents in modern agriculture and an increasingly regulated world, Annu. Rev. Entomol. 51 (2006) 45−66.

[13] S.M. Cook, Z.R. Khan, J.A. Pickett, The use of push-pull strategies in integrated pest management, Annu. Rev. Entomol. (2007) 52.

[14] C.E. Rutledge, A survey of identified kairomones and synomones used by insect parasitoids to locate and accept their hosts 7 (3) (1996) 121−131.

[15] D. Payton, M. Daily, R. Estowski, M. Howard, C. Lee, Pheromone robotics, Aut. Robots 11 (3) (2001) 319−324.

[16] P. Karlson, M.J.N. Lüscher, 'Pheromones': a new term for a class of biologically active substances 183 (4653) (1959) 55.

[17] T. Baker, M. Willis, K. Haynes, P.L. Phelan, A pulsed cloud of sex pheromone elicits upwind flight in male moths, Physiol. Entamol. 10 (3) (1985) 257−265.

[18] M.S.J.P.E.A. Blum, Mycotoxins, I. Pheromones, A.C. Chou, The Chemistry and Roles of Eusocial Insect Pheromones and Allomones, 1989, pp. 39−47.

[19] C. Teerling, D. Gillespie, J. Borden, Utilization of western flower thrips alarm pheromone as a prey-finding kairomone by predators, Can. Entomol. 125 (3) (1993) 431−437.

[20] M.S. Blum, Semiochemical parsimony in the Arthropoda, Annu. Rev. Entomol. 41 (1) (1996) 353−374.

[21] J. Farine, O. Bonnard, R. Brossut, J.L. Le Quere, Chemistry of pheromonal and defensive secretions in the nymphs and the adults of Dysdercus cingulatus Fabr.(Heteroptera, Pyrrhocoridae), J. Chem. Ecol. 18 (1) (1992) 65−76.

[22] O. Anderbrant, C. Ryne, P. Olsson, E. Jirle, K. Johnson, C. Lofstedt, Pheromones and kairomones for detection and control of indoor pyralid moths, JIWB, IOBC/WPRS Bull. 30 (2) (2007) 73.

[23] A.R. Horowitz, P.C. Ellsworth, I. Ishaaya, Biorational pest control−an overview, in: Biorational Control of Arthropod Pests, Springer, 2009, pp. 1−20.

[24] L. Sarles, et al., Semiochemicals of Rhagoletis fruit flies: potential for integrated pest management, Crop Protect. 78 (2015) 114−118.

[25] D.L.P. Schorkopf, et al., Combining attractants and larvicides in biodegradable matrices for sustainable mosquito vector control, PLoS Negl. Trop. Dis. 10 (10) (2016) e0005043.

[26] A.M. El-Sayed, et al., Potential of "lure and kill" in long-term pest management and eradication of invasive species, J. Econ. Entomol. 102 (3) (2009) 815−835.

[27] S.J. Seybold, et al., Management of western North American bark beetles with semiochemicals, Annu. Rev. Entomol. 63 (2018) 407−432.

[28] P.C. Gregg, A.P. Del Socorro, P.J. Landolt, Advances in attract-and-kill for agricultural pests: beyond pheromones, Annu. Rev. Entomol. 63 (2018) 453−470.

[29] N. Bakthavatsalam, Chapter 19 − semiochemicals, in: Ecofriendly Pest Management for Food Security, Omkar, Academic Press, San Diego, 2016, pp. 563−611.

[30] M.A. Ganai, Z.H. Khan, M.A. Dar, Pheromones in lepidopteran insects: types, production, reception and its application, J. Pharmacogn. Phytochem. 6 (5) (2017) 2552−2558.

[31] R.K. Anish, et al., A novel assembly pheromone trap for tick control in dog kennels, Vet. Parasitol. 235 (2017) 57−63.

[32] H.S. Rathore, S. Mittal, L.M.L. Nollet, Biochemical Pesticides, Biopesticides Handbook, 2015, p. 155.

[33] A.L. Knight, et al., Evaluation of novel semiochemical dispensers simultaneously releasing pear ester and sex pheromone for mating disruption of codling moth (Lepidoptera: Tortricidae), J. Appl. Entomol. 136 (1-2) (2012) 79−86.

[34] A.M. Bueno, O. Jones, Alternative methods for controlling the olive fly, Bactrocera oleae, involving semiochemicals, IOBC-WPRS Bull. 25 (9) (2002) 147−156.

[35] S. Heuskin, et al., The use of semiochemical slow-release devices in integrated pest management strategies, Biotechnol. Agron. Soc. Environ. 15 (3) (2011) 459−470.

[36] D.M. Suckling, A.M. El-Sayed, J.T.S. Walker, Regulatory innovation, mating disruption and 4-Play™ in New Zealand, J. Chem. Ecol. 42 (7) (2016) 584−589.

[37] A. Lucchi, et al., Disrupting mating of Lobesia botrana using sex pheromone aerosol devices, Environ. Sci. Pollut. Control Ser. (2018) 1−9.

[38] B. Czarnobai De Jorge, et al., Novel nanoscale pheromone dispenser for more accurate evaluation of Grapholita molesta (Lepidoptera: Tortricidae) attract-and-kill strategies in the laboratory, Pest Manag. Sci. 73 (9) (2017) 1921−1926.

[39] J.H. Borden, et al., Potential for nonhost volatiles as repellents in integrated pest management of ambrosia beetles, Integr. Pest Manag. Rev. 6 (3−4) (2001) 221−236.

[40] R.S. Mann, P.E. Kaufman, J.F. Butler, Evaluation of semiochemical toxicity to houseflies and stable flies (Diptera: Muscidae), Pest Manag. Sci. 66 (8) (2010) 816−824.

[41] S. Schlaeger, J.A. Pickett, M.A. Birkett, Prospects for management of whitefly using plant semiochemicals, compared with related pests, Pest Manag. Sci. 74 (2018).

[42] B. O'Shea, et al., Disruption of host-seeking behaviour by the salmon louse, Lepeophtheirus salmonis, using botanically derived repellents, J. Fish Dis. 40 (4) (2017) 495−505.

[43] T.V. Bhaskaran, et al., Drying fish preference assessment and efficacy of semiochemicals as repellents to blow fly Chrysomya megacephala (F.)(Diptera: Calliphoridae) during sun drying of fish, Entomon 43 (3) (2018) 157−164.

[44] D.L. Kline, Semiochemicals, traps/targets and mass trapping technology for mosquito management, J. Am. Mosq. Control Assoc. 23 (sp2) (2007) 241−251.

[45] A.M. El-Sayed, et al., Potential of mass trapping for long-term pest management and eradication of invasive species, J. Econ. Entomol. 99 (5) (2006) 1550−1564.

[46] J.G. Logan, M.A. Birkett, Semiochemicals for biting fly control: their identification and exploitation, Pest Manag. Sci. 63 (7) (2007) 647−657.

[47] M.J. Domingue, et al., Trapping of European buprestid beetles in oak forests using visual and olfactory cues, Entomol. Exp. Appl. 148 (2) (2013) 116−129.

[48] B.S. Brodie, et al., Acquired smell? Mature females of the common green bottle fly shift semiochemical preferences from feces feeding sites to carrion oviposition sites, J. Chem. Ecol. 42 (1) (2016) 40−50.

[49] J. Hernandez-Cumplido, et al., Tempo-spatial dynamics of adult plum curculio (Coleoptera: Curculionidae) based on semiochemical-baited trap captures in blueberries, Environ. Entomol. 46 (3) (2017) 674−684.

[50] L.M. Torres-Vila, et al., Mark-recapture of Monochamus galloprovincialis with semiochemical-baited traps: population density, attraction distance, flight behaviour and mass trapping efficiency, Forestry 88 (2) (2014) 224−236.

[51] P.D. Cox, Potential for using semiochemicals to protect stored products from insect infestation, J. Stored Prod. Res. 40 (1) (2004) 1−25.

[52] M. Fountain, et al., The Use of Semiochemical Mass Traps to Reduce Mirid Damage in Strawberry Crops, 2017.

[53] A. Wibe, et al., Semiochemical-based Pest Insect Management in Strawberry and Raspberry, 2016.

[54] S. Broughton, D.A. Cousins, T. Rahman, Evaluation of semiochemicals for their potential application in mass trapping of Frankliniella occidentalis (Pergande) in roses, Crop Protect. 67 (2015) 130−135.

[55] J. Gross, New strategies for phytoplasma vector control by semiochemicals. 40 years of the IOBC WPRS working group, Pheromones Semiochem Integr. Prod. 126 (2017) 12−17.

[56] L. Vaníčková, A. Canale, G. Benelli, Sexual chemoecology of mosquitoes (Diptera, Culicidae): current knowledge and implications for vector control programs, Parasitol. Int. 66 (2) (2017) 190−195.

[57] Y. Kebede, et al., Unpacking the push-pull system: assessing the contribution of companion crops along a gradient of landscape complexity, Agric. Ecosyst. Environ. 268 (2018) 115−123.

[58] Q. Xu, et al., A push−pull strategy to control aphids combines intercropping with semiochemical releases, J. Pest. Sci. 91 (1) (2018) 93−103.

[59] M.J. Hassemer, et al., Development of pull and push−pull systems for management of lesser mealworm, *Alphitobius diaperinus, in Poultry Houses Using Alarm and Aggregation Pheromones*, Pest Manag. Sci. (2018).

[60] J.A. Byers, et al., Inhibitory effects of semiochemicals on the attraction of an ambrosia beetle Euwallacea nr. fornicatus to quercivorol, J. Chem. Ecol. 44 (6) (2018) 565−575.

[61] M. Nawaz, J.I. Mabubu, H. Hua, Current status and advancement of biopesticides: microbial and botanical pesticides, J. Entomol. Zool. Stud. 4 (2) (2016) 241−246.

CHAPTER 8

Fungistatic Properties of Lectin-Containing Extracts of Medicinal Plants

SERGEY V. POSPELOV, PHD • ANNA D. POSPELOVA, PHD •
VALENTINA V. ONIPKO, DOCTOR OF SC. • MAXIM V. SEMENKO

INTRODUCTION

Agricultural practice is associated with the loss of yield due to biotic factors, such as plant damage by pests, diseases, and competing weeds. It is proved that the losses on the average make up 30%−50%, but can reach 100%. Pathogenic microorganisms that develop on plants cause infectious diseases, the treatment of which by pesticides mainly leads to the development of resistance of pathogenic microflora and the accumulation of metabolic products in plant production and the environment. To avoid such negative moments, the search for substances of plant origin with antimicrobial activity is being carried out. Natural biologically active substances that have such an effect include plant antibiotics, phytoncides, essential oils (see Chapter 6 for details), balsams, resins, tannins, organic acids, alkaloids, and glycosides. All of them are formed during the life of different groups of plants from the simplest to the highest to protect their living tissues from the reproduction of microorganisms in them. In addition, they activate vital functions of plants, destroy insects, frighten rodents, stimulate the growth of some plants, and inhibit the growth of others. This is evidenced by the numerous experimental materials of recent decades [1−3].

Prospects for the Use of Medicinal Plants in Plant Protection

Natural plant compounds—a source of biopesticides

A review of various studies suggests that in many scientific centers, there is a systematic screening of plants—producers of compounds, which would have the inherent antimicrobial and antiviral properties. It should be noted that in this plan, there are results that give hope that in the next decade new products with such properties will appear on the market of biological products.

According to Lesnikov [4], the flora of Europe is extremely rich in medicinal plants, which contain natural antifungal components. More than 879 species of higher plants from 128 families are producers of fungicides. The list of plants is constantly updated and increased, especially in recent decades.

In Germany, experiments were conducted to study more than 250 extracts derived from different parts of the plant (leaves, buds, flowers, fruits, roots, bulbs, etc.) for fungicidal activity. As a test object for experiments, *Fusarium nivale*, *Phoma lingam*, *Botrytis cinerea*, *Pythium ultimum*, *Rhizoctonia solani*, and others were used. In climatic chambers, extracts of 0.03%, 0.1%, 0.3%, and 1% concentrations were used on cucumbers against powdery mildew, beans against rust, and potatoes against phytophthora. In field experiments, they were tested on apple with scab, fruit rot, powdery mildew of wheat and roses, gray rot of strawberries, and others. The author has shown that some extracts have a high (up to 80%) fungicidal activity, not inferior to commercial drugs. However, it is noted that the technology of preparing on plant basis can be quite expensive [5].

The activity of 24 extracts from 131 species of plants in relation to the burn agent *Erwinia amylovora* was studied using the diffusion method in agar. The potency of suppression of *Juglans nigra*, *Berberis vulgaris*, and *Rhus typhina* at concentrations of 5.2% and 1.25% was comparable to that of streptomycin inhibition (17 mg dm^{-3}) [6].

In Slovakia, in the Petri dishes, the effect of extracts from 500 species of 280 genera and 74 families on

Natural Remedies for Pest, Disease and Weed Control. https://doi.org/10.1016/B978-0-12-819304-4.00008-7

the growth of *Fusarium merismoides corda* was studied with the disk method. The activity of extracts was compared with the standard preparations SPOFA (synonym of nystatin), microfungal. Inhibitory activity was detected in 180 species (36%) of the families Ranunculaceae, Brassicaceae, and Liliaceae [7].

According to Japanese researchers, among 53 species of medicinal plants, 12 had the ability to suppress phytopathogens. The extracts of roots of *Geranium pratense*, *Sanguisorba officinalis*, and *Eupatorium fortune* had the highest activity [8,9].

Among the 30 studied plants, *Potentilla erecta* and *Salvia officinalis* extracts significantly inhibited conidia germination and increased mycelium growth of *Phytophthora infestans* on tomato and potato [10].

Studies by Singh and Dwivedi [11] revealed fungi staticity of some plants in relation to root rot of barley agent *Sclerotium rolfsii*. The dry weight of mycelium and the products of sclerotia decreased under the influence of extracts of bark of *Acacia arabica*, onions, and leaves of *Allium sativum* and *A. cepa* and leaves of other plants. The authors recommend the use of a number of extracts for screening in field conditions.

An effective means of controlling *Pseudoperonospora cubensis* is the spraying of plants with the extract of *Reynoutria sachalinensis*. At the same time, the level of disease was maintained at 8% of plants, while in the untreated variants the disease was 73% [12].

Hot and cold extracts of *Carica papaya* leaves effectively reduced the distribution of *Leveillula taurica* on peppers in field conditions [13]. Infusions of *Artemisia absinthium* grass and *Tanacetum vulgare* increased the resistance of tomato plants to fungi diseases [14].

The incrustation of onion seeds with dusts and medicinal plant extracts (*Chamomilla recutita*, *Equisetum arvense*, *Salvia officinalis*, *Allium sativum*, *Menta piperita*) reduced the number of pathogens on them, and also stimulated the growth of young plants [15].

In the fight against the brown leaf spot disease of rice—*Drechslera oryzae* is proposed to use extracts of *Lawsonia inermis* leaves. At dilution, 1:40 inhibition of the development of hyphae and decrease of the absorption of oxygen were observed. A compound that has an antifungal activity was identified as 2-hydroxy-1,4-naphthoquinone (lawson) [16].

Chinese scientists estimated the extracts of 247 plant species from 85 families in relation to the pathogenic fungi of rice *Pyricularia oryzae*. Among them, ethanolic extracts of 42 species caused deformation of mycelium and inhibited the germination of pathogenic spores. A correlation was found between the increase in the antifungal action of rice leaf diffusates in the case of damage by *Pyricularia oryzae* and the level of plant resistance to the disease [17].

There are data on the use of extracts of some African plants to suppress the development of smut diseases [18] and the use of extracts with fungicidal properties for the treatment of vascular diseases of wood species [19].

Characteristically, plants that have a different degree of resistance to diseases contain in tissues some substances that have a different fungitoxic effect. At least, they are synthesized in varieties of wheat resistant to stem and brown rust [20].

On the eve of landing, potato tubers are recommended to spray with water infusion of needles (pine and spruce). This measure significantly reduces the infestation of planting tubers and sprouts with the fungus *Rhizoctonia solani*. In addition, plant preparations did not inhibit the growth and development of plants in comparison with chemical fungicides [21].

In Germany, Biofa Tema (a blend of extracts of brown algae and horsetail) and Humis Vital 80 [PL] (vegetable extracts and 80% of humic acid) have been developed. Their application on potatoes reduces the damage of phytophthora. *Rheum rhabarbarum* and *Solidago canadensis* extracts are also known to be effective [22].

The oil of medicinal plant *Cymbopogon citratus* caused complete inhibition of growth of mycelium of *Fusarium solani* f.sp.*phaseoli*, *Sclerotinia sclerotiorum* and *Rhizoctonia solani* on legumes [23].

An important point is to protect the crop from diseases during storage. It has been established that the processing of tomato fruits with extracts from the leaves of *Lantana camara* and *Adenocalymna alliaceae* reduces and retards the development of fruit rot caused by *Aspergillus niger* [24,25]. Before laying the root vegetables of carrots for storage, it is expedient to process them with extract of yellow horn poppy (*Gaudium flavum*). This helps to reduce fungi diseases and improve the quality of storage [26].

The leaf extracts of *Rauwolfia serpentina*, *Datura stramonium*, *Eucalyptus globulus*, and others are capable of inhibiting the development of root rot of barley (*Sclerotium rolfsii*) [24,27]. Water extracts of *Croton sparsiflorus*, *Datura metal*, and *Solanum nigrum* effectively inhibited the germination of uredospores of *Melampsora lini*. In this case, water extracts were more effective than decoctions [28]. The compounds isolated from oil and onion of *Allium sativum*, ginger leaves suppressed germination of spores and mycelium growth of *Helminthosporium*, *Pyricularia grisea*. In this case, the acetone solutions of these extracts were more effective compared to water [29]. For a long time at the Kharkiv Agrarian University named after V. V. Dokuchaev, research on extracts from germinating seeds of field crops was carried out [30,31].

It is noteworthy that protein fractions that have been removed from a stable maize variety have shown an antifungal property in relation to *Aspergillus flavus* [32]. Antifungal activity may be inherent in proteinaceous compounds, for example, removed from the roots of *Phytolacca americana*. The analysis allowed to characterize them as lectins. Protective functions of lectins of *Urtica dioica* are known [30]. Belgian scientists believe that their presence in plants helps to protect against pathogenic organisms. The mechanism of this phenomenon will be considered by us further. It should only be noted that pathogens of diseases contain a significant number of receptors that specifically interact with different lectins and perform regulatory functions [32]. Phenolic compounds contained in the bark of *Erythrina berteroana* showed significant activity against the *Cladosporium cucumerinum* fungus [33]. From the plants *Allium sativum* the compound—adjoen was obtained, which inhibits the germination of pathogens of dangerous diseases of grain crops—*Alternaria solani*, *A. triticina*, *Colletotrichum* sp., *Curvularia* sp., *Fusarium lini*, *F. oxysporum*, *F. semitectum*, and others. Complete inhibition was achieved at concentrations of 100 mg/dm^3. The authors discuss the possibility of using this substance against diseases in field conditions [34].

Extracts from plants have antimicrobial properties, and this is due to the presence of biologically active components: essential oils, coumarins, and flavonoids. Polar extracts of various parts of *Eupomatia laurina* [35], *Pistacia lentiscus* extracts, and decoctions have a significant antibacterial activity [36].

Traditionally, medicinal plants have significant antimicrobial activity. In the study of species of the genus *Helichrysum*, derivatives of floroglucin and acetophenone that have activity against gram-positive bacteria were isolated and identified. Both compounds also inhibited the growth of some other fungi (*Phytophthora capsici*, *Penicillium italicum*) [37]. Among the Ukrainian flora the genus *Astragalus* was studied. Representatives of this genus produce substances that inhibit in different degrees the growth of gram-positive bacteria, in particular—*Corynebacterium michiganense* [38].

Thanks to biotechnology methods, tomato plants that produce a significant amount of proteins from the root system have been created. Their task is to protect plants from bacteria and other pathogens. When growing such plants in a hydroponic culture, protein compounds can be collected and used in phytopathology [39]. Research on microbial test systems suggests that plant extracts have a significant antimutagenic effect. In this case, *Hypericum perforatum*, *Linaria vulgaris*, and *Tussilago farfara* [40,41] are pointed out.

One of the alternative means of controlling plant diseases is the use of phytosanitary properties of different cultures [42,43]. First of all, it concerns crucifers—*Brassica juncea*, *Spinacia oleraceae*, *Lepidium sativum*, and *Lactuca sativa*. During the decomposition of the green mass of these plants in the soil, compounds that act as natural fungicides and purify the soil from pathogens are released.

Thus, various chemical components that are present in plants can influence as inhibitors on pathogens of fungal, bacterial, or viral nature. Using this feature, nowadays we already have experience in the manufacture and use of biological agents for disease control.

Mechanisms of resistance and susceptibility of plants to pathogens

The problem of intercellular recognition of pathogenic microorganisms by plants at the level of receptor—ligand interaction and the inclusion of the genetic mechanism of resistance of plants to infectious diseases is a central problem of phytoimmunity [44]. It corresponds to the theory of the combined evolution of parasites and plants and is considered as the main one [45]. In evolution, there really is a selection to increase the level of complementarity. The glycoprotein is immunocompetent, in which the carbohydrate part recognizes, and the protein—determines the immunological affinity [46].

In the 50s of the 20th century, Krüpe [47] found that phytolectins recognize slight differences in the structure of carbohydrates. This became the basis for the assumption of including them in the processes of cellular recognition—the interaction was not only between the cells of the plant, but also between the cells of the microorganism and the plant. Most researchers associate endogenous lectins with the display of protective mechanisms of the organism, which are shown in the blockage of the pathogen carried out by the lectins of the host [48].

Studies that show the role of lectins in the display of resistance to phytopathogenic fungi, as well as the significance of other compounds that are closely related to lectins, in the early stages of pathogenesis, are quite controversial.

Despite this, researchers come to the general conclusion that the process of pathogenesis begins with the attachment of bacteria and fungi to animal and plant cells. As a rule, on the surface of the host cells there are receptors with a help of which they bind and recognize the pathogen. In this case, highly specific connections such as "lectin-carbohydrate" may appear [49,50]. Carbohydrate receptors are localized on the cell surface of hyphae, vesicles, sporangia, and bacteria [51].

It is believed that the level of proteins in the cell wall increases with damage. On the basis of available information, it can be assumed that the main ones are oxyproline-rich glycoproteins (ORG)—structural proteins that enhance the cell wall and have lectin-like activity [52,53]. In this case, ethylene acts as a mediator in the interaction of cell surfaces of plants and fungi, and the biosynthesis of ORG. It is proved that the launch of biosynthesis of ORG in the walls of diseased plants is mediated by the recognition of the host cell of a pathogenic fungus.

Interestingly, some plants in their roots synthesize agglutinins, rich in oxyproline, which causes the accumulation of *Pseudomonas putida* on the surface of the roots, which suppress the pathogenic microflora around itself, thus "protecting" the plant [54].

Obviously, in the primary processes of recognition a certain value is played by phenolic compounds. Monocyclic compounds have been shown to induce virulence genes in *Agrobacterium*, and flavanoid—genes of modulation in *Rhizobium* [55,56]. In parasitic plants, the ability to react to the lifetime excretions of host plants, which contain phenolic compounds, is detected. They stimulated the growth of gaustories and served as a signal for the biochemical recognition of the host plant [57].

Interaction of Lectins with Pathogenic Organisms

As already noted, plant lectins can bind to the cell wall of the pathogenic organism [58]. However, so far, scientists cannot with complete confidence put forward a unified theory that would determine the physiological role of lectins in protecting plants from pathogens.

It has been established that bacterial cells of pathogens show a characteristic reaction of agglutination with plant lectins. This way lectin-containing wheat extract interacts with *Azotobacter indicum*, *Bacillus polymyxa*, *Micrococcus citreus*, and *Mycobacterium citreum* [59]. Virulent strains of *Erwinia stewartii*, a causative agent of bacterial wilt of maize, when introduced into a plant form a polysaccharide capsule protects the bacterium from corn lectin and thus allows it to procreate. The virulence of noncapsule mutants was very low even in relation to plants of susceptible maize variety [60]. A similar result was obtained during the interaction of

virulent cells of *Pseudomonas phaseolicola* with lectin-containing bean extracts [61].

Thus, on the surface of bacterial cells are receptors that specifically interact with lectins of various plants [62]. At the same time, a large number of bacteria produce hemagglutinins in the cultural environment, which can also react with the cell structures of the host plant and participate in recognition systems [63,64].

As the study shows, the degree of resistance of plants to fungi diseases can also be associated with a specific reaction of pathogens and plant lectins. In this regard, wheat germ agglutinin is sufficiently studied [65]. The authors believe that WGA is a major component of the plant's protective response to fungal diseases.

During germination of fungal spores, the interaction of their surfaces with plant lectins is going on. The nature of this interaction depends on the polysaccharide complex, characteristic of the hyphae of the fungus. On the surface of the germinating structures of the pathogenic fungi *Phytophthora megasperma f.* sp *glycinea i ph. infestans* are areas that bind galactose and fructose, which provides adhesion to the plasma membrane of host cells [66,67]. Lectins of wheat germs, soybean seeds, and peanuts have different carbohydrate specificities, and interact in different ways with the young spores of *Penicillia* and *Aspergilli* [68]. Interaction with lectins often leads to the suppression of the growth of some fungi and the lysis of their spores [69].

Studies show that lectins can not only cause agglutination of the cell walls of mycelium and spores [70], but also the growth of hyphae [71–74]. This pattern is noted not only when the host plant is damaged by the fungus. A very active agglutinin was isolated from common nettle, which inhibited the growth of hyphae of *Trichoderma hamatum*, *Phycomyces blakesleeanus*, and *Botrytis cinerea* [69]. Extracts from leaves and fruits of oak, in which there are agglutinins, inhibited the growth of fungi *Penicillium*, *Aspergillus*, *Fusarium*, and *Mucor* [75]. According to M. Etzler [52], P. Agrawal [76], and A. Mahadevan [77], the content, localization, activity, and specificity of lectins vary in plants during ontogenesis. In addition, the fungal damage of plants causes an increase in the content of glycoproteins in vegetative organs.

Research Methodology

For research, extracts of medicinal plants were used. They were obtained from grass of common St. John's wort (*Hypericum perforatum* L.), dwarf everlast (*Helichrysum arenarium* L.), common marigold (*Calendula officinalis* L.), common yarrow (*Achillea millefolium* L.), and seeds of common sea buckthorn (*Hippophae rhamnoides* L.) by infusing them in a solution of 0.9% NaCl for 2 h in the ratio of raw material—extract 1:10 [78]. Evaluation of the content of lectins in plant extracts was carried out by the method of hemagglutination of human erythrocytes. To do this, 0.05 mL of distilled water was added to the wells of the immunological plates with a dispenser. Then, for each variant, a series of successive twofold dilutions of the extract was prepared and 0.05 mL of 2% suspension of human erythrocytes was added to each well [78]. After 2-h incubation at a temperature of 27°C, a visual assessment of the activity of lectins was conducted on the degree of erythrocyte sedimentation on a 5-point scale [79].

The primary sedimentation of lectins from the extracts was carried out by extraction using low-temperature ethanol fractionation [80] to a final concentration of 76%. After 2 h of cooling, the precipitate (lectins) was separated by centrifugation (15 min at 3000 rpm), and then dissolved in a physiological solution to an initial volume. The supernatant was dried by lyophilization, the residue was dissolved in a physiological solution to an initial volume. Thus, the lectins of the extract and the extract without lectins were obtained in volumes and concentrations corresponding to the initial native extract.

Experiments on the study of biological activity of plant extracts and their components were carried out by the method of biotesting on cress-salad sprouts [42].

Experiments on the study of the influence of lectin-containing extracts on the germination of teliospores of loose smut of barley were conducted in laboratory conditions. The spores of loose smut were sprouted by the method of hanging drop in a humid chamber at a temperature of +22°C during the day. Nutrient medium was prepared on the basis of 20% sucrose with the addition of lectin-containing extracts, lectins, and extracts without lectins at dilution in the range of 1%−0.000001%. Control—pure 20% sucrose. Repetition of the experiment is fourfold. After 24 h, the number of germinated teliospores was counted using an eyepiece micrometer. The average was derived from 40 measurements [81,82].

Studies of Biological Activity of Lectin-Containing Extracts of Medicinal Plant
Medicinal plants—a source of lectins

According to literary sources [83], lectins were found in representatives of 124 genera and 40 families of both wild and cultivated plants of Ukraine. The maximum amount of lectin-containing species is gathered in the bean family (Fabaceae). Next, in descending order:

Poaceae, Asteraceae, Lamiaceae, Solanaceae, Apiaceae, Brassicaceae, Rosaceae, Rutaceae, Euphorbiaceae and Liliaceae. These families has from 3 to 10 genera, which representatives contains lectins. The remaining 28 families had from 1 to 2 genera with lectin-containing plants.

Thus, the search for donor plants among representatives of the Ukrainian flora presents a wide field of activity for screening. Along with the presence of lectins, according to our working hypothesis, plants contain components that have the property to suppress the development of pathogenic organisms, or other plants.

According to numerous studies [84,85], substances of phenol nature, alkaloids, glycosides, and lipids may have similar properties. The plants, selected according to these criteria, were evaluated by the method of hemagglutination with erythrocytes of 0 (I) human blood type. Thus, we tested 26 medicinal plants of local flora (Table 8.1).

After a more detailed analysis, further research was devoted to the study of lectin-containing extracts of common St. John's wort, common marigold, common sea buckthorn, common agrimony, dwarf everlast, and common yarrow.

It has been proven that proteinaceous compounds that had hemagglutinating activity were lectins. For this, the isolation of proteins with the method of low-temperature ethanol fractionation was carried out. Subsequently, they were kept in a thermostat at temperature 105°C for 3 h. As a result, hemagglutinating activity decreased by 22.6%−80.0% compared to control. Incubation of the native extract with the proteolytic enzyme pronase (Pronase E) also caused a decrease in agglutinating activity of extracts by 33.3%−63.8% (Table 8.2), which proves the proteinaceous nature of agglutinins.

Evaluation of biological activity of lectin-containing extracts

Experiments indicate that biotesting of native extracts of medicinal plants enables one to quickly screen their biological activity. However, at the same time, the biotest is influenced by the whole amount of substances isolated from plants. It is impossible to conclude what determines the effectiveness of testing—lectins or a complex of other compounds. That is why for the study of the influence of biological activity of lectins, they were isolated from extracts. After that, the evaluation of the native extract and its components was carried out.

The presented data give grounds to conclude that the inhibitory effect of the native extract of St. John's wort is due to the complex of nonlectin compounds (Fig. 8.1). Lectins of the extract in all dilutions, except for the first (10%), stimulated the test object. It should also be

TABLE 8.1
Activity of Lectins of Plant Extracts.

No	Latin Name	Activity, Points
1.	*Crataegus sanguinea* Pall.	10.0
2.	*Sambucus nigra* L.	5.0
3.	*Hypericum perforatum* L.	22.0
4.	*Fragaria vesca* L.	21.0
5.	*Calendula officinalis* L.	6.0
6.	*Viburnum opulus* L.	24.0
7.	*Urtica dioica* L.	2.0
8.	*Tilia cordata* Mill.	9.0
9.	*Tussilago farfara* L.	3.5
10.	*Agrimonia eupatoria* L.	24.0
11.	*Chelidonium majus* L.	2.5
12.	*Salvia officinalis* L.	5.5
13.	*Hippophae rhamnoides* L.	24.0
14.	*Betula pendula* Roth.	4.0
15.	*Glechoma hederacea* L.	2.0
16.	*Zea mays* L.	4.0
17.	*Mentha piperita*L.	3.5
18.	*Alnus glutinosa* (L.) Gaertn.	9.5
19.	*Plantago major* L.	2.0
20.	*Polygonum aviculare* L.	3.0
21.	*Equisetum arvense* L.	5.5
22.	*Eucalyptus globulus* L.	12.5
23.	*Helichrysum arenarium* (L.) Moench.	7.5
24.	*Centaurium erythraea* Rafn.	2.0
25.	*Rubus caesius* L.	14.0
26.	*Achillea submillefolium* Klok. et Krytzka	8.0

noted that, possibly, lectins and other components interact in a native extract, which increases the action of the latter in comparison with the extract without lectins (Fig. 8.1, concentration 0%.1%—0.001%). In addition, the extract without lectins acted oppositely than lectins in relation to the test.

The results of biotesting of dwarf everlast (Fig. 8.2) indicate that the biological activity of the extract is mainly due to the activity of lectins. In all dilutions,

there was a direct correlation between the indicators. The extract without lectins had a weak inhibitory effect and somewhat increased the inhibitory effect of the native extract. In the concentrations of 10%—1%, the extract and its lectins showed inhibitory activity, and at further dilution (0.1%—0.0001%), the growth of the cress salad was stimulated.

Analysis of the data of biotesting of common agrimony (Fig. 8.3) indicates the presence of fairly strong inhibitory components in it. The native extract in all dilutions suppressed the test system. Lectins and other components inhibited biotest in dilutions 10%—0%.01%. Only in the 0.001%—0.0001% concentrations, they acted positively. Characteristically, in higher concentrations (10%—0.1%), the extract without lectins and the native extract practically did not differ in activity. In lower dilutions (0.01%—0.0001%), they jointly intensified the inhibitory effect of native extract. Even when the components themselves stimulated the test system, the native extract suppressed it.

The biological activity of the native extract of common marigold in high concentrations (10%—1%) was lower than control and then slowly increased to 0.01% dilution with subsequent decrease (Fig. 8.4). In accordance with lectins and nonlectin compounds, they in all concentrations inhibited the test object. It is important that each of the components itself exhibits inhibitory activity, while together (native extract)—stimulating. In this case, the activity of the extract more correlated with the activity of lectins than with nonlectin component. It is possible that in extracts, lectins bind or inhibit the activity of compounds, resulting in the loss of inhibitors' activity and stimulation prevails. The fact that lectins are characterized by such an effect is evidenced by classical studies on the specificity of the action of lectins and carbohydrates [86].

According to the test of common sea buckthorn, the biological activity of the native extract is determined by the activity of lectins (Fig. 8.5). In 10%—1% concentrations, the solution suppressed the test system, and with further dilution, stimulation was observed. Lectins inhibited biotest only in 10% concentration, in others—stimulated from 1.92% to 38.46%. Extracts without lectins in 10% and 1% inhibited, and at further dilution positively influenced the test system (14.7%). Due to the high activity of lectins of sea buckthorn, the extract exhibited stimulating activity, although nonprotein components in the native extract reduced the activity. This indicates the complex chemical composition of the extract, which is confirmed by the studies of many authors [87—89].

TABLE 8.2
Activity of Agglutination of Native Extracts Under the Influence of Various Factors (In Points).

Variants of Experiment	WARMING UP FOR 3 H AT TEMPERATURE 105°C			INCUBATION WITH PRONASE FOR 24 H		
	Control	Experiment	Before Control (%)	Control	Experiment	Before Control (%)
Common sea buckthorn	24.0	14.5	60.4	22.5	10.5	46.7
Common St. John's wort	15.5	12.0	77.4	16.0	9.0	56.3
Common agrimony	20.5	9.0	43.9	23.5	8.5	36.2
Common marigold	4.5	3.0	66.7	3.0	2.0	66.7
Dwarf everlast	5.0	2.0	40.0	5.5	3.5	63.6
Common yarrow	7.5	1.5	20.0	6.0	4.0	66.7

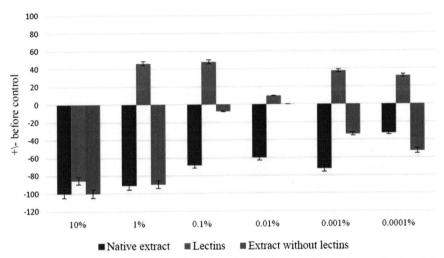

FIG. 8.1 Biological activity of the native extract of St. John's wort (*Hypericum perforatum* L.) and its components.

Biotesting of common yarrow was also carried out (Fig. 8.6). Its results indicate that the native extract at high concentrations (10%–1%) inhibited biotest, and in low (0.1%–0.0001%)—stimulated it. Lectins have a positive effect on the growth of the roots of cress salad with 1% dilution and below. The nonlectin fraction activity at concentrations of 10%–1% was higher than that of the native extracts in the same dilutions, and then (0.1%–0.0001%) was weak and did not exceed 8.54%. It can be noted that the activity of the native extract was determined by the interaction of lectins and other compounds; moreover, nonlectin components of the extract play an important role.

Comparing the obtained data, it can be concluded that the activity of the native extract in plants depended on its components. In this case, the nature of the biological activity of extracts can have three main types.

Lectin. The biological activity of native extracts is mainly due to the action of lectins in them. The nonproteinaceous component may slightly affect, depending on the concentration. In our experiments, extracts of dwarf everlast and common sea buckthorn had this type.

Nonlectin, when the biological activity of native extracts is mainly due to the action of nonlectin components, although lectins in different dilutions can

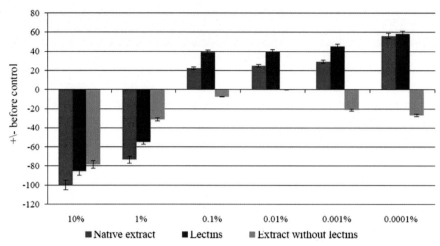

FIG. 8.2 Biological activity of the native extract of dwarf everlast (*Helichrysum arenarium* L.) and its components.

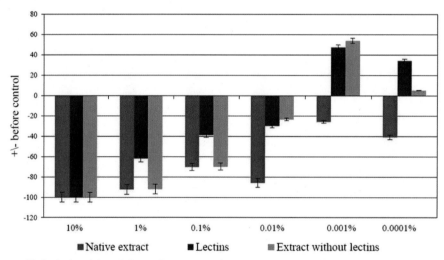

FIG. 8.3 Biological activity of the native extract of common agrimony (*Agrimonia eupatoria* L.) and its components.

reduce or increase activity. Extracts of St. John's wort and common yarrow are of this type.

Combined. In this case, biological activity was a result of the interaction between the lectin and nonlectin fractions of the extract. These components can in all cases suppress, while the native extract stimulates a test object that is a characteristic of common marigold. Otherwise, the native extract exhibited only an inhibitory effect, and its components stimulated in low concentrations (common agrimony).

The reason for the aforementioned may be the property of lectins to be selectively and inversely inhibited by carbohydrates [77]. In a native extract, the activity of lectins can be controlled by haptens of different origin, which are contained therein. If they are not in the extract, then the biological activity will mainly be determined by the activity of lectins. In that case, when lectins are, to varying degrees, suppressed by the carbohydrates contained in the extract, then there is a different type of interaction.

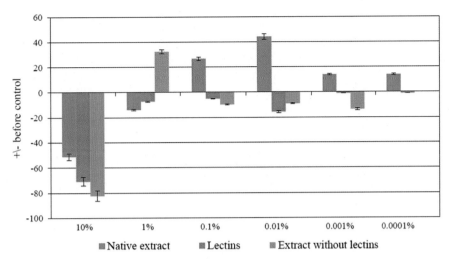

FIG. 8.4 Biological activity of the native extract of common marigold (*Calendula officinalis* L.) and its components.

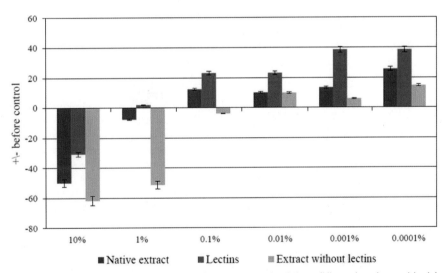

FIG. 8.5 Biological activity of the native extract of common sea buckthorn (*Hippophae rhamnoides* L.) and its components.

Influence of Medicinal Plants Extracts on Germination of Loose Smut (Ustilago nuda (Jens.) Kell. Et Sw.) Teliospores

The study of the interaction of fungal pathogens and lectins shows their influence on the processes of germination and growth of spores. It has been established that lectins, due to their unique properties to bind to the carbohydrate components of the cell membrane, can specifically interact with germinating spores and cause various breaks in the growth and development of the hyphae [90]. The authors studied antifungal activity of wheat germ lectins, seeds of *Datura stramonium*, *Chelidonium majus*, potato tubers, and *Phytolacca americana* roots. However, none of them could be compared to lectin of *Urtica dioica*. Processing with extracts of plants and preparations of proteinaceous nature of spores of the fungus *Colletotrichum lindemuthianum* caused agglutination of the latter [70].

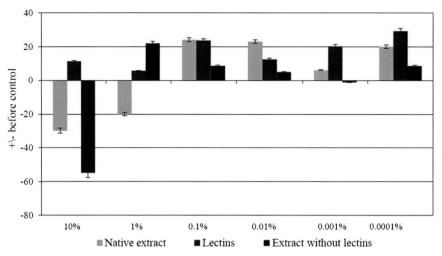

FIG. 8.6 Biological activity of the native extract of common yarrow (*Achillea millefolium* L.) and its components.

Extracts from leaves and fruits of oak significantly inhibited the growth of fungi of the genera *Penicillium, Aspergillus, Fusarium,* and *Mucor* [75]. Lectins of the seeds of eight legumes caused deformation of the conidia of *Colletotrichum capsici* and *Fusarium solani* [76]. According to the famous lectinologist M. Ettsler [52], lectin compounds are one of the components of integrated plant protection against fungal diseases. The same conclusions were made by other researchers [65,68]. The urgency of this is confirmed with the development of methods for assessing the resistance of plants by determining the amount of lectins [91]. Taking into account that in the initial period of growth plants are the most susceptible to pathogens, and the main role of lectins of seeds is considered as protective [52]. In favor of this, authors suggest that the amount of lectins sharply decreases in 2–3 days after germination of seeds.

Taking into account this from the standpoint of the biological cycle of loose smut of barley, it can be assumed that lectins of seeds, when interacting with the mycelium that is in the embryo, can inhibit its growth during the period of germination of the seeds.

To study the nature of the interaction of the pathogen with lectins, and the search for substances that would inhibit the development of the pathogen, we studied the influence of native extracts and their components on the germination of teliospores of loose smut of barley. The data obtained (Fig. 8.7) indicate that the native extract of St. John's wort suppressed the growth of teliospores at concentrations of 1% and

at dilutions of 0.00001%–0.000001%, and at concentrations of 0.1%–0.0001% stimulated the process (+8.6%–+35.7%). Lectins of the extract in all dilutions inhibited the germination of teliospores (−13.9% to −64.6%). In contrast, the extract without lectins in all variants stimulated the germination process (+17.4%–+124.9%). The most stimulating effect was observed in 0.1%–0.001% dilutions. It is worth noting that, inhibition of germination occurred more at 0.00001%–0.000001% concentrations.

Germination of teliospores on media with native extract of common yarrow showed stimulating effects at concentrations of 1%–0.001% and inhibition with further dilution (0.0001%–0.0001%) (Fig. 8.8). In the effect of lectins on the germination of teliospores, there was no clear pattern. Stimulation was noted only in 0.1% concentration (+38.4%). In the 0.01% and 0.001% concentrations, their effects were at the control level, while in other dilutions, inhibition of the germination process was observed. A similar pattern was characteristic to the effect of the extract without lectins. Only in 1% concentration, significant inhibition was observed (−80.2%). Lower inhibition (−24.8% to −33.5%) was noted at dilution of 0.0001%–0.000001% (Fig. 8.8).

Experiments with extracts of dwarf everlast indicate that the inhibitory effect was characteristic only in dilutions of 1%–0.1% (−25.0%–41.3%) (Fig. 8.9). In other concentrations, the extract stimulated the growth of teliospores (+3.5%–+24.5%). Lectins, which were removed from the extract, only inhibited the studied process in the first two dilutions (−38.4%

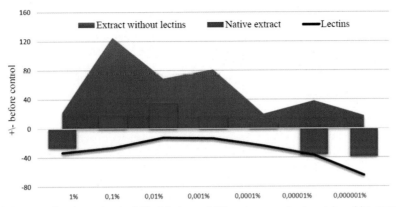

FIG. 8.7 Influence of native extract of St. John's wort *(Hypericum perforatum* L.) and its components on germination of Ustilago nuda teliospores.

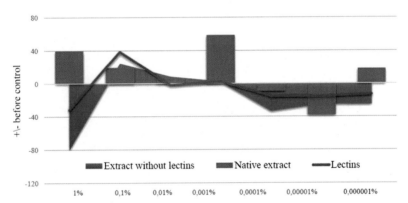

FIG. 8.8 Influence of native extract of common yarrow (*Achillea millefolium* L.) and its components on germination of Ustilago nuda teliospores.

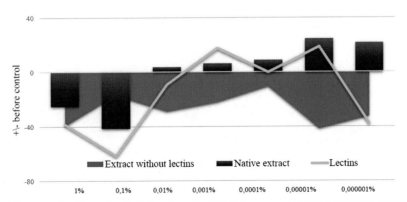

FIG. 8.9 Influence of native extract of dwarf everlast (*Helichrysum arenarium* L.) and its components on germination of Ustilago nuda teliospores.

to −61.7%). Extracts without lectins in all concentrations inhibited germination of teliospores by −10.5% to −41.6% compared to control.

Testing of extracts of common marigold (Fig. 8.10) indicates a significant inhibitory effect of native extract on germination of teliospores. Depending on the concentration, it varied from −32.2% to −70.9%. Lectins isolated from the extract also exhibited an inhibitory effect, although to a significantly smaller extent: from −3.2% to −31.2%. Extract without lectins behaved unstable. At concentrations of 0.1%; 0.01%; 0.000,001% it inhibited the development of teliospores (correspondingly −15.9%; −34.6%; −10.5%), and in other dilutions stimulated the studied object.

Conducting testing of native extracts of common sea buckthorn (Fig. 8.11) showed its significant inhibitory effect on germination of teliospores—by −28.8% to −98.8% to control. Lectins of sea buckthorn, isolated from the extract, have not shown clear patterns.

Stimulant activity was observed in concentrations of 1% and 0.001% (+7.2% and +17.7%, respectively), while in other dilutions—inhibition—from −0.9% to −35.4%. The extract, purified from lectins, only inhibited the growth of teliospores in the 1% concentration (−47.8%). In all other cases, stimulation was observed from +7.6% to +79.0%.

Established patterns have become the basis for developing a method for preliminary assessment of the effectiveness of reducing the harmfulness of loose smut with phytoextracts on grain crops, which was protected by the patent of Ukraine No. 13550 [92].

Conclusion

Study of biological activity of native extracts of medicinal plants, extracted lectins from them and extract without lectins in the system of biotests allowed us to establish the presence of three types of biological activity. The first one—the biological activity of native

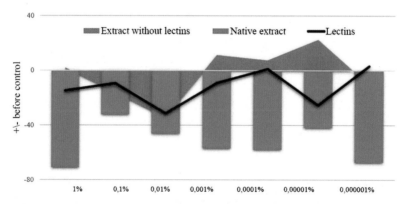

FIG. 8.10 Influence of native extract of common marigold (*Calendula officinalis* L.) and its components on germination of Ustilago nuda teliospores.

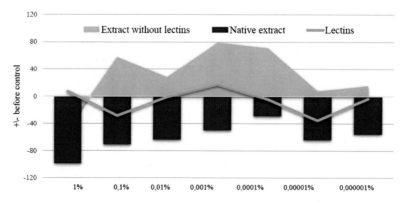

FIG. 8.11 Influence of native extract of common sea buckthorn (*Hippophae rhamnoides* L.) and its components on germination of Ustilago nuda teliospores.

extracts is determined by lectins; the second one is determined by a nonprotein fraction; the third one is the result of the interaction of lectins with other components of the extract.

Study of the germination of teliospores of loose smut of barley on media with native extracts, lectins, and lectins without extracts indicates the prospect of further research of some extracts to inhibit the development of the pathogen. Among the investigated plants, significant inhibition of growth of teliospores was recorded when using the native extract of common marigold and common sea buckthorn. Lectins, isolated from plant extracts, acted on germination of spores diversely. Only lectins of common St. John's Wort in all dilutions inhibited the studied process. Extracts without lectins in the majority caused insignificant with further stimulation of the germination of teliospores. Among all plants, extract of dwarf everlast showed a steady inhibitory effect at all concentrations.

In summary, medicinal plant extracts with lectins, could serves as a promising source of fungistatic agents for the creation of new biopesticides which can be used in systems of biological and organic cultivation of agricultural plants, the creation of model systems for studying the interaction of lectins—pathogens, and new directions of application of medicinal plants.

REFERENCES

[1] I. Berdicevsky, Promising broad spectrum antifungal activity of a plant extract and its mode of action, Mycoses 42 (3) (1999) 166–167.

[2] K. Gucwa, S. Milewski, T. Dymerski, P. Szweda, Investigation of the antifungal activity and mode of action of *Thymus vulgaris, Citrus limonum, Pelargonium graveolens, Cinnamomum cassia, Ocimum basilicum,* and *Eugenia caryophyllus* essential oils, Molecules 23 (5) (2018) 1116, https://doi.org/10.3390/molecules23051116.

[3] C.L. Cespedes, J.R. Salazar, A. Ariza-Castolo, L. Yamaguchi, J.G. Avila, P. Aqueveque, I. Kubo, J. Alarcon, Biopesticides from plants: *Calceolaria integrifolia* s.l. Environ. Res. 132 (2014) 391–406, https://doi.org/10.1016/j.envres.2014.04.003.

[4] E.P. Lesnikov, Higher Plants—Fungicide Producers. Microorganisms and Green Plant, 1967, pp. 71–80.

[5] B.D. Eggler, Fungizide Wirkung verschiedener Pelanzenextrakte Ergebnisse aus Laborscreening, Klima-kammer und Fleilandversuchen, Fungizide. Meded. Fac. landbouwwetensch. rijksuniu. Gent. 52 (3A) (1987) 971–980, 3.

[6] J. Mosh, In vitro — Untersuchungen uber die Wirkung von Pflanzenextrakten auf den Erreger des Feuerbrandes, *Erwinia amylovora* (Burrill) Winslow et al. F. Klingauf. Nachrichtenbl, Dtsch. Pflanzenschutzdienst. (BRD) 41 (8–9) (1989) 121–123.

[7] L.I. Linevich, Lectins and carbohydrate-protein recognition at different levels of the organization of living, *Uspekhi Biologicheskoi Khimii* 20 (1979) 71–94.

[8] J. Ushiki, Medicinal plants for suppressing soil-borne plant diseases. I. Screening for medicinal plants with antimicrobial activity in roots, Soil Sci. Plant Nutr. 42 (2) (1996) 423–426.

[9] J. Ushiki, Medicinal plants controlling soil-borne plant diseases, in: 15th Int. Bot. Congr., Yokodama: Abstr. — Yokodama, 1993, p. 395.

[10] P. Blaeser, Fungizide Wirkstoffe aus Pflanzenatrakten, Mitt. Biol. Bundesanst. Land und Forstwirt. 357 (1998) 167.

[11] N. Sharon, Lectins. Sci. Amer. 236 (1978) 108–119.

[12] G. Herger, Die Wirkung von Auszugen aus dem Sachalin-Staudenknoterich, *Reynoutria sachalinensis* (F. Schmidt) Nakai, gegen Pilzkrankheiten, insbesondere Echte Mehltau-Pilze, Nachrichtenbl. Dtsch. Pflanzenschutzdienst. (BRD) 40 (4) (1988) 56–60.

[13] A.C. Amadioha, V.I. Obi, Fungitoxic activity of extracts from *Azadirachta indica* and *Xylopia aethiopica* on *Colletotrichum lindemuthianum* in Cowpea, J. Herbs Spices Med. Plants 6 (2) (1998) 33–40.

[14] T.N. Filippovich, The influence of root excretions on the supply of nutrients to plants, Thethis. diss... Biol. Science, 1966, p. 18.

[15] Y. Druet, Produits et procedes de protection de semences par voie biologique. Pat 2766060 France, МПК⁶ A01N 65/00, A01N 25/08.

[16] M.R. Natarajan, D. Lalithakumar, Antifungal activity of the leaf extract of *Lawsonia inermis* on *Drechslera orezae*, Indian Phytopathol. 40 (3) (1987) 390–395.

[17] J.C. Hughes, T. Swain, Scopolin production in potato tubers infection with *Phytophthora infestans*, Phytopathology 50 (1960) 398–400.

[18] R.T. Awuah, Fungitoxic effects of extracts from some West African plants, Ann. Appl. Biol. 115 (3) (1989) 451–453.

[19] C. Desevedavy, M. Amoros, L. Girre, Properties antifongiques in vivo de la Paquerette (*Bellis perennis*) et du Mouron Rouge (*Anagallis arvensis*). Application au traitement de la graphiose de lÓrme, Fitoterapia 58 (4) (1987) 229–233.

[20] T.M. Sidorova, L.P. Evstratova, Fungitoxid substances and the resistance of wheat plants to brown and stem rust, Agric. Biol. 5 (2002) 92–94.

[21] A.G. Smirnov, Studying the fungicidal properties of spruce and pine needles. Biological bases of studying, mastering and protecting animal and plant world, soil cover of Eastern Fennoscandia, Intern. Conference. Petrozavodsk. (1999) 47–53.

[22] S. Meinck, A. Schmitt, Der Einfluss von alternativen Mitteln auf den Krankheitsbefall vin Kartoffeln mit *Phytophthora infestans* und auf den Ertrag. 51, Dtsch, Pflanzenschutztag. 357 (1998) 99.

[23] P.J. Valarini, R.T.S. Frighetto, C.A. Spadotto, Potencialde uso erva medicinal *Cymbopogon citrates* no controlede fitopatogenos do feijoeiro e plantas daninhas em areas irrigadas, Cientifica (Jaboticabal) 24 (1) (1996) 199–214.

[24] P. Sinha, S.K. Saxena, Effects of treating tomatoes with leaf extract of Adenocalymna alliacea on development of fruit rot caused by *Aspergillus niger* in presence of *Drosophila busckii*, Natl. Acad. Sci. Lett. 10 (5) (1987) 161–165.

[25] S.A. Shetty, H.S. Prakash, Efficace of certain plant extracts against seed-borne infection of Trichoconiella padwickii in paddy (*Oryza sativa*), Can. J. Bot. 67 (7) (1989) 1956–1958.

[26] E.A. Golovko, T.M. Bilyanovskaya, I.I. Vorobey, Allelo pathy of cultivated plants in the aspect of problems of agrophytocenology, Physiologist. Biochem. Cultivated Plants 31 (2) (1999) 103–113.

[27] S.N. Tewari, S. Shukla, K.M. Biswal, Fungitoxic properties of some leaf extracts, Natl. Acad. Sci. Lett. 11 (12) (1988) 369–373.

[28] S. Verma, I.B. Randey, Antifungaleffect of some medicinal angiosperms on *Melampsora lini* (Ehrneb), Proc. Natl. Acad. Sci. India 56 (3) (1986) 254–257.

[29] R. Jagannathan, V. Narasimhan, Effect of plant extracts/products on two fungal pathogens of finger-millet, Indian J. Mycol. Plant Pathol. 18 (3) (1988) 250–254.

[30] G.F. Naumov, P.V. Podoba, Allelopathic activity of extracts from seeds of various grain crops when treating seeds of spring barley, Coll. scientific tr. Kharkiv Agricultural Institute 310 (1984) 92–100.

[31] V.I. Nikolaychuk, N.P. Courtin, Influence of water infusions of some kinds of meadow plants on the laboratory similarity of seeds, Nauk. visn. Uzhhorod univ, Ser. Biology. 4 (1997) 218–223.

[32] N.V. Lyubimova, E.G. Salkova, Phytolectins in the relationship of plants and pathogenic microorganisms. Study and application of lectins, T.2. Uch.-zap. Tartu University. 869 (1989) 3–9.

[33] M. Maillard, M.P. Gupta, K. Hostettmann, A new antifungal prenylated flavanone from *Erythrina berteroana*, Planta Med. 53 (6) (1987) 563–564.

[34] U.P. Singh, V.N. Pandey, K.G. Wagner, Antifungal activity of ajoene, a constituent of garlic (*Allium sativum*), Can. J. Bot. 6 (1990) 1354–1356.

[35] M.R. Khan, M. Kihara, A.D. Omoloso, Antimicrobial activity of *Eupomatia laurina*, Pharm. Biol. 39 (4) (2001) 297–299.

[36] L. Iauk, S. Ragusa, A. Rapisarda, In vitro animicrobial activity of *Pistacia lentiscus* L. extracts: preliminary report, J. Chemother. 8 (3) (1996) 207–209.

[37] P. L Thomas, Barley smut in the prairie provinces of Canada, 1978-82, Canad. J. Plant Patol. 6 (1) (1984) 78–80.

[38] V.V. Smirnov, V.A. Prikhodko, A.A. Meshcheryakov, Antimicrobial properties of plants of the genus Astragalus, Phytoncids. Bacterial Plant Diseases: Conf. Lviv 1 (1990) 16–17.

[39] J. Travis, Getting to the root of protein production, Sci. News 155 (18) (1999) 279.

[40] V.Y. Kornienko, Sage diseases of the cereal, *Zashchita i karantin rasteniy* 3 (1998) 2–3.

[41] A.P. Dmitriev, Signal systems of plant immunity, Cytol. Genet. 3 (2002) 58–67.

[42] A. M Grodzinsky, E.A. Golovko, S.A. Gorobet, Exper. Allelopathy 236 (1987), 5-12-001928-5.

[43] W.R. Jarvis, Allelopathic control of *Fusarium oxysporum f. sp. radicis-lycopersici*. Vasc. Wilt Diseases Plants, in: Proc. NATO Adv. Res. Workshop interact. Genet. And Environ., 1989, pp. 479–486.

[44] I.K. Karanyan, Fruits of sea buckthorn – the most valuable source of biologically active substances, Agrar. Russia. 6 (2001) 65–66.

[45] F.K. Mitsue, A. Jun-ichi, Induction of callus from mistletoe and interaction with its host cells, Bull. Kyoto univ. Forests. 62 (1990) 261–269.

[46] A.S. Dvornik, T.P. Pererva, V.A. Kunakh, The antimutagenic effect of plant extracts in the test system *Escherichia coli* - bacteriophage λ in vitro, in: Reports of the National Academy of Sciences of Ukraine, 7, 2000, pp. 188–190.

[47] M. Krüpe, V. Enke, Blutgruppespezifische Pflenzliche Eiweiskörper (Phytoagglutinine), 1956, p. 131. Stuttgart.

[48] B.B. Gromova, M.V. Patrikeeva, N.A. Zhitlova, Method for Determining the Nonspecific Resistance of Potatoes to Late Blight Pathogens by Leaf Lectins, All-Union Res. Institute of Plant Protection, 1990, p. 12.

[49] B. Cifuentes, R. Urrialde, Lectins in recognition systems, Physiol. Plant. 79 (2) (1990) 97.

[50] K.H. Kogel, S. Ehrlich-Rogozinski, H.I. Reisener, Surface galactolipids of wheat protoplasts as receptors forsoybean agglutinin and their possible relevance to host-parasite interaction, Plant Physiol. 76 (4) (1984) 924–928.

[51] A. Ghnekar, M.C. Perombelon, Interactions between potato lectin and some phytobacteriain relation to potato tuber decay caused by *Erwinia caratovora*, Phytopathol. Z. 98 (2) (1980) 137–149.

[52] M.E. Etzler, Are lectins involved in plant–fungus interactions? Phytopathology 71 (7) (1981) 744–746.

[53] J. Kratka, Vliv *Fusarium oxysporum* f. sp. pisi na zmeny v obsahu hydroxyprolinu ve stene bunecne hrachu, Sb. UVTIZ (Ustav Vedeckotech. Inf. Zemed.) Ochr. Rostl. 24 (4) (1988) 275–280.

[54] A. Anderson, Understanding plant-microbe communication, Utah Sci. 47 (1) (1986) 6–8.

[55] W. Feucht, Phenolische Substanzen und Resistenzgegen Schädlinge, Erwerbsobstbau 32 (7) (1990) 188–190.

[56] N.K. Peters, D.P.S. Verma, Phenolic compounds as regulators of gene expression in plant-microbe interactions, Mol. Plant Microbe Interact. 1 (1990) 4–8.

[57] H. Rüdiger, Preparation of plant lectins, Adv. Lectin Res. v.1 (1988) 26–72. Berlin.

[58] M.T. Esquerre-Tugave, D. Mazau, A. Toppan, Hydroxyproline-rich glycoproteinsin the cell wall of diseased plants as a defense mechanism, Post-Harvest Physiol. And Crop. Preserv. Proc. NATO Adv. Study Inst., Sounion (1983) 287–298.

[59] N. Semak, S.Y. Schegolev, L.I. Pozdnyakova, Agglutination of bacterial cells with wheat lectin. N questions biochemistry, Fisiol. Microorganisms. 11 (1987) 36–42.

[60] J.J. Bradshaw-Rouse, M.N. Whatley, D.L. Coplin, Agglunation of Erwinia stewartii strains with a corn agglutinin: correlation with extracellular polysaccharide production

and pathogenicity, Appl. Environ. Microbiol. 42 (2) (1981) 344—350.

[61] F.E. El-Banoby, K. Rudolph, Agglutination of Pseudomonasphaseolicolaby bean leaf extracts (*Phaseolus vulgaris*), Phytopathol. Z. 98 (1) (1980) 91—95.

[62] D.L. Smalley, M.E. Bradley, Lectin-binding activity and antibody response with *Pseudomonas paucimobilis* and *Flavobacterium multivorum*, Can. J. Microbiol. 29 (5) (1983) 619—621.

[63] M.J. Chrispeels, N.V. Raikhel, Lectin, lectin genes and their role in plant deferense, Plant Cell 3 (1) (1991) 1—9.

[64] L. Sequera, Lectins and their role in host-pathogen specificity, Annu. Rev. Phytopathol. 16 (1978) 453—481.

[65] A. Chaboud, M. Lalonde, Lectin binding on surfaces of *Frankia* strains, Can. J. Bot. 61 (11) (1983) 2889—2897.

[66] H.R. Hohl, S. Balsiger, Surface glycosyl receptors of *Phytophthora megasperma f. sp.glicinea* and its soybean host, Bot. Helv. 98 (2) (1988) 271—277.

[67] N. Furuichi, J. Suzuki, Isolation of proteins related to the β-lectin from potato which bind to hyphal wall components of *Phytophtora infestans*, Phytopathology 127 (4) (1989) 281—290.

[68] R. Barkai-Golan, D. Mirelman, N. Sharon, Stadies on growth inhibition by lectins of Penicillia and Aspergilli, Arch. Mikrobiol. 116 (2) (1978) 119—124.

[69] N.A. Garas, J. Kus, Potato lectin lyses zoospores of *Phytophthorainfestans andprecipitates* elicitors of terpenoidaccumulation produced by the fungus, Physiol. Plant Pathol. 18 (1981) 227—237.

[70] D.H. Yong, H. Kauss, Agglutination of mycelia cell wall fragments and spores of *Colletotrichum lindemuthianum* by plant extracts, and by variousproteins, Physiol. Plant Pathol. 20 (3) (1982) 285—297.

[71] M. Mishkind, K. Keegstra, B.A. Palevitz, Distribution of wheat germ agglutinin in young wheat plants, Plant Physiol. 66 (5) (1980) 950—955.

[72] W.J. Fu, J. Liu, M. Zhang, J.Q. Li, J.F. Hu, L.R. Xu, G.Y. Dai, Isolation, purification and identification of the active compound of turmeric and its potential application to control cucumber powdery mildew, J. Agric. Sci. 3 (156) (2018) 358—366, https://doi.org/10.1017/S0021859618000345.

[73] M. Ghosh, Purification of a lectin-like antifungal protein from the medicinal herb, *Withania somnifera*, Fititerapia 2 (80) (2009) 91—95, https://doi.org/10.1016/j.fitote.2008.10.004.

[74] J.F.D. da Silva, S.P. da Silva, P.M. da Silva, A.M. Vieira, L.C.C. de Araújo, T. de Albuquerque Lima, A.P.S. de Oliveira, L.V. do Nascimento Carvalho, M./G. da Rocha Pitta, M.J.B. de Melo Rêgo, I. O Pinheiro, R.B. Zingali, M. do Socorro de Mendonça Cavalcanti, T. H Napoleão, P.M.G. Paiva, Portulaca elatior root contains a trehalose-binding lectin with antibacterial and antifungal activities, Int. J. Biol. Macromol. 126 (2019) 291—297, https://doi.org/10.1016/j.ijbiomac.2018.12.188.

[75] J. Krahe-Urban, H.R. Kanitz, W. Punin, Uber den nachweis von Agglutinen an Fruchten und Blattern der Stiel — und Traubeneiche (*Quercuspedunculata* Ehrh. und

Guercusstssiliflora Salisb.), Z. Forstgenet. und Forstpflanzenzucht. 4 (1) (1955) 18—20.

[76] P. Agrawal, A. Mahadevan, Lectins in germinatingseeds and their effect on fungi, Acta Phytopathol. Acad. Sci. Hung vol. 19 (3—4) (1984) 219—235.

[77] A. Mahadevan, G. Muthukumar, A. Arunakumari6 Lectins, Are involved in recognition of parasites by plants, Curr. Sci. 51 (6) (1982) 262—268.

[78] M.D. Lutsik, E.N. Panasyuk, A.D. Lutsik, Lectins 156 (1981). ISBN 978-5-7992-0733-5.

[79] S.V. Pospelov, Lectins of representatives of the genus *Echinacea* (*Echinacea* Moench.). 1. Methodological aspects of activity assessment, Chem. Plant Raw Mater. 4 (2012) 143—148.

[80] W. Boyd, R. Requera, Hemagglutinating substances in various plants, J. Immunol. 62 (1949) 333—339.

[81] Y.L. Golynskaya, Phytohemagglutinins of the generative organs of plants and their possible participation in the recognition reaction in the interaction of pollen and pistila, Mol. Biol. (1979) 34—41.

[82] A.S. Stepanovskikh, Smut barley diseases, Thesis. dis. Dr. Sci. (1991) 44.

[83] S.V. Pospelov, V.N. Samorodov, Lectins in flowering plants of Ukraine, in: III Ukrainian Conference on Medical Botany: Tez. report, 1992, pp. I19—120.

[84] L.T. Rasulov, N.S. Kholodetsky, Stimulating supplements and pepper harvest, Potatoes Vegetables 2 (2001) 44.

[85] H.K. Dong Aijun, S. Qishi, Y. Xinsheng, Bioactivity of 247 traditional Chinese medicines against *Pyricularia oryzae*, Pham. Biol. 39 (1) (2001) 47—53.

[86] N.V. Lyubimova, V.M. Lakhtin, E.P. Shuvalova, Lectin carbohydrate interaction in the potato system — pathogen of late blight at the plant plasmolemma level, Fiziol. Rast. 35 (5) (1988) 870—878.

[87] S.M. Nikolaev, D.T.s. Tsybikova, O.T.s. Tsyrenzhapova, To the question of pharmacological studies of preparations from sea buckthorn wastes, in: 3rd International Symposium on the Sea Buckthorn, 1998, pp. 111—113.

[88] V.F. Turetskova, N.V. Tregubenko, O.N. Borisova, Study of the possibility of complex processing of the bark and shoots of sea buckthorn and bark of the pine tree, Actual. quest.med. science. (1997) 666—667.

[89] V.F. Turetskova, O.V. Azarova, On the possibility of rational use of the waste of buckthorn cultivation, 3rd International. Symposium. on Sea-buckthorn (1998) 107—108.

[90] D. Mirelman, E. Galun, N. Sharon, Inhibition of fungal growth by WGA, Nature (1975) 256.

[91] V.M. Lakhtin, Z.M. Yakovleva, Binding of a wheat germ lectin to the surface of the mycelium and spores of Helminthosporium sativum, Izv. Academy of Sciences of the USSR. Ser. Biol. 5 (1987) 792—795.

[92] G.D. Pospelova, V.M. Pisarenko, S.V. Pospelov, Method of Preliminary Estimation of the Effectiveness of Reducing the Harmfulness of Dusty Headwith Phytoextraction on Grain Crops, 2006, p. 1. Patent 13550 Ukraine, A01G 7/00. 4.

CHAPTER 9

Biological Control of Weeds by Allelopathic Compounds From Different Plants: A BioHerbicide Approach

MOHAMMAD MEHDIZADEH, PHD • WASEEM MUSHTAQ, PHD

INTRODUCTION

The growing population of the world has led to an increase in the demand for agricultural products. Therefore, efforts to reduce yield losses of crops caused by pests and especially weeds are essential. During the long human history of cultivation, farmers have tried to manage weeds to increase crop production, and since then, weed control methods have been developed along with the advancement of science and technology. Hand weeding, mechanical, physical, cultural, biological, chemical, and integrated weed control are the main methods that have been implemented by farmers to reduce the negative impacts of these plants in agricultural systems. Generally, choosing an effective method for weed management in an agro ecosystem depends on different factors such as the type of weeds and cultivated plant, financial situation of farmers, availability of technology, damage level, as well as crop rotation. Overall, in spite of all the issues and consequences associated with the application of chemical methods, it is still considered as one of the most important and effective weed management methods in the world. Nowadays, a great number of agrochemicals (herbicides, insecticides, fungicides, nematicides, etc.) are being used in agro ecosystems for the management of weeds and other pests [1]. This extensive application of pesticides is of great concern to the general public because of the negative impacts on the environment and human health. Herbicides as the most commonly used pesticides can alter the agricultural environments, and influence the water and soil resources [2,3], human and animal health [4],

food safety [5], nontarget plants [6—8], and ecosystem function and structure [9]. In recent years, numerous incidents have been reported about the negative impacts of herbicides and their residues on different crops like cereals, vegetables, oil plants, etc. The direct damaging role of herbicides on agricultural crops is reported by some researchers in different areas and study conditions. Depending on chemical structure of herbicides, crop species, environmental condition, and soil properties, the negative impacts of herbicide residues could be categorized as complete destruction of the plant, decreased biomass with or without recovery, and enormous crop development. However, germination, flower and seed production, as well as vitality of plants could be affected by herbicides residues during the critical growing steps [10]. This calls for a holistic intervention through the development of innovative products from natural origin that are safe to man and his immediate environment, renewable with fast biodegradation on the crops in order to minimize residues on agricultural products. Producing these kinds of products could especially be useful for crops where the harvest time is long and plant protection is needed during it. Therefore, the development of applied research in order to introduce the biopesticides to control different pests in agricultural ecosystems is very important.

THE PROBLEM OF WEEDS

Many plant species grow in the agro ecosystems. Plants that interfere with human activities in different ways

Natural Remedies for Pest, Disease and Weed Control. https://doi.org/10.1016/B978-0-12-819304-4.00009-9

are called "weeds." These plants are not wanted by humans and cause major problems in agricultural production systems. However, the weeds have an important role for increasing the plant diversity in natural ecosystems, there are many important problems associated with the presence of weeds in agricultural fields, including reduction in the yield and quality of crops [11,12], causing poisoning in humans and animals [13], pollution of lakes and reservoirs and blocking channels and irrigation waterways [14], interference with crop harvesting [15], increasing the cost of agricultural production [16], etc. The damage caused by weeds is higher than that of other pests and plant diseases, as in worldwide scale the damage rate is estimated by 43% [17]. Hence, the control of weeds imposes enormous costs on farmers and producers. Weeds, due to features such as abundant seed production, the existence of seed dormancy, preservation of seed vigor for a long time, high adaptability to environmental conditions, can survive and compete with crops [18]. Crops and weeds interact with each other due to common needs such as water, light, space, and nutrients, and this interference may appear as competition or allelopathy.

WEEDS CONTROL METHODS

Weed control is one of the most important aspects of crop production in any agricultural system. Development of human knowledge in various fields, such as the accurate identification of weeds species, life cycle, methods of reproduction and distribution, seed dormancy, understanding the critical period of weed damage in different crops, are important factors that can help us to achieve sustainable weed management. Currently, different strategies for weed management are explained. Efforts should be made to use effective tools in order to minimize weed density in agricultural fields and reduce the damage level. Based on the economic damage threshold, achieving complete control of weeds is unexpected. Mainly five methods including cultural, mechanical, biological, chemical, and integrated weed management (IWM) are widely used for weed control in agro ecosystems which are briefly discussed in the following sections [19].

Cultural Method

Cultural practices are widely used throughout crop's active growth period. These methods include living mulches and crop residues [20,21], crop rotation [22],

planting dates [23], use of competitive genotypes [24], planting patterns, and plant densities [25]. These methods are used to improve the growth and development of crops in competition with weeds. Cultural methods have been used for a long time in agricultural ecosystems. Currently, with the development of new technologies and herbicides, the application of cultural methods has decreased. However, to achieve a sustainable agricultural ecosystem, the integration of cultural methods with new approaches is essential [26].

Mechanical and Physical Method

The mechanical weed control is one of the oldest weed management methods that require the physical removal of weeds by mechanical equipment before planting the main crops or during the crop growing season. One of the most important factors in mechanical control of weeds is the management of soil cultivation practices. The use of different tillage systems in weed management can play a significant role in the competitive pattern between crops and weeds. Deep tillage can stimulate seed germination in the lower layers of the soil and reduce the seed bank of weeds in the soil. However, based on sustainable agricultural foundations, today, the use of minimum tillage and nontillage conservation methods has been developed to protect the soil environment. In the mechanical method, a wide range of implements are used for management of weeds, from manual to smart and advanced tools. Generally there are some different commonly used weed control equipment and practices such as blind cultivation, between row cultivation [27], tine weeders, rotary hoes [28], basket weeders, finger weeders, row crop cultivators, rolling cultivators, robotic weeders [29], and flame weeders [30]. Control of weeds by using the harrow is a common method for managing weeds in cereals and legumes [31]. In row crops such as corn, cotton, soybeans, and vegetables, the use of between row and interrow cultivators is very efficient to weed control [32]. Hand weeding is a primitive low capacity method to weed control in some vegetables, organic, and small-scale crops [33].

Biological Method

Biological control of weeds refers to the use of living organisms including insects, fungi, bacteria, and nematodes to control weeds and reduce their population in crops. In biological management of weeds, the prevention or reduction of growth, development, and reproduction of weeds are followed through stimulation of herbivory and pathogenicity of natural enemies such as herbivores, insects, and pathogens. In this method the weed population is reduced to below the economic damage threshold [34]. There are two major strategies in the biological control of weeds. The first strategy relates to the release of biological control agents in the environment that is known as the *classical* strategy. In this strategy, biological control agents have the ability to produce new generations and control the target weeds over a long period of time in a sustainable manner [35]. The second strategy is called *bioherbicide*, and involves the use of some fungi and bacteria to control the weeds during the growing season [36]. This method is similar to the application of chemical herbicides, but the active ingredient in the bioherbicide method is the living organisms. The effectiveness of the biological control of weeds requires the presence of host-specific biological control agents, high and rapid reproductive potential of control agent for weed suppression, and high flexibility and adaptability to new environments [37]. Mehdizadeh et al. [38] reported a seed beetle (*Spermophagus sericeus* Geoffroy) as an effective biological control agent for field bindweed (*Convolvulus arvensis* L.) in Iran. Biological weed control is a time-consuming method; however, due to high stability and reduction of environmental pollution, many efforts have been made to develop this method in agricultural ecosystems. It is very difficult to control some noxious weeds and herbicide-resistant weeds by mechanical, chemical, or cultural methods. In such a situation, one of the best options for weed management is the biological method [39].

Chemical Method

The use of a group of pesticides known as herbicides to control weeds is one of the most effective methods for managing these plants in agricultural ecosystems. Improving the efficiency of herbicides and production of herbicide-resistant crops created a fundamental transformation in chemical management of weeds and made herbicides as an important tool in modern weed management. In recent decades, this has led to the extensive use of herbicides in developed and developing countries. Herbicides are classified according to different factors such as chemical families, chemical structure, methods of application (soil-applied herbicides and foliage-applied herbicides), selectivity (selective or nonselective), mobility and translocation (contact herbicide or systemic herbicide), mode of action, site of action, application time (preplant/preemergence/postemergence), [40,41]. Some of the most important herbicide modes of action include the lipid synthesis inhibitors, amino acid synthesis inhibitors, growth regulators (synthetic auxins), photosynthesis

inhibitors, nitrogen metabolism inhibitors, pigment inhibitors, cell membrane disruptors, and seedling growth inhibitors [42,43]. Based on mechanism of action, the most important chemical families of herbicides include acetyl CoA carboxylase inhibitors, acetolactate synthase or acetohydroxy acid synthase (AHAS) inhibitors, photosystem II inhibitors, photosystem I inhibitors, protoporphyrinogen oxidase (PPG oxidase or protox) inhibitors, carotenoid biosynthesis inhibitors, enolpyruvylshikimate-3-phosphate synthase inhibitors, glutamine synthetase inhibitors, dihydropteroate synthetase inhibitors, mitosis inhibitors, cellulose inhibitors, oxidative phosphorylation uncouplers, fatty acid and lipid biosynthesis inhibitors, synthetic auxins, and auxin transport inhibitors [44]. Hence, large groups of herbicides were identified, produced, and introduced to target markets. These herbicides had major differences in terms of weed control level, activity, mode of action and application, crop safety, toxicity, and environmental impacts and were accepted by farmers. However, in recent years, with the occurrence of environmental issues and major problems such as the resistance of some weeds to herbicides, the use of these chemical compounds has been widely criticized. Hence, efforts are focused on finding low-risk compounds and bioherbicides [45].

Integrated Weed Management

From weed scientists' point of view, IWM is an effective approach to sustainable weed control and reducing damage caused by weeds-crops competition, along with reducing environmental pollution and preserving the life of microorganisms [46–48]. Achieving an effective and sustainable weed management practice requires the integration approach of different methods. According to Thill et al. [49], IWM refer to combining effective, environmentally friendly, and socially acceptable methods to reduce weed interference to below the economic threshold. The common strategies in IWM are cultural methods (crop rotation, resistant cultivars, cover crops, mixed cropping, fertilizer management), mechanical methods, biological methods, and chemical methods. In integrated approach and sustainable weed management, controlling weeds during the competition period is very important, however, the reduction of viable seeds in the seed bank should be considered. Therefore, the main factors such as tolerance threshold, damage threshold, critical periods of interference, and critical competition periods play a significant role in integrated control of weeds [50]. There is much evidence about the effect of integrative methods and simultaneous application of different strategies on successful weed control. O'Donovan et al. [51] reported that combinations of different methods including crop rotation, higher seeding rates, and herbicide rates resulted in higher barley yields and reduced wild oat (*Avena fatua*) biomass. Danmaigoro et al. [52] evaluated the performance of upland rice varieties as affected by different weed management treatments and found that application of pendimethalin along with one hoe weeding at 6 weeks after sowing resulted in significantly greater plant height, leaf area index, panicle weight per plant, biological yield, and the grain yield of rice [52].

ENVIRONMENTAL RISKS ASSOCIATED WITH HERBICIDE PERSISTENCE

Environmental pollution is one of the most important issues concerning the quality of the environment and human safety. Herbicides have been used for decades for weed management and prevent the yield loss of agricultural products. In recent years, the quality of the environment is greatly considered due to enormous application of different herbicides in agro ecosystems. The presence of herbicide residues in the environment has become an important problem in many countries. Some herbicides persist for a long time in different environments and have the potential for displacement far away from the location of application. Herbicide residue could be easily moved through the surface water as well as groundwater. The presence and persistence of herbicides in the environment should be considered due to their potential to pollute water resources and aquatic life [53], endangering human health [54], causing damage on nontarget rotational crops in agro ecosystems [55], creation of superweeds and herbicides-resistant weeds [56], and the imposition of irreversible financial loss to farmers and crop producers [57]. The persistence of a single herbicide in the environment depends on chemical structure of herbicide, application rate, degradation process, solubility in water, soil texture and properties, temperature, etc. In the study of herbicide persistence in the environment, the factors such as leaching of herbicides to the lower layers of soil [58], runoff of herbicides through surface water [59], photodegradation of herbicide molecules [60], biodegradation of herbicides by soil microorganisms [61], drift of the herbicides to other locations [62], and binding of herbicide molecules to soil particles [63] are very important. Determining the fate of herbicides in the environment can help to understand the behavior of these compounds and their effects on biological ecosystems. Generally, there are some different

methods to detection and quantification of herbicides from various media (soil, water, plant tissues, and human body). Generally, analytical methods such as HPLC, GC, TLC, and bioassays are those important methods that are commonly used by researchers all over the world [64,65].

ALLELOPATHY

Concerned consequences related to commonly used herbicide in the environment have encouraged the farmers and agricultural producers to take advantage of technologies that manage weeds and be available to farmers [66]. The use of allelopathy in various agricultural systems can be considered as one of these options [67]. According to Zimdahl [68], allelopathy can be divided into two categories which include true allelopathy and functional allelopathy. Generally, direct release of allelochemical compounds in the environment, by donor species, can be considered as true allelopathy [69]. However, the functional allelopathy is referred to the use of plants with allelopathic properties in the form of mulch or cover crops in agricultural fields, which the phytotoxic compounds produced and released after their decomposition through chemical and biochemical processes. Typically, the use of allelopathic cover crops involves the application of their residues as mulch at the surface of the field or spraying of their aqueous extracts similar to postemergence herbicides [70].

The use of allelopathic phenomenon as an eco-friendly strategy for weed management can be considered. Unlike the common herbicides, allelochemicals are of plant origin and do not cause residual effects and environmental pollution. Plants with allelopathic properties are a source of various allelochemical compounds that can be used to discover and produce new bioherbicides. Allelopathy refers to any positive or negative effects of a living organism on another living organism through direct or indirect release of allelochemicals into the environment [71]. Actually, in plant allelopathic phenomenon, there is always a plant as a donor and a plant as a receptor of allelochemicals. In agro ecosystems, the receptor and donor plants may be crops or weeds. Allelochemicals are referred to chemical substances, which are responsible for allelopathic effects. These compounds are synthesized in plants as secondary metabolites and play an important role in the relationships between plants and the surrounding environment. The allelochemicals produced by the plants are composed of various compounds such as alkaloids, cyanohydrins, lactones, organic acids, amino acids, fatty acids, flavonoids, tannins, and other compounds. Allelopathic compounds are synthesized in different plant tissues such as leaves, flowers, fruits, roots, rhizomes, stems, and pollen. However, the concentration and efficiency of allelopathy depends on the age and plant tissue type. Nowadays, the allelopathic phenomenon is used to weed management and produce bioherbicides [72].

Allelopathic Compounds

As mentioned earlier, allelochemicals produced by plants are composed of various compounds such as alkaloids, cyanohydrins, lactones, organic acids, amino acids, fatty acids, flavonoids, tannins, and other compounds. Shao et al. [73] extracted the alkaloids compounds from a medicinal plant (*Peganum harmala* L.) and reported significant growth inhibitory effect on different plants (lettuce, amaranth, wheat, and ryegrass). Hussain and Reigosa [74] studied the effects of flavonoid compounds isolated from the *Artemisia annua* plant on germination and seedling growth of *Arabidopsis thaliana* and reported the significant effect of this allelochemical at concentration of 1000 μM on germination and growth of *A. thaliana* [74]. The allelochemicals are released in the environment through evaporation, leaching, root leakage, and plant decay. Generally, the allelochemicals are environmentally friendly compounds and their presence in the environment is not a threat to human health and bio-ecosystems. Allelochemical compounds can affect processes such as germination, growth and development, nutrient uptake, cell division, membrane permeability, enzyme activity, and fatty acid metabolism [75,76].

Potential Allelochemicals and Secondary Metabolites

Generally, the allelochemicals depending on their chemical similarities are classified in different groups, including alkaloids, flavonoids, phenols, tannins, cyanohydrins, amino acids, peptides, terpenes, terpenes, ketones, cinnamic acids, benzoic acid, water-soluble organic acids, fatty acids, unsaturated lactones, quinones, polyacetylenes, coumarin, steroids, and benzoquinones [76]. Nishihara et al. [77] reported that the allelopathic process in velvet bean plant is influenced by L-3-[3, 4-dihydroxyphenylalanine (L-DOPA)], which is an amino acid compound. Some of these allelochemicals are commercially produced or in the process of commercialization by various companies. For example, the Cinch is produced from an allelochemical named Cineole and Callisto is commercially produced from the leptospermone [78]. Another allelochemical that

is successfully used as a natural herbicide for controlling weeds is sorgoleone, extracted from the sorghum plant [79]. The phytotoxins derived from different microbial population have also been used as allelochemicals for weed management. Bialaphos is a natural synthesized product from *Streptomyces viridochromeogenes* and *Streptomyces hygroscopicus* that has been widely used to control weeds [80]. Although the physiological nature of allelochemicals has not yet been specifically addressed, it has been found that these compounds can interact the vital processes of plants, such as photosynthesis, cell division, enzymatic processes, and respiration [81,82]. Understanding the biochemical and physiological effects of allelochemicals and identifying and purifying the chemical compounds responsible for phytotoxic effects on target plants is one of the most important factors in the development of new herbicides [83]. An effective allelopathic compound of cucumber (*Cucumis sativus* L.), that is, (9-hydroxy-4,7 megastigmadien-9-one) HMO was isolated by Kato-Noguchi et al. [84]. They found that this compound had an inhibitory effect on *Echinochloa crus-galli* L. and *Lepidium sativum* L. at low concentrations [84].

Allelopathic Interactions Between Crops and Weeds

As previously mentioned, allelopathy refers to the interaction between living organisms and especially plant species in the environment that has occurred over centuries in agricultural ecosystems. Hence, allelopathy has an undeniable role in plant communications, including, crop-crop, crop-weed, and weed-weed interactions. Also, allelopathy plays a significant role in intra- and interspecific competition in the plant population. Mhlanga et al. [85] reported that allelopathy phenomenon using cover crops has led to the successful control of different weeds including foxtail (*Setaria* spp.), morning glory (*Ipomoea* spp.), and yellow nutsedge (*Cyperus esculentus* L.) in field. The occurrence of allelopathy among various plant species may inhibit the growth and development of neighboring plant species and affect seed germination [86,87]. Although the protection of donor plant species from adverse biotic conditions can also be considered as one of the consequences of allelopathy [88]. Today, in sustainable agricultural systems and especially organic farming and conservation agriculture, which minimize the use of agricultural pesticides, research on allelopathy and the use of allelochemicals for weed management are of great importance [89].

Methods to Study the Allelopathy

During recent years, weed management researchers have been seeking to develop practical methods for using allelopathic phenomena to control weeds in agro ecosystems. Accordingly, three major methods have been reported for the use of this biological phenomenon including (1) the use of a donor crop with allelopathic properties to suppress weeds, (2) the use of aqueous or hydro alcoholic extracts of allelopathic plants for postemergence sprays similar to herbicides, and (3) the use of living mulches or crop residues from plants with allelopathic properties in crop rotations. Plant allelopathic effects are also determined by various analytical methods such as high-performance liquid chromatography [90], gas chromatography [91], gas chromatography-mass spectrometry [92], and different bioassay methods in field or controlled conditions [93,94]. Verdeguer et al. [95] evaluated the allelopathic effects of different plants such as ambilateral (Lantana camara), red river gum (Eucalyptus camaldulensis), and wild rosemary (Eriocephalus africanus) through analytical methods (gas chromatography-mass spectrometry) in order to their herbicidal activity on green amaranth (Amaranthus hybridus) and purslane (Portulaca oleracea) [95]. They reported that red river gum extracts were the most effective on inhibition of germination and seedling growth of both weeds. Charoenying et al. [96] conducted a laboratory bioassay method to evaluate the allelopathic effects of Zanthoxylum limonella extracts on seed germination and seedling growth of barnyard grass (Echinochloa crus-galli (L.)) and Chinese amaranth (Amaranthus tricolor L.) and found that extract at concentration of 2500 μM had a significant effect on germination and seedling growth of Chinese amaranth [96]. Batish et al. [97] conducted a laboratory bioassay to determine allelopathic effects of Anisomeles indica and found the significant inhibitory effects on growth of four weeds including Amaranthus viridis, Cassia occidentalis, Bidens pilosa, and Echinochloa crus-galli [97].

Use of Allelopathy for Weed Management

Undoubtedly, the use of herbicides has a major impact on improving crop yield in agro ecosystems. However, the negative environmental effects associated with these chemicals and the threat to human health is not negligible. Therefore, allelopathy and the use of allelochemicals can be considered as an alternative to herbicides in agricultural systems. Some of these allelochemicals, due to their herbicidal properties, can be classified as a new group of herbicides called bioherbicides. Planting a crop with allelopathic properties in agronomic rotations can improve the quality of crop and increase the

yield as well as management of weeds [98]. Lin et al. [99] evaluated the allelopathic potential of Houttuynia cordata for weed management in rice paddy field of Japan and reported that the germination and seedling growths of two major weed species (Monocharia and Echinochloa) have been influenced by the aqueous extracts of H. cordata. Ramachandran [100] studied the efficacy of different botanical extracts on management of Parthenium hysterophorus (L.) and found that germination, seedling length, seedling vigor index, and seedling biomass of P. hysterophorus was significantly reduced due to application of all botanical extracts; however, the effect was more pronounced with botanicals extract in the order of Datura metel, Mangifera indica, Azadirachta indica, Tagetes erectus, Helianthus annuus, and Sorghum bicolor [100].

Some plant species can apply toxicity effects to suppress some weeds by producing secondary metabolites and releasing them in the environment. Inhibition of seeds germination and growth of some weeds have been reported in various studies due to some crops with allelopathic properties. Naeem et al. [101] studied the allelopathic effects of sorghum, sunflower, and mulberry to control wild oats (Avena fatua), little seed canary grass (Phalaris minor), lamb's quarters (Chenopodium album), and swine cress (Coronopus didymus). They found that application of water extract at the rate of $12-21$ L. ha^{-1} reduced the total weed density and dry biomass production in all weed species [101]. Sturm et al. [102] tested different cover crops with allelopathic effects to weed suppression and reported that three cover crops including Raphanus sativus, Fagopyrum esculentum, and Avena strigosa had the highest allelopathic weed suppression with up to 28%. Alam et al. [103] evaluated the allelopathic effects of rice plant on germination and seedling growth of different weeds and reported that the rice cultivars caused inhibitory effects on seed germination and seedling shoot-root length of Echinochloa crus-galli, Cyperus difformis, Cyperus iria, Fimbristylis miliacea, and weedy rice [103].

Medicinal plants can be considered as donor species in plant allelopathic interactions due to their high diversity and potential for the production of secondary metabolites for weed management in agro ecosystems [104]. Some researchers confirmed the inhibitory and allelopathic effects of various medicinal plants on different weeds. Algandabya and El-Darier [105] conducted bioassay experiments to manage Medicago polymorpha L. via allelopathy of some medicinal plants and found that the aqueous extracts of four medicinal plants including Achillea santolina L., Artemisia monosperma Del., Pituranthos tortuosus L., and Thymus capitatus

L. had a great inhibitory allelopathic effect on germination, plumule, and radicle lengths, leaf area index, total photosynthetic pigment, and chlorophyll *a* (Chl a) of this weed [105]. Petrova et al. [106] confirmed that the extracts of three medicinal plants such as lavender (*Lavandula angustifolia* Mill.), horse mint (*Mentha longifolia* (L.), and peppermint (*M. piperita* L.) significantly reduced the germination and growth of Johnson grass (*Sorghum halepense* (L.) and curly dock (*Rumex crispus* L.). Teerarak et al. [107] studied the allelopathic effects of Spanish jasmine (*Jasminum officinale* L.) for management of weeds and found a glucoside named oleuropein as the main active allelopathic compound for this medicinal plant. They reported that germination and growth of *Phaseolus lathyroides* L. and *Echinochloa crus-galli* L. weeds are significantly affected by Spanish jasmine extract [107].

CONCLUSION

Application of biological-based herbicides can be considered as an appropriate alternative method to weed management, and can achieve sustainable agriculture and reduce environmental hazards due to synthetic herbicides in agro ecosystems. Hence, the allelochemicals and secondary metabolites from some potential crops and medicinal plants can be considered as a possible option for the production of safe and environmentally friendly herbicides. Determination of chemical and biochemical mechanisms for biological management of weeds via allelopathy phenomenon requires coherent studies at laboratory levels and field conditions. Generally, the ability to suppress weeds through some of the natural compounds produced by plants can open a new horizon in the management of weeds. Although it is not possible to complete control of weeds through the allelopathy and the use of allelochemicals, it could be achieved to minimize the weed populations in agro ecosystems by combining the allelopathy phenomena with other methods such as crop rotation, using competitive cultivars, water and soil management, and other agricultural and mechanical methods. It should be noted that in each region, there are a large number of plants with secondary metabolites and allelopathic potential that can be evaluated in order to produce effective and reliable bioherbicides. On the other hand, it may be possible to manage weeds by transferring the genes responsible for production of allelopathy to different crops via biotechnological and plant breeding methods.

REFERENCES

[1] N. Ghimire, R.T. Woodward, Under- and over-use of pesticides: an international analysis, Ecol. Econ. 89 (2013) 73–81. https://doi.org/10.1016/j.ecolecon.2013.02.003.

[2] W.T. Tsai, Trends in the use of glyphosate herbicide and its relevant regulations in Taiwan: a water contaminant of increasing concern, Toxics 7 (1) (2019) 4. https://doi.org/10.3390/toxics7010004.

[3] X. Xu, R. Zarecki, S. Medina, S. Ofaim, X. Liu, C. Chen, S. Hu, D. Brom, S. Gat, S. Porob, H. Eizenberg, Z. Ronen, J. Jiang, S. Freilich, Modeling microbial communities from atrazine contaminated soils promotes the development of biostimulation solutions, ISME J. 13 (2) (2019) 494–508. https://doi.org/10.1038/s41396-018-0288-5.

[4] A.H.C. Van-Bruggen, M.M. He, K. Shin, V. Mai, K.C. Jeong, M.R. Finckh, J.G. Morris, Environmental and health effects of the herbicide glyphosate, Sci. Total Environ. 616–617 (2018) 255–268. https://doi.org/10.1016/j.scitotenv.2017.10.309.

[5] T. Torretta, I.A. Katsoyiannis, P. Viotti, E.C. Rada, Critical review of the effects of glyphosate exposure to the environment and humans through the food supply chain, Sustainability 10 (2018) 950. https://doi.org/10.3390/su10040950.

[6] M. Mehdizadeh, M.T. Alebrahim, M. Roushani, J.C. Streibig, Evaluation of four different crops' sensitivity to sulfosulfuron and tribenuron methyl soil residues, Acta Agric. Scand. Sect. B Soil Plant Sci 66 (8) (2016) 706–713. https://doi.org/10.1080/09064710.2016.1212919.

[7] M. Mehdizadeh, F. Gholami-Abadan, Negative effects of residual herbicides on sensitive crops: impact of rimsulfuron herbicide soil residue on sugar beet, J. Res. Weed Sci. 1 (1) (2018) 1–6. https://dx.doi.org/10.26655/jrweedsci.2018.6.1.

[8] M. Mehdizadeh, Sensitivity of oilseed rape (*Brassica napus* L.) to soil residues of imazethapyr herbicide, Int. J. Agric. Environ. Food Sci. 3 (1) (2019) 46–49. https://dx.doi.org/10.31015/jaefs.2019.1.10.

[9] M.T. Rose, T.R. Cavagnaro, C.A. Scanlan, T.J. Rose, T. Vancov, S. Kimber, I.R. Kennedy, R.S. Kookana, L. Van-Zwieten, Impact of herbicides on soil biology and function, Adv. Agron. 136 (2016) 133–220. https://doi.org/10.1016/bs.agron.2015.11.005.

[10] C. Boutin, B. Strandberg, D. Carpenter, S.K. Mathiassen, P.J. Thomas, Herbicide impact on non-target plant reproduction: what are the toxicological and ecological implications? Environ. Pollut. 185 (2014) 295–306. https://doi.org/10.1016/j.envpol.2013.10.009.

[11] M.R. Fernandez, R.P. Zentner, M.P. Schellenberg, J.Y. Leeson, O. Aladenola, B.G. McConkey, M.S. Luce, Grain yield and quality of organic crops grown under reduced tillage and diversified sequences, Agron. J. (2019). https://doi.org/10.2134/agronj2018.01.0029.

[12] N. Fernando, S.K. Florentine, M. Naiker, J. Panozzo, B.S. Chauhan, Annual ryegrass (*Lolium rigidum* Gaud) competition altered wheat grain quality: a study under elevated atmospheric CO_2 levels and drought conditions, Food Chem. 276 (2019) 285−290. https://doi.org/10.1016/j.foodchem.2018.09.145.

[13] L. Lei, W. Sun, L. He, H. Jiang, M. Zhang, W. He, Z. Hu, Y. Gu, H. Song, Y. Zhang, Cardiotoxicity of Consolida rugulosa, a poisonous weed in Western China, Ecotoxicol. Environ. Saf. 170 (2019) 141−147. https://doi.org/10.1016/j.ecoenv.2018.11.109.

[14] W. Abtew, S.B. Dessu, Aquatic weeds and the grand Ethiopian renaissance dam, in: The Grand Ethiopian Renaissance Dam on the Blue Nile, Springer Geography. Springer, Cham, 2019, pp. 147−159.

[15] I. Nosratti, S. Almaleki, B.S. Chauhan, Seed germination ecology of soldier thistle (Picnomon acarna): an invasive weed of rainfed crops in Iran, Weed Sci. (2019). https://doi.org/10.1017/wsc.2018.74.

[16] Y. Gharde, P.K. Singh, R.P. Dubey, P.K. Gupta, Assessment of yield and economic losses in agriculture due to weeds in India, Crop Protect. 107 (2018) 12−18. https://doi.org/10.1016/j.cropro.2018.01.007.

[17] E.C. Oerke, Crop losses to pests, J. Agric. Sci. 144 (1) (2006) 31−43. https://doi.org/10.1017/S0021859605005708.

[18] R.L. Zimdahl, Fundamentals of Weed Science, third ed.tion, Academic Press, 2007, 666 pp.

[19] K.N. Harker, J.T. O'Donovan, Recent weed control, weed management, and integrated weed management, Weed Technol. 27 (1) (2013) 1−11. https://doi.org/10.1614/WT-D-12-00109.1.

[20] A.O. Ikeh, E.I. Udoh, A.C. Opara, Effect of mulching on weed, fruit yield and economic returns of garden egg (*Solanum melogena*) in Okigwe southeastern Nigeria, J. Res. Weed Sci. 2 (1) (2019) 52−64. https://dx.doi.org/10.26655/jrweedsci.2019.1.5.

[21] C.L. Mohler, A.G. Taylor, A. DiTommaso, R.R. Hahn, Effects of incorporated rye and hairy vetch cover crop residue on the persistence of weed seeds in the soil, Weed Sci. 66 (3) (2018) 379−385. https://doi.org/10.1017/wsc.2017.80.

[22] J.L. Gonzalez-Andujar, M.J. Aguilera, A.S. Davis, L. Navarrete, Disentangling weed diversity and weather impacts on long-term crop productivity in a wheat-legume rotation, Field Crop. Res. 232 (2019) 24−29. https://doi.org/10.1016/j.fcr.2018.12.005.

[23] P.J.W. Lutman, S.R. Moss, S. Cook, S.J. Welham, A review of the effects of crop agronomy on the management of *Alopecurus myosuroides*, Weed Res. 53 (5) (2013) 299−313. https://doi.org/10.1111/wre.12024.

[24] D. Lemerle, D.J. Luckett, P. Lockley, E. Koetz, H. Wu, Competitive ability of Australian canola (*Brassica napus*) genotypes for weed management, Crop Pasture Sci. 65 (12) (2014) 1300−1310. https://doi.org/10.1071/CP14125.

[25] C. Marin, J. Weiner, Effects of density and sowing pattern on weed suppression and grain yield in three varieties of maize under high weed pressure, Weed Res. 54 (5) (2014) 467−474. https://doi.org/10.1111/wre.12101.

[26] M.D. Owen, H.J. Beckie, J.Y. Leeson, J.K. Norsworthy, L.E. Steckel, Integrated pest management and weed management in the United States and Canada, Pest Manag. Sci. 71 (3) (2014) 357−376. https://doi.org/10.1002/ps.3928.

[27] M.D. Frost, E.R. Haramoto, K.A. Renner, D.C. Brainard, Tillage and cover crop effects on weed seed persistence: do light exposure and fungal pathogens play a role? Weed Sci. 67 (1) (2019) 103−113. https://doi.org/10.1017/wsc.2018.80.

[28] J.E. Maul, M.A. Cavigelli, B. Vinyard, J.S. Buyer, Cropping system history and crop rotation phase drive the abundance of soil denitrification genes nirK, nirS and nosZ in conventional and organic grain agroecosystems, Agric. Ecosyst. Environ. 273 (2019) 95−106. https://doi.org/10.1016/j.agee.2018.11.022.

[29] S.A. Fennimore, M. Cutulle, Robotic weeders can improve weed control options for specialty crops, Pest Manag. Sci. (2019). https://doi.org/10.1002/ps.5337.

[30] E. Kanellou, G. Economou, M. Papafotiou, N. Ntoulas, D. Lyra, E. Kartsonas, S. Knezevic, Flame weeding at archaeological sites of the mediterranean region, Weed Technol. 31 (3) (2017) 396−403. https://doi.org/10.1017/wet.2016.31.

[31] S. Stenerud, K. Mangerud, H. Sjursen, T. Torp, L.O. Brandsæter, Effects of weed harrowing and undersown clover on weed growth and spring cereal yield, Weed Res. 55 (5) (2015) 493−502. https://doi.org/10.1111/wre.12163.

[32] A. Peruzzi, L. Martelloni, C. Frasconi, M. Fontanelli, M. Pirchio, M. Raffaelli, Machines for non-chemical intra-row weed control in narrow and wide-row crops: a review, J. Agric. Eng. 48 (2) (2017) 57−70. https://doi.org/10.4081/jae.2017.583.

[33] F.A. Nwagwu, I.A. Udo, Effects of integrated weed management on tuber yield of cassava (*Manihot esculenta* Crantz), J. Res. Weed Sci. 2 (1) (2019) 1−15. https://dx.doi.org/10.26655/jrweedsci.2019.1.1.

[34] F.D. Panetta, B. Gooden, Managing for biodiversity: impact and action thresholds for invasive plants in natural ecosystems, NeoBiota 34 (2017) 53−66. https://doi.org/10.3897/neobiota.34.11821.

[35] A.W. Sheppard, R.H. Shaw, R. Sforza, Top 20 environmental weeds for classical biological control in Europe: a review of opportunities, regulations and other barriers to adoption, Weed Res. 46 (2) (2006) 93−117. https://doi.org/10.1111/j.1365-3180.2006.00497.x.

[36] C.D. Boyette, R.E. Hoagland, K.C. Stetina, Bioherbicidal enhancement and host range expansion of a mycoherbicidal fungus via formulation approaches, Biocontrol Sci. Technol. 28 (3) (2018) 307−315. https://doi.org/10.1080/09583157.2018.1445199.

[37] M. Schwarzländer, V.C. Moran, S. Raghu, Constraints in weed biological control: contrasting responses by implementing nations, BioControl 63 (3) (2018) 313−317. https://doi.org/10.1007/s10526-018-9888-2.

[38] M. Mehdizadeh, G.A. Asadi, A. Delobel, Biological control potential of *Spermophagus sericeus* Geoffroy, 1785 (Coleoptera: chrysomelidae) against field bindweed as the first report from Iran, J. Res. Weed Sci. 1 (1) (2018) 40−47. https://dx.doi.org/10.26655/jrweedsci.2018.6.5.

[39] I. Gnanavel, S.K. Natarajan, Eco-friendly weed control options for sustainable agriculture, Sci. Int. 35 (3) (2014) 172−183. https://doi.org/10.17311/sciintl.2015.37.47.

[40] S. Varshney, S. Hayat, M.N. Alyemeni, A. Ahmad, Effects of herbicide applications in wheat fields: is phytohormones application a remedy? Plant Signal. Behav. 7 (2012) 570−575. https://doi.org/10.4161/psb.19689.

[41] F. Torrens, G. Castellano, Molecular classification of pesticides including persistent organic pollutants, phenylurea and sulphonylurea herbicides, Molecules 19 (6) (2014) 7388−7414. https://doi.org/10.3390/molecules19067388.

[42] G.R.B. Webster, Herbicides: chemistry, degradation and mode of action, J. Fish. Res. Board Can. 34 (12) (1977), 2426-2426. https://doi.org/10.1139/f77-331.

[43] R.R. Schmidt, Classification of Herbicides According to Mode of Action, Bayer Ag, Leverkusen, 1998, p. 8.

[44] A. Forouzesh, E. Zand, S. Soufizadeh, S. Samadi Foroushani, Classification of herbicides according to chemical family for weed resistance management strategies-an update, Weed Res. 55 (4) (2015) 334−358. https://doi.org/10.1111/wre.12153.

[45] J. Kao-Kniffin, S.M. Carver, A. DiTommaso, Advancing weed management strategies using metagenomic techniques, Weed Sci. 61 (2) (2013) 171−184. https://doi.org/10.1614/WS-D-12-00114.1.

[46] K.N. Harker, J.T. O'Donovan, R.E. Blackshaw, H.J. Beckie, C. Mallory-Smith, B.D. Maxwell, Our view, Weed Sci. 60 (2) (2012) 143−144. https://doi.org/10.1614/WS-D-11-00177.1.

[47] D.L. Shaner, Lessons learned from the history of herbicide resistance, Weed Sci. 62 (2) (2014) 427−431. https://doi.org/10.1614/WS-D-13-00109.1.

[48] M. Liebman, B. Baraibar, Y. Buckley, D. Childs, S. Christensen, R. Cousens, H. Eizenberg, D. Loddo, A. Merotto, M. Renton, M. Riemens, Ecologically sustainable weed management: how do we get from proof-of-concept to adoption? Ecol. Appl. 26 (5) (2016) 1352−1369. https://doi.org/10.1002/15-0995.

[49] D.C. Thill, L.M. Lish, R.H. Callihan, E.J. Bechinski, Integrated weed management−A component of integrated pest management: a critical review, Weed Technol. 5 (1991) 648−656. https://doi.org/10.1017/S0890037X00027500.

[50] M.B. Bertucci, K.M. Jennings, D.M. Monks, J.R. Schultheis, F.J. Louws, D.L. Jordan, C. Brownie, Critical period for weed control in grafted and nongrafted watermelon grown in plasticulture, Weed Sci. 67 (2018) 221−228. https://doi.org/10.1017/wsc.2018.76.

[51] J.T. O'Donovan, K.N. Harker, T.K. Turkington, G.W. Clayton, Combining cultural practices with herbicides reduces wild oat (*Avena fatua*) seed in the soil seed bank and improves barley yield, Weed Sci. 61 (2) (2013) 328−333. https://doi.org/10.1614/WS-D-12-00168.1.

[52] O. Danmaigoro, A.G. Halilu, A.U. Izge, Growth and yield of direct seeded upland rice varieties as influenced by weed management and organic manure application, J. Res. Weed Sci. 2 (2) (2019) 103−114. https://dx.doi.org/10.26655/jrweedsci.2019.3.2.

[53] C.S. Machado, R.I.S. Alves, B.M. Fregonesi, K.A.A. Tonani, B.S. Martinis, J. Sierra, M. Nadal, J.L. Domingo, S. Segura-Munoz, Chemical contamination of water and sediments in the pardo river, sao paulo, Brazil, Procedia Eng. 162 (2016) 230−237. https://doi.org/10.1016/j.proeng.2016.11.046.

[54] A. Morris, E.G. Murrell, T. Klein, B.H. Noden, Effect of two commercial herbicides on life history traits of a human disease vector, *Aedes aegypti*, in the laboratory setting, Ecotoxicology 25 (5) (2016) 863−870. https://doi.org/10.1007/s10646-016-1643-9.

[55] P. Janaki, S. Bhuvanadevi, M. Dhananivetha, P. Murali-Arthanari, C. Chinnusamy, Persistence of quizalofop ethyl in soil and safety to ground nut by ultrasonic bath extraction and HPLC- DAD detection, J. Res. Weed Sci. 1 (2) (2018) 63−74. https://dx.doi.org/10.26655/jrweedsci.2018.9.1.

[56] C. Bain, T. Selfa, T. Dandachi, S. Velardi, 'Superweeds' or 'survivors'? Framing the problem of glyphosate resistant weeds and genetically engineered crops, J. Rural Stud. 51 (2017) 211−221. https://doi.org/10.1016/j.jrurstud.2017.03.003.

[57] D. Bourguet, T. Guillemaud, The hidden and external costs of pesticide use, in: E. Lichtfouse (Ed.), Sustainable Agriculture Reviews, vol. 19, Springer, Cham, 2016.

[58] S. Manna, N. Singh, Biochars mediated degradation, leaching and bioavailability of pyrazosulfuron-ethyl in a sandy loam soil, Geoderma 334 (2019) 63−71. https://doi.org/10.1016/j.geoderma.2018.07.032.

[59] J. Boulange, F. Malhata, P. Jaikaew, K. Nanko, H. Watanabe, Portable rainfall simulator for plot-scale investigation of rainfall-runoff, and transport of sediment and pollutants, Int. J. Sediment Res. 34 (1) (2019) 38−47. https://doi.org/10.1016/j.ijsrc.2018.08.003.

[60] A. Chenchana, A. Nemamcha, H. Moumeni, J.M. Dona-Rodriguez, J. Arana, J.A. Navio, O. Gonzalez-Diaz, E. Pulido-Melian, Photodegradation of 2,4-dichlorophenoxyacetic acid over TiO2(B)/anatase nanobelts and Au-TiO2(B)/anatase nanobelts, Appl. Surf. Sci. 467−468 (2019) 1076−1087. https://doi.org/10.1016/j.apsusc.2018.10.175.

[61] L. Carles, M. Joly, F. Bonnemoy, M. Leremboure, F. Donnadieu, I. Batisson, P. Besse-Hoggan, Biodegradation and toxicity of a maize herbicide mixture: mesotrione, nicosulfuron and S-metolachlor, J. Hazard Mater. 354 (2018) 42−53. https://doi.org/10.1016/j.jhazmat.2018.04.045.

[62] N.R. Steppig, J.K. Norsworthy, R.C. Scott, G.M. Lorenz, Insecticide seed treatments as safeners to drift rates of herbicides in soybean and grain sorghum, Weed

Technol. 32 (2) (2018) 150−158. https://doi.org/10.1017/wet.2017.102.

[63] R. Jiang, M. Wang, W. Chen, Characterization of adsorption and desorption of lawn herbicide siduron in heavy metal contaminated soils, Chemosphere 204 (2018) 483−491. https://doi.org/10.1016/j.chemosphere.2018.04.045.

[64] M. Mehdizadeh, M.T. Alebrahim, M. Roushani, Determination of two sulfonylurea herbicides residues in soil environment using HPLC and phytotoxicity of these herbicides by lentil bioassay, Bull. Environ. Contam. Toxicol. 99 (1) (2017) 93−99. https://doi.org/10.1007/s00128-017-2076-8.

[65] S. Sondhia, P.K. Singh, Bioefficacy and fate of pendimethalin residues in soil and mature plants in chickpea field, J. Res. Weed Sci. 1 (1) (2018) 28−39. https://dx.doi.org/10.26655/jrweedsci.2018.6.4.

[66] B. Mhlanga, S. Cheesman, B.S. Chauhan, C. Thierfelder, Weed emergence as affected by maize (*Zea mays* L.) cover crop rotations in contrasting arable soils of Zimbabwe under conservation agriculture, Crop Protect. 81 (2016) 47−56. https://doi.org/10.1016/j.cropro.2015.12.007.

[67] K. Jabran, G. Mahajan, V. Sardana, B.S. Chauhan, Allelopathy for weed control in agricultural systems, Crop Protect. 72 (2015) 57−65. https://doi.org/10.1016/j.cropro.2015.03.004.

[68] R.L. Zimdahl, Fundamentals of Weed Science, 4rd Edition, Academic Press, London, United Kingdom, 2013.

[69] T.D. Baratelli, A.C. Candido-Gomes, L.A. Wessjohann, R.M. Kuster, N.K. Simas, Phytochemical and allelopathic studies of *Terminalia catappa* L. (Combretaceae), Biochem. Syst. Ecol. 41 (2012) 119−125. https://doi.org/10.1016/j.bse.2011.12.008.

[70] R. Ashraf, B. Sultana, S. Yaqoob, M. Iqbal, Allelochemicals and crop management: a review, Curr. Sci. 3 (2017) 1−13.

[71] E.L. Rice, Allelopathy, second ed., Academic Press, Inc., Orlando, F.L., 1984.

[72] U.A. Abd El-Razek, R.A. El-Refaey, S.M. Shebl, S.S.M. Abd El-Naby, Integrating allelopathy, plant population and use of herbicides for weed control of six rice genotypes, Asian J. Crop Sci. 6 (1) (2014) 1−14. https://doi.org/10.3923/ajcs.2014.1.14.

[73] H. Shao, X. Huang, Y. Zhang, C. Zhang, Main alkaloids of *Peganum harmala* L. And their different effects on dicot and monocot crops, Molecules 18 (3) (2013) 2623−2634. https://doi.org/10.3390/molecules18032623.

[74] M.I. Hussain, M.J. Reigosa, Evaluation of herbicide potential of sesquiterpene lactone and flavonoid: impact on germination, seedling growth indices and root length in *Arabidopsis thaliana*, Pak. J. Bot. 46 (2014) 995−1000.

[75] Z.H. Li, Q. Wang, X. Ruan, C.D. Pan, D.A. Jiang, Phenolics and plant allelopathy, Molecules 15 (12) (2010) 8933−8952. http://doi.org/10.3390/molecules15128933.

[76] F. Cheng, Z. Cheng, Research progress on the use of plant allelopathy in agriculture and the physiological and ecological mechanisms of allelopathy, Front. Plant Sci. 6 (2015) 1020. https://dx.doi.org/10.3389/2Ffpls.2015.01020.

[77] E. Nishihara, M.M. Parvez, H. Araya, S. Kawashima, L-3-(3, 4-Dihydroxyphenyl) alanine (L-DOPA) an allelochemical exuded from velvet bean (*Mucuna pruriens*) roots, Plant Growth Regul. 45 (2) (2005) 113−120. https://doi.org/10.1007/s10725-005-0610-x.

[78] G. Mitchell, D.W. Bartlett, T.E. Fraser, T.R. Hawkes, D.C. Holt, J.K. Townson, R.A. Wichert, Mesotrione: a new selective herbicide for use in maize, Pest Manag. Sci. 57 (2) (2001) 120−128. https://doi.org/10.1002/1526-4998(200102)57:2%3c120::AID-PS254%3e3.0.CO;2-E.

[79] L.A. Weston, M.A. Czarnota, Activity and persistence of sorgoleone, a long-chain hydroquinone produced by Sorghum bicolor, J. Crop Prod. 4 (2) (2001) 363−377. https://doi.org/10.1300/J144v04n02_17.

[80] F.E. Dayan, S.O. Duke, Natural compounds as next-generation herbicides, Plant Physiol. 166 (3) (2014) 1090−1105. https://doi.org/10.1104/pp.114.239061.

[81] S.L. Peng, W. Jun, G.Q. Feng, Mechanism and active variety of allelochemicals, Acta Bot. Sin. 46 (7) (2004) 757−766.

[82] I.D.S. Mendes, M.O.O. Rezende, Assessment of the allelopathic effect of leaf and seed extracts of *Canavalia ensiformis* as post emergent bioherbicides: a green alternative for sustainable agriculture, J. Environ. Sci. Health Part B 49 (5) (2014) 374−380. https://doi.org/10.1080/03601234.2014.882179.

[83] T. Abbas, M.A. Nadeem, A. Tanveer, R. Ahmad, Identifying optimum herbicide mixtures to manage and avoid fenoxaprop-p-ethyl resistant *Phalaris minor* in wheat, Planta Daninha 34 (4) (2016) 787−793. https://doi.org/10.1590/s0100-83582016340400019.

[84] H. Kato-Noguchi, H.L. Thi, T. Teruya, K. Suenaga, Two potent allelopathic substances in cucumber plants, Sci. Hortic. 129 (4) (2011) 894−897. https://doi.org/10.1016/j.scienta.2011.04.031.

[85] B. Mhlanga, S. Cheesman, B. Maasdorp, T. Muoni, S. Mabasa, E. Mangosho, C. Thierfelder, Weed community responses to rotations with cover crops in maize-based conservation agriculture systems of Zimbabwe, Crop Protect. 69 (2015) 1−8. https://doi.org/10.1016/j.cropro.2014.11.01.

[86] T. Abbas, M.A. Nadeem, A. Tanveer, B.S. Chauhan, Can hormesis of plant-released phytotoxins be used to boost and sustain crop production? Crop Protect. 93 (2017) 69−76. https://doi.org/10.1016/j.cropro.2016.11.020.

[87] R. Fakhari, A. Tobeh, P. Sharifi-Vizeh, G. Didehbaz-Moghanlo, B. Khalil-Tahmasbi, Effects of cover crop residue management on corn yield and weed control, J. Res. Weed Sci. 1 (1) (2018) 7−17. https://dx.doi.org/10.26655/jrweedsci.2018.6.2.

[88] C.H. Kong, Ecological pest management and control by using allelopathic weeds (*Ageratum conyzoides, Ambrosia*

trifida, and *Lantana camara*) and their allelochemicals in China, Weed Biol. Manag. 10 (2) (2010) 73−80. https://doi.org/10.1111/j.1445-6664.2010.00373.x.

[89] H.M. Kruidhof, L. Bastiaans, M.J. Kropff, Ecological weed management by cover cropping, effects on weed growth in autumn and weed establishment in spring, Weed Res. 48 (6) (2008) 492−502. https://doi.org/10.1111/j.1365-3180.2008.00665.x.

[90] F. Fuentes-Gandara, A. Torres, M.T. Fernandez-once, L. Casas, C. Mantell, R. Varela, E.J. Ossa-Fernandezc, F.A. Macias, Selective fractionation and isolation of allelopathic compounds from *Helianthus annuus* L. leaves by means of high-pressure techniques, J. Supercrit. Fluids 143 (2019) 32−41. https://doi.org/10.1016/j.supflu.2018.08.004.

[91] A. Ismail, H. Mohsen, J. Bassem, H. Lamia, Chemical composition of *Thuja orientalis* L. essential oils and study of their allelopathic potential on germination and seedling growth of weeds, Arch. Phytopathol. Plant Prot. 48 (1) (2015) 18−27. https://doi.org/10.1080/03235408.2014.882107.

[92] X. Zhang, Q.X. Cui, Y. Zhao, H.Y. Li, Allelopathic potential of *Koelreuteria bipinnata* var. *integrifoliola* on germination of three turf grasses, Russ. J. Plant Physiol. 65 (6) (2018) 833−841. https://doi.org/10.1134/S1021443718060146.

[93] S.J. Meiners, K.K. Phipps, T.H. Pendergast, T. Canam, W.P. Carson, Soil microbial communities alter leaf chemistry and influence allelopathic potential among coexisting plant species, Oecologia 183 (4) (2017) 1155−1165. https://doi.org/10.1007/s00442-017-3833-4.

[94] A. Scavo, C. Rial, J.M. Molinillo, R.M. Varela, G. Mauronicale, F.A. Macias, The extraction procedure improves the allelopathic activity of cardoon (*Cynara cardunculus* var. *altilis*) leaf allelochemicals, Ind. Crops Prod. 128 (2019) 479−487. https://doi.org/10.1016/j.indcrop.2018.11.053.

[95] M. Verdeguer, M.A. Blazquez, H. Boira, Phytotoxic effects of *Lantana camara*, *Eucalyptus camaldulensis* and *Eriocephalus africanus* essential oils in weeds of Mediterranean summer crops, Biochem. Syst. Ecol. 37 (4) (2009) 362−369. https://doi.org/10.1016/j.bse.2009.06.003.

[96] P. Charoenying, M. Teerarak, C. Laosinwattana, An allelopathic substance isolated from *Zanthoxylum limonella* Alston fruit, Sci. Hortic. 125 (3) (2010) 411−416. https://doi.org/10.1016/j.scienta.2010.04.045.

[97] D.R. Batish, M. Kaur, H.P. Singh, R.K. Kohli, Phytotoxicity of a medicinal plant, *Anisomeles indica*, against

Phalaris minor and its potential use as natural herbicide in wheat fields, Crop Protect. 26 (7) (2007) 948−952. https://doi.org/10.1016/j.cropro.2006.08.015.

[98] T.D. Khanh, M.I. Chung, T.D. Xuan, S. Tawata, The exploitation of crop allelopathy in sustainable agricultural production, J. Agron. Crop Sci. 191 (3) (2005) 172−184. https://doi.org/10.1111/j.1439-037X.2005.00172.x.

[99] D. Lin, Y. Sugitomo, Y. Dongm, H. Terao, M. Matsuo, Natural herbicidal potential of saururaceae (*Houttuynia cordata* Thunb) dried powders on paddy weeds in transplanted rice, Crop Protect. 25 (2006) 1126−1129. https://doi.org/10.1016/j.cropro.2006.02.004.

[100] A. Ramachandran, Efficacy of different botanical extracts on the management of *Parthenium hysterophorus* (L.), Journal of Research in Weed Science 2 (1) (2019) 16−32. https://dx.doi.org/10.26655/jrweedsci.2019.1.2.

[101] M. Naeem, Z.A. Cheema, M.Z. Ihsan, Y. Hussain, A. Mazari, H.T. Abbas, Allelopathic effects of different plant water extracts on yield and weeds of wheat, Planta Daninha 36 (2018) 1−8. https://doi.org/10.1590/s0100-83582018360100094.

[102] D.J. Sturm, G. Peteinatos, R. Gerhards, Contribution of allelopathic effects to the overall weed suppression by different cover crops, Weed Res. 58 (5) (2018) 331−337. https://doi.org/10.1111/wre.12316.

[103] M.A. Alam, M. Hakim, A.S. Juraimi, M. Rafii, M.M. Hasan, F. Aslani, Potential allelopathic effects of rice plant aqueous extracts on germination and seedling growth of some rice field common weeds, Ital. J. Agron. 13 (2) (2018) 134−140. https://doi.org/10.4081/ija.2018.1066.

[104] J.R. Qasem, Allelopathic effects of selected medicinal plants on *Amaranthus retroflexus* and *Chenopodium murale*, Allelopathy J. 10 (2) (2002) 105−122.

[105] M.M. Algandabya, S.M. El-Darierb, Management of the noxious weed; *Medicago polymorpha* L. via allelopathy of some medicinal plants from Taif region, Saudi Arabia, Saudi J. Biol. Sci. 25 (7) (2018) 1339−1347. https://doi.org/10.1016/j.sjbs.2016.02.013.

[106] S.T. Petrova, E.G. Valcheva, I.G. Velcheva, A case study of allelopathic effect on weeds in wheat, Ecol. Balk. (2015) 121−129.

[107] M. Teerarak, C. Laosinwattana, P. Charoenying, Evaluation of allelopathic, decomposition and cytogenetic activities of *Jasminum officinale* L. f. var. *grandiflorum* (L.) Kob. on bioassay plants, Bioresour. Technol. 101 (14) (2010) 5677−5684. https://doi.org/10.1016/j.biortech.2010.02.038.

CHAPTER 10

Microbial Control of Pests and Weeds

DEEPA VERMA, MSC, B.ED, M.A, M.PHIL, MBA, PHD • TEMITOPE BANJO, PHD •
MADHVI CHAWAN, MSC • NIKHIL TELI, MSC, PHD •
ROHAN GAVANKAR, MSC, B.ED, PHD

INTRODUCTION

Within the next 20 years, crop production will have to increase significantly to meet the needs of the rising human population. This has to be done without destructing the environment and human health. Over the past 50 years, crop safety has relied comprehensively on synthetic chemical pesticides, but their accessibility is now declining as a result of different legislation and the evolution of resistance in pest populations. There will not be a "silver bullet" solution to the forthcoming food production challenge. Rather, a sequence of transformations must emerge to counter the diverse needs of farmers according to their local circumstances [1]. One way to amplify food availability is to develop effective management programs for pests. Around 67,000 divergent crop pest species including plant weeds, pathogens, invertebrates, and a few vertebrate species jointly cause about a 40% decline in the world's crop production [2]. Crop losses caused by pests undermine food safety alongside other constraints, such as poor soils, inclement weather as well as farmer's inadequate access to technical awareness [3]. Weeds constitute a severe risk to crop production mainly responsible for the reduction in the quality and yields of crops [4]. Over the years, different approaches to weed control have been adopted. The introduction and adoption of selective herbicides by farmers resulted in the control of specific weeds while the yield and quality of crops significantly increased [5]. Furthermore, genetic modification of crops was introduced as an improvement over the selective herbicides employed in weed control. This newer approach was found to control a wider group of weeds, thus making it more effective than the selective herbicides which only target a specific weed each time [6]. However, this approach resulted in resistance in the weed population. More so, there is a rising concern on the negative effects of chemical residues present in herbicides on human health [7]. There is need to develop a safer, eco-friendly, and more effective approach of controlling weed population. This review therefore considered the applications of biological agents in the management of pests and weeds of agricultural crops.

BIOLOGICAL CONTROL OF PESTS AND WEEDS OF AGRICULTURAL CROPS

Biological control is a process of suppressing or controlling the unwanted population of insects, other animals, or plants by the introduction or artificial increase of their natural enemies to reasonably nonimportant levels.

Biological Control of Pests

The biological control of pest is a significant aspect of integrated pest management (IPM) [8]. Research on microbiological pathogens of insects is growing significantly in recent times to find out eco-friendly alternatives to harmful chemical insecticides. These microbial pesticides accounts for 1.3% of the world's entire pesticide market of which, 90% of them are used as insecticides [9]. The pathogens that are responsible to cause diseases in insects generally fall into four main categories: bacteria, fungi, viruses, and protozoans. Among a range of bio-products, *Bacillus thuringiensis* (Bt), *Metarhizium* spp., *Trichoderma viride*, *Beauveria bassiana*, and Nuclear polyhedrosis virus are prevalently used in protection of plants [10,11].

Biological Control of Weeds

The management of weeds has been exercised over the years with the use of bacteria and fungi [12]. However, the use of viruses in weed control has been limited [13]. The application of bioherbicide in the management of weeds is fast gaining acceptability among ecologists, farmers, and the consumers. It has been stated that the cost of developing and application of bioherbicides is cheaper compared to chemical herbicides [12]. Furthermore, the preference of consumers for crops produced using bioherbicides rather than chemical herbicides is on the increase. This is due to the negative perception of consumers about food products grown with synthetic chemical herbicides [14].

BIOLOGICAL CONTROL MECHANISMS

Parasitism

Parasitism is a coexisting association in which two microorganisms that are not evolutionarily linked to each other live together for a long period of time and as a result of this kind of association, usually one of these two types of microorganisms which is physically smaller and is called the parasite benefits from the other one which is partially damaged and is called the host. The most direct type of competition in biological control is hyperparasitism, in which obligate parasites are used to eliminate the same plant pathogen [15].

Competition

Competition is a kind of mutual coexistence between two living microorganisms, which occurs when different microorganisms within a population effort to achieve something similar. The objective may be a place or food. This type of mechanism represents a negative relationship between two microbial population in which both the population are unfavorably affected with respect to their growth and survival. Microbial population generally competes for any growth limiting resources such as nitrogen source, carbon source, vitamins, phosphorus, growth factors [16].

PREDATION

This primarily refers to animals that at the higher stage of nourishment and in the macroscopic world are predators. It is a widespread occurrence when one organism (predator) engulf or attack another organism (prey). The prey can be larger or smaller than the predator and this usually results in the death of the prey. In general predator-prey interaction is of short duration [16].

ANTAGONISM

When one microbial population synthesizes components, which exert inhibitory effect to other microbial population, then this type of relationship is known as antagonism. It is the negative effect that a microorganism has on another one. This is a one-way process and is due to the genesis of certain compounds such as antibiotics or bacteriocins produced by a microorganism. The first population which synthesizes inhibitory components is unaffected or may gain a competition and survive in the habitat while other population gets inhibited. This chemical inhibition is generally known as antibiosis [16].

BIOPESTICIDES

As defined by the US Environmental Protection Agency (EPA), biopesticides are natural pesticides that are sourced from natural materials, such as plants, animals, bacteria, and certain minerals. In commercial terms, biopesticides include microorganisms that control insect pests (microbial pesticides), naturally occurring constituents that control insect pests (biochemical pesticides), and pesticidal materials formed by plants containing added genetic material (plant-incorporated protectants).

Classes of Biopesticides

Biopesticides mainly fall into three major classes which include biochemical pesticides, microbial pesticides, and bacterial biopesticides.

Biochemical pesticides

These are naturally occurring constituents that control pests by nontoxic mechanisms. Traditional pesticides, by contrast, are usually synthetic materials that particularly kill or inactivate the pest. Biochemical pesticides comprise substances that obstruct with mating, such as insect sex pheromones, as well as several scented plant extracts that attract insect pests to traps. Biochemical pesticides, with the exception of pheromones, tend to have much less species-specificity and are broader-spectrum pesticides than the microbial pesticides. Biochemical pesticides are mainly prepared from naturally occurring constituents such as diatomaceous earth, tea tree oil, neem oil, canola oil, baking soda, cayenne pepper, and other compounds to kills pests. These substances are naturally formed by a plant or an organism. They are nontoxic as well as biodegradable in nature. Lethal but nontoxic biochemical pesticides include abrasives (e.g., diatomaceous earth), desiccants (e.g., acetic acid), suffocating agents (e.g., soybean oil).

Microbial pesticides

This class of biopesticides comprises of microorganisms (e.g., a fungus, bacterium, virus, or protozoan) as the active component. Microbiological pesticides can control several dissimilar kinds of pests, although each separate active constituent is reasonably specific for its target pest. For example, there are some fungi that control certain weeds and other fungi that kill definite insects. The most extensively used microbial pesticides are subspecies and strains of *B. thuringiensis*, or Bt, and each strain of this bacterium yields a different mix of proteins that specifically kills one or a few associated species of insect larvae. Some Bt ingredients control moth larvae found on plants while other Bt ingredients are specific for larvae of flies and mosquitoes. The target insect species are determined by whether the particular Bt produces a protein which can bind to a larval gut receptor and thereby causing the insect larvae to starve [16].

Bacterial biopesticides. Several bacterial pathogens of different insects are being used as insecticides. These are *Pseudomonas, Clostridium, Bacillus, Proteus, Enterobacter, Serratia*, etc. *B. thuringiensis*, or Bt, is one of the most broadly used bacterial biopesticides against an extensive variety of more than 150 insects. The bacterium basically kills insect larva by generating a toxin that binds to the larval stomach cells. With the help of recombinant DNA technology, the gene having insecticidal characteristics of *B. thuringiensis* has been transferred to the crop plants like tomato in 1987, with the help of *Agrobacterium tumefaciens*. Similar accomplishment has been achieved in different crop plants like cotton, tobacco etc., by using the related technique. Nowadays, Bt cotton is very common among the farmers [4,5]. Table 10.1 presents different examples of bacterial biopesticides.

Bacterial bioherbicides. The application of bacteria as bioherbicide in the control of weeds has received more attention due to the many advantages it has over other bioherbicides. Bacteria as agents of biocontrol are believed to multiply rapidly, possess genetic

makeup, can be easily modified, and require simple and easily available nutrients for their growth. There are different genera of bacteria that have been reported as potential bioherbicides in the control of weeds of agricultural crops. However, the genera *Pseudomonas* and *Xanthomonas* have received great attention [19]. Table 10.2 shows that two members of the genus *Pseudomonas*: *Pseudomonas fluorescens* strain D7 and *P. fluorescens* strain BRG100 have been reported to have suppressed the growth of *Bromus tectorum* and *Setaria viridis*, respectively [21,22]. Imaizumi et al. [23] also reported the use of *Xanthomonas campestris* cv. *poae* JT-P482 and *X. campestris* LVA-987 as biocontrol agent of weeds, *Poa annua* and *Conyza canadensis*, respectively. The mechanism of operation of most of these bacteria used as bioherbicide is their production of metabolites which upsets the germination and growth of plants [21]. Table 10.2 presents different examples of bacterial bioherbicides.

Fungal biopesticides

Entomopathogenic fungi are reflected to play a vital role as biocontrol agent of insect populations. A very diverse range of fungal species is found from several different groups that infect insects. These biopesticides do not have to be ingested to inhibit or kill target pests; physical contact is enough. Fungal biopesticides typically consist of fungal spores and are very simple to apply. Soil fungus *Beauveria bassiana* controls a varied array of pests, killing them within a few days. Fungi in the genus *Trichoderma* is used to manage plant diseases by parasitizing harmful fungi in the root zone of plants and improving plant growth and resistance. At the recent period, about 90 genera and roughly above 700 species are considered as insect infecting fungi [24,25]. A group of fungi that basically kill or inactivate an insect by attacking and infecting its insect host is known as entomopathogenic fungi [26]. The key path

TABLE 10.1
Examples of Bacterial Biopesticides

Bt Array	Object Pest
Bacillus sphaericus	Mosquito larvae
Bacillus thuringiensis subsp. *kurstaki*	Lepidopteran larvae
Bacillus moritai	Diptera
Bacillus thuringiensis subsp. *galleriae*	Lepidopteran larvae

Source: Y. Kunimi. Current status and prospects on microbial control in Japan. J. Invertebr. Pathol. (95) (2007) 181−186; J.T. Kabaluk, A.M. Svircev, M.S. Goette, S.G. Woo. The use and regulation of microbial pesticides in representative jurisdictions worldwide. IOBC Global (2010) 99.

TABLE 10.2
Examples of Bacterial Bioherbicides

Bacterial Array	Object Weed
Pseudomonas fluorescens strain D7	*Bromus tectorum*
P. fluorescens strain BRG100	*Setaria viridis*
Xanthomonas campestris cv. *poae* (JT-P482)	*Poa annua* and *Poa attenuate*
X. campestris LVA-987	*Conyza canadensis*

Source: D.P. Harding and M.N. Raizada. Controlling weeds with fungi, bacteria and viruses: a review. Front. Plant Sci. 6 (2015) 659.

TABLE 10.3
Examples of Fungal Biopesticides

Fungal Array	Target Organism/Weed
Hirsutella thompsonii	Phytophagous mites
Fusarium lateritium	Velvet leaf (weed)
Verticillium lecanii	Aphid, whiteflies, and scales
Alternaria cassia	Sickle pod (weed)
Phytophthora palmivora	Milk weed vine (weed)

Source: G. Zimmermann. The entomopathogenic fungus Metarhizium anisopliae and its potential as a biocontrol agent. Pestic. Sci. 37 (1993) 375—379.

TABLE 10.5
Examples of Viral Biopesticides

Viral Array	Host Object
Pox virus	Amsacta moorei
Granulosis virus	Cnaphalocrocis medinalis, Pericallia ricini, Achaea janata, Phthorimaea operculella, and Chilo infuscatellus
Cytoplasmic polyhedrosis virus	Helicoverpa armigera

Source: N. Ramakrishnan and S.K. Kumar. Biological control of insects by pathogen and nematodes, Pesticides. (1976) 32—47.

of entry of the entomopathogen is through integument and it may also infect the insect by ingestion or through the trachea or wounds [27]. Table 10.3 presents different examples of fungal biopesticides.

Fungal bioherbicides

Some fungi have been employed as biocontrol agents in the control of weeds. The use of fungi as bioherbicides has been receiving attention over the years in the control of weeds of crops and turfs (Table 10.4). Some fungi of the genus *Colletotrichum* have been used as commercial formulations in the control of certain weeds. Mortensen [29] and Menaria [30] reported the use of *Colletotrichum gloeosporioides* f. sp. *malvae* and *C. gloeosporioides* f. sp. *aeschynomene* as effective agent for the control of *Malva pusilla* and *Aeschynomene virginica*, respectively. Furthermore, the application of *C. truncatum* and *C. orbiculare* in the control of *Sesbania*

TABLE 10.4
Examples of Fungal Bioherbicides

Fungal Array	Object Weed
Colletotrichum gloeosporioides f. sp. Cuscuta	Cuscuta species
C. gloeosporioides f. sp. Malvae	Malva pusilla
Phoma herbarum	Taraxacum officinale
Puccinia canaliculata	Cyperus esculentus
Puccinia thlaspeos	Isatis tinctorial
Alternaria destruens	Dodder species
Phytophthora palmivora	Morrenia odorata

Source: D.P. Harding and M.N. Raizada. Controlling weeds with fungi, bacteria and viruses: a review. Front. Plant Sci. 6 (2015) 659.

exaltata and *Xanthium spinosum* has been established [31]. In turf management, some fungi of the genus *Sclerotinia* has been found very useful in the control of weeds associated with turfs. It was reported that *Sclerotinia minor* was effective in the control *Taraxacum officinal*, *Trifolium repens*, and *Plantago major* [32]. The mechanism of action of these bioherbicides is the production of some acids by the bioherbicides which break down the cell wall and the defense mechanism of the weeds. In addition, these acids suppress the release of polyphenol oxidase which plays an important role in the suppression of the defense mechanism of weeds [33]. Table 10.4 presents different examples of fungal bioherbicides.

Viral biopesticides

Viruses have been isolated from more than 1000 species of insects from at least 13 different insect orders [34]. Endogenous viruses mainly fall into two classes, viz. inclusion viruses (IV) generating inclusion bodies in the host cells and noninclusion viruses (NIV) which do not generate inclusion bodies. The insect viruses found in nature belonging to the baculovirus family (*Baculoviridae*) were considered for the improvement of most commercial viral biopesticides [35—37]. Members of this family are considered as safe for vertebrates and no cases of pathogenicity of a baculovirus to a vertebrate have been reported till date [38,39]. Table 10.5 presents different examples of viral biopesticides.

Viral bioherbicides

The application of viruses as bioherbicides in the management of weeds of agricultural crops has been established (Table 10.6). Studies have shown that Tobacco mild green mosaic possesses the ability to curb the growth and activities of *Solanum vivarium* [13]. Furthermore, Elliot et al. [41] reported on the use of Araujo

TABLE 10.6
Examples of Viral Bioherbicides

Virus	Object Weed
Tobacco mild green mosaic tobamovirus	*Solanum vivarium*
Araujia mosaic virus	*Araujia hortorum*
Tobacco rattle-like virus	*Impatiens glandulifera*
Obuda pepper virus	*Solanum nigrum*
Pepino mosaic virus	*S. nigrum*

Source: D.P. Harding and M.N. Raizada. Controlling weeds with fungi, bacteria and viruses: a review. Front. Plant Sci. 6 (2015) 659.

TABLE 10.7
Examples of Nematodes Biopesticides

Nematodes Array	Host Organism
Steinernema scapterisci	Mole crickets (*Scapteriscus* spp.)
Heterorhabditis megidis	Weevils
Heterorhabditis zealandica	Scarab grubs
Heterorhabditis indica	Fungus gnats, root mealy bug, grubs

Source: N. Tofangsazi, S.P. Arthurs, R.M.G. Davis. Entomopathogenic Nematodes (Nematoda: Rhabditida: Families Steinernematidae and Heterorhabditidae). One of a Series of the Entomology and Nematology Department, UF/IFAS Extension (2015) 1—5.

mosaic virus in the effective control *Araujo moratorium*. Another virus of interest that is being considered as a bioherbicide in the control of *Impatiens glandulifera* is Tobacco rattle virus [42]. The report of Kazinczi et al. [43] also showed the efficacy and potentials of Obuda pepper virus and Pepino mosaic virus in the control of *Solanum nigrum*. These viruses resulted in a significant reduction of the weed populations which would have otherwise competed with the agricultural crops of interest. The result of the application of these viral bioherbicides is the prevention of losses and resultant increase in yields and quality of agricultural produce (Table 10.6).

Nematodes biopesticides
Entomopathogenic nematodes are soft bodied, no segmented roundworms that are obligate or sometimes facultative parasites of insects. Entomopathogenic nematodes occur naturally in soil and trace their host in response to carbon dioxide, vibration, and other chemical cues [44]. Species in two families (*Heterorhabditidae* and *Steinernematidae*) have been efficiently used as biological insecticides in pest management programs [45]. Entomopathogenic nematodes fit agreeably into integrated pest management programs since they are considered nontoxic to human beings, comparatively precise to their target pests, and can be effectively applied with standard pesticide equipment [46]. Entomopathogenic nematodes have been exempted from the US EPA pesticide registration. Table 10.7 presents different examples of nematodes biopesticides.

Plant-incorporated-protectants
These are certain pesticidal constituents that are produced by plants from genetic material that has been added to the plant. For example, scientists can take the gene for the Bt pesticidal protein and introduce

the gene into the plant's own genetic material. Then the plant, instead of the Bt bacterium, manufactures the substance that destroys the pest. The protein and its genetic material, but not the plant itself, are regulated by EPA.

Commercialization of Biopesticide and Bioherbicides
There are about 1400 biopesticides products produced worldwide [48]. To put this into context, these biopesticide products signify just 2.5% of the total pesticide market [7]. However, the market may need to increase considerably more than this if biopesticides and bioherbicides are to play a full role in minimizing our overreliance on synthetic chemicals for control of pest and weeds. Similarly, the decision for a farmer whether or not to adopt a novel technology can be thought of in economic terms as a cost-benefit comparison of the profits to be made from using then oval versus the incumbent technology. Several different features of the agricultural economy make it very complicated for companies to invest in developing new biopesticide and bioherbicide products and at the same time, make it difficult for farmers to decide about accepting and adopting the new technology. Although biopesticide and bioherbicide use at a global scale is increasing by almost 10% every year, it appears that the global market must increase further in the future if such kind of biological control agents are to play a noticeable role in replacement for synthetic chemicals and reducing the current overdependence on them [49]. It should be observed, however, that biopesticides and bioherbicides are assessed in the EU by the same regulations used for the assessment of synthetic active

substances, and this condition required numerous new provisions in the current legislation, as well as the preparation of new guidelines facilitating the registration of prospective biopesticides and bioherbicides products. Currently, there are fewer biopesticide and bioherbicide active substances registered in the EU than in the United States, India, Brazil, or China. The relatively low level of research on these biological control agents in the EU is related to the greater complexity of EU-based regulations on these products [49]. The development of biopesticide and bioherbicide market in the future is strongly related to research on biological control agents. A number of scientists from diverse research institutes have done some primary research in the field, but complete and systematic reports are scarce. Therefore, it is essential to strengthen the collaboration of enterprises and research institutes on this topic. It seems that biopesticides and bioherbicides cannot as yet fully replace chemical pesticides and herbicides, so the agricultural and forestry sector can and should benefit from the coexistence of biopesticides with chemical pesticides. In this regard, accelerating practical application of research results is anticipated to facilitate large-scale industrial development.

CONCLUSION

Negative effects of chemical pesticides and herbicides on nature and natural resources like pollution, pesticide and herbicide residue, pest and weed resistance, etc. have forced many to shift focus onto more reliable, sustainable, and eco-friendly agents of biocontrol—the biopesticides and bioherbicides. Microbial control agents of pests and weeds are being familiarized in this new scenario of crop protection and presently several valuable microorganisms are the active constituents of a new generation of microbiological pesticides and herbicides. Entomopathogenic viruses, bacteria, and fungi are currently used as alternatives to traditional insecticides and herbicides. The microorganism provides certain distinct advantages over many other control agents and methods. The major advantage of exploiting microorganism for the control of pests and weeds is their environmental safety primarily due to the host specificity of these pathogens. Microorganisms have natural potential of causing disease at epizootic levels due to their persistence in soil and efficient transmission. Governments are likely to carry on imposing strict safety standards on conventional chemical control agents, and this will result in fewer products on the market. This will create a real opportunity for biopesticide and bioherbicide companies to help fill the gap, although there will also

be critical challenges for these companies, most of which are small and medium enterprises with inadequate resources for Research and Development, product registration, and promotion. The future of microbial pesticides is not only in improving new active ingredients based on microorganisms advantageous to plants, but in generating self-protected plants by transforming agronomically high-value crop plants with genes from biocontrol agents.

LIST OF ABBREVIATION

IPM	integrated pest management
Bt	*Bacillus thuringiensis*
EPA	Environmental Protection Agency
DNA	Deoxyribonucleic acid
IV	inclusion viruses
NIV	noninclusion viruses
PIPs	plant-incorporated protectants

REFERENCES

[1] L. Bastian's, R. Paolini, D.T. Baumann, Focus on ecological weed management: what is hindering adoption? Weed Res. 48 (2008) 481–491.

[2] E.C. Oerke, H.W. Dehne, F. Schoenbeck, A. Weber, Crop Production and Crop Protection: Estimated Losses in Major Food and Cash Crops, Elsevier Science Publishers B.V., Amsterdam, The Netherlands, 1994.

[3] C.I. Speranza, B. Kiteme, U. Wiesmann, Droughts and famines: the underlying factors and the mcausal links among agro-pastoral households in semiarid Makueni district, Kenya, Glob. Environ. Chang. 18 (2008) 220–233.

[4] G. Gadermaier, M. Hauser, F. Ferreira, Allergens of weed pollen: an overview on recombinant and natural molecules, Methods 66 (2014) 55–66.

[5] J. Mithila, J.C. Hall, W.G. Johnson, K.B. Kelley, D.E. Riechers, Evolution of resistance to euxinic herbicides: historical perspectives, mechanisms of resistance, and implications for broadleaf weed management in agronomic crops, Weed Sci. 59 (2011) 445–457.

[6] J.M. Green, M.D.K. Owen, Herbicide-resistant crops: utilities and limitations for herbicide-resistant weed management, J. Agric. Food Chem. 59 (2011) 5819–5829.

[7] K.L. Bailey, S.M. Boyetchko, T. Langle, Social and economic drivers shaping the future of biological control: a Canadian perspective on the factors affecting the development and use of microbial biopesticides, Biol. Control 52 (2010) 221–229.

[8] C.R. Weeden, A.M. Shelton, M.P. Hoffman, Biological Control: A Guide to Natural Enemies in North America, Cornell University College of Agriculture and Life Sciences, 2007.

[9] J.J. Menn, F.R. Hall, Biopesticides — present status and future prospects, in: J.J. Menn, F.R. Hall (Eds.), Methods in Biotechnology, Biopesticides Use and Delivery, vol. 5, Humana Press, Totowa, NJ, 2001, pp. 1–9.

[10] L.A. Falcon, R.T. Hess, Electron microscope observations of multiple occluded virions in the granulosis virus of the codling moth, *Cydia pomonella*, Hess. J. Invertebr Pathol 45 (1985) 356–359.

[11] J.A. Jehle, G.W. Blissard, B.C. Bonning, J.S. Cory, E.A. Herniou, G.F. Rohrmann, D.A. Theilmann, S.M. Thiem, J.M. Vlak, On the classification and nomenclature of baculoviruses: a proposal for revision, Arch. Virol. 15 (2006) 1257–1266.

[12] Y.Q. Li, Z.L. Sun, X.F. Zhuang, L. Xu, S.F. Chen, M.Z. Li, Research progress on microbial herbicides, Crop Protect. 22 (2003) 247–252.

[13] R. Diaz, V. Manrique, K. Hibbard, A. Fox, A. Roda, D. Gandolfo, Successful biological control of tropical soda apple (*Solanales: Solanaceae*) in Florida: a review of key program components, Fla. Entomol. 97 (2014) 179–190.

[14] P. Bazoche, P. Combris, E. Giraud-Heraud, A. Pinto, F. Bunte, E. Tsakiridou, Willingness to pay for pesticide reduction in the EU: nothing but organic? Eur. Rev. Agric. Econ. 41 (2014) 87–109.

[15] A. Mitra, A review on the biological control of plant diseases using various microorganisms, J. Res. Med. Dent. Sci. 6 (4) (2018) 30–35.

[16] G.R. Stirling, Biological control of plant-parasitic nematodes, in: Diseases of Nematodes, CRC Press, 2017, pp. 103–150.

[17] Y. Kunimi, Current status and prospects on microbial control in Japan, J. Invertebr. Pathol. 95 (2007) 181–186.

[18] J.T. Kabaluk, A.M. Svircev, M.S. Goette, S.G. Woo, The use and regulation of microbial pesticides in representative jurisdictions worldwide, IOBC Global (2010) 99.

[19] G.M. Banowetz, M.D. Azevedo, D.J. Armstrong, A.B. Halgren, D.I. Mills, Germination-Arrest Factor (GAF): biological properties of a novel, naturally-occurring herbicide produced by selected isolates of rhizosphere bacteria, Biol. Control 46 (2008) 380–390.

[20] D.P. Harding, M.N. Raizada, Controlling weeds with fungi, bacteria and viruses: a review, Front. Plant Sci. 6 (2015) 659.

[21] J.W. Quail, N. Ismail, M.S.C. Pedras, S.M. Boyetchko, Pseudophomins A and B, a class of cyclic lipodepsi peptides isolated from a *Pseudomonas* species, Acta Crystallogr. C Crystal Struct. Commune 58 (2002) 268–271.

[22] A.C. Kennedy, B.N. Johnson, T.L. Stubbs, Host range of a deleterious rhizobacterium for biological control of downy brome, Weed Sci. 49 (2001) 792–797.

[23] S. Imaizumi, T. Nishino, K. Miyabe, T. Fujimori, M. Yamada, Biological control of annual bluegrass (*Poa annua* L.) with a Japanese isolate of *Xanthomonas campestris* cv. *poae* (JT-P482), Biol. Control 8 (1997) 7–14.

[24] E.R. Moorhouse, A.T. Gillespie, E.K. Sellers, A.K. Charnley, Influence of fungicides and insecticides on the endogenous fungus *Metarhizium anisopliae* a pathogen of the vine weevil, *Otiorhynchus sulcatus*, Biocontrol Sci. Technol. 2 (1) (1992) 49–58.

[25] A.E. Hajek, R.J. Leger, Interactions between fungal pathogens and insect hosts, Annu. Rev. Entomol. 39 (1) (1994) 293–322.

[26] S. Singkaravanit, H. Kinoshita, F. Ihara, T. Nihira, Cloning and functional analysis of the second geranylgeranyl diphosphate syntheses gene influencing helvellic acid biosynthesis in *Metarhizium anisopliae*, Appl. Microbiol. Biotechnol. 87 (3) (2010) 1077–1088.

[27] D.J. Holder, N.O. Keyhani, Adhesion of the entomopathogenic fungus *Beauveria* (*Cordyceps*) *bassiana* to substrata, Appl. Environ. Microbiol. 71 (9) (2005) 5260–5266.

[28] G. Zimmermann, The entomopathogenic fungus *Metarhizium anisopliae* and its potential as a biocontrol agent, Pestic. Sci. 37 (1993) 375–379.

[29] K. Mortensen, The potential of an endemic fungus, *Colletotrichum gloeosporioides*, for biological control of round-leaved mallow (*Malva pusilla*) and velvetleaf (*Abutilon theophrasti*), Weed Sci. 36 (1988) 473–478.

[30] B.L. Menaria, Bioherbicides: an eco-friendly approach to weed management, Curr. Sci. 92 (2007) 10–11.

[31] B.A. Auld, M.M. Say, H.I. Ridings, J. Andrews, Field applications of *Colletotrichum orbiculare* to control *Xanthium spinosum*, Agric. Ecosyst. Environ. 32 (1990) 315–323.

[32] PMRA, *Sclerotinia Minor* Strain IMI 344141. RD2010-08. Health Canada, Health Canada, Ottawa, ON, 2010, p. 50.

[33] S.G. Cessna, V.E. Sears, M.B. Dickman, P.S. Low, Oxalic acid, a pathogenicity factor for *Sclerotinia sclerotiorum*, suppresses the oxidative burst of the host plant, Plant Cell 12 (2000) 2191–2199.

[34] K.P. Srivastava, G.S. Dhaliwal, A Textbook of Applied Entomology, Kalyani Publishers, New Delhi, 2010, p. 113.

[35] P.F. Entwistle, H.F. Evans, Viral control, in: L.I. Gilbert, G.A. Kerkut (Eds.), Conprehensive Insect Physiology. Biochemistry and Pharmacology, vol. 12, Pergamon Press, Oxford, UK, 1985, pp. 347–412.

[36] R.R. Granados, B.A. Federici, Introduction: historical perspective, in: The Biology of Bocaviruses, Biological Properties and Molecular Biology, CRC Press, Boca Raton, FL, USA, 1986, pp. 1–35, 1.

[37] N.F. Moore, L.A. King, R.D. Posse, Viruses of insects, Insect Sci. Appl. 3 (1987) 275–289.

[38] A. Krieg, J.M. Franz, A. Groner, J. Huber, H.G. Miltenburger, Safety of entomopathogenic viruses for control of insect pests, Environ. Conserv. 7 (1980) 158–160.

[39] P.F. Entwistle, Viruses for insect pest control, Span 26 (1983) 59–62.

[40] N. Ramakrishnan, S.K. Kumar, Biological control of insects by pathogen and nematodes, Pesticides (1976) 32–47.

[41] M.S. Elliott, B. Massey, X. Cui, E. Hiebert, R. Charudattan, N. Waipara, Supplemental host range of *Araújo mosaic virus*, a potential biological control agent of moth plant in New Zealand, Australas. Plant Pathol. 38 (2009) 603–607.

[42] J. Kollmann, M.J. Banuelos, S.L. Nielsen, Effects of virus infection on growth of the invasive alien *Impatiens glandulifera*, Preslia 79 (2007) 33–44.

[43] G. Kazinczi, D. Lukacs, A. Takacs, J. Horvath, R. Gaborjanyi, M. Nadasy, Biological decline of *Solanum nigrum* due to virus infections, J. Plant Dis. Prot. 32 (2006) 5–330.

[44] H.K. Kaya, R. Gauged, Entomopathogenic nematodes, Annu. Rev. Entomol. 38 (1993) 181–206.

[45] P.S. Grewal, R.U. Ehlers, D.I. Shapiro Ilan, Nematodes as Biocontrol Agents, CABI, New York, 2005, pp. 79–90.

[46] D. Shapiro-Ilan, D.H. Gough, S.J. Piggott, F.J. Patterson, Application technology and environmental considerations for use of entomopathogenic nematodes in biological control, Biol. Control 38 (2006) 124–133.

[47] N. Tofangsazi, S.P. Arthurs, R.M.G. Davis, Entomopathogenic Nematodes (Nematoda: Rhabditida: Families Steinernematidae and Heterorhabditidae), in: One of a Series of the Entomology and Nematology Department, UF/IFAS Extension, 2015, pp. 1–5.

[48] P.G. Marrone, Barriers to Adoption of Biological Control Agents and Biological Pesticides, in: CAB Reviews: Perspectives in Agriculture, Veterinary Science, Nutrition and Natural Resources, vol. 2(15), CABI publishing, Wallingford, UK, 2017.

[49] S. Kumar, A. Singh, Biopesticides: present status and the future prospects, J Fertil. Pestic 6 (2015) 129.

CHAPTER 11

Biological Control of Plant Pests by Endophytic Microorganisms

ALLOYSIUS CHIBUIKE OGODO, BSC, MSC, PHD

INTRODUCTION

The term "endophyte" is used to refer to interior colonization of plants by microorganisms (bacteria or fungi). In other words, endophytic microorganisms grow in the intercellular spaces of higher plants [1]. They have been found to exist in all parts of plant, including xylem and phloem [2]. Endophytic microorganisms comprise mainly bacteria and fungi that while colonizing the internal tissues plants do not cause visible damage to their host. However, they may contribute to the well-being of the plant while forming colonies and biofilms within the plant tissues. The host plant provides nutrients to the endophytic microorganisms, while the microbes may produce substances that help protect the host plant from attack by other microbes, animals, or insects [3−5].

Organisms that cause plant disease including fungi pose serious threat to sustainable food production. Moreover, the use of chemicals such as pesticides, fungicides, and other chemicals leaves residues on either the leaves, fruits, or other parts of plant meant for consumption. On this regard, there is need for safer and more effective alternative. Endophytic microorganisms have received much attention recently as an alternative to chemical control of pest. It is an option to host-plant resistance to pesticide-based pest and pathogen control and has been recognized as one of the most promising group of microorganisms in terms of diversity and pharmacological potentials [1,6]. Variable bioactive compounds (metabolites) such as antiviral, anticancer, antidiabetic, and antibacterial compounds have been found in endophytes [2].

Endophytes penetrate and become disseminated in the host plant and colonize the apoplast, conducting vessels, and intracellular space. Their colonization presents an ecological niche similar to that occupied by plant pathogens. Therefore, the endophytic microorganisms can act as biological control of the pathogens by suppressing their growth [1,7]. To achieve this, the endophytes act directly on the pathogens by antibiosis; competition for food, nutrients, and space; or inducing plant resistance response as well as production of bioactive compounds [8−11].

The use of bacterial endophytes in agriculture has immense potential to reduce the environmental adverse effects of chemical fertilizers. Bacterial endophyte synthesizes phytohormones (e.g., indole-3-acetic acid, cytokinins, and gibberellins) and siderophores to promote plant growth or it can promote growth through the regulation of internal hormones in the plant [12,13]. Moreover, endophytic strains of *Bacillus*, *Burkholderia*, *Enterobacter*, *Pseudomonas*, and *Serratia* were found to be effective in suppressing the growth of pathogenic microorganisms in *vivo* and in *vitro* while endophyte strains of the genera *Bacillus*, *Enterobacter*, *Pseudomonas*, *Azotobacter*, *Arthrobacter*, *Streptomyces*, and *Isoptericola* alleviate drought, heat, and salt stress in different crops [14−16].

ENDOPHYTIC MICROORGANISMS

Microorganism that at least for one period of their life cycle inhabit the interior of a higher plant may be considered as an endophyte. These organisms are different from epiphytes (organisms that live on the surface of plants) and phytopathogens (organisms that cause plant diseases to) in that they inhabit the interior of plants and do not cause any noticeable damage to the plant [17]. Nitrogen-fixing bacteria and mycorrhizae that relate intimately with their hosts could be regarded as endophytes, although mycorrhizae fungi and some nitrogen-fixing bacteria such as *Rhizobium* possess hyphae and nodules, respectively, as external structures. Endophytic microorganisms started receiving more considerable attention when their capacity to protect their host plants against insects-pests, phytopathogens

Natural Remedies for Pest, Disease and Weed Control. https://doi.org/10.1016/B978-0-12-819304-4.00011-7

as well as domestic herbivores such as sheep and cattle was recognized. The continued study of these endophytic microorganisms (fungi and bacteria) made it clear that they could confer some characteristics to plants, including greater resistance to stress conditions (i.e., water), alteration in physiological properties, production of phytohormones and other compounds of biotechnological interest (such as enzymes and pharmaceutical drugs) [17].

Several endophytic bacteria and fungi have been isolated from different plants including coffee, beans, potato, cotton, tomato, maize, cocoa, banana, date palm, rice, lettuce, lemon, etc. Some of the fungal and bacteria genera include *Acremonium, Beauveria, Cladosporium, Clonostachys, Paecilomyces, Lecanicillium, Verticillium, Isaria, Achromobacter, Acinetobacter baumannii, Alcaligenes, Moraxella, Arthrobacter, Bacillus, Burkholderia, Citrobacter, Corynebacterium, Curtobacterium, Enterobacter, Methylobacterium, Pantoea,* and *Pseudomonas* [1,18−21]. These organisms have been shown to be potential biocontrol endophytes.

Biological Control Potential of Endophytic Microorganisms

According to Nair and Padmavathy [22], endophytic microorganisms are regarded as an effective biocontrol agent, alternative to chemical control reports of many researchers bothering on the potential use of endophytic microorganisms as biocontrol agent against insect-pests. In the review, endophytic fungi play an important role in controlling insect herbivory in grasses and conifers. For example, *Beauveria bassiana,* an endophytic fungus, controls borer insect in coffee seedlings and sorghum; *Bacillus subtilis,* an endophytic bacterium isolated from *Speranskia tuberculata* (Bunge.) Baill, was found to exhibit antagonistic activity against the fungal pathogen, *Botrytis cinerea in-vitro*; a new strain of *Burkholderia pyrrocinia* JK-SH007 and *Burkholderia cepacia* were identified as potential biocontrol agent against poplar canker.

Naturally occurring endophytes are used as biocontrol agents. However, they are also engineered genetically to express antipest proteins like lectins. Attempts have been made to introduce heterologous gene into an endophytic microorganism for insect control [23]. Fungal endophyte of *Chaetomium globosum* YY-11 with antifungal activities, isolated from rape seedlings, and bacterial endophytes of *Enterobacter* species and *B. subtilis* isolated from rice seedlings were used to express *Pinellia ternata* agglutinin (PtA) gene. These recombinant endophytes expressing PtA gene were found to effectively control the population of sap-

sucking pests in several crop seedlings. Similarly, in a different study, recombinant endophytic bacteria *Enterobacter cloacae* expressing PtA gene proved to be a bioinsecticide against white backed planthopper, *Sogatella furcifera* (Horváth). Use of recombinant endophytes as biocontrol agents expressing different antipest proteins becomes a promising technique for control of plant pests as these endophytes can easily colonize within different crop plants successfully [22,24,25].

Control of Insect-pests by Endophytic Microorganisms

The first report indicating that endophytic microorganisms especially fungi could play an important role in reduction of insect attack to their host plants has been published as early as 1980s. In 1981, Webber [26] reported the protection of elm trees against the beetle, *Physocnemum brevilineum* by endophytic fungi, *Phomopsis oblonga*. In this report, *P. oblonga* was considered to be responsible for reducing the spread of *Ceratocystis ulmi,* the causative agent of elm Dutch disease due to its ability to control the vector *P. brevilineum*. Here there was a correlation between the repellant effects observed toward the insect to toxic substances produced by the fungi [17]. Four years later, Claydon et al. [27] reported that secondary metabolites from endophytic fungi of the family *Xylariaceae, in the host, Fagus* are effective against the larvae of beetles. Moreover, high fungi infection of ryegrass results to decrease in the attack of *Listronotus bonariensis* (Argentine stem weevil) with aphid and grasshopper being affected by endophytic fungi *Acremonium* [28−30].

The process of plant protection by endophytes is complex. It may involve a mutualistic interaction or relationship between the host plant and the endophyte with resultant negative effect on the insect-pests or it may involve neutral, symbiotic, trophobiotic, and in some cases antagonistic interactions [17,31]. Plants infected with endophytic fungi have less infection of billbug *Sphenophorus parvulus* [32]. The reduction in attach of aphid, *Diuraphis noxia* in *Lolium* and *Festuca* by *Acremonium* has been reported [33]. Similarly, Soria et al. [34] demonstrated the use of endophytic bacteria from *Pinus taeda* L., as biocontrol agents of *Fusarium circinatum* (pitcher canker fungus). It was recorded from their study that *B. subtilis* strains and one *Burkholderia* species isolated as *P. taeda* endophytes showed antagonistic effect on the growth of *Fusarium circinatum,* arresting mycelia at 1 cm of the bacteria colony while the thermostable metabolites from the bacteria reduced the fungal growth by more than 50%.

Endophytic Bacteria in the Control of Pest Insects

Isolates of *Kluyvera ascorbata* and *Alcaligenes piechaudii* reduce the viability of *Plutella xylostella* to about 80% to 50%, respectively. Similarly, strains of *Bacillus thuringiensis* from Brazil caused 100% mortality in caterpillars in the 3° instar of *P. xylostella* after 48 h, more than 75% mortality against *Diatraea saccharalis* (Lepidoptera: Crambidae), more than 50% mortality against cotton aphid (*Aphis gossypii*), more than 75% mortalities against caterpillars of *Spodoptera frugiperda* [35−39].

Endophytes can be combined with other microorganisms to control pest insects in Agriculture. For example, Broderick et al. [40] reported an increase of 35% in the mortality of *Lepidopterous lymantria* dispar (L) when *B. thuringiensis* and intermixing of *Bacillus cereus* were combined to produce synergistic effect. A synergistic effect of a commercial product of *B. thuringiensis* and the fungus *B. bassiana* were used simultaneously in *Leptinotarsa decemlineata*. The effect was attributed to the intoxication caused by entomopathogen, which inhibits insect feeding leading to stress and physiological effects that aid fungal penetration in the insect [41]. Similar effect has been reported against the larvae of *Ostrinia furnacalis* [42].

Effects of Entophytic Microorganisms toward Pathogens

Endophytic microorganisms have the capacity to control plant pathogens and nematodes. Several bacterial endophytes have been reported to support growth and improve the health of plants and therefore may be important sources of biocontrol agents. For example, *Erwinia carotovora* is inhibited by endophytic bacteria, including strains of *Pseudomonas* sp., *Curtobacterium luteum*, and *Pantoea agglomerans* while *B. subtilis* strains isolated from the xylem sap of healthy chestnut trees exhibited antifungal effects against *Cryphonectria parasitica* that causes chestnut blight disease. Moreover, endophytes from potato stem tissues have been shown to be effective against *Clavibacter michiganensis* [9,43]. This is a very important discovery for phytopathologists, and above all for agriculture, because it was proved for the first time that this factor can limit potato quarantine disease.

Biologically active endophytes and some root-colonizing microorganisms belong to the phylum, actinobacteria, and genus *Streptomyces*. A number of endophytic actinobacteria belonging to the genera *Streptomyces*, *Microbispora*, *Micromonospora*, and *Nocardioides* are capable of suppressing fungal pathogens of wheat such as *Rhizoctonia solani*, *Pythium* species, and *Gaeumannomyces graminis var tritici* in vitro. These show their potential use as biocontrol agents [44,45].

Endophytic Microorganisms Affecting Parasites of Domestic Animals

Therefore, it is important to verify if endophytes that reduce insects-pest can also interfere with the latter's biological controls. The report of Azevedo et al. [17] indicated that *A. lolii* infecting *Lolium perenne* did not affect *M. hyperbolae*, facilitating its performance as a biological control agent of the insect-pest that was fed also on infected plants. The development of the larvae is in hosts fed with artificial diet containing alkaloids and diterpene produced by *A. loci*. Similarly, Bultman et al. [46] reported that in the interaction of *F. arundinacea*, *A. coenophialum*, alkaloids produced by the fungus interfered on the fall armyworm *S. frugiperda* and also on two parasites that control this pest, *Euplectrus comstockii* and *Euplectrus plathypenae*. The fungus has a moderate negative effect on the parasites that feed on *S. frugiperda* larvae.

Haematobia irritans larvae of horn fly, which is a cattle ectoparasite, are reported to have been killed when cattle manure was amended with seed extracts containing lolines from plants infected with *N. monophylum*. These effects will probably also occur on insects that feed on manure from animal fed on *F. arundinacea* infected with the endophyte *N. coenophialum* [17,47].

MECHANISMS OF DISEASES/PEST CONTROL DISPLAYED BY ENDOPHYTIC MICROORGANISMS

The suppression of plant diseases by endophytic microorganisms are achieved through two major mechanisms: directly on the pathogen inside the plant by antibiosis and competition for nutrients, or indirectly by induction of plant resistance response [1].

Endophytes are source of bioactive natural products due to the fact that there are so many of them occupying different niches in higher plants growing in so many unusual environments. These natural products are important in the selection of plant as they are responsible biological activity of the products associated with endophytic microorganisms. Terpene production and Peppermint growth of *Mentha piperita* generated in-vitro in response to inoculation with a leaf fungal endophyte is an indication of essential oil profile by fungal infection. Moreover, the weight of roots, seedlings, and terpenoid production of *Euphorbia pekinensis* are shown to have increased after inoculation with the endophyte, *Phomopsis* sp [1]. Microbial elicitor obtained from endophytic fungi induces the biosynthesis and production of terpenoids (artemisinin) in plant suspension cells. More so, endophytic actinomycetes may also affect plant growth either by nutrient assimilation or enhanced secondary metabolites (anthocyanin) synthesis [48,49].

Endophytes can have microbiocidal property against certain fungi and bacteria by producing certain mixture volatile compounds which can be identified, synthesized, and made into artificial mixture with antimicrobial effect mimicking volatile compounds produced by the fungus [50]. In the Northern Territory of Australia, newly described *Muscod* was obtained from tree species and was effective against other microorganisms *in-vitro*. Similarly, the endophytic *Streptomycete* (NRRL 30566), from a fern-leaved *Grevillea pteridifolia*, produces a novel antibiotics in-vitro called kakadumycins which are shown to contain alanine, serine, and an unknown amino acid [51,52]. A metabolite of *Colletotrichum gloeosporioides*, an endophytic fungus in *Artemisia mongolica*, known as colletotric acid, displays antimicrobial activity against bacteria as well as against the fungus *Helminthosporium sativum* [53]. A number of compounds produced by cultures of *Epichloe* and *Neotyphodium* species including indole derivatives, indole-3-acetic acid, and indole-3-ethanol, a sesquiterpene and a diacetamide, have been identified and shown to have antifungal activity against the chestnut blight fungus *Cryphonectria parasitica* and suggest that they may play a similar role against other pathogens [54]. Indirect plant disease control by endophytic microorganisms is achieved by mechanisms modulating the plant immune response, including the induction of systemic acquired resistance [1,55].

Entry of Bacterial Endophytes into the Host Plant

Bacterial endophytes attach to the plant root surface or rhizoplane and access the internal plant tissues through openings where the root hairs or lateral roots emerge, stomata, wounds, and hydathodes in the shoot as well as the natural discontinuities in the plant body [16]. Moreover, some endophytes may secrete cell wall cellulolytic enzymes such as cellulases, xylanases, pectinases, and endoglucanases, to modify the cell wall of the plant and facilitate bacterial entry and spread within the plant tissues. In addition, some bacteria use root apex and root hairs as entry points followed by endophytic colonization in root cortex and vascular tissues [56,57].

INFLUENCE OF GENETIC AND ENVIRONMENTAL MODIFICATIONS IN DISEASE/PESTS' CONTROL BY ENDOPHYTES

Endophytic microorganisms are identified mainly by culture methods. However, the analysis of ribosomal ribonucleotide (rRNA) using molecular techniques for identification eliminates the set-backs mostly encountered using culture techniques. The means of assessing the variations in the genetic composition of the endophytes and the host plants are provided by molecular markers and in turn provide an insight into the relationship between variation in endophyte and host plants and the variability of agronomic traits. Molecular techniques such as terminal restriction fragment length polymorphism (T-RFLP) analysis or denaturing gradient gel electrophoresis in combination with sequence analysis of rRNA genes allow rapid characterization of microbial communities [58,59]. PCR-denaturing gradient gel electrophoresis has been used to monitor endophytic populations in potato, which revealed that different range of endophytic organisms fall into different distinct phylogenetic groups. The result also revealed the presence of nonculturable endophytic microorganisms in potato [60].

The expression of insect resistance by plants infected with endophytic microorganisms may be affected by several factors such as active amounts of allelochemicals, plant genotype, endophyte concentration, soil fertility, and endophyte genotype. Moreover, hydric stress, temperature, soil pH, insect-pest resistance, etc. may also affect the endophyte concentration and toxin production [17].

BIOTECHNOLOGICAL POTENTIAL OF ENDOPHYTIC MICROORGANISMS

It is known that endophytic microorganisms inhabit the interior of plants and provide protection against insect-pests. This property is an indication of their biotechnological potentials. Endophytic microorganisms are used as genetic vectors and are potentially useful to agriculture and industry such as food, pharmaceutical, and agrochemical industries. Endophytes are capable of producing antimicrobials such as toxins and antibiotics and can confer resistance to stress conditions as well as producing phytohormones [17,61].

The induction of systematic resistance against a wide variety of pests is acquired after appropriate stimulation which explains the control of insects-pests in agriculture. The ability of the endophytes to modify the cell wall structure (i.e., deposition of lignin) and the biological and physiological changes account for the synthesis of chemical compounds and proteins involved in plant defense mechanisms [62]. The potential to use plant growth promoting bacteria in insect-pest control has been related to stimuli generated in the plant itself, through different metabolic routes such as salicylic acid, jasmonic acid, and ethylene. As elicitors, these compounds induce defense and/or resistance, which are not activated in their absence with resultant production or increase in the production of proteinase-inhibitor compounds like serine, cysteine and aspartate (to inactivate extracellular proteinases produced by the pathogen), glycoalkaloids, polyphenols, and other compounds for resistance. Also, the response induced by the plant can involve accumulation of secondary metabolites including synthesis of siderophores, phytoalexins, and phenylpropanoids as well as biosynthesis of pathogenesis-related proteins (PR-protein) [62]. According to Neto et al. [63], the elucidation of the ecological functions of these endophytes can bring benefits while exploring their biotechnological potential as plant-growth promoting agents which can be analyzed through molecular techniques such as quantification of genes of desired characteristics.

BENEFITS OF ENDOPHYTES IN PRACTICAL AGRICULTURE

Plant growth promotion by endophytic microorganisms can be achieved through direct or indirect mechanisms. Endophytes that start their journey as rhizosphere bacteria may retain their attributes inside plant tissues. Moreover, most endophytes can be cultured and can survive outside host in rhizosphere [64].

Endophytic microorganisms can directly provide benefits to plants through the production of antimicrobial substances, secretion of insecticidal by-products (such as pyrrolopyrazine alkaloid per amine, ergot alkaloid ergo valine, and pyrrolizidine loline alkaloids), iron chelators, phosphate solubilizing compounds, and the ability to fix nitrogen. They can also oxidize elemental sulfur to sulfate which can be used by plants and produce phytochemicals which are antagonistic to plant pathogens as well as active secondary metabolites thereby contributing to plant metabolite production. Additionally, endophytic microorganisms can directly promote plant growth through the production of phytohormones (such as indole acetic acid, gibberellic acid, ethylene, and auxins), siderophores (such as catecholate, hydroximate and/or phenolate types), induced systemic tolerance through 1-aminocyclopropane-1-carboxylase deaminase production, induced systemic resistance, and antagonism [64–70].

Endophytes indirectly help plants to overcome series of unfavorable environmental and biotic stresses like drought, cold, hypersaline condition, heavy metal accumulation, pathogenesis, and radiations. Some endophytes which may have evolved from plant pathogens could induce plant defense responses through induced systemic responses (ISR). Endophytes can help in bioremediation by reducing heavy metal stress to plants, degrading toxic metabolites secreted by plants, and removing greenhouse gases from air as well as controlling the growth of pests on plants. Moreover, endophytes help in phytoremediation by decreasing metal phytotoxicity and their role in metal translocation and accumulation [71,72].

Endophytic microorganisms can also serve as biocontrol agents (e.g., *Bacillus and Pseudomonas*) through natural antagonism. They compete for nutrients and produce antibiotics such as pseudomycins, xiamycins, ecomycins, and munumbicins. Moreover, endophytes produce metabolites such as flavonoids which are applicable in agrochemical and pharmaceutical industries due to their antimicrobial properties [73].

FUTURE PROSPECTS

Endophyte which are to be exploited for agricultural purposes must not induce plant disease, should be capable to spread inside plant parts, should be culturable, and must colonize plant parts naturally obligately with species specifies [64]. Novel endophytes with desirable traits can be screened from plants, especially those growing in extreme environments (temperature, salinity, etc.). Molecular techniques such as gene manipulation and recombinant DNA technology can be used to develop new traits such as pesticide and herbicide resistance in plants. There is need for bulk plant-specific inoculum production of the endophytes in large scale for commercial use. However, research is required to understand host specific population dynamics. Optimized and enhanced production of host-specific endophyte inoculum will contribute to the reduction over dependent on chemical fertilizers, pesticides, and fungicides. Moreover, discovery of pesticides with synergistic effect on endophyte bioinoculant may be used to control different pathogens and the development of endophytes that can be sprayed along with chemical pesticides can contribute to effective integrated pest management [64].

George et al. [31] asserted major and promising area of research for future studies is developing endophytes (and rhizobacteria) to promote the sustainable production of biomass and bioenergy crops in conjunction with phytoremediation of soil contamination. The research on the role of endophytes in insect control in specific hosts requires more new findings. The use of endophytes and biological control agents in combination with commercial pesticides could have synergistic antimicrobial and pesticidal on one or multiple pathogens and pests.

CONCLUSION

Endophytes have antimicrobial properties against pathogens and biocontrol potentials against insects-pest. They are recognized to be sources of bioactive compounds and metabolites that are quite novel which can potentially be exploited in medicine, agriculture, and various industries. They play important role in protecting plants against diseases and pests. However, these group organisms are poorly investigated. It is essential to know endophyte diversity, their presence, distribution frequency, and functions. This understanding will create an avenue to expand scope of the use of endophytes as a biotechnological tool with the sole aim of increasing yield of crops and decrease the use of chemical in the control of pest thereby providing solutions to damages caused by pest insects in agriculture. Therefore, more studies are needed in the role of endophytes in insect-pest control.

REFERENCES

[1] W.M. Haggag, Role of entophytic microorganisms in biocontrol of plant diseases, Life Sci. J. 7 (2) (2010) 57–62.

[2] K.G. Arun, A.A. Robert, V.R. Kannan, Exploration of endophytic microorganisms from selected medicinal plants and their control potential to multi drug resistant pathogens, J. Med. Plants Stud. 3 (2) (2015) 49–57.

[3] O. Petrini, Fungal endophytes of tree leaves, in: J.H. Andrews, S.S. Hirano (Eds.), Microbial Ecology of Leaves, Springer-Verlag, New York, USA, 1991, pp. 179–198.

[4] X. Yang, G. Strobel, A. Stierle, W.M. Hess, J. Lee, J. Clardy, A fungal endophyte-tree relationship: *Phoma* sp. in *Taxus wallachiana*, Plant Sci. 102 (1994) 1–9.

[5] K. Ulrich, A. Ulrich, D. Ewald, Diversity of endophytic bacterial communities in poplar grown under field conditions, FEMS Microbiol. Ecol. 63 (2008) 169–180.

[6] M.M. Wagenaar, J. Clardy, Dicerandrols, new antibiotics and cytotoxic dimmers produced by the fungus *Phomopsis longicolla* isolated from an endangered mint, J. Nat. Prod. 64 (2001) 1006–1009.

[7] J. Hallmann, A. Quadt-Hallmann, W.F. Mahaffee, J.W. Kloepper, Bacterial endophytes in agricultural crops, Can. J. Microbiol. 43 (1997) 895–914.

[8] P. M'piga, R.R. Bélanger, T.C. Paulitz, N. Benhamou, Increased resistance to *Fusarium oxysporum f.* sp. *Radicis lycopersiciin* tomato plants treated with the endophytic bacterium *Pseudomonas fluorescens* strain 63-28, Physiol. Mol. Plant Pathol. 50 (1997) 301–320.

[9] A.V. Sturz, B.R. Christie, B.G. Matheson, W.J. Arsenault, N.A. Buchanan, Endophytic bacterial communities in the periderm of potato tubers and their potential to improve resistance to soil-borne plant pathogens, Plant Pathol. 48 (1999) 360–369.

[10] M. Puentea, Y.C. Li, C. Yoav Bashana, Endophytic bacteria in cacti seeds can improve the development of cactus seedlings, Environ. Exp. Bot. 66 (2009) 402–408.

[11] Y. Huang, J. Wang, G. Li, Z. Zheng, W. Su, Antitumor and antifungal activities in endophytic fungi isolated from pharmaceutical plants Taxus Maier, *Cephalotaxus fortunei* and *Torreya grandis*, FEMS Immunol. Med. Microbiol. 34 (2001) 163–167.

[12] S. Spaepen, J. Vanderleyden, Auxin and plant-microbe interactions, Cold Spring Herb Perspective Biol. 3 (2011) 1–13.

[13] G. Santoyo, G. Moreno-Hagelsieb, M. del Carmen Orozco-Mosqueda, B.R. Glick, Plant growth-promoting bacterial endophytes, Microbiol. Res. 183 (2016) 92–99.

[14] Q. Esmaeel, M. Pupin, N.P. Kieu, G. Chataigné, M. Béchet, J. Deravel, F. Krier, M. Höfte, P. Jacques, V. Leclère, *Burkholderia* genome mining for no ribosomal peptide synthetases reveals a great potential for novel siderophores and lipopeptides synthesis, Microbiol. Open (2016) 1–5.

[15] S.L. Kandel, A. Firrincieli, P.M. Joubert, P.A. Okubara, N.D. Leston, K.M. McGeorge, G.S. Mugnozza, A. Harfouche, S.H. Kim, S.L. Doty, An in vitro study of bio-control and plant growth promotion potential of *Salicaceae* endophytes, Front. Microbiol. 8 (2017a) 1–16.

[16] S.L. Kandel, P.M. Joubert, S.L. Doty, Bacterial endophyte colonization and distribution within plants, Microorganisms 5 (2017b) 77. https://doi.org/10.3390/microorganisms5040077.

[17] J.L. Azevedo, W. Maccheroni Jr., J.O. Pereira, W. Luiz de Araújo, Endophytic microorganisms: a review on insect control and recent advances on tropical plants, Electron. J Biotechnol. 3 (1) (2000) 40–65.

[18] W.L. Araújo, H.O. Saridakis, P.A.V. Barroso, C.I. Aguilar-Vildoso, J.L. Azevedo, Variability and interactions between endophytic bacteria and fungi isolated from leaf tissues of citrus rootstocks, Can. J. Microbiol. 47 (2001) 229–236.

[19] M.A.C. Zawadneak, H.R. Vidal, B. Santos, Lagarta-da-coroa, *Duponchelia fovealis* (Lepidoptera: *Crambidae*), in: E. Vilela, R.A. Zucchi (Eds.), Pragas Introduzidas: Insetos Eácaros, ESALQ/FEALQ, Piracicaba, 2015, pp. 216–231.

[20] S. Parsa, A.M. García-Lemos, K. Castillo, V. Ortiz, L.A.B. López-Lavalle, J. Braun, F.E. Vega, Fungal endophytes in germinated seeds of the common bean, *Phaseolus vulgaris*, Fungal Biol. 120 (5) (2016) 783−790.

[21] R.F. Amatuzzi, N. Cardoso, A.S. Poltronieri, C.G. Poitevin, P. Dalzoto, M.A. Zawadeneaka, I.C. Pimentel, Potential of endophytic fungi as biocontrol agents of *Duponchelia fovealis* (Zeller) (Lepidoptera: *Crambidae*), Braz. J. Biol. 78 (3) (2018) 429−435.

[22] D.N. Nair, S. Padmavathy, Impact of endophytic microorganisms on plants, environment and humans, Sci. World J. (2014) 1−12. https://doi.org/10.1155/2014/250693.

[23] J.W. Fahey, M.B. Dimock, S.F. Tomasino, J.M. Taylor, P.S. Carlson, Genetically engineered endophytes as biocontrol agents: a case study in industry, in: J.H. Andrews (Ed.), Microbial Ecology of Leaves, Springer, New York, NY, USA, 1991, pp. 402−411.

[24] X. Zhao, G. Qi, X. Zhang, N. Lan, X. Ma, Controlling sap sucking insect pests with recombinant endophytes expressing plant lectin, Nature Preceding 21 (21) (2010).

[25] X. Zhang, J. Li, G. Qi, K. Wen, J. Lu, X. Zhao, Insecticidal effect of recombinant endophytic bacterium containing *Pinellia ternate agglutinin* against white backed planthopper, *Sogatella furcifera*, Crop Protect. 30 (11) (2011) 1478−1484.

[26] J. Webber, A natural control of Dutch elm disease, Nature 292 (1981) 449−451. London.

[27] N. Claydon, J.F. Grove, M. Pople, Elm bark beetle boring and feeding deterrents from *Phomopsis oblonga*, Phytochem. 24 (1985) 937−943.

[28] D.L. Gaynor, W.F. Hunt, The relationship between nitrogen supply, endophytic fungus and Argentine stem weevil resistance in ryegrass, Proc. N. Z. Grassl. Assoc. 44 (1983) 257−263.

[29] M.C. Johnson, L.D. Dahlman, M.R. Siegel, L.P. Bush, G.C.M. Latch, D.A. Potter, D.R. Varney, Insect feeding deterrents in endophyte-infected tall fescue, Appl. Environ. Microbiol. 49 (1985) 568−571.

[30] G.C.M. Latch, M.J. Christensen, Artificial infections of grasses with endophytes, Ann. Appl. Biol. 107 (1985) 17−24.

[31] A. George, Role of endophytes in insect control, Acta Sci. Agric. 1 (4) (2017) 1−3.

[32] S. Ahmad, J.M. Johnson-Cicalese, W.K. Dickson, C.R. Funk, Endophyte-enhanced resistance in perennial ryegrass to the bluegrass billbug *Sphenophorus parvulus*, Entomol. Exp. Appl. 41 (1986) 3−10.

[33] S.L. Clement, K.S. Pike, W.J. Kaiser, A.D. Wilson, Resistance of endophyte-infected plants of tall fescue and perennial ryegrass to the Russian wheat aphid (Homoptera: *Aphidiae*), J. Kans. Entomol. Soc. 63 (1990) 646−648.

[34] S. Soria, R. Alonso, L. Bettucci, Endophytic bacteria from *Pinus taedal* as biocontrol agents of *Fusarium circinatum* Nirenberg & O'Donnell, Chil. J. Agric. Res. 72 (2) (2012) 281−284.

[35] R.B. Thuler, R. Barros, R.L.R. Mariano, J.D. Vendramim, Effect of plant growth-promoting bacteria (BPCP) in developing *Plutella xylostella* (L.) (Lepidoptera: *Plutellidae*) in cabbage, Science 34 (2) (2006) 217−222.

[36] V.M. Melatti, L.B. Praça, E.S. Martins, E. Sujii, C. Berry, R.G. Monserrat, Selection of *Bacillus thuringiensis* strains toxic against cotton aphid, *Aphis gossypii* Glover (Hemiptera: *Aphididae*), Bio Assay 5 (2) (2010) 1−4.

[37] E.B. Campanini, C.C. Davolos, E.C.C. Alves, M.V.F. Lemos, Characterization of new isolates of *Bacillus thuringiensis* for control of important insect pests of agriculture, Bragantia 71 (3) (2012) 362−369.

[38] C.L. Macedo, E.S. Martins, L.L.P. Macedo, A.C. Santos, L.B. Praça, L.A.B. Góis, R.G. Monnerat, Selection and characterization of *Bacillus thuringiensis* strains effective against *Diatraea saccharalis* (Lepidoptera: *Crambidae*), Pesqui. Agropecuária Bras. 47 (12) (2012) 1759−1765.

[39] L.B. Praça, Interactions between Bacillus Thuringiensis Strains and Hybrids of Cabbage for the Control of Plutella Xylostella and Plant Growth Promotion. Brasília: UNB, 2012, Thesis (doctoral), Faculty of Agronomy and veterinary medicine, University of Brasília, Brasília, 2012, p. 141.

[40] N.A. Broderick, R.M. Goodman, K.F. Raffa, J. Handelsman, Synergy between zwittermicin A and *Bacillus thuringiensis* subsp. *kurstaki* against gypsy moth (Lepidoptera: *Lymantriidae*), Biol. Control 29 (1) (2000) 101−107.

[41] S.P. Wraight, M.E. Ramos, Synergistic interaction between *Beauveria bassiana* and *Bacillus thuringiensis* tenebrionids based biopesticides applied against field populations of Colorado potato beetle larvae, J. Invertebr. Pathol. 90 (3) (2005) 139−150.

[42] X.M. Ma, X.X. Liu, X. Ning, B. Zhang, X.M. Guan, Y.F. Tan, Q.W. Zhang, Effects of *Bacillus thuringiensis* toxin Cry 1Ac and *Beauveria bassiana* on Asiatic corn borer (Lepidoptera: *Crambidae*), J. Invertebr. Pathol. 99 (2) (2008) 123−128.

[43] E. Wilhelm, W. Arthofer, R. Schafleitner, Bacillus subtilis, an endophyte of chestnut (*Castanea sativa*), as antagonist against chestnut blight (*Cryphonectria parasitica*), in: A.C. Cassells (Ed.), Pathogen and Microbial Contamination Management in Micropropagation, Kluwer Academic Publishers, Dortrecht, The Netherlands, 1997, pp. 331−337.

[44] K. Xaio, L.L. Kinkel, D.A. Samac, Biological control of *Phytophthora* root rots on alfalfa and soybean with Streptomyces, Biol. Control 23 (2002) 285−295.

[45] J.T. Coombs, C.M.M. Franco, Isolation and identification of actinobacteria isolated from surface-sterilized wheat roots, Appl. Environ. Microbiol. 69 (2003) 5303−5308.

[46] T.L. Bultman, K.L. Borowicz, R.M. Scneble, T.A. Couldron, L.P. Bush, Effect of a fungal endophyte on the growth and survival of two *Euplectrus* parasitoids, Oikos 78 (1997) 170−176.

[47] C.T. Dougherty, F.W. Knapp, L.P. Bush, J.E. Maul, J. Van Willigen, Mortality of horn fly (Diptera: *Muscidae*) larvae in bovine dung supplemented with loline alkaloids from tall fescue, J. Med. Entomol. 35 (1998) 798−803.

[48] S. Hasegawa, A. Meguro, M. Shimizu, T. Nishimura, H. Kunoh, Endophytic actinomycetes and their

interactions with host plants, Actinomycetologica 20 (2006) 72–81.

[49] J.W. Wang, L.P. Zheng, R.X. Tan, The Preparation of an elicitor from a fungal endophyte to enhance artemisinin production in hairy root Cultures of *Artemisia annua* L, Chin. J. Biotechnol. 22 (2006) 829–834.

[50] G.A. Strobel, E. Dirksie, J. Sears, C. Markworth, Volatile antimicrobials from a novel endophytic fungus, Microbiology 147 (2001) 2943–2950.

[51] J. Worapong, G.A. Strobel, B. Daisy, U. Castillo, G. Baird, W.M. Hess, *Muscodor roseusanna*. Nov. an endophyte from *Grevillea pteridifolia*, Mycotaxon 81 (2002) 463–475.

[52] U. Castillo, J.K. Harper, G.A. Strobel, J. Sears, K. Alesi, E. Ford, J. Lin, M. Hunter, M. Maranta, H. Ge, D. Yaver, J.B. Jensen, H. Porter, R. Robison, D. Millar, W.M. Hess, M. Condron, D. Teplow, Kanamycin's, novel antibiotics from *Streptomyces* sp. NRRL 30566, an endophyte of Grevillea pteridifolia, FEMS Lett. 224 (2003) 183–190.

[53] W.X. Zou, J.C. Meng, H. Lu, G.X. Chen, G.X. Shi, T.Y. Zhang, R.X. Tan, Metabolites of *Colletotrichum gloeosporioides*, an endophytic fungus in *Artemisia mongolica*, J. Nat. Prod. 63 (2000) 1529–1530.

[54] Q. Yue, C.J. Miller, J.F. White, M.D. Richardson, Isolation and characterization of fungal inhibitors from *Epichloe festucae*, J. Agric. Food Chem. 48 (2000) 4687–4692.

[55] S.C. van Wees, M. Luijendijk, I. Smoorenburg, L.C. van Loon, C.M. Pieterse, Rhizobacteria mediated induced systemic resistance (ISR) in *Arabidopsisis* not associated with a direct effect on expression of known defense-related genes but stimulates the expression of the jasmonate-inducible gene Atvspoon challenge, Plant Mol. Biol. 41 (1999) 537–549.

[56] D. Compant, B. Duffy, J. Nowak, C. Clément, E.A. Barka, Use of plant growth promoting bacteria for biocontrol of plant diseases: principles, mechanisms of action, and future prospects, Appl. Environ. Microbiol. 71 (2005) 4951–4959.

[57] A.L.S. Rangel de Souza, S.A. De Souza, M.V.V. De Oliveira, T.M. Ferraz, F.A.M.M.A. Figueiredo, N.D. Da Silva, P.L. Rangel, C.R.S. Panisset, F.L. Olivares, E. Campostrini, G.A. De Souza Filho, Endophytic colonization of *Arabidopsis thaliana* by *Gluconacetobacter diazotrophicus* and its effect on plant growth promotion, plant physiology, and activation of plant defense, Plant Soil 399 (2016) 257–270.

[58] K. Smalla, G. Wieland, A. Buchner, A. Zock, J. Parzy, S. Kaiser, N. Roskot, H. Heuer, G. Berg, Bulk and rhizosphere soil bacterial communities studied by denaturing gradient gel electrophoresis: plant-dependent enrichment and seasonal shifts revealed, Appl. Environ. Microbiol. 67 (2001) 4742–4751.

[59] H.A. Gamper, J.P.W. Young, D.L. Jones, A. Hodge, Real-time PCR and microscopy: are the two methods measuring the same unit of arbuscular mycorrhizal fungal abundance, Fungal Genet. Biol. 45 (2008) 581–596.

[60] P. Garbeva, L.S. van Overbeek, J.W.L. van Vuurde, J.D. van Elsas, Analysis of endophytic bacterial communities of potato by plating and denaturing gradient gel electrophoresis (DGGE) of 16S rDNA-based PCR fragments, Microb. Ecol. 41 (2001) 369–383.

[61] A.Q.L. Souza, S. Astolfi Filho, M.L. Belem Pinheiro, M.I.M. Sarquis, J.O. Pereira, Antimicrobial activity of endophytic fungi isolated from toxic plants in the Amazon: *Palicourea longiflora* (aubl.) rich and *Strychnos cogens* Bentham, Acta Amazonica 34 (2) (2004) 185–195.

[62] E.O. Araújo, Rizobacteria in the control of pest insects in agriculture, Afr. J. Plant Sci. 9 (9) (2015) 368–373.

[63] P.A.S. Neto, J.L. Azevedo, W.L. Araujo, Endophytic microorganisms, Biotechnol. Sci. Develop. 29 (2002) 62–77.

[64] A. Yadav, K. Yadav, Exploring the potential of endophytes in agriculture: a minireview, Adv. Plants Agric. Res. 6 (4) (2017) 102–106.

[65] H.H. Wilkinson, Contribution of fungal loline alkaloids to protection from aphids in a grass–endophyte mutualism, Mol. Plant Microbe Interact. 13 (10) (2000) 1027–1033.

[66] M. Rajkumar, Potential of siderophore–producing bacteria for improving heavy metal phytoextraction, Trends Biotechnol. 28 (3) (2010) 142–149.

[67] E.A.A. Pinheiro, J.M. Carvalho, D.C.P. dos Santos, A.O. Feitosa, P.S.B. Marinho, G.M.P.S. Guilhon, L.S. Santos, A.L.D. de Souza, A.M.R. Marinho, Chemical constituents of *Aspergillus* sp. EJC08 isolated as endophyte from *Bauhinia guianensis* and their antimicrobial activity, Ann. Brazilian Acad. Sci. 85 (4) (2013) 1247–1252.

[68] G. Brader, Metabolic potential of endophytic bacteria, Curr. Opin. Biotechnol. 27 (2014) 30–37.

[69] J.L. Knoth, Biological nitrogen fixation and biomass accumulation within poplar clones as a result of inoculations with diazotrophic endophyte consortia, New Phytol. 201 (2) (2014) 599–609.

[70] H. Nisa, Fungal endophytes as prolific source of phytochemicals and other bioactive natural products: a review, Microb. Pathog. 82 (2015) 50–59.

[71] X. Zhang, C. Li, Z. Nan, Effects of cadmium stress on seed germination and seedling growth of *Elymus dahuricus* infected with the *Neotyphodium endophyte*, Sci. China Life Sci. 55 (9) (2012) 793–799.

[72] Z. Stepniewska, A. Kuźniar, Endophytic microorganisms–promising applications in bioremediation of greenhouse gases, Appl. Microbiol. Biotechnol. 97 (22) (2013) 9589–9596.

[73] A. Christina, V. Christapher, S.J. Bhore, Endophytic bacteria as a source of novel antibiotics: an overview, Pharmacogn. Rev. 7 (13) (2013) 11–16.

CHAPTER 12

Techniques for the Detection, Identification, and Diagnosis of Agricultural Pathogens and Diseases

AJAY KUMAR GAUTAM, PHD • SHASHANK KUMAR, PHD, MSC, BSC

INTRODUCTION

Food crops are essential sources of energy to the body. Nowadays, the demand for supply of food crops is on the increase. To maintain the balance of food, we need to protect the crops from pathogens which could be ectoparasites, viruses, bacteria, oomycetes, fungi, protozoans, or nematodes. The production of vegetable crops represents an important economic section of the agricultural field. These days, diseases, insects, and weeds cause ~25% of crop loss and ultimately affects economic values according to the Food and Agriculture Organization (FAO). A notable example is the destruction of rice by blast disease every year [1]. Vegetables are prone to pathogens; soil-borne pathogens are considered as severe threat in vegetable production. These pathogens can survive for a long time without a host crop which makes it challenging to control them. Some viruses can only survive inside a living plant tissues, and are difficult to be predicted, detected, and diagnosed [2].

In the past, farmers use supernatural practices for the destruction of pathogens to protect their crops. By this method, there are problems associated with proper identification, false result interpretation, and incorrect disease diagnosis [3]. However, in the present time, the detection of plant disease is based on modern plant pathology. The most reported pathogens were treated using chemical and biological pesticides, herbicides, and fungicides. Before treating the crop with disinfectant, there is need to study the type of pathogen and disease condition. To achieve this, few chemical, biological, and physical methods need to be employed. Nowadays, methods for the detection and identification of the plant pathogens have been used based on nucleic acid sequencing using PCR amplification and DNA/RNA probe technology [4,5].

CLASSIFICATION OF AGRICULTURAL PATHOGENS

As noted in the previous chapters, a pathogen is an infectious disease–causing agent that invades and causes disease to the host. These pathogens may be zoonotic and capable of causing disease in both animals and humans. These organisms mainly include bacteria, viruses, fungus, nematodes, and protozoans. Invasive species of the pathogens present noteworthy threats to worldwide agriculture; it is still unclear how the magnitude and distribution of the threats vary between countries and regions.

Bacteria

Plant pathogenic bacteria are extremely diverse in the environment and require precise identification diagnostic test. Bacteria are the most commonly identified organism, in which some of them are highly pathogenic to plants. The most commonly found pathogenic bacteria are bacilli (rod-shaped). These bacteria act on different parts of the plants, and localize and inhibit few functions of the plants. Some are cell wall–degrading bacteria species such as *Erwinia*. Some species can hinder the plant hormone levels such as soft rot *Agrobacterium*.

Fungi

There are over 1.5 million species of fungi (known mainly as decomposers) in the planet [6]. Most fungi are harmful and act as pathogens- cause diseases. Various studies have shown overwhelming effects of these fungi on crop yield. Approximately half of the world population depends on the rice crop which is considered as staple food, but 10%–30% of rice crops are lost per year by the invasion of *Pyricularia oryzae*, causing rice blast disease [7]. The fungi plant pathogens can be classified as *Ascomycetes* and *Basidiomycetes*. The most common types in *Ascomycetes* are *Fusarium* spp., *Verticillium* spp., *Sclerotinia sclerotiorum* commonly called cottony rot, and *Magnaporthe grisea* commonly called rice blast, and *Basidiomycetes* are classified as *Rhizoctonia* spp., *Puccinia* spp., and *Ustilago* commonly called smuts.

Nematodes
Nematodes are wormlike animals present in the soil which affect the plant roots. The best example is *Globodera pallida* and *Globodera rostochiensis* commonly called potato cyst nematodes. Nematodes are involved in the transmission of few viruses and some of the nematodes are root-feeders and omnivorous.

Viruses
The ability of viruses to infect a host is both time- and condition-dependent [8]. Common types of viruses that severely affect agriculture crops are Tobacco mosaic virus, Tomato spotted wilt virus, Tomato yellow leaf curl virus, Cucumber mosaic virus, Potato virus Y, Cauliflower mosaic virus, African cassava mosaic virus, Plum pox virus, Brome mosaic virus Potato virus, etc. Plant pathogens like viruses and viroids are silent killer of plants and lead to economic losses in various crops, especially those for which no virus-resistant varieties are available. Thus, there is need for the early detection of pathogens to control the expansion of the disease [9,10].

Protozoa
The most common plant species hosting protozoa *Phytomonas* belong to Euphorbiaceae and Asclepiadaceae family. Members of the Apocynaceae, Urticaceae, Moraceae, Compositae, Sapotaceae, and Solanaceae infect fruits [11,12]. The single-celled protozoa feed on bacteria and fungus present in the soil which digests the organic matter. Some of the protozoa are *Mastigophora*, *Ciliophora*, and *Sarcodina*.

MOLECULAR METHODS FOR DISEASE DIAGNOSIS
Molecular methods are conventional methods used to determine the disease conditions through DNA and RNA sequence analysis. To analyze these sequences, there are a few methods like nucleic acid hybridization and polymerase chain reaction. The most accurately and

recently used type of molecular detection technique is fluorescence in-situ hybridization (FISH), which is applied for bacterial detection in combination with microscopy and hybridization.

MICROSCOPY METHODS FOR DISEASE DIAGNOSIS
General microscopes, fluorescence microscopes, and scanning electron microscope (SEM) are commonly used for the identification of different types of pathogens through the observation of their morphology and imaging. Some organisms cannot be observed using compound microscopes, so they require more sophisticated types of microscopes such as fluorescence microscopes. Diagnostic electron microscopy allows rapid morphologic identification and differential diagnosis of various infectious agents in the specimen [13]. Fluorescence imaging is a widely used technique to analyze pathogen infections in the leaves on the basis of changes in the photosynthetic apparatus and photosynthetic electron transport reactions [14–16].

SPECTROSCOPY-BASED DISEASE DIAGNOSIS
Spectroscopy methods are most effective ways of detecting and analyzing plant pathogens for diseases due to their flexibility and cost-efficiency. Spectroscopy methods are classified as imaging spectroscopy and nonimaging spectroscopy. Some of the most common types of spectroscopy are visible wavelength spectroscopy (VIS), near-infrared resonance (NIR), short-wave infrared wavelength (SWIR), nuclear magnetic resonance (NMR), and selected ion flow tube mass spectroscopy (SIFT-MS).

Visible (VIS) and infrared (IR) spectroscopy are the most promising methods accepted in many areas of agriculture fields [17]. There are so many studies done using VIS/IR to resolve pathogen identification issues in plant disease [18]. The identification and quantification of the olive leaf spot disease are usually performed by the use of visible/near-infrared spectroscopy and multivariate data analysis [19,20]. These techniques can be used in the detection of plant stress due to variation in the absorption of incident rays in VIS and IR spectrum [19,20].

Literature studies reveal the use of VIS/IR spectroscopy technique for detecting and monitoring plant diseases at early stage such as yellow rust in wheat, parley, and olive leaves; spot in wheat; HLB and CVC disease in citrus; leaf-roll in grape (*Grapevine leafroll-associated virus*); verticillium wilt in cotton; scab in apple; Fiji leaf gall in sugarcane; pathogen in tomato; crown rot in tomato; fungal infection in corn; and leaf folder infestation in rice [21−23].

NMR is routinely used in the plant metabolite profiling and disease metabolomics due to its high fidelity and speed. It is a very fast method for the metabolite fingerprinting of plants, convenient, and an effective tool for discriminating between groups of related samples. This technique needs no prior sample purification. It detects hydrophilic and lipophilic metabolites, and requires less amount of sample. The NMR techniques can also be used for disease metabolomics studies, the easiest being one-dimensional 1H NMR that is able to detect compounds above micromolar concentrations [24]. From earlier study, the approaches of NMR spectroscopy and chemometrics for the analysis of metabolic biomarkers from *Citrus sinensis* leaves enabled successful metabolic profile differentiation of disease early stages [25].

SIGNIFICANCE OF ADVANCED DETECTION METHODS

New and emerging technologies for the detection and analysis of plant pathogens are becoming more pronounced, for example, nanotechnology-based technology using gold nanoparticles and quantum dots which helps in detecting biological markers with high accuracy and calibrated data generated for high-throughput. Biosensors and quantum dots for nano-imaging and DNA sequencing tools, which are sensitive, highly specific, and have speed outcome in detecting pathogens, have found wide application.

Colony-Forming Units

Colony-forming unit (CFU) is commonly used to count the number of viable or live fungal or bacterial cells in the unknown sample. It requires significant growth of microbial culture and microscope for the colonies counting of viable or dead cells and expresses the results as colony-forming units. This method has some drawbacks: counting is time-consuming and errors might emerge from incomplete separation of clumping spores [26].

Polymerase Chain Reaction

The polymerase chain reaction (PCR) is a simple, original method to amplify target sequences which involves three essential steps: melting of the target, annealing denatured strands, and primer extension using Taq DNA polymerase. The frequent use of PCR grew speedily in plant pathology research and diagnosis. This method provides several advantages compared to more traditional methods of diagnosis [27]. Its productivity mainly depends on the concentration of dNTPs, polymerase type, purity of template and cycling parameters, etc. [28]. Therefore, nowadays it is considered a standard tool in diagnosis, alone or in combination with other methods.

Nested Polymerase Chain Reaction

The sensitivity and specificity of PCR and RT-PCR for the detection of viral and bacterial pathogens can be eliminated by the use of nested PCR−based methods based on their two rounds of amplification [29,30]. In this method, first PCR amplification product is transferred to another tube and then nested PCR is performed using one or two internal primers (heminested or nested amplification). There are many studies that have been reported related to the detection of the viral and bacterial pathogens in the plant [31−33]. Olmos et al. [34] reported the detection of Citrus tristeza virus by using single-tube nested PCR.

Multiplex Polymerase Chain Reaction

Multiplex PCR methods are usually used for the simultaneous detection of two or more target pathogenic DNA or RNA molecules by using several specific primers in a single PCR reaction. It is a precious and frequently used method for the detection and identification of bacterial or viral pathogens which shows infection in plants [28,35]. Few examples have been reported for the detection of viruses such as CLRV, CMV, SLRSV, Arabis mosaic virus (ArMV), Olive latent virus-1, and Olive latent virus-2 affecting olive trees [36] and simultaneous detection of nine viruses in grapevine tree [37].

Multiplex nested Polymerase Chain Reaction

Multiplex nested RT-PCR method is the combination of multiplex PCR and nested PCR. It consists of sensitivity and reliability with time-saving and also decreases the reagent cost because two reactions are performed in a single reaction mixture. It conducts the simultaneous detection of target DNA and RNA and also avoids primer-dimer and hairpin formation. Some examples have been reported for the detection of virus, fungi, and phytoplasm as in plants [38−40].

Real-time Polymerase Chain Reaction

Real-time PCR is one of the useful tools for the detection and accurate quantification of the PCR products [41,42]. It helps reduce the detection time and uses less concentration of targeted DNA or RNA [43]. It has no necessity of gel electrophoresis to confirm the PCR product and has no risk of contamination [44]. The running process of the real-time PCR can be monitored directly by the exponential curve analysis on a computer screen [45]. The procedure initially require reverse transcriptase enzyme to convert the single strand RNA to cDNA through primary PCR [46]. Experimental approach to this was reported in studies involving Citrus tristeza virus, TMV, Citrus leaf blotch virus (CLBV) [42,47,48] and in the detection of potato pathogenic bacteria [49].

ENZYME-LINKED IMMUNOSORBENT ASSAY

Enzyme-linked immunosorbent assay (ELISA) is the most commonly used diagnostic technique for the detection of antigen by using specific antibody [50]. In ELISA assay, specific antibodies bind to the target antigen present in the sample which coats with microtiter plate wells. The recognition of antigen can be easily possible by the enzyme-mediated color reaction. All the process happens in an appropriate 96 well plastic plate where the color change reaction measurement is done with the help of ELISA plate reader or spectrophotometer to quantify a load of pathogens [51,52]. ELISA techniques are usually more sensitive and specific due to using the polyclonal and monoclonal antibodies [53,54]. The broad spectrum of the pathogen strains can be distinguished by the use of polyclonal antibodies [55]. However, monoclonal antibodies can be used to identify the specific one [56]. The combination of both antibodies could be used for the detection of many pathogenic strains [57]. The technique is sensitive enough to detect the bacteria at a lower concentration of nearly 10^6 CFU/mL and usually cheaper and more samples can be processed in one plate [58,59].

Lateral Flow Microarrays

Lateral flow microarrays (LFM) is a recently used method for the rapid detection of nucleic acids based on hybridization and finally visualized by colorimetric signal [60]. It is inexpensive in the laboratory and assembled on the lateral flow chromatography using nitrocellulose membrane; detection is based on the availability of strong and reliable host and pathogen biomarkers [61]. This method is widely used for the identification of plant metabolites as biomarkers in different plants during pathogen infections or stresses [62–64].

Direct Tissue Blot Immunoassay

Direct tissue blot immunoassay (DTBIA) is an antibody-based technique like ELISA. It is performed on the nylon and nitrocellulose membranes in the place of ELISA plate, so it is called DTBIA [65]. It requires a specific antibody for binding with pathogens and a large amount of virus concentration to decrease the false negative result. It has excellent benefits over ELISA in relation to less detection time, low cost, and more sensitive for the diagnosis of viral diseases, detection of citrus disease agents, mainly Citrus tristeza virus, Citrus psorosis virus [66–68], and *Xylella fastidiosa* in olive trees [69].

Fluorescence In-situ Hybridization

Fluorescence *in-situ* hybridization (FISH) is a technique useful for the detection of bacteria with the use of microscopy in combination with hybridization of DNA probes and target gene from plant samples [70]. The mechanism is based on the hybridization of DNA probes to species-specific regions of bacterial ribosomes. The ribosomal RNA contains specific sequence to all species, and this is the basic requirement for FISH technique to recognize the unique sequence of target organism. FISH can distinguish even single cells and has high sensitivity and affinity for DNA probe and can detect at least 10^3 cells/L. In practice, FISH has been employed in the diagnosis of bacteria as plant pathogens [71].

Immunofluorescence

Immunofluorescence (IF) is a technique used for the detection of microbes with the help of the fluorescence microscope. The principle of this technique, the exposure of antigen present inside the cell, is only possible by the use of a specific antibody to target fluorescent dyes and allows visualization. The detection is based on two methods: a direct method using single antibody linked with a fluorophore, and an indirect method using two antibodies; first antibody binds specifically with antigen and second antibody carries fluorophore that binds to the primary antibody which can be detected by fluorescence microscope [52]. Nowadays, IF has been used in the field of agriculture for the detection of pathogens, for example, detection of fungi *Puccinia striiformis* f. sp. *tritici* and *Puccinia triticina* in wheat rust [72], detection of *Botrytis cinerea* fungus in onion crop [73].

Flow Cytometry

Flow cytometry (FCM) is a technique used for the counting of a single cell in the fluid sample. Also, it is used in the cell cycle kinetics and antibiotic susceptibility study for plant disease. It comprises three components: fluidics, optics (laser-based), and electronics, commonly used in the research field for the differencing between viable and nonviable bacteria. FCM is a rapid (30 min) and accurate (10^4 CFU) detection technique with the use of fluorescent probes for the detection of foodborne bacterial pathogens and *Bacillus subtilis* in mushroom [74−76].

Electron Microscopy for Viral Pathogen Detection

It is the basic technique for the detection of virus by acquiring the morphological detail about virus shape, size, and another surface feature. Generally, the detection of filamentous and rod-shaped viruses occurred by the use of negative staining leaf-dip technique. One more method is also famous for rapid detection; in this method the epidermis is removed from the leaf with the help of negative strain on the electron microscope grid. Nowadays, the immunosorbent electron microscopy (combination with serology) method has been used for the increase in the efficiency of virus detection [77].

Gas Chromatography-Mass Spectrometry

Gas chromatography-mass spectrometry (GC-MS) is a nonoptical-based indirect method, in which the separation of the compounds is based on the flow of inert helium or nitrogen gas. It involves the profiling and signature of released specific volatile organic compounds (VOCs) in the stressed plant indicative during the infection [78]. The identification of the unknown VOCs could be possible by the use of gas-chromatography combined with mass-spectrometry (GC-MS) to enhance the performance [79−81]. For example, the release of p-ethylguaiacol and p-ethylphenol compound by the infection of fungi *Phytophthora cactorum* and causing crown rot diseases in strawberry plants [82].

MATRIX-ASSISTED LASER DESORPTION/IONIZATION-TIME-OF-FLIGHT MASS SPECTROMETRY

Matrix-assisted laser desorption/ionization-time-of-flight mass spectrometry (MALDI-TOF) is a useful technique for the analysis of macromolecules such as glycoproteins, proteins, and nucleic acids. It is also used for the identification of bacteria, bacterial spores, fungi, and yeasts [83]. The identification of the molecules is based on the comparison of observed spectrum with the reference database [84]. Initially, the spore suspension or target fragments are mixed with a matrix which becomes crystals after drying during the process.

The crystallized sample tends to be ionized through short laser pulse, then ions move and accelerate in a vacuum flight tube and measure the time of flight [85]. It is a powerful tool for the differentiation of genotype and is gaining widespread application as a diagnostic tool in genetic testing and plant pathology [86], and the identification of bacteria [87,88] and viruses [89].

NANOPORE SEQUENCING

The detection and identification of plant pathogens is possible by the use of sequencing-based methods. The earliest possibility of pathogens diagnosis examined by the use of single-molecule sequencing platform of Oxford Nanopore as a general method for diagnosis of plant diseases. The testing of the plant pathogens is initiated by isolation of DNA or RNA (e.g., viruses, bacteria, fungi, phytoplasma) from infected plants and finally sequencing of the target ones. The benefits of Nanopore include fast runtimes, long read lengths, portability, low cost, and the possibility of use. In this process, the pathogens can be identified by using real time within short time of running the nanopore sequencer and can be classified [90].

REFERENCES

[1] R.A. Dean, N.J. Talbot, D.J. Ebbole, M.L. Farman, T.K. Mitchell, M.J. Orbach, et al., The genome sequence of the rice blast fungus Magnaporthegrisea, Nature 434 (2005) 980−986.

[2] S. Koike, K. Subbarao, R.M. Davis, T. Turini, Vegetable Diseases Caused by Soilborne Pathogens, University of California, Division of Agriculture and Natural Resources, 2003, ISBN 978-1-60107-273-3, pp. 1−13.

[3] F. Martinelli, R. Scalenghe, S. Davino, S. Panno, G. Scuderi, P. Ruisi, et al., Advanced methods of plant disease detection. A review, Agron. Sustain. Dev. 35 (1) (2014) 1−25.

[4] J. Duncan, L. Torrance, Techniques for the Rapid Detection of Plant Pathogens, Black Well Scientific Publications, 1992.

[5] R.R. Martin, D. James, C.A. Lévesque, Impacts of molecular diagnostic technologies on plant disease management, Annu. Rev. Phytopathol. 38 (1) (2000) 207−239.

[6] N. Capote, A. Aguado, A.M. Pastrana, P. Sánchez-Torres, Molecular Tools for Detection of Plant Pathogenic Fungi and Fungicide Resistance, INTECH Open Access Publisher, Valencia, 2012.

[7] N.J. Talbot, On the trail of a cereal killer: exploring the biology of Magnaporthe grisea, Annu. Rev. Microbiol. 57 (1) (2003) 177−202.

[8] C.P. Gerba, Applied and theoretical aspects of virus adsorption to surfaces, Adv. Appl. Microbiol. 30 (1984) 133−168.

[9] M.M. López, E. Bertolini, A. Olmos, P. Caruso, M.T. Gorris, P. Llop, et al., Innovative tools for detection of plant pathogenic viruses and bacteria, Int. Microbiol. 6 (2003) 233–243.

[10] N. Boonham, J. Kreuze, S. Winter, R. van der Vlugt, J. Bergervoet, J. Tomlinson, et al., Methods in virus diagnostics: from ELISA to next generation sequencing, Virus Res. 186 (2014) 20–31.

[11] R.B. Harvey, S.B. Lee, Flagellates of laticiferous plants, Plant Physiol. 18 (1943) 633–655.

[12] F.O. Holmes, Herpetomonas bancrofti N. sp. from the latex of a Ficus in Queensland, Contrib. Boyce Thompson Inst. 3 (1931) 375–383.

[13] P.R. Hazelton, H.R. Gelderblom, Electron microscopy for rapid diagnosis of emerging infectious agents, Emerg. Infect. Dis. 9 (3) (2003) 294–303.

[14] K. Bürling, M. Hunsche, G. Noga, Use of blue-green and chlorophyll fluorescence measurements for differentiation between nitrogen deficiency and pathogen infection in winter wheat, J. Plant Physiol. 168 (2011) 1641–1648.

[15] J. Kuckenberg, I. Tartachnyk, G. Noga, Temporal and spatial changes of chlorophyll fluorescence as a basis for early and precise detection of leaf rust and powdery mildew infections in wheat leaves, Precis. Agric. 10 (2009) 34–44.

[16] L. Chaerle, S. Lenk, I. Leinonen, H.G. Jones, D. Van Der Straeten, C. Buschmann, Multi-sensor plant imaging: towards the development of a stress-catalogue, Biotechnol. J. 4 (2009) 1152–1167.

[17] D.W. Sun, Infrared Spectroscopy for Food Quality Analysis and Control, Academic Press, Massachusetts, United States, 2009.

[18] N. Abu-Khalaf, M. Salman, Visible/near infrared (VIS/NIR) spectroscopy and multivariate data analysis (MVDA) for identification and quantification of olive leaf spot (OLS) disease, Palest. Tech. Univ. Res. J. 2 (1) (2014) 1–8.

[19] G.H. Mohammed, T.L. Noland, P.H. Irving, P.H. Sampson, P.J. Zarco-Tejada, J.R. Miller, Natural and Stress-Induced Effects on Leaf Spectral Reflectance in Ontario Species, Ontario Ministry of Natural Resources, 2000.

[20] M. Zhang, Z. Qin, X. Liu, S.L. Ustin, Detection of stress in tomatoes induced by late blight disease in California, USA, using hyperspectral remote sensing, Int. J. Appl. Earth Obs. Geoinf. 4 (4) (2003) 295–310.

[21] S.A. Hawkins, B. Park, G.H. Poole, T. Gottwald, W.R. Windham, K.C. Lawrence, Detection of citrus huanglongbing by Fourier transform infrared - Attenuated total reflection spectroscopy, Appl. Spectrosc. 64 (1) (2010) 100–103.

[22] S. Sankaran, A. Mishra, J.M. Maja, R. Ehsani, Visible-near infrared spectroscopy for detection of hangdogging in citrus orchards, Comput. Electron. Agric. 77 (2) (2011) 127–134.

[23] M.C.D.B. Cardinali, P.R. Villas Boas, D.M.B.P. Milori, E.J. Ferreira, M.F.E. Silva, M.A. MacHado, B.S. Bellete, M.F.D.G.F. Da Silva, Infrared spectroscopy: a potential tool in huanglongbing and citrus variegated chlorosis diagnosis, Talanta 91 (2012) 1–6.

[24] M.F. Bhat, R. Hassan, M.H. Masoodi, Nuclear magnetic resonance (NMR) for plant profiling and disease metabolomics-fast tracking plant-based drug discovery from northern India, Int. J. Chem. Sci. 2 (1) (2018) 08–09.

[25] J.G.M. Pontes, W.Y. Ohashi, A.J.M. Brasil, P.R. Filgueiras, A.P.D.M. Espanola, J.S. Silva, R.J. Poppi, H.D. Coletta-Filho, L. Tasic, Metabolomics by NMR spectroscopy in plant disease diagnostic: huanglongbing as a case study, Chemistry Select 6 (2016) 1176–1178.

[26] E. Goldman, L.H. Green, Practical Handbook of Microbiology, Second Edition (Google eBook)), second ed., CRC Press, Taylor and Francis Group, USA, 2008, ISBN 978-0-8493-9365-5, p. 864.

[27] J.M. Henson, The polymerase chain reaction and plant disease diagnosis, Annu. Rev. Phytopathol. 31 (1993) 81–109.

[28] M.M. López, E. Bertolini, E. Marco-Noales, P. Llop, M. Cambra, Update on molecular tools for detection of plant pathogenic bacteria and viruses, in: J.R. Rao, C.C. Fleming, J.E. Moore (Eds.), Molecular Diagnostics: Current Technology and Applications, Horizon Bioscience, 2006, pp. 1–46. Wymondham, UK, ISBN: 13: 978-1904933199.

[29] P. Simmonds, P. Balfe, J.F. Peutherer, C.A. Ludlam, J.O. Bishop, A.J. Brown, Human immunodeficiency virus-infected individuals contain provirus in small numbers of peripheral mononuclear cells and at low copy numbers, J. Virol. 64 (1990) 864–872.

[30] K. Porter-Jordan, E.I. Rosenberg, J.F. Keiser, J.D. Gross, A.M. Ross, S. Nasim, C.T. Garrett, Nested polymerase chain reaction assay for the detection of cytomegalovirus overcomes false positives caused by contamination with fragmented DNA, J. Med. Virol. 30 (1990) 85–91.

[31] P.D. Roberts, Survival of Xanthomonas fragariae on strawberry in summer nurseries in Florida detected by specific primers and nested polymerase chain reaction, Plant Dis. 80 (1996) 1283–1288.

[32] A. Olmos, M. Cambra, M.A. Dasi, T. Candresse, O. Esteban, M.T. Gorris, M. Asensio, Simultaneous detection and typing of plum pox potyvirus (PPV) isolates by Heminested-PCR and PCR-ELISA, J. Virol. Methods 68 (1997) 127–137.

[33] P.M. Pradhanang, J.G. Elphinstone, R.T.V. Fox, Sensitive detection of Ralstonia solanacearum in soil: a comparison of different detection techniques, Plant Pathol. 49 (2000) 414–422.

[34] A. Olmos, O. Esteban, E. Bertolini, M. Cambra, Nested RT-PCR in a single closed tube, in: J.M.S. Bartlett, D. Stirling (Eds.), PCR Protocols, second ed.Methods in Molecular Biology, vol. 226, Humana Press, Totowa, USA, 2003, pp. 156–161.

[35] Z.K. Atallah, W.R. Stevenson, A methodology to detect and quantify five pathogens causing potato tuber decay using real-time quantitative polymerase chain reaction, Phytopathology 96 (2006) 1037–1045.

[36] E. Bertolini, A. Olmos, M.C. Martínez, M.T. Gorris, M. Cambra, Single-step multiplex RTPCR for simultaneous and colorimetric detection of six RNA viruses in olive trees, J. Virol. Methods 96 (2001) 33–41.

[37] G. Gambino, I. Gribaudo, Simultaneous detection of nine grapevine viruses by multiplex reverse transcription polymerase chain reaction with amplification of an RNA as internal control, Phytopathology 96 (2006) 1223–1229.

[38] C.I. Dovas, N.I. Katis, A spot multiplex nested RT-PCR for the simultaneous and generic detection of viruses involved in the aetiology of grapevine leaf roll and rugose wood of grapevine, J. Virol. Methods 109 (2003) 217–226.

[39] E.H. Stukenbrock, S. Rosendahl, Development and amplification of multiple codominant genetic markers from single spores of arbuscular mycorrhizal fungi by nested multiplex PCR, Fungal Genet. Biol. 42 (2005) 73–80.

[40] D. Clair, J. Larrue, G. Aubert, J. Gillet, G. Cloquemin, E. Boudon-Padieu, A multiplex nested-PCR assay for sensitive and simultaneous detection and direct identification of phytoplasma in the Elm yellows group and Stolbur group and its use in survey of grapevine yellows in France, Vitis 42 (2003) 151–157.

[41] A.H. McCartney, S.J. Foster, B.A. Fraaige, E. Ward, Molecular diagnostics for fungal plant pathogens, Pest Manag. Sci. 59 (2003) 129–142.

[42] S. Ruiz-Ruiz, S. Ambros, M. del Carmen Vives, L. Navarro, P. Moreno, G. Jose, Detection and quantitation of Citrus leaf blotch virus by TaqMan real-time RT-PCR, J. Virol. Methods 160 (2009) 57–62.

[43] C.A. Heid, J. Stevens, K.J. Livak, M.P. Williams, Real time quantitative PCR, Genome Res. 6 (2011) 986–994.

[44] M.M. Lopez, P. Lop, A. Olmos, E. Marco-Noales, M. Cambra, E. Bertolini, Are molecular tools solving the challenges posed by detection of plant pathogenic bacteria and viruses, Curr. Issues Mol. Biol. 11 (2008) 13–45.

[45] U.E.M. Gibson, C.A. Heid, P.M. Williams, A novel method for real time quantitative RT-PCR, Genome Res. 6 (1996) 995–1001.

[46] A. Olmos, E. Bertolini, M. Gil, M. Cambra, Real-time assay for quantitative detection of non-persistently transmitted Plum pox virus RNA targets in single aphids, J. Virol. Methods 128 (2005) 151–155.

[47] S. Ruiz-Ruiz, P. Moreno, G. Jose, S. Ambros, A real-time RT-PCR assay for detection and absolute quantitation of Citrus tristeza virus in different plant tissues, J. Virol. Methods 145 (2007) 96–105.

[48] J.G. Yang, F.L. Wang, D.X. Chen, L.L. Shen, Y.M. Qian, J.Y. Liang, W.C. Zhou, T.H. Yan, Development of a one-step immunocapture real-time RT-PCR assay for detection of Tobacco Mosaic Virus in Soil, Sensors 12 (2012) 16685–16694.

[49] X.S. Qu, L.A. Wanner, B.J. Christ, Multiplex real-time PCR (TaqMan) assay for the simultaneous detection and discrimination of potato powdery and common scab diseases and pathogens, J. Appl. Microbiol. 110 (2011) 769–777.

[50] M.F. Clark, A.N. Adams, Characteristics of the microplate method of enzyme-linked immune sorbent assay for the detection of plant viruses, J. Gen. Virol. 34 (1977) 475–483.

[51] J.P. Paulin, Erwinia amylovora: general characteristics, biochemistry and serology, in: J.L. Vanneste (Ed.), Fireblight, the Disease and its Causative Agents, Erwinia Amylovora, CABI Wallingford, UK, 2000, pp. 87–115.

[52] E. Ward, S.J. Foster, B.A. Fraaiji, H.A. McCartney, Plant pathogen diagnostics: immunological and nucleic acid-based approaches, Ann. Appl. Biol. 145 (2004) 1–16.

[53] I. Pankova, B. Kokoskova, Sensitivity and specificity of monoclonal antibody Mn-Csl for detection and determination of Clavibacter michiganensis subsp. sepedonicus, the casual agent of bacterial ring rot of potato, Plant Prot. Sci. 38 (2002) 17–124.

[54] A.A.G. Westra, S.A. Slack, J.L. Drennan, Comparison of some diagnostic assay for bacterial ring rot of potato: a case study, Am. Potato J. 71 (1994) 557–565.

[55] R. Zielke, A. Schmidt, K. Naumann, Comaprison of different serological methods for the detection of the fire blight pathogen, Erwinia amylovora (Burrill) Winslow et al, Zentralbl. Mikrobiol. 148 (1993) 379–391.

[56] C.P. Lin, T.A. Chen, J.M. Wells, T. van der Zwet, Identification and detection of Erwinia amylovora with monoclonal antibodies, Phytopathology 77 (1987) 367–380.

[57] M.T. Gorris, E. Camarasa, M.M. Lopez, M. Cambra, J.P. Paulin, R. Chartier, Production and characterization of monoclonal antibodies specific for Erwinia amylovora and their use in different serological techniques, Acta Hortic. 411 (1996) 47–52.

[58] S.H. De Boer, A. Boucher, T.L. De Haan, Validation of thresholds for serological tests that detect Clavibacter michiganensis subsp. sepedonicus in potato tuber tissue, Bull ORPP/EPPO Bull 26 (1996) 391–398.

[59] B. Kokoskova, I. Mraz, J.D. Janse, J. Fousek, R. Jerabkova, Reliability of diagnostic techniques for determination of Clavibacter michiganensis subsp.sepedonucus, Pfl Krankh 1 (2005) 1–16.

[60] D.J. Carter, R.B. Cary, Lateral flow microarrays: a novel platform for rapid nucleic acid detection based on miniaturized lateral flow chromatography, Nucleic Acids Res. 35 (2007) e74. https://doi.org/10.1093/nar/gkm269.

[61] F. Martinelli, R.L. Reagan, S.L. Uratsu, M.L. Phu, U. Albrecht, et al., Gene regulatory networks elucidating Huanglongbing disease mechanisms, PLoS One 8 (2013a) e74256. https://doi.org/10.1371/journal.pone.0074256.

[62] F.M. Rizzini, C. Bonghi, L. Chkaiban, F. Martinelli, P. Tonutti, Effects of postharvest partial dehydration and prolonged treatments with ethylene on transcript profiling in skins of wine grape berries, Acta Hortic. 877 (2010) 1099–1104.

[63] R. Tosetti, F. Martinelli, P. Tonutti, Metabolomics approach to studying minimally processed peach (Prunuspersica) fruit, Acta Hortic. 934 (2012) 1017–1022.

[64] A.M. Ibanez, F. Martinelli, S.L. Uratsu, A. Vo, M.A. Tinoco, M.L. Phu, Y. Chen, D.M. Rocke, A.M. Dandekar, Transcriptome and metabolome analysis of Citrus fruit to elucidate puffing disorder, Plant Sci. 217 (2014) 87–98.

[65] C.G. Webster, J.S. Wylie, M.G.K. Jones, Diagnosis of plant viral pathogens, Curr. Sci. 86 (2004) 1604–1607.

[66] S.M. Garnsey, T.A. Permar, M. Cambra, C.T. Henderson, Direct tissue blot immunoassay (DTBIA) for detection of Citrus tristeza virus (CTV), in: Proceedings 12th Conference of International Organization of Citrus Virologists, New Delhi, India 1992, 1993, pp. 39–50.

[67] M. Cambra, M.T. Gorris, M.P. Román, E. Terrada, S.M. Garnsey, E. Camarasa, A. Olmos, M. Colomer, Routine detection of citrus tristeza virus by direct immunoprinting-ELISA method using specific monoclonal and recombinant antibodies, in: Proceedings 14th Conference of International Organization of Citrus Virologists, Campinas, Brazil 1998, 2000, pp. 34–41.

[68] A.M. D'Onghia, K. Djelouah, D. Frasheri, O. Potere, Detection of citrus psorosis virus by direct tissue blot immunoassay, J. Plant Pathol. 83 (2001) 139–142.

[69] K. Djelouah, D. Frasheri, F. Valentini, A.M.D.M. Digiaro, Direct tissue blot immunoassay for detection of *Xylella fastidiosa* in olive trees, Phytopathol. Mediterr. 53 (3) (2014) 559–564.

[70] A. Volkhard, J. Kempf, K. Trebesius, I.B. Autenrieth, Fluorescent in situ hybridization allows rapid identification of microorganisms in blood cultures, J. Clin. Microbiol. 38 (2000) 830–838.

[71] B.A. Wullings, A.R. Beuningen, J.D. van Janse, A.D.L. Akkermans, A.R. Van Beuningen, Detection of *Ralstonia solanacearum*, which causes brown rot of potato, by fluorescent in situ-hybridization with 23S rRNA-targeted probes, Appl. Environ. Microbiol. 64 (1998) 4546–4554.

[72] L. Gao, W. Chen, T. Liu, B. Liu, An immunofluorescence assay for the detection of wheat rust species using monoclonal antibody against urediniospores of *Puccinia triticina*, J. Appl. Microbiol. 115 (2013) 1023–1028.

[73] F. Dewey, G. Marshall, Production and use of monoclonal antibodies for the detection of fungi, in: Proceeding of British Crop Protection Council Symposium, Farnham, UK, 18–21 November 1996, 1996.

[74] L.G. Chitarra, R.W. van den Bulk, The application of flow cytometry and fluorescent probe technology for detection and assessment of viability of plant pathogenic bacteria, Eur. J. Plant Pathol. 109 (2003) 407–417.

[75] J. Diaper, C. Edwards, Flow cytometric detection of viable bacteria in compost, FEMS Microbiol. Ecol. 14 (1994) 213–220.

[76] W.S. Kaneshiro, A.M. Alvarez, Specificity of PCR and ELISA assays for hypovirulent and avirulent *Clavibacter michiganensis* subsp. *Michiganensis*, . Phytopathology 91 (2001) S46.

[77] F.J. Louws, J.L.W. Rademaker, F.J. De Bruijn, The three Ds of PCR based genomic analysis of phytobacteria: diversity, detection, and disease diagnosis, Annu. Rev. Phytopathol. 37 (1999) 81–125.

[78] Y. Fang, Y. Umasankar, R.P. Ramasamy, Electrochemical detection of p-ethylguaiacol, a fungi infected fruit volatile using metal oxide nanoparticles, Analyst 139 (2014) 3804–3810.

[79] V. Isidorov, I. Zenkevich, B. Ioffe, Volatile organic compounds in the atmosphere of forests, Atmos. Environ. 19 (1985) 1–8.

[80] J. Kesselmeier, M. Staudt, Biogenic volatile organic compounds (VOC): an overview on emission, physiology and ecology, J. Atmos. Chem. 33 (1999) 23–88.

[81] R.M. Perera, P.J. Marriott, I.E. Galbally, Headspace solid-phase microextraction—comprehensive two-dimensional gas chromatography of wound induced plant volatile organic compound emissions, Analyst 127 (2002) 1601–1607.

[82] M. Ellis, G. Grove, Fruit rots cause losses in Ohio strawberries, Ohio Rep. Res. Dev. 67 (1982) 3–4.

[83] C. Lacroix, A. Gicquel, B. Sendid, J. Meyer, I. Accoceberry, N. Francois, F. Morio, G. Desoubeaux, J. Chandenier, C. Kauffmann-Lacroix, C. Hennequin, J. Guitard, X. Nassif, E. Bougnoux, Evaluation of two matrixassisted laser desorption ionization-time of flight mass spectrometry (MALDI-TOF MS) systems for the identification of *Candida* species, Clin. Microbiol. Infect. 20 (2) (2014) 153–158.

[84] M. Trevino, P. Areses, M.D. Penalver, S. Cortizo, F. Pardo, L. Pérez del Molino, C. García-Riestra, M. Hernández, J. Llvo, B.J. Regueiro, Susceptibility trends of *Bacteroides fragilis* group and characterisation of carbapenemase-producing strains by automated REP-PCR and MALDITOF, Anaerobe 18 (1) (2012) 37–43.

[85] L.F. Marvin, M.A. Roberts, L.B. Fay, Matrix-assisted laser desorption/ionization time-of-flight mass spectrometry in clinical chemistry, Int. J.Clin. Chem. Diagnos. Lab. Med. 337 (1–2) (2003) 11–21.

[86] J. Leushner, N.H.L. Chiu, Automated mass spectrometry: a revolutionary technology for clinical diagnostics, Mol. Diagn. 5 (2000) 341–348. https://doi.org/10.1054/modi.2000.19574.

[87] C.A. Lowe, M.A. Diggle, S.C. Clarke, A single nucleotide polymorphism identification assay for the genotypic characterisation of *Neisseria meningitidis* using MALDI-TOF mass spectrometry, Br. J. Biomed. Sci. 61 (2004) 8–10.

[88] G.W. Jackson, R.J. McNichols, G.E. Fox, R.C. Willson, Universal bacterial identification by mass spectrometry of 16S ribosomal RNA cleavage products, Int. J. Mass Spectrom. 26 (2007) 218–226. https://doi.org/10.1016/j.ijms.2006.09.021.

[89] Y.J. Kim, S.O. Kim, H.J. Chung, M.S. Jee, B.G. Kim, et al., Population genotyping of hepatitis C virus by matrix-assisted laser desorption/ionization time-of-flight mass spectrometry analysis of short DNA fragments, Clin. Chem. 51 (2005) 1123–1131. https://doi.org/10.1373/clinchem. 2004.047506.

[90] A.D. Chalupowicz, V. Gaba, N. Luria, M. Reuven, A. Beerman, O. Lachman, O. Dror, G. Nissan, S. Manulis-Sasson, Diagnosis of plant diseases using the Nanopore sequencing platform, Plant Pathol. 68 (2019) 229–238.

CHAPTER 13

Nucleic Acid—Based Methods in the Detection of Foodborne Pathogens

HAKIYE ASLAN, PHD • AYTEN EKINCI, PHD • İMRAN ASLAN, ASSOC. PROF.

INTRODUCTION

The consumption of food or water contaminated with chemicals or microbial agents such as pathogens or their toxins leads to foodborne illnesses. Despite advances in food safety, foodborne outbreaks are a significant public health problem worldwide [1]. According to the World Health Organization (WHO) report, 31 foodborne threats (bacteria, viruses, parasites, chemicals, and toxins) led to nearly 600 million cases of disease and 420,000 deaths in 2010 [2]. Among the foodborne diseases, the highest incidence is caused by microbial factors and the number of deaths from them is high [3]. It has been reported that each year, nearly 30% of the population in developed countries suffer from foodborne illnesses caused by microorganisms [1]. Furthermore, it is estimated that pathogens cause the 9.4 million foodborne disease, 55,961 hospitalizations, and 1351 deaths in the United States annually. *Campylobacter* spp., nontyphoidal *Salmonella* spp., *Clostridium perfringens*, and *Norovirus* are the most common pathogens of them causing the foodborne diseases [3,4]. *Salmonella* spp., *Listeria monocytogenes*, *Toxoplasma gondii*, and Norwalk-like viruses cause the highest number of deaths related to foodborne illnesses [5].

Foodborne diseases, as well as threatening public health, cause a significant economic loss, influence international trading of food products, and damage consumer confidence [6]. Furthermore, foodborne outbreaks induce loss of billions of dollars each year due to medical costs, lower productivity, and product recall. The total annual medical cost related to foodborne illness caused by pathogens is $6.5—9.4 billion according to the forecast of researchers at the Economic Research Service of the US Department of Agriculture [7]. To provide the safety of food, protect the public health, and prevent economic losses, many countries established early warning systems. These systems provide controls at every department of the food production chain. The critical step in these systems is the utilization of high-efficiency food safety analyzing techniques, which have to be correct and susceptible enough for detection of low levels of contamination in foods for avoiding pathogen transmission through the food chain to human [8].

Conventional or traditional cultural (phenotypic) methods for the detection of pathogen microorganisms in foods are standardized although they are time consuming and labor intensive [9,10]. Conventional cultural methods require nonselective preenrichment/ selective enrichment, morphological, biochemical, and serological tests for the identification and characterization of pathogens [11]. Generally, traditional methods require 2—3 days for preliminary identification of pathogens, while more than 1 week is required for detection at the strain level [9]. For example, the detection of *Campylobacter* by the conventional cultural method, takes 4—9 days to obtain negative results and between 14 and 16 days for verification of positive results [12]. Furthermore, traditional cultural methods are gold standards for the detection of pathogens; they are sometimes insufficient to the classification of pathogens at strain level or subtyping. For example, nearly 15% of *Mycobacterium avium* strains are not characterized by phenotypic methods because of the unavailability of proper antiserum [13]. Moreover, the detection of viable but nonculturable microorganisms by conventional cultural methods have some restrictions [14,15]. Therefore, it is necessary to apply novel methods for the rapid, sensitive, and specific identification and detection of foodborne pathogens as a complementary or an alternative to the conventional techniques. Nucleic acid—based methods (genotypic) have been developed and implemented as a fast, accurate, and precision technique for supplying microbial food safety. Some foodborne viruses such as adenoviruses or caliciviruses are not culturable because there are no standardized methods for cultivation or detection exist. So nucleic acid–based techniques remain the only choice for their detection [16,17].

NUCLEIC ACID–BASED METHODS

In recent years, significant development has been made in nucleic acid–based methods used for foodborne pathogen detection. These techniques depend on the detection of specific DNA or RNA sequences in the target pathogen microorganisms by two fundamental techniques: in vitro amplification and hybridization of nucleic acids (no amplification). In the following section, the basis of nucleic acid amplification techniques (thermal cycling amplification and isothermal amplification methods) and nucleic acid hybridization are overviewed. Furthermore, the advanced nucleic acid–based methods derived from these two fundamental techniques are discussed.

Nucleic Acid Amplification Methods

Compared to the conventional methods, nucleic acid amplification methods are fast, highly sensitive, specific, and simple so they are performed as a complement or as an alternative to the traditional cultivation methods for the detection of pathogens in foodstuffs. In this section, nucleic acid amplification techniques are categorized as thermal cycling amplification and isothermal amplification methods.

Thermal cycling amplification methods

Thermal cycling amplification methods can be called polymerase chain reaction (PCR)–based assays. PCR is one of the most common methods used to detect and identify foodborne pathogen microorganisms by molecular techniques. PCR was discovered in the early 1980s and in theory, it allows the determination of a single pathogenic cell in foods by amplifying a specific target DNA sequence [1]. The principles of PCR were described by Kleppe et al. [18] for the first time [19]. After the discovery of thermostable DNA polymerase enzyme in 1985, the first scientific study on PCR was published by Saiki and colleagues working in the field of human genetics at Cetus Corporation [20–22]. The discovery of PCR was attributed to a colleague of Saiki, Kry B. Mullis won the Nobel Prize in chemistry in 1993 [19,23]. After the invention of PCR, molecular methods for the detection and identification of microbial pathogens have been revolutionized. One of the earliest applications of PCR assays was to develop rapid techniques for the detection of foodborne pathogens [24,25]. At the present time, PCR is a widely used

molecular tool for the determination of microbial food safety and new PCR platforms have been developed to enhance the specificity, accuracy, and sensitivity of this method. In the following section, the most used PCR-based assays for the detection of microbial pathogens in foods such as conventional PCR, nested PCR, real-time PCR (qPCR), live/dead cell detection by qPCR (reverse-transcription PCR (RT-PCR) and PMA/EMA qPCR), multiplex PCR, and digital PCR are discussed later.

Conventional PCR

PCR is a nucleic acid–based molecular technique that is used for in vitro amplification of specific DNA or RNA sequences in the target pathogens [21]. A PCR reaction requires the following constituents for the transcription of target DNA or RNA sequence:

 i. *The template DNA (cDNA) containing the DNA (RNA) region to be amplified.*
 ii. *Two primers (oligonucleotides) to be connected to the beginning (forward primer) and end (reverse primer) sites of the DNA region to be amplified.*
 iii. *Heat-stable DNA polymerase enzyme that replicates the DNA region desired to be amplified by polymerization.*
 iv. *Four deoxynucleoside triphosphates (dNTPs: dATP, dTTP, dGTP, and dCTP), containing the bases necessary to synthesize the new DNA chain from the template DNA.*
 v. *Buffer solution supplying an appropriate chemical medium for optimum efficiency and stability of DNA polymerase.*
 vi. *The Mg^{+2} ion, which acts as a cofactor for the polymerase enzyme, thereby allowing the bases to be recognized by the polymerase enzyme* [26].

The temperature control required to perform DNA amplification in vitro by the PCR method is carried out in devices that allow automated temperature control called "thermal cycler." The amplification of the specific DNA sequence by PCR is carried out by two stages: holding (initialization) stage and cycling stage. Holding (initialization) stage is required for the activation of thermostable polymerase enzyme and applied by holding for 1–10 min at high temperature (usually between 94°C and 98°C) in a single step. In this step, the template DNA is efficiently denatured [27]. Cycling stage contains a series of 20–40 repetitive cycles of three steps operating at different temperatures [28]. Each PCR cycle contains three steps: denaturation, annealing, and elongation (synthesis). In the denaturation step of DNA, the temperature is generally raised to usually between 93°C and 96°C (usually > 90°C), the weak hydrogen bonds between the bases, which hold the DNA in the double-stranded helical structure, are broken, and the two single-stranded DNAs are obtained. The covalent bonds between deoxyribose and phosphate in the structure of DNA are not affected by the applied temperature because they are strong. In the second step, the annealing step, the temperature is reduced to usually between 40°C and 65°C, so that the primers bind to the regions to be complementary on the single-stranded template DNA. The temperature applied during this process is called the annealing temperature (Ta) of the primers and varies depending on the length and the sequence of the primers. In the elongation step, by using free dNTPs from the reaction mixture, the DNA polymerase enzyme synthesizes a new DNA strand complementary to the DNA template strand. The temperature at this step should be brought to 72°C, which is the optimum temperature for the activity of *Thermus aquaticus* (Taq) polymerase enzyme. The time of the elongation phase can vary depending on the length of the region to be amplified. In the PCR analysis, generally, 1 min is sufficient for the synthesis of new DNA strand [29]. As the synthesized new products contain primer binding sites, they behave as template DNA in the next cycle and so that a series of chain reactions are carried out to geometrically replicate the specific regions of DNA. So, it is possible to obtain billions or trillions of copies of the selected region on the genome as a result of applying a series of 20—40 repetitive cycles during a PCR reaction.

PCR products, amplicons, can be detected through several methods including electrophoresis, DNA hybridization and nongel methods such as enzyme-linked immunosorbent assay [4]. Among them, gel electrophoresis methods are widely used to evaluate the amplification products formed after the PCR process. For this purpose, amplified DNA fragments are separated according to their length by agarose or polyacrylamide gel electrophoresis and visualized under ultraviolet light by staining with double-stranded DNA dyes such as ethidium bromide, SYBR GREEN, and Nancy-520 [26].

PCR-based assays have been used as powerful diagnostic tools for the detection of microorganisms present in foods [30,31]. At present, the International Organisation for Standardisation (ISO) has developed ISO 22,174 "Microbiology of food and animal feeding stuffs-polymerase chain reaction for the detection of foodborne pathogens — general requirements and definitions." In this standard, principal criteria and parameters for PCR performance as diagnostic tools are described. PCR assay has been used for the detection

and validation of various foodborne bacteria due to its specificity, sensitivity, and rapidity [9]. Several PCR-based protocols have been developed over last decades to detect different foodborne pathogens such as *Escherichia coli* [32], *Listeria monocytogenes* [33], *Salmonella* spp. [34], *Campylobacter jejuni* [35], *Staphylococcus aureus* [36], *Shigella* spp. [37], *Yersinia enterocolitica* [38], *Bacillus cereus* [39], and *Vibrio cholerae* [40]. The efficiency of PCR to detect microorganisms in foods depends on the purity and sufficient amount of the template used as a target [31,41]. Furthermore, the sensitivity of PCR may be affected by the presence of PCR inhibitors in food samples that cause the production of false-negative results [42]. Therefore, efficient sample treatment methods for preparing template DNA are needed to remove the inhibitory substance and improve the utility of PCR in the detection of foodborne pathogenic microorganisms [31,43]. To overcome these limitations, several methods have been suggested including enrichment of food samples in a suitable medium specific for the target pathogen(s) [44], immunomagnetic separation [43], buoyant density centrifugation [45], aqueous two-phase system [46], and different DNA extraction procedures from food samples [47]. The major drawbacks of conventional PCR are the requirement for postamplification handling steps that is time consuming and has contamination risk of amplicons and the initial concentration of pathogens in food samples cannot be determined [48].

Nested PCR

Nested PCR, a modified conventional PCR method, was developed to increase the amount of amplified products in the PCR reaction and the sensitivity of PCR-based detection [31]. In this method, two PCRs have been applied by two sets of primers (P1F, P1R, and P2F, P2R). In the first PCR, a long region is amplified using two outer primers (P1F and P1R) specific to the outer region of the target DNA sequence. Then, a small aliquot of the primary PCR amplification products, perhaps 1 μL of a 1-in-10 or 1-in-100 dilution can be used as the template for the secondary PCR [27]. The secondary PCR is performed using the second set of primers (P2F and P2R) on the DNA sequence obtained as a result of primary PCR amplification. The application of secondary amplification process improves the sensitivity and specificity of PCR detection [49].

Nested PCR is predicted to increase sensitivity by 10^4-fold in amplifying the target DNA sequence but great attention must be taken to avoid contamination when pipetting the primary PCR products. To prevent any contamination problems, both the primary and secondary PCR reactions are suggested to be performed using two different reaction temperatures in a single tube [27,50,51].

Minarovičová et al. [52] developed a single tube nested real-time PCR method for the sensitive detection of *Cryptosporidium parvum* in environmental and food samples by reducing the risk of laboratory contamination. Saroj et al. [53] reported a rapid, sensitive, and validated the method for detection of *Salmonella typhimurium* in food by enrichment in lactose broth culture and nested PCR combination assay. The results of the research showed that the PCR amplification are not affected by background microflora and food contents, so the combination of an enrichment step and nested PCR assay increases the sensitivity and decreases the generation of false-negative results for detection of *Salmonella typhimurium* in food samples.

Real-time PCR

The real-time PCR or quantitative PCR (qPCR) technique can be defined as a PCR method in which the amplification process of target DNA or cDNA are simultaneously monitored and the accumulated amplicon, quantified by measuring the fluorescence produced during each cycle of the PCR reaction. In the qPCR method, different from conventional PCR, real-time monitoring of the amplification process reduces analysis time and the risk of contamination by eliminating the need for post-PCR analysis, such as gel electrophoresis. Furthermore, the qPCR technique can be determined by the initial DNA concentration with accuracy and high sensitivity, in contrast to conventional PCR [54]. Another important advantage of qPCR methods is that the use of various fluorescent dyes and probes at different wavelengths allow the development of multiple assays with an automated system [25,55]. In recent years, qPCR technology has become an increasingly choice technique for research and diagnostic applications alternative to conventional PCR. In microbial food safety, qPCR assays have been widely used for the detection of pathogenic microorganisms and virulence factors such as *Listeria monocytogenes* [56], *Salmonella* spp. [57], *Escherichia coli* O157: H7 [58], *Campylobacter jejuni* [59], *Staphylococcus aureus* [60], and noroviruses [61].

In qualitative (indicating the presence or absence of the target DNA sequence) and quantitative analysis of samples by qPCR, the fluorescence signal increases proportionally to the amount of PCR product [62,63]. The measured fluorescence signals by the detector of the qPCR machine are converted into the numerical values by the software program. The most crucial numerical variable in the qPCR method is the threshold cycle (CT) value. The CT value is the cycle at which fluorescence reaches a defined threshold. The CT value indicates the number of cycles in which the first significant increase in the amplification product. During qPCR analysis, an amplification curve is obtained (Fig. 13.1). An amplification curve contains three phases: initiation phase, exponential phase, and plateau phase. Initiation phase takes place during the initial cycles of PCR where the intensity of emitted fluorescence is indistinguishable to the baseline. In exponential or log phase, the qPCR products double at every cycle.

The last phase is the plateau phase that the reagents are exhausted and the amplification reaction stops. The quantitative determination of amplification products of target DNA sequence in qPCR method can be determined by interpolating the obtained CT value of sample analyzed in a linear standard curve of CT values

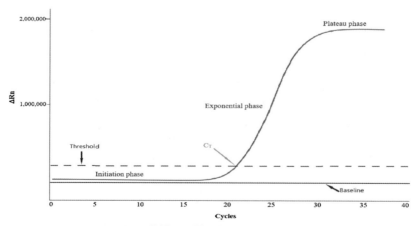

FIG. 13.1 qPCR amplification curve and CT value.

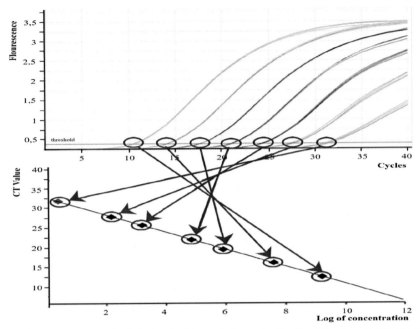

FIG. 13.2 Standard curve method. A standard curve is formed using the CT values obtained against the logarithmic concentration values of the serial dilution of target nucleic acid molecules (DNA, RNA, cDNA, etc.).

obtained from serially diluted known-amount standards (Fig. 13.2) [19].

Various fluorescent detection systems have been developed for qPCR analysis. These can be classified as follows: (1) nonspecific fluorescent dyes such as SYBR-green I, ethidium bromide (EtBr), Eva Green, BEBO, YOYO-1, and TOTO-1, which are binding any double-stranded DNA, (2) sequence-specific probes that are oligonucleotide labeled with a fluorescent reporter called TaqMan, Molecular Beacons, fluorescence resonance energy transfer (FRET) hybridization probes, Scorpion primers, etc. [64].

Nonspecific Fluorescent Dyes Connected to Double-Stranded DNA: qPCR amplification products can be detected by using nonspecific fluorescent dyes. The specialty of these dyes is to give increasing fluoresces when binding double-stranded DNA during the analysis. There are several commercial dyes as SYBR Green, EtBr, Eva Green, BEBO, YOYO-1, TOTO-1, etc. SYBR Green I dye is the most used among them [65]. SYBR Green is a fluorescent dye that fluorescence when it binds to double-stranded DNA and is visible under UV light. This dye is oxidized at a wavelength of 480 nm and reduced at a wavelength of 520 nm. The SYBR Green fluorescence intensity increases by 1000-fold when it bounds to double-stranded DNA

(dsDNA), so measuring the increase in SYBR Green fluorescence allows the detection and quantitation of PCR product after each PCR cycle [66]. PCR products are monitored by measuring the increase in fluorescence in real time during PCR reaction (Fig. 13.3) [67]. SYBR Green used in qPCR analysis, the fluorescence is measured after the each extension stage of every cycle [68].

SYBR Green I method for detecting foodborne pathogen microorganisms by qPCR is cost-effective because it does not require fluorescence-labeled probes and is more practical as there is no need to design probes. However, as SYBR Green binds to all double-stranded DNA molecules, it does not always indicate that the target DNA region is duplicated, and a false positive result may be obtained. Fluorescence emission can also occur as a result of the formation of short double-stranded DNA regions called "primer dimers" as a result of the binding of nonspecific amplicons and primers in the environment [69]. Whether or not the amplification product is in the desired target region of DNA can be determined by performing a melting curve analysis in qPCR devices [70]. The melting curve analysis allows the identification of the difference between nonspecific amplicons and primer dimers with the main PCR product [71].

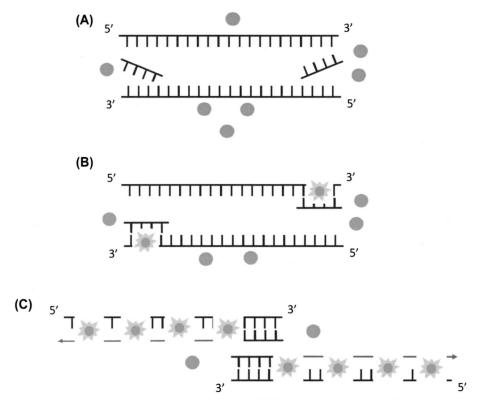

FIG. 13.3 SYBR Green I fluorescence mechanism: **(A)** SYBR Green I Dye is free when DNA is denatured, **(B)** a small amount of SYBR Green fluorescent appears when primers bind to the target DNA molecule **(C)** Fluorescence increases when SYBR Green Dye binds to newly synthesized double-stranded DNA molecule.

Target Specific Probes: DNA probe hybridization methods analyze the presence or the absence of a specific part of genes. So, it is possible for the identification of a target organism. These methods are also used for the detection and identification of pathogenic microorganisms. In this method, the short, specific single DNA strands called probe are being used. These probes are sequence-specific to target genes and labeled with reporter dyes [72]. The most important advantage of probes is that they increase the specificity of the analysis and allow the application of multiplex studies based on the amplification of different species in the same tube. A probe should give strong fluorescence when it is bound to the target while giving a weak fluorescence in free form and should be high specificity. The most common fluorescence-labeled probes used in the qPCR technique are "*TaqMan probe, Molecular beacons, Fluorescence Resonance Energy Transfer (FRET), Hybridization Probes and Scorpion Primers.*"

TaqMan Probe: TaqMan probes are developed for qPCR analysis to eliminate the difficulties in SYBR Green I and called as "double-dye oligonucleotides, dual-labeled probes, or 5′ nuclease probes." TaqMan probes are single-chain oligonucleotides that contain a fluorescent reporter dye at the 5′-end and a nonfluorescent quencher dye at the 3′-end [64]. The reporter dye and quencher dye are close to each other in an intact probe, so the quencher prevents the emitted fluorescence of the reporter [68]. During amplification, TaqMan probe connected to specific nucleotide sequence in one of the strands of amplicon internal to both primers on the denatured target DNA molecule and is cleaved by the 5′ nuclease activity of Taq DNA polymerase. Then, Taq DNA polymerase extends the primer to synthesize new DNA molecule and breakdowns the TaqMan probe from its 5′ end. The breakdown of the probe separates the reporter dye from quencher dye, and increases the reporter dye fluorescence intensity (Fig. 13.4). TaqMan probe is used in qPCR analysis, and the fluorescence is measured following each extension stage of every cycle [68].

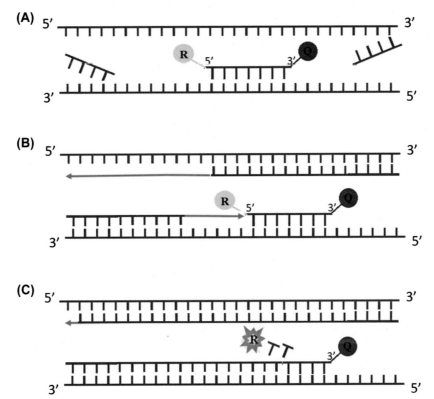

FIG. 13.4 TaqMan probe mechanism: (A) TaqMan probe binds a specific region between the primers target region on the denatured DNA molecule, (B) during the synthesizing of each new DNA molecule, the TaqMan probe begins to break down from its 5′ end, and (C) the reporter dye is released at the 5′ end, and it increases the fluorescence signal.

Molecular Beacons: Molecular beacons are single-stranded short oligonucleotide probes with hairpin or stem-and-loop shaped. The sequence complementary to a target sequence is present in the loop portion of molecular beacons [73]. The stem part of molecular beacons has five to seven base pairs that are complementary. A reporter dye is linked to one end of the stem, and a quencher dye is linked to the other end of the stem. These two dyes are near, so the quencher prevents the emission of fluorescence-labeled reporter dye during the real-time PCR reaction, the probe comes across the target DNA, and the molecular beacons undergoes conformational changes and becomes a single linear chain. Hence, the reporter and quencher dye are separated from each other. In this way, the reporter dye emits intense fluorescence. The fluorescence signal is measured after the annealing step of each cycle [68].

FRET Probes: FRET probes contain two fluorescently labeled oligonucleotides. Each probe has a single label with different fluorescent dye. The probe labeled with a fluorophore at the 3′ end is called donor while the other probe labeled with a fluorophore at the 5′ end is called acceptor. The sequences of the two probes are designated to hybridize the same strand of an amplicon in a head-to-tail orientation. When the FRET probes hybridize the target DNA, the distance between donor and acceptor dye is critical and must be in near (generally 1—5 nucleotides distance). During the annealing step, the two probes hybridize to the adjacent regions of the target DNA. The donor fluorophore dye is excited by an external light source, and then the excitation energy of donor is transferred to the acceptor fluorophore. The excited acceptor fluorophore emits light at a different wavelength that can be detected and measured after annealing step of each cycle [68].

Scorpion Primers: Scorpion primers term consists of a specific probe sequence that is held in a hairpin loop configuration and a target-specific PCR amplification primer covalently attached to this probe sequence. On the hairpin probe, there is a fluorophore attached at the end of 5′-end and a quencher dye (generally methyl red) joined to the 3′-end. This probe is linked to the 5′-end of a target-specific PCR amplification primer with a PCR stopper. During the PCR, the primers perform replication of amplicons, then the probe opens and hybridizes to this newly synthesized amplicon; thus, fluorophore separates from the quencher dye and fluorescence. The amount of PCR product is determined by measuring this fluorescent increase [74].

Finally, the qPCR is a quite well nucleic acid–based method for detecting, quantifying, or typing microorganisms that are important for food safety. The qPCR techniques described earlier are common methods for the detection of foodborne pathogens. In many research, these methods have been used successfully (Table 13.1).

Determination of live/dead cells using qPCR with PMA and EMA, reverse-transcription qPCR

The use of rapid and reliable methods for detecting live pathogenic bacteria at the species level has great importance for the food industry and public health [89,90]. In the food industry, the efficiency of the technological processes used in processing food stuff on pathogenic microorganisms is generally assessed by the ability of cultured cells to multiply as colonies in agar environments with cultural-based methods. Detection of pathogenic bacteria alive in foods by conventional cultural-based methods is simple and inexpensive methods that are standardized, but time consuming and labor intensive. In addition, some pathogenic microorganisms provide resilience to changes in the structural, physiological, biochemical, and molecular levels to protect against adverse environmental conditions (e.g., nutrient deprivation, pressure, and temperature application, etc.). In encountering such environment, they are metabolically active although they lose their ability to dissociation, growth, and colony formation [91,92]. In fact, some pathogenic bacteria have been found to survive for several months [93,94]. Therefore, when cultivating live pathogenic bacteria in foods, the presence of cells that cannot be developed in culture but in the environment will cause false-negative results (viable but nonculturable bacteria). Therefore, other techniques have been developed as an alternative to classical cultural methods in the detection of pathogenic bacteria found live in food. Rapid, reliable, precise, and reproducible results can be obtained by qPCR, one of the modern techniques alternative to classical cultural methods, in the detection of foodborne

pathogens. Furthermore, quantitative detection of microorganisms in different origins is also possible [30]. However, this method is inadequate in the distinction of living/dead cells in a mixed population of microorganisms consisting of living and dead cells [95,96]. Due to the fact that DNA protects its structure from several days to several weeks after the death of the cell, the number of pathogenic microorganisms may be higher than the true value in PCR-based methods. To overcome this problem, the most commonly used method in the past has been the detection of mRNA, an indicator that bacterial metabolism is active.

Reverse-transcriptase (RT)-qPCR is a rapid and sensitive method that can be performed in gene expression analysis as a result of the RNA molecules isolated from cells by performing complementary DNA (cDNA) synthesis with the help of reverse-transcriptase enzyme isolated from retroviruses. With this method, very small amounts of RNA can be detected and the amount of expression can be determined. RT-qPCR method using mRNA as the target nucleic acid has been applied to different microorganisms [97]. However, RT-qPCR method is a complex and expensive technique that requires extensive optimization. Furthermore, RNA degradation can occur during processing and storage of the sample especially in complex matrix food samples. RNA degradation leads to false-negative results [98].

In recent years, it has been recommended that samples to be analyzed before DNA isolation should be treated with nucleic acid intercalating dyes such as ethidium monoazide (EMA) and/or propidium monoazide (PMA) in the detection of live cells by qPCR [99].

EMA and PMA are a type of nucleic acid intercalating dye that is a derivative of propidium iodide [100,101]. EMA and PMA penetrate cells with damaged membranes and bind into the cell's DNA. Before DNA isolation, the samples containing EMA and PMA are incubated for a while in a dark place at room temperature and later exposed to strong visible light to provide photolysis of EMA and PMA. During photolysis, it is indicated that the azeotropic group in EMA and PMA structure turns into a highly reactive nitric radical and the resulting nitric radical modifies the DNA structure by forming carbon–nitrogen covalent bond with the C atom in the DNA molecule, thus preventing DNA amplification during PCR analysis [102].

Nocker and Camper [103] found that the DNA to which the EMA was linked turned into insoluble form and was removed with cell debris during DNA isolation. Furthermore, EMA and thereby, DNA in the living cells with undamaged cell membrane/wall is not affected by

TABLE 13.1
Real-Time PCR-Based Methods Reported for Detection of Pathogenic Microorganisms in Several Food Samples

Pathogenic Microorganisms	Real-Time PCR Method	Target Genes	Detection Limit	Food Sample	References
Bacillus cereus	SYBR green	pc-plc	3 CFU per reaction or 60 CFU/mL of food	Liquid egg and reconstituted infant formula	[75]
Campylobacter jejuni and Campylobacter coli	TaqMan-based multiplex qPCR	hipO, ccoN, and cadF genes	10^2–10^3 CFU/mL	Chicken juice and tap water	[76]
Vibrio cholera V. parahaemolyticus V. vulnificus	TaqMan	hlyA, ctxA tlh, tdh hyl A, vvhA	0.6 ± 0.3 CFU/reaction for V. vulnificus and V. Cholerae and 0.7 ± 0.4 CFU for V. parahaemolyticus	Several seawater fishs (Alaska pollock, Theragra chalcogramma; striped catfish, Pangasianodon hypophthalmus, etc.)	[77]
Escherichia coli O157: H7	TaqMan	uidA gene	10^2 CFU/mL	Raw milk cheeses and raw ground meat	[78]
		rfb E. coli O157 and stx2	10^4–10^9 CFU/g (CFU/mL)	Pasteurized milk, beef, and apple juice	[79]
		uidA gene	10^4 CFU/g	Ground beef	[80]
Listeria monocytogenes	Hybridization probe	hlyA gene	<100 CFU/g	Sausage, ground meat, processed cheese, processed milk, infant formula	[81]
Listeria spp.	SYBR green	23S rDNA gene	4.4 log CFU/g	Spiked soybean sprout samples	[82]
Salmonella enterica and Salmonella bongori	TaqMan	ttrBCA genes	10^3–10^4 CFU/mL	Chicken rinses, minced meat, fish, and raw milk	[83]
Salmonella enterica		ssaN	1 cfu/10 g sample	Chicken, liquid egg, and peanut butter samples	[84]
Salmonella spp.		invA gene	100 fg/PCR reaction	Chicken, minced meat, salmon, milk chocolate, tiramisu, vanilla, chocolate and walnut ice cream, black pepper, paprika, garlic, green and fennel tea	[85]
Staphylococcus aureus		Sa0836 gene	10^1–10^7 CFU/mL	Milk	[86]
Staphylococcus aureus, Salmonella enterica, Bacillus cereus and Vibrio parahaemolyticus	TaqMan-based multiplex qPCR	invA, hbl, and tlhA genes	$10^{2.5}$ CFU/mL for S. aureus, S. enterica and V. parahaemolyticus and $10^{3.5}$ CFU/mL for B. cereus	Mixtures of pure cultures (S. aureus, S. enterica, B. cereus and V. parahaemolyticus)	[87]
Staphylococcus aureus	TaqMan	Acr genes (AcrB/AcrD/AcrF)	6.8×10^1 and 3.4×10^1 CFU/mL	Several food samples (meat products, fish products, dairy products, delicatessen, etc.)	[88]

acr, Acriflavine resistance protein gene; bip A, Bvg-intermediate-phase gene; cadF, cadF protein gene; ctxA, Cholera enterotoxin A gene; ccoN, ccoN protein gene; hbl, Hemolysin BL gene; hipO, hippurate hydrolase; hlyA, Hemolysin A gene; invA, invasion protein invA gene; pc-plc, phosphatidylcholine-specific phospholipase C gene; $rfb_{E.\ coli/O157}$, E.coli O antigen gene; Sa0836, putative transcriptional regulator gene of S. aureus; ssaN, secretion ATP synthase gene; stx 2, Shiga-like toxin two gene; tdh, Thermostable direct hemolysin; tlh, Thermolabile hemolysin; ttrBCA, tetrathionate reductase gene; uidA, beta-glucuronidase gene; vvhA, gene for the cytolysin and contained the N-terminal amino acid sequence.

EMA or PMA during the DNA isolation process [104,105]. As a result, EMA/PMA application to bacterial cultures of living and dead cells provides selective removal of dead cell DNA. Methods of EMA-qPCR and PMA-qPCR were tested in many bacteria types as in *E. coli* O157: H7, *S. typhimurium* [99], *L. Monocytogenes* [106,107], *Campylobacter jejuni* Rudi et al. [106], and *Vibrio vulnificus* [96].

The EMA-qPCR method was first used by Nogva et al. [99] to detect viable *E. coli* O157: H7, *L. Monocytogenes* and *Salmonella* live cells. However, Nocker and Camper [103] reported that the EMA application resulted in the removal of genomic DNA of dead cells in *E. coli* O157: H7 culture also could cause a loss of about 60% in the genomic DNA of live cells. In addition, in this study, it was determined that EMA penetrated to other living cells of bacterial species and caused partial DNA loss. Similar results were obtained in another study where *C. jejuni* and *L. monocytogenes* as model organisms were used and depending on the bacterial cells and used concentration of EMA (1−100 μg/mL) in the sample, EMA application affected not only the dead cells but also the living cells and it was not a successful indicator of survival [108].

We suggest that the disadvantages of EMA in live-dead cell differentiation may be reduced by the use of PMA [102]. Compared to EMA, PMA molecule has a higher charge (EMA has one, PMA has two positive charges) and is successfully applied to a wide range of cell types and has a very high selectivity [109,110].

Digital PCR

Digital polymerase chain reaction (dPCR) is an emerging and powerful technique that allows determining the "absolute quantification" of target nucleic acid sequences by counting positive amplification signals derived from amplification of a single DNA template from minimally diluted samples [111,112]. The dPCR method was developed by Sykes et al. [113] and Vogelstein & Kinzler [114] to expand the applications of PCR. Currently, qPCR is a widespread method for the nucleic acid measurement due to the lower cost [115]. It has been used extensively in the diagnostic laboratory for the detection of genetically modified organisms, foodborne pathogens and others [116]. In qPCR, the amplification reactions are monitored along the PCR process, and the results detected as cycle threshold (CT) value at which fluorescence produced from fluorescent DNA-intercalating dyes or fluorescent dye-labeled probes during PCR reach a certain threshold. The "relative quantification" of target nucleic acid can be calculated by comparing the CT of the analyzing

sample to the standard curve that generated from a known amount of target [117]. The quantification accuracy of qPCR depends on the careful calibration of the standard curve. Furthermore, the same source material must be used as a reference to establish the standard curve because of preventing the differences in parallel testing results among laboratories [116]. In contrast to qPCR, the quantification of nucleic acid in dPCR does not base on the CT values, standard curves, and internal controls. Although the same fluorescent dyes are used for the nucleic acid amplification reactions in both techniques, highly sensitive, specific, precise, and reproducible measurements are obtained by dPCR [118]. Furthermore, it has been shown that dPCR is more resistant to PCR inhibitors than qPCR [119]. In dPCR, the sample volume used in the amplification reaction is separated into many small volumes, so thousands or millions of partitions (pL−nL range) are generated by using microwell plates, capillaries, oil emulsions, or arrays. Each partition is individual reaction mixture contains either a few target nucleic acid (ideally a single target nucleic acid in each partition) or none at all. Then, the amplification of partitioned samples is completed to the end point and the number of positive (fluorescent is present) and negative partitions are counted to determine the target copy number in the sample. In cases where the target concentration is high, the relevant partition probably contains two or more targets. So, Poisson's Law is used to correctly enumerate the number of target nucleic acids per partition and the copy number in the analyzing sample [117,120−122]. Commercially, several dPCR systems are available and the reactions occur either in a droplet-based dPCR system (QX100 Droplet Digital PCR; Bio-Rad) or two chip-based dPCR systems (Biomark HD, Fluidigm; QuantStudio 3D Digital PCR, Thermo Fisher Scientific) [123]. The trade-off between the degree of precision, throughput and the cost of the system and the assay affects the choice of dPCR systems [115,116].

dPCR has been extensively used especially in clinical microbiology and cancer research in medicine and determination of genetically modified organisms in food science [117,124−126]. There is a limited number of research on the availability of dPCR for the detection of foodborne pathogens. Wang et al. [127] compared the effect between qPCR and digital droplet PCR (ddPCR) in detecting *Salmonella typhimurium* in milk samples. They reported that ddPCR had stronger resistance to inhibitors than qPCR. Furthermore, they stated that the ddPCR technique has advanced in sensitivity, resistance, and decreasing preculturing time, representing its great applicability in detecting food-borne pathogens, especially for zero tolerance. Cremonesi et al. [128] developed the first ddPCR method without selective enrichment to simultaneously determine *Listeria* spp., *L. monocytogenes*, *Salmonella* spp., verocytotoxin-producing *E.coli* and *Campylobacter* spp. in soft cheese. Monteiro and Santos [129] improved a new protocol able to detect Norovirus using a single-step digital PCR reaction (RT-dPCR) and compared the performance of the RT-dPCR with RT-qPCR. Although the quantitative data determined by both methods were not significantly different, single-step absolute quantification of dPCR is a useful method to minimize the analyzing time to amplify viral RNA. According to the results of the scientific research, dPCR technology showed good performance at detection of foodborne pathogens; hence, further research is needed to increase the applicability of this technique in the routine analysis of food products.

Isothermal-based amplification methods

The isothermal amplification of nucleic acids at a constant temperature has a practical and promising alternative method to PCR [130]. Although PCR is the best-known nucleic acid amplification technology, it requires expensive and high-precision thermal cycler device that mostly restrict the application of PCR in resource-limited settings and for point-of-care analysis. Furthermore, it needs highly purified nucleic acids to maintain the activity of Taq DNA polymerase used in PCR assay [131,132]. Isothermal nucleic acid amplification techniques were developed to overcome these drawbacks of PCR by eliminating the need for a thermal cycler device. Besides, isothermal amplification techniques are easier to use, cost effective, and more tolerant to the inhibitory components compared to PCR [8,25,133]. Isothermal amplification techniques contain loop-mediated isothermal amplification (LAMP), nucleic acid sequence—based amplification (NASBA), helicase-dependent amplification (HDA), strand displacement amplification (SDA), rolling circle amplification, recombinase polymerase amplification (RPA), hybridization chain reaction, isothermal exponential amplification reaction, single primer isothermal amplification, isothermal and chimeric primer-initiated amplification of nucleic acids, and isothermal multiple displacement amplification [133]. Currently, various platforms where isothermal amplification of nucleic acids can be made are commercially available. Those are GenieII (OptiGene) and Illumigene (Meridian Bioscience Inc.) based on LAMP amplification, OligoC-TesT (Coris BioConcept) and NucliSENS easyQ

(BioMerieux) based on NASBA, Twista (TwistDx,UK) based on RPA, AmpliVue (Quidel Corporation) based on HDA, and BD Probetec ET System (Becton Dickinson, USA) based on SDA 119]. In this section, some isothermal amplification techniques frequently investigated for the detection of foodborne pathogens are discussed later.

Loop-mediated isothermal amplification

The LAMP is the most widely applied isothermal amplification method developed by Notomi et al. [134] and based on auto-cycling strand displacement nucleic acid amplification carried out by *Bst* DNA polymerase large fragment enzyme at a constant temperature of 60−65°C for 60 min [63]. In the LAMP method, a set of four specially designed primers including two inner primers and two outer primers are used to amplify six specific regions of target DNA [134]. The LAMP products are mixtures of many different sizes of stem-loop DNAs with several inverted repeats of the target nucleic acid sequence and cauliflower-like structures with multiple loops [135−137]. A comprehensive definition of LAMP reaction mechanism is available on the Eiken Chemical Co., Ltd. Website (http://loopamp.eiken.co.jp/e/index.html) [138].

The LAMP amplification products can be determined by either end-point methods or real-time assays. Agarose gel electrophoreses or lateral flow devices are used for end-point detection [133]. Turbidity [139] or fluorescence determination methods [140] are used for real-time detection or visual inspection of LAMP products. In the turbidity detection method, magnesium pyrophosphate that is the by-product of DNA synthesis is used as an indicator to assess the appearance of white turbidity. Furthermore, the instruments for real-time monitoring of LAMP products that rely on the turbidimetric measurement of magnesium pyrophosphate precipitation are commercially available (Eiken Chemical Co., Ltd.,Tochigi, Japan). The fluorescence detection methods use fluorescent chelation dyes such as calcein [140], hydroxyl naphthol blue [141], or SYB green dye l [142] and visual determination of LAMP products can be performed by naked eye assessment through the color changes [133].

Many studies have been shown that detection of foodborne pathogens by LAMP methods provides more specific and sensitive results than PCR assays [143,144]. Moreover, the amount of amplicon produced by LAMP within 1 hour is 10^3-fold or higher level than traditional PCR, so LAMP reaction is more rapid than conventional PCR [63]. LAMP methods have been used for the investigation of various foodborne pathogens due to its rapidity and sensitivity. Further-more, commercial LAMP kits are available for the detection of *Listeria*, *Salmonella*, *Campylobacter*, *Legionella*, and verotoxin-producing *Escherichia coli* [63,145].

Nucleic acid sequence–based amplification
NASBA is an isothermal (41°C) and sensitive method used to amplify RNA. The multiplication of RNA is achieved by two specially designed primers to amplify complementary sequences of the target RNA and an enzyme mix consisting of avian myeloblastosis virus to reverse transcriptase for cDNA synthesis, RNase H to degrade the RNA strand of the RNA:DNA heteroduplex, and T7 RNA polymerase to synthesize RNA from the T7 promoter [4,146,147]. NASBA products can be detected by various methods such as gel electrophoresis [148], colorimetry [149], electrochemiluminescence [150], or real-time detection by molecular beacons [151]. A molecular beacon-based real-time NASBA assay provides the opportunities that no contamination during the amplification of nucleic acids amplifies the nucleic acids at constant temperature (41°C) without the need for a thermal cycler, requires less time than 1 h for the reaction, and has higher sensitivity compared to reverse-transcriptase qPCR [152–154]. This technique has been performed for the detection of foodborne pathogens such as *Salmonella* genus [155], *Listeria monocytogenes* [151], *Campylobacter jejuni* [156], *Bacillus* spp. [157], hepatitis A virus, and rotavirus [158]. Furthermore, NASBA has been applied to detect viable microorganisms in food samples [159].

Nucleic Acid Hybridization Methods
Nucleic acid hybridization is a screening technique for the detection of foodborne pathogens. This technique provides quite fast detecting of foodborne pathogens without needing enrichment medium. Furthermore, this technique enables the identification of closely related nucleic acid molecules in two different populations [48,160].

In this technique, labeled DNA probes are required for hybridization. Probs are nucleic acids that are labeled with a radioactive or nonradioactive marker, complementary to the investigated nucleic acid sequence. By the agency of the marker that probe contains, the double-stranded molecule formed by the hybridization reaction becomes visible. Thus, the labeled probe makes it possible to identify the target DNA or RNA in an unlabeled complex nucleic acid mixture containing the DNA or RNA molecule of interest [160]. The DNA probes are usually double-stranded DNA and generally consist of 15–30 nucleotides [161]. Although single-stranded

oligonucleotides and RNA probes should not be denatured, the double-stranded DNA probes should be denatured before hybridization. After denaturation, DNA probes annealed to sample DNA that has also been denatured. Hybridizations can be performed in all DNA–DNA, DNA–RNA, or RNA–RNA combinations, and the nucleotide sequence of the probe controls the specificity of the hybridization application [48]. In the hybridization technique, first of all, the cell is lysed and purified to obtain the free nucleic acid. Then the target nucleic acid is denatured by high temperature (above 95°C) or alkaline pH (above 12). After denaturation, labeled nucleic acid probes bind to the target nucleic acid and synthesize the complement sequence of the single-chain target nucleic acid to form double-stranded hybrids [48,160,161].

In addition, RNA molecules are often preferred as the target nucleic acid for hybridization because rRNA molecules copy number is higher than DNA in the target bacterial cell. Thus, using rRNA molecules as a target increased the detection sensitivity of hybridization products. However, the low specificity of the rRNA-based method restricts the use of this method. Because there is a very similar rRNA sequence between closely related species (e.g., *Listeria monocytogenes* and *Listeria innocua*) so in these conditions using rRNA for hybridization is insufficient to distinguish these species [160,161].

The hybridization products can be detected directly or indirectly. For direct detection, radioactive and fluorescent probes can be used. Indirect detection is possible with using enzyme reporters on solid media membranes (nitrocellulose or nylon) and polymer particles. Southern blot and dot blot are most common techniques for indirect detection [161].

Fluorescent in situ hybridization
The fluorescent in situ hybridization (FISH) method is a molecular method based on the principle of marking the target with fluorescently labeled probes targeting the rRNA and imaging it in a fluorescent microscope [48]. The probes used in the FISH technique can be labeled in two ways. If the fluorescent nucleotides are used for labeling, it is called direct labeling. However, if probes are combined with reporter molecules that are defined by fluorescent antibodies or other affinity molecules, it is called indirect labeling [162].

To identify specific pathogen microorganisms in a complex species, medium FISH is an original microscopic technique. Thus, featuring new methods and innovative approaches make FISH technique more critical to research microorganisms in food matrices. It is based on hybridization with 16S or 23S rRNA and does not need to the cultivation of target microorganisms so that it provides a large work area for microbial identification. Studies have shown that FISH technique can be used in the identification of bacteria even at low cultivable capability [48,163]. Furthermore, there are several studies about pathogen detection and identification in foods via FISH method [164,165]. For example in a study for detection and enumeration of *Salmonella enterica* in food samples, it is reported that FISH method is more rapid and specific according to other methods [164]. In another study, FISH method used for detection of *Listeria* genus, *Listeria monocytogenes* and *Listeria ivanovii* species showed that this method is suitable for identification of pathogenic *Listeria* spp in food and food-like samples [165].

DNA microarray

Arrays obtained by binding identification ligands such as various oligonucleotides, cDNAs, proteins, peptides, antibodies, carbohydrates, tissues, or aptamers, to discrete points on a solid matrix, are called microarray. In other words, DNA microarrays (also known as gene chips, DNA chips or biochips) are microscopic DNA spots that are sequenced to monitor the expression level of thousands of genes simultaneously on a solid surface such as glass, plastic, or silicon chip [160]. The spots where DNA microarrays are generated can be composed of short oligonucleotides (20 ± 25 bp), specific gene sequences (500 ± 2000 bp), or cDNAs. The DNA fragments (usually $20-100$ nucleotides in length) attached to the surface were identified as probes [48,160,166]. In this technique, the target DNA or protein molecule can be analyzed via labeled and hybridized identification probe on the array. The probe's position on the array is called spot or future. Thousands of these spots can be found in a microarray [166].

Microarray technology allows simultaneous analysis of multiple DNA sequences and a large number of samples at the same time. In a microbial analysis, DNA microarray is usually used to analyze genotyping and gene expression. In a rutin gene expression analysis, at first, the RNA molecule was obtained from target samples and made proper for hybridization (e.g., cDNA by reverse transcriptase), and then it is hybridized to a microarray after labeled with a fluorescent dye. On the other hand, microarray genotyping is used to determine the presence of a specific sequence in a genomic sample [167].

DNA microarray is a new technology for identifying and characterizing pathogens. The first microarray application about food microbiology was the

identification of enteric food pathogens based on their various virulence factors [167]. In a research, six different genes encoding antigenic agents and virulence factors of 15 *Salmonella*, *Shigella*, and *E. coli* strains were determined by gene-specific oligonucleotide probes on a microarray [168]. In another example, a microarray application designed for the identification of *B. anthracis* from *Bacillus* species mixed culture via using *Bacillus anthracis* virulence factor genes. In this research, it is determined that to identify *B. anthracis* from mixed culture, the microarray is a successful method [169].

CONCLUSION

The detection of foodborne pathogen microorganisms in food samples by traditional cultural methods are considered as "gold standard" methods. However, these methods are time consuming, labor intensive, and unable to detect viable but nonculturable pathogen microorganisms. To overcome the limitations of traditional cultural methods, several detection methods have been developed like nucleic acid–based detection methods. Nucleic acid–based methods are more specific, sensitive, time saving and possible in the detection of viable but nonculturable pathogen microorganisms. However, these methods have some restrictions such as required trained personnel, specialized instruments, and high cost. Reducing the analyzing time and increasing the specificity and selectivity of analysis cause the nucleic acid–based methods valuable in the detection of foodborne pathogens in control laboratories and scientific research. It is critically important that validation and standardization of these methods should be developed in order to use nucleic acid based methods instead of cultural methods for detecting foodborne pathogens.

REFERENCES

[1] V. Velusamy, K. Arshak, O. Korostynska, K. Oliwa, C. Adley, An overview of foodborne pathogen detection: in the perspective of biosensors, Biotechnol. Adv. 28 (2010) 232−254.

[2] World Health Organization (WHO), WHO Estimates of the Global Burden of Foodborne Diseases, 2015. https://www.who.int/foodsafety/publications/foodborne_disease/fergreport/en/.

[3] E. Scallan, R.M. Hoekstra, F.J. Angulo, R.V. Tauxe, M.-A. Widdowson, S.L. Roy, et al., Foodborne illness acquired in the United States-major pathogens, Emerg. Infect. Dis. 17 (1) (2011) 7−15.

[4] V. Fusco, G.M. Quero, Nucleic acid-based methods to identify, detect and type pathogenic bacteria occurring in milk and dairy products, in: A.A. Eissa (Ed.), Structure

and Function of Food Engineering, IntechOpen, Rijeka, 2012, pp. 371–404.

[5] L. Jackson, Prevalence of foodborne pathogens, in: R.H. Schmidt, G.E. Rodrick (Eds.), Food Safety Handbook, John Wiley & Sons, Canada, 2003, pp. 127–136.

[6] B. Possé, G. Rasschaert, M. Heyndrickx, L. De Zutter, L. Herman, Detection, identification and typing of bacterial pathogens in animal food products, in: C. Van Peteghem, S. De Saeger, E. Daeseleire (Eds.), Towards a Safer Food Supply in Europe, Belgian Science Policy, Belgium, 2007, pp. 10–34.

[7] E.C. Alocilja, S. Pal, Microbiological detectors for food safety applications, in: J.G. Voeller (Ed.), Food Safety and Food Security, John Wiley & Sons, Canada, 2014, pp. 1–26.

[8] A. Souii, M.B. M' hadheb-Gharbi, J. Gharbi, Nucleic acid-based biotechnologies for food-borne pathogen detection using routine time-intensive culture-based methods and fast molecular diagnostics, Food Sci Biotechnol 25 (1) (2016) 11–20.

[9] X. Zhao, C.W. Lin, J. Wang, D.H. Oh, Advances in rapid detection methods for foodborne pathogens, J. Microbiol. Biotechnol. 24 (3) (2014) 297–312.

[10] E.A. Mothershed, A.M. Whitney, Nucleic acid-based methods for the detection of bacterial pathogens: present and future considerations for the clinical laboratory, Clin. Chim. Acta 363 (1–2) (2006) 206–220.

[11] D. Rodrìguez-Lazàro, B. Lombard, H. Smith, A. Rzezutka, M. D'Agostino, R. Helmuth, et al., Trends in analytical methodology in food safety and quality: monitoring microorganisms and genetically modified organisms, Trends Food Sci. Technol. 18 (6) (2007) 306–319.

[12] B.W. Brooks, J. Devenish, C.L. Lutze-Wallace, D. Milnes, R.H. Robertson, G. Berlie-Surujballi, Evaluation of a monoclonal antibody-based enzyme-linked immunosorbent assay for detection of *Campylobacterfetus* in bovine preputial washing and vaginal mucus samples, Vet. Microbiol. 103 (1–2) (2004) 77–84.

[13] A.Y. Tsang, J.C. Denner, P.J. Brennen, J.K. McClatchey, Clinical and epidemiological importance of typing *Mycobacterium avium* complex isolates, J. Clin. Microbiol. 30 (2) (1992) 479–484.

[14] D.M. Rollins, R.R. Colwell, Viable but nonculturable stage of *Campylobacter jejuni* and its role in survival in the natural aquatic environment, Appl. Environ. Microbiol. 52 (1986) 531–538.

[15] J.L. Tholozan, J.M. Cappelier, J.P. Tissier, G. Delattre, M. Federighi, Physiological characterization of viable-but-nonculturable *Campylobacter jejuni* cells, Appl. Environ. Microbiol. 65 (1999) 1110–1116.

[16] S. Ceuppens, D. Li, M. Uyttendaele, P. Renault, P. Ross, M.V. Ranst, et al., Molecular methods in food safety microbiology: interpretation and implications of nucleic acid detection, Compr. Rev. Food Sci. Food Saf. 13 (4) (2014) 551–577.

[17] C.M.A.P. Franz, H.M.W. Den Besten, C. Böhnlein, M. Gareis, M.H. Zwietering, V. Fusco, Microbial food safety in the 21st century: emerging challenges and foodborne pathogenic bacteria, Trends Food Sci. Technol. 81 (2018) 155–158.

[18] K. Kleppe, E. Ohtsuka, R. Kleppe, I. Molineux, H.G. Khorana, Studies on polynucleotides: XCVI. Repair replication of short synthetic DNA's as catalyzed by DNA polymerases, J. Mol. Biol. 56 (2) (1971) 341–361.

[19] D. Rodríguez-Lazáro, M. Hernàndez, Introduction to the real-time polymerase chain reaction, in: D. Rodríguez-Lazáro (Ed.), Real-time PCR in Food Science Current Technology and Applications, Caister Academic Press, Norfolk, 2013, pp. 3–19.

[20] R.K. Saiki, S. Scharf, F. Faloona, K.B. Mullis, G.T. Horn, H.A. Erlich, et al., Enzymatic amplification of beta-globin genomic sequences and restriction site analysis for diagnosis of sickle cell anemia, Science 230 (4732) (1985) 1350–1354.

[21] R.K. Saiki, D.H. Gelfand, S. Stoffel, S.J. Scharf, R. Higuchi, G.T. Horn, et al., Primer-directed enzymatic amplification of DNA with a thermostable DNA polymerase, Science 239 (1988) 487–491.

[22] K. Mullis, F. Faloona, S. Scharf, R. Saiki, G. Horn, H. Erlich, Specific enzymatic amplification of DNA in vitro: the polymerase chain reaction, Cold Spring Harbor Symp. Quant. Biol. 51 (1986) 263–273.

[23] S. Malhotra, S. Sharma, N.J.K. Bhatia, P. Kumar, C. Hans, Molecular methods in microbiology and their clinical applications, J. Mol. Genet. Med. 8 (4) (2014) 1–9.

[24] W.E. Hill, The polymerase chain reaction: a for the detection of foodborne pathogens, Crit. Rev. Food Sci. Nutr. 36 (1–2) (1996) 123–173.

[25] Q. Yan, K.A. Lampel, Molecular methods to characterize foodborne microbial pathogens, in: V. Ravishankar Rai, A. Jamuna Bai (Eds.), Trends in Food Safety and Protection, CRS Press Taylor & Francis Group, Boca Raton, 2018, pp. 17–50.

[26] A.M.A. Martinon, Development of innovative PCR and fluorescense activated cell sorting methodologies for the detection of a range of foodborne pathogens, University of Limerick, Doctor of Philosophy, Bordeaux, 2011, p. 247.

[27] M.J. McPherson, S.G. Møller, PCR, second ed., Taylor & Francis Group, Bodmin

[28] P.K. Mandal, A.K. Biswas, K. Choi, U.K. Pal, Methods for rapid detection of foodborne pathogens: an overview, Am. J. Food Technol. 6 (2) (2011) 87–102.

[29] D. Rodriguez-Lazáro, N. Cook, M. Hernández, Current challenges in real-time PCR diagnostic in food science, in: D. Rodriguez-Lazáro (Ed.), Real-time PCR in Food Science Current Technology and Applications, Caister Academic Press, Norfolk, 2013, pp. 21–26.

[30] B. Malorny, P.T. Tassios, P. Radstrom, N. Cook, M. Wagner, J. Hoorfar, Standardization of diagnostic PCR for the detection of foodborne pathogens, Int. J. Food Microbiol. 83 (1) (2003) 39–48.

[31] R.M.U.S.K. Rathnayaka, R.K. Devappa, S.K. Rakshit, Rapid detection of food pathogens using molecular mathods, in: A.F. El Sheikha, R. Levin, J. Xu (Eds.), Molecular Techniques in Food Biology, John Wiley & Sons, Pondicherry, 2018, pp. 343—359.

[32] J. Meng, S. Zhao, M.P. Doyle, S.E. Mitchell, S. Kresovich, Polymerase chain reaction for detecting *Escherichia coli* O157:H7, Int. J. Food Microbiol. 32 (1996) 103—113.

[33] R. Aznar, B. Alarcón, On the specificity of PCR selection of *Listeria monocytogenes* in food: a comparison of published primers, Syst. Appl. Microbiol. 25 (1) (2002) 109—119.

[34] S.G. Pathmanathan, N. Cardona-Castro, M.M. Sánchez-Jiménez, M.M. Correa-Ochoa, S.D. Puthucheary, K.L. Thong, Simple and rapid detection of *Salmonella* strains by direct PCR amplification of the *hilA* gene, J. Med. Microbiol. 52 (9) (2003) 773—776.

[35] D.K. Winters, M.F. Slavik, Evaluation of a PCR based assay for specific detection of *Campylobacter jejuni* in chicken washes, Mol. Cell. Probes 9 (5) (1995) 307—310.

[36] B. Alarcón, B. Vicedo, R. Aznar, PCR based procedures for detection and quantification of *Staphylococcus aureus* and their application in food, J. Appl. Microbiol. 100 (2) (2005) 352—364.

[37] E. Villalobo, A. Torres, PCR for detection of *Shigella* spp. in mayonnaise, Appl. Environ. Microbiol. 64 (4) (1998) 1242—1245.

[38] S.G. Fenwick, A. Murray, Detection of pathogenic *Yersinia enterocolitica* by the polymerase chain reaction, Lancet 337 (8739) (1991) 496—497.

[39] Y.M. Hsieh, S.J. Sheu, Y.L. Chen, Enterotoxigenic profiles and polymerase chain reaction detection of *Bacillus cereus* group cells and *B.cereus* strain from foods and food-borne outbreaks, J. Appl. Microbiol. 87 (4) (2001) 481—490.

[40] W.E. Hill, S.P. Keasler, M.W. Trucksess, P. Feng, C.A. Kaysner, K.A. Lampel, Polymerase chain reaction identification of Vibrio vulnificus in artificially contaminated oysters, Appl. Environ. Microbiol. 57 (3) (1991) 707—711.

[41] C.S.M.L. Estrada, L.D.C. Velázquez, S.D. Genaro, A.M.S. de Guzmán, Comparison of DNA extraction methods for pathogenic *Yersinia enterocolitica* detection from meat food by nested PCR, Food Res. Int. 40 (5) (2007) 637—642.

[42] L. Rossen, P. Nørskov, K. Holmstrøm, O.F. Rasmussen, Inhibition of PCR by components of food samples, microbial diagnostic assays and DNA-extraction solutions, Int. J. Food Microbiol. 17 (1992) 37—45.

[43] G. Jeníková, J. Pazlarová, K. Demnerová, Detection of *Salmonella* in food samples by the combination of immunomagnetic separation and PCR assay, Int. Microbiol. 3 (4) (2000) 225—229.

[44] S. Bhaduri, B. Cottrell, Sample preparation methods for PCR detection of *Escherichia coli* O157:H7, *Salmonella typhimurium*, and *Listeria monocytogenes* on of beef chuck shoulder using a single enrichment medium, Mol. Cell. Probes 15 (2001) 267—274.

[45] R. Lindqvist, B. Norling, S.T. Thisted-Lambert, A rapid sample preparation method fort he PCR detection of food pathogens based on buoyant density centrifugation, Lett. Appl. Microbiol. 24 (1997) 306—310.

[46] P.-G. Lantz, F. Tjerneld, E. Borch, B. Hahn-Hägerdal, P. Rådström, Enhanced sensitivity in PCR detection of *Listeria monocytogenes* in soft cheese through use of an aqueous two-phase system as a sample preparation method, Appl. Environ. Microbiol. 60 (1994) 3416—3418.

[47] C. Soumet, G. Ermel, P. Fach, P. Colin, Evaluation of different DNA extraction procedures for the detection of *Salmonella* from chicken products by polymerase chain reaction, Lett. Appl. Microbiol. 19 (1994) 294—298.

[48] G. Rusul, L.O. Chuah, Molecular methods for the detection and characterization of food-borne pathogens, in: V. Ravishankar Rai (Ed.), Advances in Food Biotechnology, John Wiley & Sons, Noida, 2016, pp. 471—497.

[49] T. Liu, K. Liljebjelke, E. Bartlett, C. Hofacre, S. Sanchez, J.J. Maurer, Application of nested polymerase chain reaction to detection of *Salmonella* in poultry environment, J. Food Prot. 65 (8) (2002) 1227—1232.

[50] A. Stankevicius, D. Wasyl, A. Jablonski, M. Stankeviciene, Z. Pejsak, One-tube Nested PCR for the detection of *Salmonella* sp. in swine faeces, Bull. Vet. Inst. Pulawy 50 (2006) 35—39.

[51] E. Sevindik, Z.T. Abacı, Nested PCR and applications area, Turkish J. Sci. Rev. 6 (2) (2013) 22—26.

[52] J. Minarovičová, E. Kaclíková, K. Krascsenicsová, P. Siekel, T. Kuchta, A single-tube nested real-time polymerase chain reaction for sensitive contained detection of *Cryptosporidium parvum*, Lett. Appl. Microbiol. 49 (2009) 568—572.

[53] S.D. Saroj, R. Shashidhar, M. Karani, J.R. Bandekar, Rapid, sensitive, and validated method for detection of *Salmonella* in food by an enrichment broth culture — nested PCR combination assay, Mol. Cell. Probes.

[54] E. Navarro, G. Serrano-Heras, M.J. Castaño, J. Solera, Real-time PCR detection chemistry, Clin. Chim. Acta 439 (2015) 231—250.

[55] K. Smith, M.A. Diggle, S.C. Clarke, Automation of a fluorescence-based multiplex PCR fort he laboratory confirmation of common bacterial pathogens, J. Med. Microbiol. 53 (2004) 115—117.

[56] P. Rossmanith, M. Krassnig, M. Wagner, I. Hein, Detection of *Listeria monocytogenes* in food using a combined enrichment/real-time PCR method targeting the prfA gene, Res. Microbiol. 157 (8) (2006) 763—771.

[57] C.-M. Cheng, W. Lin, K.T. Van, L. Phan, N.N. Tran, D. Farmer, Rapid detection of *Salmonella* in foods using real-time PCR, J. Food Prot. 71 (12) (2008) 2436—2441.

[58] A.A. Bhagwat, Simultaneous detection of *Escherichia coli* O157:H7, *Listeria monocytogenes* and *Salmonella* strains by real-time PCR, Int. J. Food Microbiol. 84 (2) (2003) 217—224.

[59] A. Debretsion, T. Habtemariam, S. Wilson, D. Nganwa, T. Yehualaeshet, Real-time PCR assay for rapid detection and quantification of *Campylobacter jejuni* on chicken rinses from poultry processing plant, Mol. Cell. Probes 21 (3) (2007) 177–181.

[60] I. Hein, H.J. Jørgensen, S. Loncarevic, M. Wagner, Quantification of *Staphylococcus aureus* in unpasteurised bovine and caprine milk by real-time PCR, Res. Microbiol. 156 (4) (2005) 554–563.

[61] C. Bartsch, K. Szabo, M. Dinh-Thanh, C. Schrader, E. Trojnar, R. Johne, Comparison and optimization of detection methods for noroviruses in frozen strawberries containing different amounts of RT-PCR inhibitors, Food Microbiol. 60 (2016) 124–130.

[62] E. Omiccioli, G. Amagliani, G. Brandi, M. Magnani, A new platform for real-time PCR detection of *Salmonella* spp.,*Listeria monocytogenes* and *Escherichia coli* O157 in milk, Food Microbiol. 26 (6) (2009) 615–622.

[63] J.W. Law, N.S. Ab Mutalib, K.G. Chan, L.H. Lee, Rapid methods for the detection of foodborne bacterial pathogens: principles, applications, advantages and limitations, Front. Microbiol. 5 (2015) 770.

[64] M. Arya, I.S. Shergill, M. Williamson, L. Gommersall, N. Arya, H.R. Patel, Basic principles of real-time quantitative PCR, Expert Rev. Mol. Diagn. 5 (2) (2005) 209–219.

[65] A. Gulluce, Detection of Different Animal Species in Heat Processed Meat Mixes by Real Time PCR Method (Unpublished master's thesis), Graduate School of Natural and Applied Sciences, Erciyes University Kayseri, Turkey, 2009, 77 pp.

[66] M. Kubista, J.M. Andrade, M. Bengtsson, A. Forootan, J. Jonak, K. Lind, R. Sindelka, et al., The real-time polymerase chain reaction, Mol. Asp. Med. 27 (2006) 95–125.

[67] M. Jalali, J. Zaborowska, M. Jalali, The polymerase chain reaction: PCR, qPCR, and RT-PCR, in: M. Jalali, F. Saldanha, M. Jalali (Eds.), Basic Science Methods for Clinical Researchers, Mica Haley, London, United Kingdom, 2017, pp. 1–18.

[68] R.E. Levin, The application of real-time PCR to food and agricultural systems. A review, Food Biotechnol. 18 (1) (2004) 97–133.

[69] D.A.C. Simpson, S. Feeny, C. Boyle, A.W. Stitt, Retinal VEGF mRNA measured by SYBR gren I fluorescence, A versatile approach to quantitative PCR, Mol. Vis. 6 (2000) 178–183.

[70] K.M. Ririe, R.P. Rasmussen, C.T. Wittwer, Product differentiation by analysis of DNA melting curves during the polymerase chain reaction, Anal. Biochem. 245 (1997) 154–160.

[71] V.H. Van der Velden, A. Hochhaus, G. Cazzaniga, T. Szczepanski, J. Gabert, J.J. van Dongen, Detection of minimal residual disease in hematologic malignancies by real-time quantitative PCR: principles, approaches, and laboratory aspects, Leukemia 17 (6) (2003) 1013–1034.

[72] M. Mangal, B. Sangita, S.K. Satish, G.K. Ram, Molecular detection of foodborne pathogens: a rapid and accurate answer to food safety, Crit. Rev. Food Sci. Nutr. 56 (2016) 1568–1584.

[73] W. Tan, X. Fang, J. Li, X. Liu, Molecular beacons: a novel DNA probe for nucleic acid and protein studies, Chem. Eur J. 6 (7) (2000) 1107–1111.

[74] N. Thelwell, S. Millington, A. Solinas, J. Booth, T. Brown, Mode of action and application of Scorpion primers to mutation detection, Nucleic Acids Res. 28 (19) (2000) 3752–3761.

[75] J.F. Martinez-Blanch, G. Sanchez, E. Garay, R. Aznar, Development of a real-time PCR assay for detection and quantification of enterotoxigenic members of *Bacillus cereus* group in food samples, Int. J. Food Microbiol. 135 (2009) 15–21.

[76] N. Toplak, M. Kovač, S. Piskernik, S.S. Možina, B. Jeršek, Detection and quantification of *Campylobacter jejuni* and *Campylobacter coli* using real-time multiplex PCR, J. Appl. Microbiol. 112 (4) (2012) 752–764.

[77] E. Eschbach, A. Martin, J. Huhn, C. Seidel, R. Heuer, J.-H. Schumacher, et al., Detection of enteropathogenic, *Vibrio cholerae* and *Vibrio vulnificus*: performance of real-time PCR kits in an inter laboratory study, Eur. Food Res. Technol. 243 (2017) 1335–1342.

[78] S.D. Miszczycha, S. Ganet, L. Dunière, Novel real-time PCR method to detect *Escherichia coli* O157:H7 in raw milk cheese and raw ground meat, J. Food Prot. 75 (8) (2012) 1373–1381.

[79] C.F. Hsu, T.Y. Tsai, T.M. Pan, Use of the duplex TaqMan PCR system for detection of shiga-like toxin-producing *Escherichia coli* O157, J. Clin. Microbiol. 43 (6) (2005) 2668–2673.

[80] L. Wang, Y. Li, A. Mustapha, Detection of viable *Escherichia coli* O157:H7 by ethidium monoazide real-time PCR, J. Appl. Microbiol. 107 (5) (2009) 1719–1728.

[81] E.J. Heo, B.R. Song, H.J. Park, Y.J. Kim, J.S. Moon, S.H. Wee, et al., Rapid detection of *Listeria monocytogenes* by real-time PCR in processed meat and dairy products, J. Food Prot. 77 (3) (2014) 453–458.

[82] S. Wei, B.-J. Park, S.-H. Kim, K.-H. Seo, Y.-G. Jin, D.-H. Oh, Detection of *Listeria monocytogenes* using Dynabeads® anti-*Listeria* combined with real-time PCR in soybean sprouts, LWT - Food Sci. Technol. 99 (2019) 533–539.

[83] B. Malorny, E. Paccassoni, P. Fach, C. Bunge, A. Martin, R. Helmuth, Diagnostic real-time PCR for detection of *Salmonella* in food, Appl. Environ. Microbiol. 7 (12) (2004) 7046–7052.

[84] J. Chen, L. Zhang, G.C. Paoli, C. Shi, S.-I. Tu, X. Shi, A real-time PCR method for the detection of *Salmonella enterica* from food using a target sequence identified by comparative genomic analysis, Int. J. Food Microbiol. 137 (2010) 168–174.

[85] A. Anderson, K. Pietsch, R. Zucker, A. Mayr, E. Müller-Hohe, U. Messelhausser, et al., Validation of a duplex real-time PCR for the detection of *Salmonella* spp. in

different food products, Food Anal. Methods 4 (3) (2011) 259—267.

[86] M. Goto, H. Takahashi, Y. Segawa, Y. Hayashidani, K. Takatori, Y. Hara-Kudo, Real-Time PCR method for quantification of *Staphylococcus aureus* in milk, J. Food Prot. 70 (1) (2007) 90—96.

[87] C.-Y. Cheng, M.-J. Huang, H.-C. Chiu, S.-M. Liou, C.-C. Chou, C.-C. Huang, Simultaneous detection of food pathogens, *Staphylococcus aureus, Salmonella Enterica, Bacillus cereus* and *Vibrio parahaemolyticus* by multiplex real-time polymerase chain reaction, J. Food Drug Anal. 20 (1) (2012) 66—73.

[88] T. Trnčíková, V. Hrušková, K. Oravcová, D. Pangallo, E. Kaclíková, Rapid and sensitive detection of *Staphylococcus aureus* in food using selective enrichment and real-time PCR targeting a new gene marker, Food Anal. Methods 2 (2009) 241—256.

[89] S. Perelle, F. Dilasser, J. Grout, P. Fach, Screening food raw materials for the presence of shiga toxin-encoding *Escherichia coli* O26, O103, O111, O145 and O157, Int. J. Food Microbiol. 113 (3) (2007) 284—288.

[90] G.S. Stewart, Challenging food microbiology from a molecular perspective, An. Microbiol. 143 (7) (1997) 2099—2108.

[91] J.D. Oliver, Recent findings on the viable but nonculturable state in pathogenic bacteria, FEMS Microbiol. Rev. 34 (4) (2010) 415—425.

[92] T. Ramamurthy, A. Ghosh, G.P. Pazhani, S. Shinoda, Current perspectives on viable but non-culturable (VBNC) pathogenic bacteria, Front. Public Health 2 (103) (2014) 1—9.

[93] Y. Liu, C. Wang, G. Tyrrell, S.E. Hrudey, X.F. Li, Induction of *Escherichia coli* O157:H7 into the viable but non-culturable state by chloraminated water and river water, and subsequent resuscitation, Environ. Microbiol. Rep. 1 (2) (2009) 155—161.

[94] C. Signoretto, M.M. Lleò, M.C. Tafi, P. Canepari, Cell wall chemical composition of *Enterococcus faecalis* in the viable but nonculturable state, Appl. Environ. Microbiol. 66 (5) (2000) 1953—1959.

[95] K. Rudi, H.K. Nogva, B. Moen, H. Nissen, S. Bredholt, T. Moretro, K. Naterstad, A. Holck, Development and application of new nucleic acid-based technologies for microbial community analyses in foods, Int. J. Food Microbiol. 78 (1—2) (2002) 171—180.

[96] S. Wang, R.E. Levin, Discrimination of viable *Vibrio vulnificus* cells from dead cells in real-time PCR, J. Microbiol. Methods 64 (2006) 1—8.

[97] N.J. Morin, Z. Gong, X.F. Li, Reverse transcription-multiplex PCR assay for simultaneous detection of *Escherichia coli* O157:H7, *Vibrio cholerae* O1, and *Salmonella Typhi*, Clin. Chem. 50 (11) (2004) 2037—2044.

[98] C. Frazão, C.E. McVey, M. Amblar, A. Barbas, C. Vonrhein, C.M. Arraiano, M.A. Carrondo, Unravelling the dynamics of RNA degradation by ribonuclease II and its RNA-bound complex, Nature 443 (7107) (2006) 110—114.

[99] H.K. Nogva, S.M. Drømtorp, H. Nissen, K. Rudi, Ethidium monoazide for DNA-based differentiation of viable and dead bacteria by 5'-nuclease PCR, Biotechniques 34 (4) (2003) 804—813.

[100] E. Barbau-Piednoir, J. Mahillon, J. Pillyser, W. Coucke, N.H. Roosenes, N. Botteldoorn, Evaluation of viability-qPCR detection system on viable and dead *Salmonella* serovar Enteritidis, J. Microbiol. Methods 103 (2014) 131—137.

[101] T. Soejima, K. Iida, T. Qin, H. Taniai, M. Seki, A. Takade, et al., Photoactivated ethidium monoazide directly cleaves bacterial DNA and is applied to PCR for discrimination of live and dead bacteria, Microbiol. Immunol. 51 (8) (2007) 763—775.

[102] A. Nocker, C.Y. Cheung, A.K. Camper, Comparison of propidium monoazide with ethidium monoazide for differentiation of live vs. dead bacteria by selective removal of DNA from dead cells, J. Microbiol. Methods 67 (2006) 310—320.

[103] A. Nocker, A.K. Camper, Selective removal of DNA from dead cells of mixed bacterial communities by use of ethidium monoazide, Appl. Environ. Microbiol. 72 (3) (2006) 1997—2004.

[104] D.B. Kell, A.S. Kaprelyants, D.H. Weichart, C.R. Harwood, M.R. Barer, Viability and activity in readily culturable bacteria: a review and discussion of the practical issues, Antonie Leeuwenhoek 73 (1998) 169—187.

[105] M.C. DeTraglia, J.S. Brand, A.M. Tometski, Characterization of azidobenzamidines as photoaffinity labeling for trypsin, J. Biol. Chem. 253 (6) (1978) 1846—1852.

[106] K. Rudi, B. Moen, S.M. Drømtorp, A.L. Holck, Use of ethidium monoazide and PCR in combination for quantification of viable and dead cells in complex samples, Appl. Environ. Microbiol. 71 (2) (2005) 1018—1024.

[107] K. Rudi, K. Naterstad, S.M. Drømtorp, H. Holo, Detection of viable and dead *Listeria monocytogenes* on gouda-like cheeses by real-time PCR, Lett. Appl. Microbiol. 40 (2005) 301—306.

[108] G. Flekna, P. Štefanič, M. Wagner, F.J.M. Smulders, S.S. Možina, I. Hein, Insufficient differentiation of live and dead *Campylobacter jejuni* and *Listeria monocytogenes* cells by ethidium monoazide (EMA) compromises EMA/real-time PCR, Res. Microbiol. 158 (5) (2007) 405—412.

[109] G. Nebe-Von-Caron, P. Stephens, R.A. Badley, Assessment of bacterial viability status by flow cytometry and single cell sorting, J. Appl. Microbiol. 84 (1998) 988—998.

[110] G. Nebe-Von-Caron, P.J. Stephens, C.J. Hewitt, J.R. Powell, R.A. Badley, Analysis of bacterial function by multi-colour fluorescence flow cytometry and single cell sorting, J. Microbiol. Methods 42 (2000) 97—114.

[111] G. Pohl, L.M. Shih, Principle and applications of digital PCR, Expert Rev. Mol. Diagn. 4 (1) (2004) 41—47.

[112] P.L. Quan, M. Sauzade, E. Brouzes, dPCR: a technology review, Sensors 18 (4) (2018) 1271.

[113] P.J. Sykes, S.H. Neoh, M.J. Brisco, E. Hughes, J. Condon, A.A. Morley, Quantitation of targets for PCR by use of limiting dilution, Biotechniques 13 (3) (1992) 444–449.

[114] B. Vogelstein, K.W. Kinzler, Digital PCR, Proc. Natl. Acad. Sci. U.S.A. 96 (16) (1999) 9236–9241.

[115] I. Hudecova, Digital PCR analysis of circulating nucleic acids, Clin. Biochem. 48 (2015) 948–956.

[116] W. Xu, Development of accurate nucleic acid detection technology for target quantification, in: W. Xu (Ed.), Functional Nucleic Acid Detections in Food Safety, Springer, Singapore, 2016, pp. 143–166.

[117] J. Kuypers, K.R. Jerome, Applications of digital PCR for clinical microbiology, J. Clin. Microbiol. 55 (2017) 1621–1628.

[118] J.F. Huggett, C.A. Foy, V. Benes, K. Emslie, J.A. Garson, R. Haynes, et al., The digital MIQE guidelines: minimum information for publication of quantitative digital PCR experiments, Clin. Chem. 59 (6) (2013) 892–902.

[119] N. Rački, T. Dreo, I. Gutierrez-Aguirre, A. Blejec, M. Ravnikar, Reverse transcriptase droplet digital PCR shows high resilience to PCR inhibitors from plant, soil and water samples, Plant Methods 10 (1) (2014) 42.

[120] M. Vynck, W. Trypsteen, O. Thas, L. Vandekerckhove, W. De Spiegelaere, The future of digital polymerase chain reaction in virology, Mol. Diagn. Ther. 20 (5) (2016) 437–447.

[121] J.F. Huggett, S. Cowen, C.A. Foy, Considerations for digital PCR as an accurate molecular diagnostic tool, Clin. Chem. 61 (1) (2015) 79–88.

[122] S.R. Hall, K.R. Jerome, The potential advantages of digital PCR for clinical virology diagnostics, Expert Rev. Mol. Diagn. 14 (4) (2014) 501–507.

[123] J. Pavšič, A. Devonshire, A. Blejec, C.A. Foy, F.V. Heuverswyn, G.M. Jones, et al., Inter-laboratory assessment of different digital PCR platforms for quantification of human cytomegalovirus DNA, Anal. Bioanal. Chem. 409 (2017) 2601–2614.

[124] P. Corbisier, S. Bhat, L. Partis, V.R. Xie, K.R. Emslie, Absolute quantification of genetically modified MON810 maize. (*Zea mays* L.) by digital polymerase chain reaction, Anal. Bioanal. Chem. 396 (6) (2010) 2143–2150.

[125] D. Dobnik, T. Demšar, I. Huber, L. Gerdes, S. Broeders, N. Roosens, et al., Inter-laboratory analysis of selected genetically modified plant reference materials with digital PCR, Anal. Bioanal. Chem. 410 (1) (2018) 211–221.

[126] D. Morisset, D. Štebih, M. Milavec, K. Gruden, J. Žel, Quantitative analysis of food and feed samples with droplet digital PCR, PLoS One 8 (5) (2013) e62583.

[127] M. Wang, J. Yang, Z. Gai, S. Huo, J. Zhu, J. Lia, et al., Comparison between digital PCR and real-time PCR in detection of *Salmonella typhimurium* in milk, Int. J. Food Microbiol. 266 (2018) 251–256.

[128] P. Cremonesi, C. Cortimiglia, C. Picozzi, G. Minozzi, M. Malvisi, M. Luini, et al., Development of a droplet digital polymerase chain reaction for rapid and simultaneous identification of common foodborne pathogens in soft cheese, Front. Microbiol. 7 (2016) 1725.

[129] S. Monteiro, R. Santos, Nanofluidic digital PCR for the quantification of Norovirus for water quality assessment, PLoS One 12 (7) (2017) e0179985.

[130] T. Kuchta, R. Knutsson, A. Fiore, E. Kudirkiene, A. Höhl, D.H. Tomic, et al., A decade with nucleic acid-based microbiological methods in safety control of foods, Lett. Appl. Microbiol. 59 (2014) 263–271.

[131] Y. Zhao, F. Chen, Q. Li, L. Wang, C. Fan, Isothermal amplification of nucleic acids, Chem. Rev. 115 (2015) 12491–12545.

[132] R.R. Kundapur, V. Nema, Loop-mediated isothermal amplification: beyond microbial identification, Cogent Biol 2 (1) (2016) 1137110.

[133] H. Deng, Z. Gao, Bioanalytical applications of isothermal nucleic acid amplification techniques, Anal. Chim. Acta 853 (2015) 30–45.

[134] T. Notomi, H. Okayama, H. Masubuchi, T. Yonekawa, K. Watanabe, N. Amino, et al., Loop-mediated isothermal amplification of DNA, Nucleic Acids Res. 28 (12) (2000) E63.

[135] Z. Xu, L. Li, J. Chu, B.M. Peters, M.L. Harris, B. Li, et al., Development and application of loop-mediated isothermal amplification assays on rapid detection of various types of staphylococci strains, Food Res. Int. 47 (2) (2012) 166–173.

[136] L. Wang, L. Shi, M.J. Alam, Y. Geng, L. Li, Specific and rapid detection of food-borne *Salmonella* by loop-mediated isothermal amplification method, Food Res. Int. 41 (1) (2008) 69–74.

[137] X. Zhao, Y. Li, L. Wang, L. You, Z. Xu, L. Li, et al., Development and application of a loop-mediated isothermal amplification method on rapid detection *Escherichia coli* O157 strains from food samples, Mol. Biol. Rep. 37 (5) (2010) 2183–2188.

[138] T. Notomi, Y. Mori, N. Tomita, H. Kanda, Loop-mediated isothermal amplification (LAMP): principle, features, and future prospects, J. Microbiol. 53 (1) (2015) 1–5.

[139] Y. Mori, K. Nagamine, N. Tomita, T. Notomi, Detection of loop-mediated isothermal amplification reaction by turbidity derived from magnesium pyrophosphate formation, Biochem. Biophys. Res. Commun. 289 (1) (2001) 150–154.

[140] N. Tomita, Y. Mori, H. Kanda, T. Notomi, Loop-mediated isothermal amplification (LAMP) of gene sequences and simple visual detection of products, Nat. Protoc. 3 (5) (2008) 877–882.

[141] M. Goto, E. Honda, A. Ogura, A. Nomoto, K. Hanaki, Colorimetric detection of loop-mediated isothermal amplification reaction by using hydroxy naphthol blue, Biotechniques 46 (3) (2009) 167–172.

[142] K. Karthik, R. Rathore, P. Thomas, T.R. Arun, K.N. Viswas, R.K. Agarwal, et al., Loop-mediated isothermal amplification (LAMP) test for specific and rapid detection of *Brucella abortus* in cattle, Vet. Q. 34 (4) (2014) 174–179.

[143] H. Zhang, J. Feng, R. Xue, X.J. Du, X. Lu, S. Wang, Loop-mediated isothermal amplification assays for detecting *Yersinia pseudotuberculosis* in milk powders, J. Food Sci. 79 (5) (2014) M967—M971.

[144] P.P. Kumar, R.K. Agarwal, P. Thomas, B. Sailo, A. Prasannavadhana, A. Kumar, Rapid detection of *Salmonella enterica* subspecies *enterica* serovar Typhimurium by loop mediated isothermal amplification (LAMP) test from field chicken meat samples, Food Biotechnol. 28 (2014) 50—62.

[145] Y. Mori, T. Notomi, Loop-mediated isothermal amplification (LAMP): a rapid, accurate, and cost-effective diagnostic method for infectious diseases, J. Infect. Chemother. 15 (2) (2009) 62—69.

[146] J. Compton, Nucleic acid sequence-based amplification, Nature 350 (1991) 91—92.

[147] J. Zhong, X. Zhao, Isothermal amplification technologies for the detection of foodborne pathogens, Food Anal. Methods 11 (2018) 1543—1560.

[148] M.-C. Kao, R.A. Durst, Detection of *Escherichia coli* using nucleic acid sequence-based amplification and oligonucleotide probes for 16S ribosomal RNA, Anal. Lett. 43 (2010) 1756—1769.

[149] P. Gill, R. Ramezani, M.V.-P. Amiri, A. Ghaemi, T. Hashempour, N. Eshraghi, et al., Enzyme-linked immunosorbent assay of nucleic acid sequence-based amplification for molecular detection of *M. Tuberculosis*, Biochem. Biophys. Res. Commun. 347 (4) (2006) 1151—1157.

[150] K. Loens, T. Beck, H. Goossens, D. Ursi, M. Overdijk, P. Sillekens, et al., Development of conventional and real-time nucleic acid sequence-based amplification assays for detection of *Chlamydophila pneumoniae* in respiratory specimens, J. Clin. Microbiol. 44 (4) (2006) 1241—1244.

[151] A. Nadal, A. Coll, N. Cook, M. Pla, A molecular beacon-based real time NASBA assay for detection of *Listeria monocytogenes* in food products: role of target mRNA secondary structure on NASBA design, J. Microbiol. Methods 68 (2007) 623—632.

[152] L. Birch, C.E. Dawson, J.H. Cornett, J.T. Keer, A comparison of nucleic acid amplification techniques for the assessment of bacterial viability, Lett. Appl. Microbiol. 33 (4) (2001) 296—301.

[153] L. Heijnen, G. Medema, Method for rapid detection of viable *Escherichia coli* in water using real-time NASBA, Water Res. 43 (2009) 3124—3132.

[154] A.F. El Sheikha, R.C. Ray, Is PCR-DGGE an innovative molecular tool for the detection of microbial plant pathogens? in: N. Sharma (Ed.), Biological Controls for Preventing Food Deterioration John Wiley & Sons, 2014, pp. 409—433.

[155] H. Mollasalehi, R. Yazdanparast, An improved non-crosslinking gold nanoprobe-NASBA based on 16S rRNA for rapid discriminative bio-sensing of major salmonellosis pathogens, Biosens. Bioelectron. 47 (2013) 231—236.

[156] I. Cools, M. Uyttendaele, E. D'Haese, H.J. Nelis, J. Debevere, Development of a real-time NASBA assay for the detection of *Campylobacter jejuni* cells, J. Microbiol. Methods 66 (2006) 313—320.

[157] H.M. Gore, C.A. Wakeman, R.M. Hull, J.L. McKillip, Real-time molecular beacon NASBA reveals hblC expression from *Bacillus* spp. in milk, Biochem. Biophys. Res. Commun. 311 (2003) 386—390.

[158] J. Jean, B. Blais, A. Darveau, I. Fliss, Simultaneous detection and identification of hepatitis A virus and rotavirus by multiplex nucleic acid sequence-based amplification (NASBA) and microtiter plate hybridization system, J. Virol. Methods 105 (2002) 123—132.

[159] D. Rodríguez-Lázaro, M. Hernández, M. D'Agostino, N. Cook, Application of nucleic acid sequence-based amplification for the detection of viable foodborne pathogens: progress and challenges, J. Rapid Methods Autom. Microbiol. 14 (2006) 218—236.

[160] S.L. Foley, K. Grant, Molecular techniques of detection and discrimination of foodborne pathogens and their toxins, in: S. Simjee (Ed.), Foodborne Diseases, Humana Press, Totowa, New Jersey, 2007, pp. 485—510.

[161] A. Martinovic, H.M. Østlie, S.B. Skeie, Novel food pathogen testing technologies: molecular biology methods, in: Today Science-Tomorrow Industry. International Scientific and Professional Conference. 13th Ružička Days, Vukovar, September 16—17, 2011, pp. 55—68.

[162] Z.A. Ratan, S.B. Zaman, V. Mehta, M.F. Haidere, N.J. Runa, N. Akter, Application of fluorescence in situ hybridization (FISH) technique for the detection of genetic aberration in medical science, Cureus 9 (6) (2017) e1325.

[163] B. Bottari, A. Mancini, D. Ercolini, M. Gatti, E. Neviani, FISHing for food microorganisms, in: T. Liher (Ed.), Fluorescence in Situ Hybridization (FISH) — Application Guide, Springer, Berlin, 2017, pp. 511—530.

[164] S. Shimizu, R. Aoi, Y. Osanai, Y. Kawai, K. Yamazaki, Rapid quantitative detection of *Salmonella enterica* using fluorescence in situ hybridization with filter-cultivation (FISHFC) method, Food Sci. Technol. Res. 19 (2013) 59—67.

[165] X. Zhang, S. Wu, K. Li, J. Shuai, Q. Dong, W. Fang, Peptide nucleic acid fluorescence in situ hybridization for identification of *Listeria* genus, *Listeria monocytogenes* and *Listeria ivanovi*, Int. J. Food Microbiol. 157 (2012) 309—313.

[166] A. Ehrenreich, DNA microarray technology for the microbiologist: an overview, Appl. Microbiol. Biotechnol. 73 (2) (2006) 255—273.

[167] A. Rasooly, K. Herold, Food microbial pathogen detection and analysis using DNA microarray technologies, Foodb. Pathog. Dis. 5 (4) (2008) 531—550.

[168] A. Loy, C. Schulz, S. Lücker, A. Schöpfer-Wendels, K. Stoecker, et al., 16S rRNA gene-based oligonucleotide microarray for environmental monitoring of the beta-proteobacterial order "*Rhodocyclales*", Appl. Environ. Microbiol. 71 (3) (2005) 1373—1386.

[169] J.E. Burton, O.J. Oshota, N.J. Silman, Differential identification of *Bacillus anthracis* from environmental *Bacillus* species using microarray analysis, J. Appl. Microbiol. 101 (4) (2006) 754—763.

CHAPTER 14

Plant Biotechnology in Food Security

BARBARA SAWICKA, PHD • KRISHNAN UMACHANDRAN, BE, MS, MBA, PHD •
DOMINIKA SKIBA, PHD • PARISA ZIARATI, PHD

INTRODUCTION

Biotechnology is the application of science and technology to living organisms as well as parts, products, and models thereof, to alter living or nonliving materials for the production of knowledge, goods, and services [1]. The complexity of plants and microorganisms present a greater challenge to advance applied genetics. The successful genetic manipulation of microbes has encouraged researchers in the agricultural sciences to use plant breeding tools. The plant breeder's approach is determined for the particular biological factors of the crop being bred to allow development of new varieties and even species of plants by circumventing current biological barriers to the exchange of genetic material. Later the plant is selected, regenerated, and evaluated under field conditions to ensure that the genetic change is stable and that the attributes of the new variety meet commercial requirements [2,3]. Techniques available for manipulation of regenerating plants are then tested for the genetic basis of novel traits. Molecular breeding approach identifies and validates quantitative trait loci markers associated with for traits of interesting genes that can be introgressed in elite lines through marker-assisted backcrossing [3]. Biotechnology add significant value to genetic resources and increased remarkably the potential returns from genetic product development [4]. With progressing improvements in science, automation, and biotechnology, concerted efforts are underway to end the widespread poverty and hunger, aiming at safety, food security, and sustainability in food. The concern for biotechnological development should facilitate biosafety, health regulations, and not be regulated by the motive of corporate profit interests through associated funded research and becoming increasingly permissive to explorative trials on human health and life. The environmental protection is the crucial component in sustainable development of biotechnology either directly through remediation processes or indirectly preventive through substitutions in conventional processes [2,5–7]. Adopt R&D partnerships at all levels and create an instrument to enable incubation for agricultural developments to reach market with an entrepreneurial spirit. Thus, biotechnological products and services require an indispensable point of emerging and matching field coverage and clinical trials to precede trade. Therefore, testing, governing, and advisory mechanism are to be transparent in biotechnology development. Finally, endure developments in biotechnology research capacities though protection of biosafety guidelines be transparent and flexible [6,8]. The global biotechnology industry pragmatically should have been the consideration as a modern tool for solving the problem of global hunger; ironically, the major investments were on products of little relevance to the needs of the world's hungry, fertility, self-respect, and well-being [9]. Recently, there has been a dynamic development of biotechnology related to genetic engineering. The main achievements of biotechnology are primarily genetically modified plants (GMOs), which allowed to direct and control changes in the genome, to strengthen plants against diseases and pests and adverse environmental conditions. In addition, these plants have become more resistant to herbicides, fungi, viruses, and viroids. For example:

- transgenic potato has an increased content of starch and is resistant to herbicides, potato stalks, and viruses. A new sweet potato variety was created by adding sweet protein or thaumatin. Transgenic potatoes are resistant to enzymatic darkening and free of harmful glycoalkaloids [10];
- transgenic strawberries sweeter, more resistant to frost and longer ripening;
- transgenic oilseed rape, with a lower content of unsaturated fatty acids, resistant to herbicides [6];
- increasing plant sales through their better taste and appearance;
- reduction in the use of chemical plant protection products [6,7];

- reducing hunger by increasing the size of crops [10–12];
- the creation of modern biopharmaceuticals and the development of new therapeutic techniques, for example, gene therapy [12–14].

HISTORY OF BIOTECHNOLOGY

Processes related to biotechnology have accompanied people for a long time. Activities such as growing plants, making bread or wine are an example of the fact that despite the lack of awareness, our distant ancestors were in contact with biotechnology. The term biotechnology was first used much later in 1917. This was due to the description of the process that used living organisms to create products. One of the most important moments in the field of medical biotechnology was the use of antibiotics on a massive scale. This fact took place around 1945, when penicillin G. was produced. Another important date that gave rise to related food production;

- classical biotechnology is an interdisciplinary field of science, using biological processes on an industrial scale, associated with food production, antibiotics; does not contain biopolymers and simple chemical compounds;
- modern biotechnology—applies to molecular biology, genetic engineering, and bioengineering [15–17].

The Current State of Biotechnology

Division of biotechnology:

- Red biotechnology—concerns health protection, namely genetic diagnosis, production of new biopharmaceuticals, and gene therapy and xeno-transplantology [7,18–21];
- Biotechnology white—refers to biological systems in industrial production and environmental protection through biocatalysis and bioprocesses [6,22–25];
- Green biotechnology—concerns the improvement of agricultural and animal production [26];
- Blue biotechnology—applies to modern biology used in the production of marine food and receiving new medicines [25,27,28];
- Violet biotechnology deals with social and legal issues [29–32].

The degree of biotechnology intervention in food:

- food produced through genetic modification—containing foreign genes, for example, potato, tomato [2,16,33,34];
- food with processed GMOs, for example, frozen fries [35,36];
- foods produced using GMOs, for example, bread [23,37,38];
- GMO-derived products but not having transgenic components, for example, sugar from transgenic beets [3,37,38].

Plant Biotechnology

Agricultural biotechnology is built with competencies and technical skills that describe the fundamental use of it in plant as medicine and food and defined as a set of tools that use living organisms or parts of it to make or modify a product, improve plants, trees, or develop microorganisms for specific uses pertaining to food production as a big picture to provide food security to the people [25]. Agricultural biotechnology has products and process where in the crop improvement happens through rudimentary biotechnology following Mendelian genetics where two plant types of the same species are crossed to produce a better plant type. Crossing a plant that has a high tolerance to disease, drought tolerance, and pest resistance, with a plant that has a high fruit yield gives you a disease and drought tolerant and pest-resistant plant with a high fruit yield, which will be critical for helping farmers adapt to new growing conditions. Professional plant breeders acquiring genetic material from ex situ collections benefit extensively from free access, information, and from materials that have been selected in various breeding programmers' for its desirable characteristics [31]. Changing genetic traits of plants in a way to increase their resistance against pests or droughts [5] thereby the commercial biotechnology steps from a successful transformational event into a plant breeding stage, where the benefits of investing in research are taken by partner organizations or countries through global technology transfer [23]. Producing enough food for our growing global population comes with the added challenge of changing consumption patterns, as well as the need to cope with the dramatic effects of climate change and the increasing scarcity of water and land. The current food system requires a global response to supply issues integrated in information, trade, and technology affected with rising population and more land utilization; thus, there is a need to increase food productivity in a sustainable manner, flexible to handle exigencies with better water and soil management. All these are possible with biotechnology revolution as an integrative production system maintaining ecological diversity [12,15,38,39]. Countries will have to pursue biotechnology applications that reduce waste from food and animal production, by altering animal metabolic processes to improve the environmental impact of animal waste management and disposal [40].

BIOTECHNOLOGY OF MICROORGANISMS

Currently, biotechnology of microorganisms is the main component of global industry, primarily the pharmaceutical industry, food, chemical, and agricultural. Thanks to the ability to conduct synthesis processes and the degradation of various types of compounds, microorganisms have become the object of interest between other green, red, and white biotechnology. They play a key role in the degradation of renewable biopolymers, such as cellulose, hemicellulose, lignin, and chitin, to simple oligomers or monomeric sugars that can be used directly in industry, but they can also be further metabolized by microorganisms of these renewable raw materials, chitin is an extremely valuable biopolymer, the second after cellulose in occurrence in nature (1010−1020 tons per year), which is a rich source of carbon and nitrogen. It is a homopolymer of simple sugar (N-acetylglucosamine) and requires specific enzymes, chitinases, for its hydrolysis [41−43]. The products of hydrolysis can be used, among others in agriculture, food industry, pharmacy, medicine, or waste management. The main source of chitinases is microorganisms that use chitin as the main source of nutrients [41]. These enzymes also occur in plants where they perform different functions during growth and development [44,45]. They also have a defensive function against fungal pathogens not only in plants [46], but also in animals [6], and during infection caused by viruses, bacteria, or fungi, they are considered virulence factors [20,47].

Chitinase. Enzymatic hydrolysis of this polysaccharide is carried out in the presence of glycoside hydrolases chitinases. They are produced by microorganisms, insects, plants, and animal, but it is the bacterial chitinases that play a fundamental role in degradation of the chitin. Chitinases and their products, chitooligomers, have been of interest in recent years due to their wide range of applications in agriculture, medicine, and industry [45].

Sources of Chitin and Its Structure

Chitin with cellulose is one of the most common polymers in nature. It is a component of fungal cell walls, exoskeletons, and shells of arthropod eggs (insects and crustaceans), nematode shells, and shells of mollusks. It is mainly found in complexes with proteins or other polysaccharides, such as β-1,3-glucan, which is also the building block of fungal cell walls [20].

Crustaceans are potentially the largest source of chitin, such as crabs, shrimps, and krill. The dry weight of chitin in these wastes is from 20% to 58%, and therefore is of interest to renewable raw material, which is not only a rich source of carbon but also nitrogen [13,20,48,49].

Chitin is an unbranched polymer of β-1,4-N-acetylglucosamine (GlcNAc), linked by β-1,4-glycosidic linkages. It is similar to cellulose but instead of a hydroxyl group (−OH) at C_2 has an acetylamino group ($NH.CO.CH_3$). It is a crystalline polysaccharide occurring in nature in three forms: α-chitin, β-chitin, and γ-chitin [45,50,51]. α-chitin is the most widespread, isomorphic, and compact form due to the antiparallel distribution of chitin chains, which promotes the creation of strong hydrogen bonds. β-chitin is loosely packed, and the chains are arranged in a parallel manner, at a greater distance from each other, with weaker intermolecular forces, which make it a less stable form. Furthermore, it is not possible to synthesize β-chitin from the solution in vitro. The mixture of both forms is polymorphic γ-chitin [52−54].

Chitinases—Construction and Operation

Chitinases are enzyme proteins belonging to a larger group enzymatic—glycoside hydrolases (GH). This group includes all enzymes that hydrolyze glycosidic bonds in polysaccharides, oligosaccharides, and glycosides. It is a very large group and based for the amino acid sequences of their catalytic domains, and they were divided into more than 115 families of glycoside hydrolases [44−46].

Chitinase catalyzes the hydrolysis of β-glycosidic bonds between the C_1 and C_4 carbons of two neighboring N-acetylglucosamines in the chitin chain [13,45]. Two types of chitinases were distinguished: endochitinase (EC 3.2.1.14) and exochitinase (EC 3.2.1.52) [55]. Endochitinases cleave the bonds at random within the chitin polymer, forming soluble oligomers of N-acetylglucosamine (GlcNAc), of low molecular weight, such as chitotetrose, chitotriose, and N,N'-diacetylchitobiose dimer. Exochitinases is subdivided into two subcategories: chitobiosidase (EC 3.2.1.29), now also called N-acetylhexosaminidases (EC 3.2.1.52), which catalyze the gradual release of N,N'-diacetylchitobiose from the nonreduced end of chitin microfibrils and 1-4-β. -N-acetylglucosaminidase (EC3.2.1.30), which cleaves oligomeric products resulting from the action of endochitinases and chitobiosidase, generating on the occasion of GlcNAc monomers without mono- or oligosaccharides [40,46]. Bacteria have the ability to complete degradation of chitin. Formed in earlier ones stages of the hydrolysis of chitin molecules of N-acetylglucosamine, they are hydrolyzed by transferase acetyl glucosamine to glucosamine and then to acetate, ammonia, and fructose-6-phosphate due to the presence glucosamine-6-phosphate transferase. Many organisms have the ability to synthesize chitinolytic enzymes. Based on the similarity of amino acid sequences of chitinases from different organisms, they were divided into five basic classes and assigned to the family of 18–20 glycosyl hydrolases [13,45,46].

Bacteria produce chitinases that belong mainly to the GH 18 family and some bacteria from the genus *Streptomyces* to GH 19 [34,45]. The GH 18 family also includes all fungal chitinases), similarly as chitinases produced by insects [2]. Genes encoding chitinases from this family are widespread in viruses as well as in animals and higher plants. However, chitinases from the GH 19 family are found mainly in higher plants and some bacteria [44,46]. It is believed that the origin of these two families of hydrolases is different because they differ in amino acid sequence, three-dimensional structure (3D), and the molecular mechanism of catalytic reactions [13,34].

Acquisition of Bacterial Chitinases

Isolation from the environment of chitinase-producing bacteria is not a complicated process, because it mainly involves the addition of colloidal chitin to isolation substrates. Most often, chitinolytic bacteria are isolated from the soil. The Gram-positive actinomycetes, in particular those belonging to the genus Streptomyces, are potentially the largest source of extracellular endo- and exochitinases. In addition to actinomycetes, the presence of chitinases was also found in bacteria of the genus *Bacillus, Pseudomonas, Serratia, Enterobacter, Aeromonas*, as well as *Vibrio* or *Stenotrophomonas* [20,45,47]. Demonstration of the presence of chitinases in bacteria is the first stage in a long-term process that is the optimization of the enzyme's operating conditions, which is necessary to determine its application potential, because it is a group of enzymes very different in optimum performance.

Acquisition of Bacterial Chitinases, Optimization of Hydrolysis Conditions, and Determination of Activity

Isolation from the environment of chitinase-producing bacteria is not a complicated process, because it mainly involves the addition of colloidal chitin to isolation substrates. Most often, chitinolytic bacteria are isolated

from the soil. Gram-positive actinomycetes, especially those belonging to the genus *Streptomyces*, are potentially the largest source of extracellular endo- and exochitinases. In addition to actinomycetes, the presence of chitinases was also found in bacteria of the genus *Bacillus, Pseudomonas, Serratia, Enterobacter, Aeromonas*, as well as *Vibrio* or *Stenotrophomonas* [20,47,55]. Demonstration of the presence of chitinases in bacteria is the first stage in the long-term process of optimization of the enzyme's operating conditions, which is necessary to determine its application potential. Increased activity of bacterial chitinases can be obtained by addition of chitinase substrates, colloidal chitin, chitooligosaccharides or elements of fungal cell walls to nutrient media, and even N-acetylglucosamine [56–59]. The activity of chitinases in *Bacillus thuringiensis* and *Bacillus licheniformis* is affected by time and temperature incubation, medium pH, and substrate concentration [57]. The highest activity of this enzyme was found only on day 5 of incubation. The delay may be due to the fact that chitin is a compound with high molecular weight and the organisms they need a long time to break it down. The effect of using various carbon or nitrogen sources on chitinolytic activity was found [46,57,60].

The Role of Bacterial Chitinases in Green Biotechnology

Bacteria capable of chitin hydrolysis can be used in green biotechnology as biopesticides for the control of fungi (biofungicides) and insects (bioinsecticides) [45,59].

Biofungicides

Mushrooms are among the most dangerous pathogens causing huge losses in agriculture. To limit their development, synthetic chemical compounds called fungicides are used. An alternative to them are biofungicides, which include microorganisms. They can limit the development of fungal diseases through the production of biologically active substances, including enzymes that digest the fungal cell walls. The microorganisms using this mechanism include chitinase-producing bacteria, capable of combating fungi belonging to the genera: *Fusarium, Penicillium, Alternaria, Botrytis, Cladosporium, Aspergillus, Rhizoctonia Phytophthora, Pythium*, or *Colletotrichum*, causing large losses in agriculture. Among the most frequently used bacteria as biofungicides, there are bacteria of the genus *Streptomyces, Bacillus*, or *Pseudomonas* [10,11,46,61] (Table 14.1).

TABLE 14.1
Bacteria That Produce Chitinases and Their Use to fight Fungal Pathogens (Biofungicides)

Bacteria	Fungus Pathogen	References
Aeromonas caviae	*Sclerotium rolfsii, Rhizoctonia solani, Fusarium oxysporum*	[40,62]
Aeromonas hydrophila SBK1	*Aspergillus flavus, Fusarium oxysporum*	[62]
Alcaligenes xylosoxydans	*Fusarium udum, Rhizoctonia bataticola*	[63]
Bacillus cereus CH2	*Verticillium dahlia*	[51]
Bacillus cereus YQQ 308	*Fusarium oxysporum, F. solani, Pythium ultimum*	[64,65]
Bacillus circulans GRS 243	*Phaeoisariopsis personata*	[66]
Bacillus licheniformis NM120-17	Among others mushrooms from genera: *Rhizoctonia, Trichoderma harzianum, Fusarium, Penicillum, Aspergillus, Pythium*	[56,57]
Bacillus pumilus SG2	*Fusarium graminearum, Magnaporthe grisea, Sclerotinia sclerotiorum, Trichoderma reesei, Botrytis cinerea, Bipolaris* sp	[67,68]
Bacillus thuringiensis NM101-19	Among others mushrooms from genera: *Rhizoctonia, Trichoderma harzianum, Fusarium, Penicillum, Aspergillus, Pythium*	[57]
Bacillus thuringiensis spp. *Colmeri*	*Rhizoctonia solani, Botrytis cinerea, Penicillium chrysogenum, P. piricola, P. glaucum, Sclerotinia fuckelian*	[69,70]
Bacillus thuringiensis UM96	*Botrytis cinerea*	[71]
Enterobacter sp. NRG4	*Fusarium moniliforme, Aspergillus niger, Mucor rouxi, Rhizopus nigricans*	[72,73]
Fluorescent *Pseudomonas*	*Fusarium oxysporum fa. sp. Dianthi*	[74]
Fluorescent *Pseudomonas*	*Phytophthora capsici, Rhizoctonia solani*	[75]
Fluorescent *Pseudomonas*	*Botrytis cinerea, Rhizoctonia solani, Diaporthe phaseolorum, Colletotrichum lindemuthianum*	[47,75,76]
Pseudomonas sp.	*Pythium aphanidermatum, Rhizoctonia solani*	[76]
Rhizobium sp.	*Aspergillus flavus, Aspergillus niger, Curvularia lunata, Fusarium oxysporum, F. udum*	[77]
Serratia marcescens B2	*Botrytis cinereal*	[78]
Serratia marcescens GPS 5	*Phaeoisariopsis personata*	[66]
Stenotrophomonas maltophilia	*Fusarium oxysporum*	[67]
Streptomyces exfoliatus MT9	*Alternaria, Aspergillus niger, Botryodiplodia, Colletotrichum, Fusarium, Geotrichum, Penicillium, Phoma, Rhizopus*	[20,21]
Streptomyces griseus	*Fusarium solani, F. oxysporum, F. moniliforme, Colletotrichum dematium, Rhizopus stolonifera*	[58]
Streptomyces sp. M-20	*Botrytis cinerea*	[78]
Streptomyces sp. MT7	*Phanerochaete chrysosporium, Schizophyllum commune, Gloeophyllum trabeum, Polyporus agaricans, Polyporus friabilis, Coriolus versicolor, Postia placenta*	[20,21,79]
Streptomyces sp. NK1057	*Fusarium oxysporum*	[74,80]
Streptomyces viridificans	*Pythium aphanidermatum, Fusarium oxysporum, Rhizoctonia solani, Colletotrichum dematium*	[57]
Vibrio pacini	*Mucor racemosus, Trichoderma viride, Zygorhynchus heterognmus, Candida albicans*	[6,13,81]

Bioinsecticides

Although chitinases produced by fungi, especially those attacking insects, are most often used for this purpose, there are reports on the use of bacterial chitinases as bioinsecticides. Bacteria are a very attractive alternative to insecticide fungi due to their faster rate of multiplication and, consequently, an increase in the speed of killing insects. Insecticidal action of chitinases mainly consists of supporting and synergistic action with the toxins produced by bacteria, for example, *Bacillus sphaericus*, which has been successfully used for years against mosquitoes. Its effectiveness in destroying mosquito larvae is associated with the production of crystalline binary toxins [69,82]. Chitinase from *B. sphaericus* can support the toxicity of binary toxin against *Spodoptera exigua* larvae [70]. Bacteria are an attractive alternative to insecticide fungi due to their faster multiplication, and consequently the speed killing insects. The insecticidal action of chitinases is based on supporting and synergistic action with the toxins produced by bacteria, for example, *Bacillus sphaericus*, which is used for mosquito control. Its effectiveness in destroying mosquito larvae is associated with the production of crystalline binary toxins [82,83]. Chitinase from *B. sphaericus* may support the toxicity larvae of the spider pestle *Spodoptera exigua* [69]. Chi-AII endochitinase from *S. marcescens*, as well as the bacteria themselves in combination with the CrylC endotoxin protein from *B. thuringiensis*, also show high insecticidal activity against the larvae of this pest [71]. Protein with chitinolytic activity was isolated from extracellular vesicles produced by bacteria *Photorhabdus luminescens* var. *akhurstii* and *Xenorhab* [47,83].

The Use of Chitinases in White Biotechnology

Chitinases can also be used in industrial biotechnology called white, dealing with the processing of renewable raw materials, such as chitin. The source of chitin is, for example, seafood in the form of postproduction waste. The waste consists mainly of shells of crustaceans, such as shrimps or crabs. Their degradation reduces the pollution of the aquatic environment by the industry associated with the production of seafood [11,84]. The source of chitinolytic bacteria utilizing this waste can be both the marine environment, including shrimp waste [10] and even soil [11]. Recombinant chitinases are used for the depolymerization of chitin biomass, and bioactive mono- or oligosaccharides have a broad spectrum of biotechnology applications, for example, as biogas [46,85,86], or substrates for cosmetics [86]. Another approach to the effective use of chitin waste is the production of single-cell protein, where the chitin degradation products serve as sources of carbon or nutrients for the production of biomass [45,87]. Objects of interest in white biotechnology are also fungi, including yeast, which are used to produce protoplasts for research on the synthesis of cell walls and the synthesis and secretion of enzymes. Dahiya et al. [72,73] showed that chitinases from Enterobacter sp. NRG4 are very effective in obtaining protoplasts of *Trichoderma reesei*, *Pleurotus florida*, *Agaricus bisporus*, or *A. nicer*.

The Use of Chitinases in Red Biotechnology

Chitinases can be used in red biotechnology used in healthcare, especially in the production of new biopharmaceuticals. The product of chitinase activity is chitooligosaccharides, which are tested in medical use. The antineoplastic effect of oligosaccharides by inhibiting the growth or reduction of tumor metastases in mice after administration of N-acetylchitohexose has been confirmed [12,67,82]. Bacterial-derived chitinases also have anticancer effects. Chitinases from *Bacillus amyloliquefaciens* V656 were used to hydrolyze chitosan, and these hydrolysates (GlcNAc) 6 inhibited the proliferation of CT26 colon cancer cell lines [46,88]. These enzymes are an excellent marker of many dangerous diseases caused by bacteria such as by *Legionella pneumophila*, *Listeria monocytogenes*, or *Pseudomonas aeruginosa* [47]. Many infectious diseases are transmitted by insects, and chitinases can be used to control mosquitoes. Chitinase from *B. sphaericus* may support toxicity of the binary toxin against *Culex quinquefasciatus* [89]. Renewable biopolymers play a key role in the development of the global economy. Chitin, the second after cellulose, the most common biopolymer, is used not only as a source of carbon but also as nitrogen. Crustaceans are potentially the largest source of chitin, although it also occurs in cell walls of fungi or exoskeleton of insects. Chitin degradation is catalyzed by the enzymatic reaction by chitinases that cleave β-glycosidic bonds between the C_1 and C_4 carbons of two neighboring N-acetylglucosamines. They are commonly occurring enzymes in nature; however, bacterial chitinases play a key role in the hydrolysis of chitin, mainly due to their easier acquisition, which makes it possible to use them in agriculture, industry, or medicine. Bacterial chitinases as well as their action products, chitooligomers can be used in green biotechnology as biopesticides for combating fungi (biofungicides) and insects (bioinsecticides), as well as in white biotechnology, for degradation waste, or in red biotechnology, mainly for the production of biopharmaceuticals [45,46].

SUSTAINABLE FOOD PRODUCTION THROUGH PLANTS

In addition to sustaining populations around the world, food production methods must help preserve the environment and natural resources, and also support the livelihoods of farmers. It is also possible to improve how well-suited plants and animals are to food production by modifying their genetics, either through breeding or using techniques in the lab to alter them. There is great inequality in where food is produced, another especially important way of increasing the efficiency of the food production process is by working to decrease food waste and improve food distribution. Biotechnology plants can resist the dreaded borer, resulting in healthier plants, leading to more food, feed, fuel and fewer insecticide applications [35]. Thus, it has been proven that plants that have been modified through GM have delivered benefits for food production including resistance to insects, viruses, and herbicides. These innovations have reduced the loss of crops to pests and diseases, minimized insecticide use and helped prevent damage to the soil from deep ploughing [35,36]. The insect and disease resistance through plant biotechnology engineering identified protein and other nutritional storages with potential benefits from tissue culture, biofertilizers, biopesticides, and medicinal plants [29]. Producing biotech drugs is a complicated and time-consuming process covering years spent just identifying the therapeutic protein, determining its gene sequence, and working out a process to make the molecules using biotechnology. Devising a scaled up on these the biotech medicines when produced in large batches by growing host cells that have been transformed to contain the gene of interest in carefully controlled conditions in large stainless-steel tanks and then the cells when kept alive can stimulate to produce the target proteins through precise environmental conditions [13,35]. Bioactive ingredients from natural sources are used for their ability to promote better health and prevent chronic diseases [3,90]. Enhancing productivity in a sustainable manner is essential for future agriculture. Genetic improvement of crop plants relies on the cultivation of genotypes that possess favorable alleles/genes controlling desirable agronomic traits [80,91]. Micropropagation is useful in maintaining valuable plants, breeding otherwise difficult-to-breed species (e.g., many trees), and speeding up plant breeding and providing abundant plant material for research. In this the process includes taking small sections of plant tissue, or entire structures such as buds, and culturing them under artificial conditions to regenerate complete plants. Renewable biopolymers play a key role in the development of the global economy. Chitin, the second after cellulose, the most common biopolymer, is used not only as a source of carbon but also as nitrogen. Crustaceans are potentially the largest source of chitin, although it also occurs in cell walls of fungi or exoskeleton of insects. Chitin degradation is catalyzed by the enzymatic reaction by chitinases that cleave β-glycosidic bonds between the C1 and C4 carbons of two neighboring N-acetylglucosamines. They are commonly occurring enzymes in nature, however, bacterial chitinases play a key role in the hydrolysis of chitin, mainly due to their easier acquisition, which makes it possible to use them in agriculture, industry or medicine. Bacterial chitinases as well as their action products, chitooligomers can be used in green biotechnology as biopesticides for combating fungi (biofungicides) and insects (bioinsecticides), as well as in white biotechnology, for degradation waste, or in red biotechnology, mainly for the production of biopharmaceuticals [12,52–54].

To ensure continuity of food security for a growing world population, focus should be on improving crop productivity through the use of new genetic sequencing and advanced "genomic reproduction" technologies and proteomics. These technologies provide an improvement in the production volume of plants through the development of crop species under specific environmental conditions. These technologies also enable plant breeders to effectively and simultaneously redirect to new crop species and traits, such as resistance, quality, and crops that are key to food security. Molecular culture plays a key role in improving crops. Although genetically modified (GM) plants are promising in increasing crop yields, GM crops face many challenges related to development, development and sustainability [3,12,17].

PRESERVATION AND FOOD SAFETY

Emulsification gives food products the characteristic appearance and sensory properties that significantly influence the consumer's perception of such products. As human sensation is complex, food processors usually require numerical standards for better quality control. Bioactive compounds isolated from natural sources often cannot be added directly into food systems due to incompatibility with the food matrix, rapid degradation during food processing, and vulnerability to digestion activity in the biological system. Thus, these bioactive ingredients are encapsulated for better protection and easier incorporation into the food matrix as bioactive proteins could degrade rapidly in the gastrointestinal tract due to acidic environment and digested by proteolytic enzymes. Moreover, hydrophobic compounds, such as flavonoids, have a poor aqueous solubility that causes challenges in their incorporation into food as well as being absorbed in the digestion tract [6,11].

Biotechnological approaches increase the extraction of oil from a plant source; hence, with the depletion of world hydrocarbon reserves, in the future it is possible that herbal biodiesel could compete with oil, coal, and gas [9]. The biotechnology industries are induced to relocate to countries for preferential fast-track access to genetic resources by prohibiting onward transfer of resources to countries that have not adopted appropriate complimentary legislative measures [31]. The diversity of applications is virtually endless leading scientists to be encouraged to look DNA inside the human body, and outside in the plants and animals around us [35,36]. As part of establishing an ecosystem for the development of new biotech products by creating a strong infrastructure for R&D and commercialization, Bio Incubators and Biotech Parks are supported/established/encouraged for industrial R&D and related investment flows, bilaterally and/or regionally in the field of biotechnology and to promote transparency through exchange of information and cooperation among relevant institutions [11,65].

Plant Productivity Performance

The world is facing the major challenge of providing food security for an ever-growing world population, which will require an increase in food production that is exceeding the one from previous decades. To achieve a new green revolution that will result in high yield plants that grow in soils with very low potential, innovative application of technologies is required. Biotechnology research is currently concentrated upon self-sufficiency in food; development in crop quality by aiming on the management of drought tolerance, growth in yield reduced cost of labor and employment generation. Thus, agricultural biotechnology has prospects to increase national production and increase availability of food thereby improving agricultural incomes and also aid in the decline of consumers food prices. However, in reality, the biotechnology research activities are absorbed more on the food security objectives than others [23]. Experimental events in staple crops, banana, sorghum, brinjal, and cotton clearly establish that the convergence is more focused on food through risk mitigation, yield increases and nutritional enhancement through successful events established in food

varieties than on to others such as cotton for textiles. Biotech crops with herbicide amounts to 71% of the global crop area with soybean, maize, and canola, while the insect-resistant cotton crops were 18%, and 11% crops were the stacked genes [38]. Roots are the most underestimated parts of a plant, even though they are crucial for water and nutrient uptake and consequently growth; hence, improved and increased root system that can support the required increase in plant yield and thereon to guarantee food security through plant hormone auxin together with an increased cell cycle activity leading to a boost in root branching with two proteins that are crucial for embryo development that plays a critical role in root branching [26].

GENETIC APPLICATIONS

Plant biotechnology applications respond to increasing demands in food security, socioeconomic development, promotion of conservation, diversification, and sustainable use of plant genetic resources as basic inputs for future agriculture. Advanced biotechnologies, such as cell and tissue culture, molecular genome analysis, plant genetic transformation, molecular plant disease diagnosis, and germplasm cryoconservation, can be successfully used to cope with; genetic erosion, to reinforce ex situ collections and in situ conservation, to upgrade the supply of improved and healthy seeds and planting materials to farmers, and to integrate a new approach into the development programs for food production and feeding security [22]. Gene banks sprouted from the traditional breeding techniques are completely resourceful with new information on genes and innovative technologies. Agro biotechnology enhances existing varieties by exploiting genes of particular agronomic interest and increases the availability of gene pools [14,92]; hence, application of biotechnology in plant breeding is an innovative phase for the creation of transgenic varieties [13,38]. Therefore, this requires expertise in conducting scientific "event" experiments to insert specific genes into plants to evolve into a successful event [23]. Implementing the appropriate plant breeding methods and selection regimes for landrace improvement in participatory breeding programs through cooperation between farmers and plant breeders to identify crop needs or priorities, varieties section, and evaluation for in situ conservation and environmental control where normal crop improvement has been frustrated. On a longer run, these are likely to have negative impacts on diversity of landraces, as they tend to change local crop population structure to make it higher yielding. Management of on-farm populations should include conservation of maximum number of multilocus genotypes and maximum allelic richness; safeguarding the evolutionary processes that generate new multilocus genotypes; and improve the population performance and increase the productivity in a defined range of local environments [39]. Plant genetic resources are becoming more complex for food and agriculture in the multilateral system. Strategic resource for sustainable agriculture is facilitated to access with utmost importance in the given challenges of climate change and quickly adapts to meet future challenges in seed and plant breeding. They provide the biological options to build food and farming systems that are resilient, sustainable, and productive. Crop genetic diversity is the biological cornerstone of global food security; it is the basis for livelihood strategies and nutritional well-being, especially for poor and marginalized people. Cross-border movement and facilitated exchange of plant genetic resources is self-sufficient to crop genetic diversity [12]. Early domestication and modern plant breeding have resulted in severe genetic bottlenecks, reducing the levels of genetic diversity, landraces, and the genes controlling important traits compared to the wild relatives [33,34,63].

FOOD SUPPLY AND REACH

Vertical and horizontal integration of the food supply into ever fewer conglomerates are routinely taken for granted in agribusiness trade publications. The struggle over the loss of another measure of control by farmers who may wish to return to their former ways, having experienced the onset of super weeds and other internalizing costs during introductory promotions, discovers the stringency of the biotech webs having enveloped and trapped them into dependency. The plant biotechnology capacitated cutting and splicing genes and transferring them from one biological entity to another, thereby crossing broad species barriers; using bacterium or virus to carry genes into a plant, or by electrical shock to get pure DNA into the plant cells nucleus, or alternatively with micro projectiles coated with DNA. Regulating the use Biotechnology, GMO crops, as similar to any hybrid crop varieties, and follow food safety evaluations [63,93]. Cofounded facilitates can benefit investing developing countries or organization to skip through profiteering access to agricultural innovations in biotechnology with meager transaction costs and through sharing information synergically within them [23,93]. Biotechnology evolved products such as fuel-grade ethanol, organic acids, bulk amino acids including lactic acid and biodegradable plastics

indicate the arrival of products as alternatives to petrochemical products in market place [29]. Active agricultural biotechnology research is very important for marketing the exports of fresh fruits/vegetables and develop policies dealing with the import of genetically modified organisms [93,94]. The applications of biotechnology are innumerable ranging from energy to environment controlling all parameters of costs, quality, productivity, and regulatory factors as the innovation drivers, including those of creating plants that can be used as bioreactors [5]. Malnutrition is the related term in medicine for hunger leading to undernourished children becoming victims to the impact of every disease, including measles and malaria [9].

GENETIC ENGINEERING

Genomics and Transgenics inserted specific genes from other species into plants/animals, to increase yields or protect against pests or environmental conditions. They are tested to ensure no adverse environmental or health effects [93,95]. Induced mutation-assisted breeding (IMAB) on plant varieties are subjected to mutagenic agents (like radiation) to induce mutations and then selected for the desired new or modern traits that appeared. IMAB has resulted in the introduction of new varieties of many crops such as rice, wheat, barley, apples, citrus fruits, sugar cane, and banana. The drawback is that stringent effort is required to ensure that the mutagenic agent is not passed into the food item. Genetic Engineering/GMO has transfer of genes and twerking. Laboratory-based biotechnology developments have increased the value of genetic resources and the associated traditional knowledge of indigenous communities that provide important leads to commercially exploitable properties of the bioresources [31,38]. Insect-protected cotton with natural insecticide protein from *Bacillus thuringiensis* (Bt cotton) provides increased yields, reduced insecticide costs and fewer health risks [16,29]. Plant breeding tools of agricultural biotechnology offer a cutting-edge technique that is an easier and quicker way than any conventional plant breeding methods. They select the gene that produces the desired trait and transfer them to another plant to increase the plant yields, nutritional content, and values [2,30,91]. In the close transfer process, the gene from one plant species is inserted to another plant species (same kingdom); whereas, in the distant transfer process, the gene from one species is inserted to another species from a different kingdom (i.e., bacteria gene into a plant); moreover, in the tweaking process, the genes already present in the organism are "tweaked" to change the level of a particular protein. The processes, artificial insemination (AI) and multiple ovulation/embryo transfer (MOET), have aided the global diversity and strength of livestock. In the primary function, AI happens over transfer of sperm and MOET happens over transfer of ova or a fertilized embryo from an animal in one part of the world, to an animal in another part of the world. Genetic modification also includes biotechnological research methods such as tissue culture (TC) and marker-assisted selection in crops [23]. Intellectual property rights are acquiring more importance in biotechnology management, as the appropriate development-oriented complementary tool to genetic improvement of crops and crop production can constitute a valuable tool for the sustainable development [22]. Genetically modified seeds are controlled by intellectual property guidelines anticipated to incentivize investors, which are primarily governed by global corporations that proliferate GM technology as a corporate leader and regulate the food chain. Biodiversity-rich countries opt to adopt harmonized access legislation providing incentives to allow preferential access to genetic resources to those countries that have implemented complimentary legislative measure [31]. Golden Rice contains three new genes, two from the daffodil and one from a bacterium that helps it to produce provitamin A, available for mass distribution as a new biotech product contributing development of biotechnology to society [9,91].

Genotoxic Stress

U UR ER. Cell cycle stop. Endoreduplication. Activation of the protein response (UPR) in mammalian cells leads to cell cycle arrest in the G1 phase [14,96—98]. The way in which UPR signaling affects cell cycle arrest remains unknown in plants. The UPR and the endo reduplication in Col-0, WEE1, and ER-B-stressing plants with endoplasmic reticulum (ER) and B deficiency were tested under stress and ER DNA replication. It was found that WEE1, as an important, negative regulator of the cell cycle, is involved in maintaining ER homeostasis during genotoxic stress, and hypersensitivity to stress ER IRE1a 1b is alleviated by a mutation causing loss of cell function in WEE1. If the IRE1-bZIP60 pathway is activated during ER stress, it is required to stop the cell cycle via WEE1. In contrast, the loss of function mutation in WEE1 results in increased expression of UPR-related genes during replication DNA stress. WEE1 and IRE1 are required for endo reduplication, respectively, during stress of DNA replication and ER stress. These recent findings suggest that cell cycle regulation involves UPR activation in various ways during ER stress and *Arabidopsis* DNA replication stress. ER-

related stress weakens progression in the G1 and G2 phases of the cell cycle. G2 inhibits increased retention in p53 mutant cells, but does not increase expression of the p53/47 isoform. Thus, early inhibition (stopping) of G2 on ER stress is a response to suppression of translation [14,16,18]. Cell cycle checkpoints make proliferation only under permissive conditions, but their role in combining the availability of nutrients with cell division is incomplete. Deposition of proteins in the ER is extremely sensitive to the source of energy and sources of amino acids, because deficiencies weaken the folding of the luminal protein and consequently trigger the signaling of ER stress. As a consequence of ER stress, many types of arrest cells in G1 phase, although recent studies, have identified a new stress ER G2 checkpoint. ER stress affects the progression of the cell cycle through two signal classes: early inhibition of protein synthesis leading to G2 delay by CHK 1 and subsequent induction of G1 arrest associated with both p53 induction of target genes and loss of cyclin D1. Thomas et al. [14] showed that 47 p53/p53 substitution suppresses ER G1 control stress, suppresses protein translation recovery, undermining the NOXA induction of cell death mediator. We suggest regulating the cell cycle, in response to stress, the ER contains the excess pathways triggered sequentially the first to harm G2 progression before the final G1 stop [14,93]. Understanding the regulation of the cell cycle under ER stress has an impact on rational anticancer therapy [14,16].

PLANT BIOTECHNOLOGY BUSINESS SUPPORT

To stay ahead of the competition with the most other Biotechnology companies, they fund research and readily extend support to universities to help drive success of their organizations that give recommendations targeting their business. Biotechnology's interdisciplinary character covers a varied range of opinions in society and dependence on employees knowledge and skills [5,25,99]. Investments drive in plant biotechnology and GM crops are affected more through the interests of different stakeholders, rather than to improve food security, reduce poverty, and advance human development. Hence, there are doubts that need to be cleared in the minds of general public through strengthening the capacity for safe biotechnology management [25,99,100]. The applications of biotechnological processes in industry are very heterogeneous ranging from the use of renewable raw materials replacing fossil fuel raw materials to the embraces of all conventional nonbiological processes replaced by biological systems and processes [5,100]. International partnerships in biotechnology are important to recognize the needs of biotechnological strategies to be followed at a quicker pace. The capability to successfully use present and evolving biotechnologies will be contingent largely on the leverage of resource investments [8]. Plant biotechnology engineering has genetically modified transgenic crops leading to significant impacts on canola, cotton, maize, and soybean farming and enlarged the toolbox for plant breeding to gene functional modifications in site-directed mutagenesis, targeted deletion or insertion of genes into plant genomes, or using transgenes to facilitate the breeding process [24]. Agricultural biotechnology delivers biomass for food, feed, genetic modifications, and molecular tools to enhance the plant breeding potential, resulting in increased food supplies, farm income, and reduced damage to ecology and environment. There are various innovations in plant biotechnology that are available on the market or at the late developmental stages, and their application to agriculture, agroforestry, industrial processes, and pharmaceutical industry providing emphasis and approaches adapted to meet heterogeneous needs support inclusiveness of locals [13,16]. All recombinant DNA experiments except those that involve the molecular manipulation of human and animal and plant pathogens, which would enhance the virulence of a pathogen or render a nonpathogen virulence in plant, animal, and human pathogens, are major threats to world security [12,91,101].

The production of genetically modified food creates many opportunities and carries a huge number of potential benefits. However, it is increasingly the subject of numerous debates and controversies. So far, the negative impact of genetically modified food on the human body could not be confirmed. It has not been proven to be completely harmless. The effects that may be caused by their long-term use are still unknown. At the same time, genetically modified crops and food obtained through genetic engineering are a very promising technology. This gives hope for a better life in the future; however, for their safe use, appropriate legal standards should be developed and necessary long-term research should be carried out. Pharmaceutical and biotechnology companies striving for the largest possible and quick profit do not always remember it. Therefore, like any revolutionary idea, it raises a lot of controversy, but also hope. Unfortunately, both modern technologies and genetic engineering are not without flaws. Interfering with the genes of living organisms and creating large-scale mutants in laboratories faces ethical doubts. People are afraid to eat genetically modified food because it is not yet known how it affects the human body, and the negative effects may appear only after many years [16,91,99].

CONCLUSION

Biotechnology is using living organisms to make useful products with production carried out by using microorganisms and natural substances from organisms. Biotechnologies in food production obligated effectively through application in reclamation of wasteland by deployment of microorganisms and plants to degrade toxic compounds; genetic modification in food, nonfood crops; weed and pest control methods; minimized the consumption of energy and water in production processes to result in improved value-added activities and thus enhanced productivity. If the biotechnological tool is used with commercial interests, then intellectual property rights protection to biotechnology processes and modifications is likely and may impede benefits to poor farmers. Crop intensification has increased food production and labor productivity without putting more land under cultivation, but has led to land degradation through soil erosion, inappropriate irrigation and land management practices, loss of soil organic matter and nutrients, depletion of freshwater resources, pollution of waterways and marine environments through inappropriate use of nutrients and crop protection products, increased greenhouse gas emissions, and reduction in biodiversity and ecological resilience through dependence on a reduced number of species and varieties. Prospects for the development of biotechnology are currently enormous. The direction of research development is the creation of second-generation biopharmaceuticals, which will be more effective and less harmful. In the near future, biotechnology will focus on gene therapy, which will involve the introduction of foreign nucleic acids into cells whose presence forces cells to produce proteins encoded by the introduced genes or inhibition or modulation of gene expression. Biotechnology aims to create the possibility of treating heart disease, stroke and multiple sclerosis through stem cells that, thanks to appropriate conditions of growth and development, will be able to transform into any type of tissue or organ. In addition, industrial biotechnology is becoming more and more important from the point of view of rising oil prices and the depletion of nonrenewable energy sources. The production of bioproducts, however, encounters a number of obstacles, the removal of which in the future will be a goal for biotechnologists. Directing resources for agricultural biotechnology techniques, such as genetic alteration, tissue culture, and biofortification, can harness all options that contribute to food security and economic empowerment.

REFERENCES

[1] UNCST, The Open Forum on Agricultural Biotechnology in Africa (OFAB)-Uganda Chapter. Summary of 2009 OFAB Proceedings Kampala, 2010, p. 82. http://www.ofabafrica.org/UserFile/OFABUganda-I.pdf.

[2] H.S. Zarandi, A. Bagheri, A. Baghizadeh, N. Moshtaghi, Quantitative analysis of chitinase gene expression in chickpea, Russ. J. Plant. Physl. 58 (2011) 681–685.

[3] M.A. El-Esawi, Introductory chapter: introduction to biotechnological approaches for maize improvement, in: The Book Maize Germplasm – Characterization and Genetic Approaches for Crop Improvement, IntechOpen, 2018, p. 4. https://www.intechopen.com/books/maize-germplasm-characterization-and-genetic-approaches-for-crop-improvement/introductory-chapter-introduction-to-biotechnological-approaches-for-maize-improvement.

[4] X. Xiao-Jing, Z. Li-Qun, Z. You-Yong, T. Wen-Hua, Improving biocontrol effect of Pseudomonas fluorescens P5 on plant diseases by genetic modification with chitinase gene, Chin. J. Agric. Biotechnol. 2 (2005) 23–27.

[5] A. Simon, Corporate Social Responsibility and Biotechnology Identifying Social Aspects for European Biotechnology Companies, The International Institute for Industrial Environmental Economics, Lund, Sweden, 2002, p. 68. http://lup.lub.lu.se/luur/download?func=downloadFile&recordOId=1322494&fileOId=1322495.

[6] I. Chen, Z. Shen, J. Wu, Expression, purification and in vitro antifungal activity of acidic mammalian chitinase against Candida albicans, Aspergillus fumigatus and Trichophyton rubrum strains, Clin. Exp. Dermatol. 34 (2009) 55–60.

[7] M.U. Swiontek-Brzezinska, A. Jankiewicz, M. Burkowska, M. Walczak, Chitinolytic microorganisms and their possible application in environmental protection, Curr. Microbiol. 68 (2014) 71–81.

[8] C. Juma, I. Serageldin, Freedom to Innovate: Biotechnology in Africa's Development, A Report of the High-Level African Panel on Modern Biotechnology. African Union and New Partnership for Africa's Development, Addis Ababa and Pretoria: AU and NEPAD, 2007, p. 163.

[9] UN Chronicle, Biotechnology – a solution to hunger?, 46(3/4) (2009). https://unchronicle.un.org/article/biotechnology-solution-hunger.

[10] M. Swiontek-Brzezinska, E. Lalke-Porczyk, W. Donderski, Chitinolytic activity of bacteria and fungi isolated from shrimp exoskeletons, Oceanol. Hydrobiol. Stud. 36 (2007) 101–111.

[11] M. Swiontek-Brzezinska, E. Lalke-Porczyk, W. Donderski, Occurrence and activity of microorganisms in shrimp waste, Curr. Microbiol. 57 (2008) 580–587.

[12] K. Umachandran, B. Sawicka, B.,A. Mohammed, N.N.-B. Nasir, A. Pasqualone, Relevance of nanotechnology in food processing industries, Int. J. Agric. Sci. 7 (2018) 5730–5733. ISSN: 0975-3710&E-ISSN: 0975-9107.

[13] T. Ohnuma, T. Numata, T. Osawa, M. Mizuhara, O. Lampela, A.H. Juffer, K. kriver, T. Fukamizo, A class V chitinase from Arabidopsis thaliana: gene responses, enzymatic properties, and crystallographic analysis, Planta 234 (2011) 123–137.

[14] S.E. Thomas, E. Malzer, A. Ordóñez, L.E. Dalton, E.F. van Wout, E.F.A. Liniker, E. Crowther, D.C. Lomas, S.J. Marciniak, P53 and suppression translation regulate various control points of the cell cycle during stress of the endoplasmic reticulum (ER). Biotechnology activities for sustainable development in agriculture, J. Biol. Chem. 288 (2013) 7606–7617.

[15] A. Jansen, F. Valkema, Global change, what solutions does biotechnology offer?, chapter in: H. van den Belt, A. Jansen, F.W.J. Keulartz, F. Valkema, C.N. van der Weele (Eds.), Global Change and Biotechnology, The Hague, on Behalf of the Netherlands Commission Genetic Modification, 2008, p. 162.

[16] S. De Buck, D. de Oliveira, M. Van Montagu, Key Innovations in Plant Biotechnology and Their Applications in Agriculture, Industrial Processes, and Healthcare, International Plant Biotechnology Outreach (IPBO), Ghent University, 2016, p. 21. http://ipbo.vib-ugent.be/wp-content/uploads/2017/11/Key-innovations.pdf.

[17] J.U. Pathak, A. Pandey, S.P. Singh, R.P. Sinha, Global agriculture and the impact of biotechnology, in: Current Developments in Biotechnology and Bioengineering Publication: First Chapter 1 Publisher: Elsevier, Radarweg 29, PO Box 211, 1000 AE, Amsterdam, the Netherlands. Editors: Suresh Kumar Dubey, Ashok Pandey, Rajender Singh Sangwan, 2016, https://doi.org/10.1016/B978-0-444-63661-4.00001-3.

[18] N.K. Bhullar, Z. Zhang, T. Wicker, B. Keller, Wheat gene bank accessions as a source of new alleles of the powdery mildew resistance gene Pm3: a large scale allele mining project, BMC Plant Biol. 10 (2010) 88. https://doi.org/10.1186/1471-2229-10-88.

[19] C. Chiarolla, H. Shand, An Assessment of Private Ex Situ Seed Collections: The Private Sector's Participation in the Multilateral System of the FAO International Treaty on Plant Genetic Resources for Food and Agriculture, 2013, p. 40. https://www.publiceye.ch/fileadmin/files/documents/Biodiversitaet/20130916_private_collections.pdf.

[20] A.B. Nagpure, R.K. Choudhary, Gupta, Chitinases: in agriculture and human healthcare, Crit. Rev. Biotechnol. 34 (2013) 215–232.

[21] A. Nagpure, B. Choudhary, S. Kumar, R.K. Gupta, Isolation and characterization of chitinolytic Streptomyces sp. MT7 and its antagonism towards wood-rotting fungi, Ann. Microbiol. 64 (2013) 531–541.

[22] T. Avila, J. Izquierdo, Management of the appropriate agricultural biotechnology for small producers: Bolivia case study, Biotechnology issues for developing

countries, Electronic Journal of Biotechnology, Pontificia Universidad Católica de Valparaíso, Chile 9 (1) (2006) 1–7, https://doi.org/10.2225/vol9-issue1-fulltext-4. http://www.ejbiotechnology.info/content/vol9/issue1/full/4/4.pdf.

[23] S. Fukuda-Parr, Recapturing the narrative of development, in: R.A. Wilkinson (Ed.), Millennium Development Goals and beyond: Global Development after 2015, Routledge, Abingdon, 2012, p. 16S, 35-53.

[24] R. Ortiz, A. Jarvis, P. Fox, P.K. Aggarwal, B.M. Campbell, Plant Genetic Engineering, Climate Change and Food Security, CCAFS Working Paper 72, CGIAR Research Program on Climate Change, Agriculture and Food Security (CCAFS), Copenhagen, Denmark, 2014, p. 27.

[25] T.M. Oluwambe, S.A. Oludaunsi, Agricultural biotechnology, the solution to food crisis in Nigeria, Adv Plants Agric Res 6 (4) (2017) 219–226, https://doi.org/10.15406/apar.2017.06.00219.

[26] I. De Smet, S. Diederich, The Roots of Food Security, Max Planck Institute for Developmental Biology, Tübingen, 2018. https://www.mpg.de/617818/pressRelease20100127.

[27] SDSN, Solutions for Sustainable Agriculture and Food Systems Technical Report for the Post-2015 Development Agenda, The Sustainable Development Solutions Network, 2013, p. 108. http://unsdsn.org/wp-content/uploads/2014/02/130919-TG07-Agriculture-Report-WEB.pdf.

[28] S. Krimsky, J. Gruber, The GMO Deception, Skyhorse Publishing, 2014, ISBN 978-1-62873-660-1, p. 321.

[29] J. Nyerhovwo, A. Tonukari, Fostering biotechnology entrepreneurship in developing countries, Afr. J. Biotechnol. 3 (6) (2004) 299–301. ISSN: 1684–5315.

[30] J.Z. Ohikere, A.F. Ejeh, The potentials of agricultural biotechnology for food security and economic empowerment in Nigeria, Arch. Appl. Sci. Res. 4 (2) (2012) 906–913.

[31] A. Shamama, B.P. Abraham, Bioprospecting: promoting and regulating access to genetic resources and benefit sharing, Working Paper Series, WPS 631 (2008) 46.

[32] Anonymous, European commission. New techniques in agricultural biotechnology, independent scientific advice for policy making. Scientific advice mechanism (SAM). Directorate-general for research and innovation high level group of scientific advisors, Explanatory Note 02 (2017) 152.

[33] S.B. Brush, Genes in the Field: On-Farm Conservation of Crop Diversity, International Development Research Centre, 2000, ISBN 0-88936-884-8, p. 301. file:///C:/Users/umachandran/Downloads/IDL-32583.pdf.

[34] M. Wiweger, I. Farbos, M. Ingouff, U. Langercrantz, S. Von Arnold, Expression of Chia4-Pa chitinase genes Turing somatic and zygotic embryo development in Norway spruce (*Picea abies*): similarities and differences between gymnosperm and angiosperm class IV chitinases, J. Exp. Bot. 54 (2003) 2691–2699.

[35] Anonymous, Healing, Fueling, Feeding: How Biotechnology Is Enriching Your Life, Biotechnology Industry Oragnization, Washington, 2010, p. 86. https://www.bio.org/sites/default/files/files/ValueofBiotech.pdf.

[36] Anonymous, British Council. Global Food Security How Can We Feed a Growing Population? Commonwealth Class, 2016, p. 15. https://schoolsonline.britishcouncil.org/sites/default/files/resource/downloads/global_food_security_how_do_we_feed_a_growing_population_web.pdf.

[37] Anonymous, Biotechnology in Europe. The Tax, Finance and Regulatory Framework and Global Policy Comparison, Ernst & Young, EuropaBio, 2014, p. 232. http://www.ey.com/Publication/vwLUAssets/EY-biotechnologyin-europe-cover/$FILE/EY-biotechnology-in-europe.pdf.

[38] C. James, Global status of commercialized transgenic crops: 2005, ISAAA Briefs 34 (2005). Ithaca, N.Y.: ISAAA.

[39] R. Bhattacharjee, Harnessing biotechnology for conservation and utilization of genetic diversity in orphan crops, African Technology Development Forum, ATDF Journal 6 (3/4) (2009) 24–31.

[40] W. Quaye, Climate Change and Food Security: The Role of Biotechnology, African Journal of Food, Agriculture, Nutrition and Development (AJFAND), Eidgenössische Technische Hochschule Zürich, 2013, p. 5. http://www.css.ethz.ch/en/services/digital-library/articles/article.html/174856/pdf.

[41] R. Cohen-Kupiec, I. Chet, The molecular biology of chitin digestion, Curr. Opin. Biotechnol. 9 (1998) 270–277.

[42] Y. Itoh, K. Takahashi, H. Takizawa, N. Nikaidou, H. Tanaka, H. Nishihashi, T. Watanabe, Y. Nishizawa, Family 19 chitinase of Streptomyces griseus HUT6037 increases plant resistance to the fungal disease, Biosci. Biotechnol. Biochem. 67 (2003) 847–855.

[43] M.B. Howard, N.A. Ekborg, R.M. Weiner, S.W. Hutcheson, Detection and characterization of chitinases and other chitin-modifying enzymes, J. Ind. Microbiol. Biotechnol. 30 (2003) 627–635.

[44] A. Kasprzewska, Plant chitinases − regulation and function, Cell. Mol. Biol. Lett. 8 (2003) 809–824.

[45] A. Kisiel, E. Kepczyńska, Bacterial chitinases and their application in biotechnology, Postepy Mikrobiol. 56 (3) (2017) 306–315. http://www.pm.microbiology.pl.

[46] A. Kisiel, K. Jeckowska, E. Kepczyńska, The role of chitinases in plant development, Adv. Cell Biol. 44 (2016) 273–288 (in polish).

[47] R.F. Frederiksen, D.K. Paspaliari, T. Larsen, B.G. Storgaard, M.H. Larsen, H. Ingmer, B.G. Storgaard, M.H. Larsen, H. Ingmer, M.M. Palcic, J.J. Leisner, Bacterial chitinases and chitin-binding proteins as virulence factors, Microbiology 159 (2013) 833–847.

[48] G.W. Gooday, The ecology of chitin degradation, Adv. Microb. Ecol. 11 (1990) 387–430.

[49] Z. Novotna, K. Fliegerova, J. Simunek, Characterization of chitinases of polycentric anaerobic rumen fungi, Folia Microbiol. 53 (2008) 241–245.

[50] K. Kurita, Controlled functionalization of the polysaccharide chitin, Prog. Polym. Sci. 26 (2001) 1921–1971.

[51] J.G. Li, Z.Q. Jiang, L.P. Xu, F.F. Sun, J.H. Guo, Characterization of chitinase secreted by Bacillus cereus strain CH2 and evaluation of its efficacy against Verticillium wilt of eggplant, Biocontrol 53 (2008) 931−944.

[52] S.N. Yano, M. Rattanakit, T. Wakayama, Tachiki, A chitinase indispensable for formation of protoplast of *Schizophyllum commune* in basidiomycete-lytic enzyme preparation produced by *Bacillus circulans* KA-304, Biosci. Biotechnol. Biochem. 68 (2004) 1299−1305.

[53] S.N. Yano, A. Rattanakit, Y. Honda, M. Noda, A. Wakayama, T. Plikomol, Tachiki, Purification and characterization of chitinase A of *Streptomyces cyaneus* SP-27: an enzyme participates in protoplast formation from *Schizophyllum commune* mycelia, Biosci. Biotechnol. Biochem. 72 (2008) 54−61.

[54] Q. Yan, S.S. Fong, Bacterial chihitinase: nature and perspectives for sustainable bioproduction, Biores. Bioproc. 2 (2015) 31.

[55] Y. Kezuka, M. Ohishi, Y. Itoh, J. Watanabe, M. Mitsutomi, T. Watanabe, T. Nonaka, Structural studies of a two-domain chitinase from *Streptomyces griseus* HUT6037, J. Mol. Biol. 358 (2006) 472−484.

[56] S.L. Erb, R.S. Bourchier, K.V. Frankenhuyzen, S.M. Smith, Sublethal effects of *Bacillus thuringiensis* berliner subsp. *kurstaki* on *Lymantria dispar* (*Lepidoptera: Lymantriidae*) and the tachinid parasitoid *Compsilura concinnata* (*Diptera: Tachinidae*), Environ. Entomol. 30 (2001) 1174−1181.

[57] E.Z. Gomaa, Chitinase production by *Bacillus thuringiensis* and *Bacillus licheniformis*: their potential in antifungal biocontrol, J. Microbiol. 50 (2012) 103−111.

[58] R. Gupta, R.K. Saxena, P. Chaturvedi, J.S. Viridi, Chitinase production by *Streptomyces viridificans*: its potential in fungal cell wall lysis, J. Appl. Bacteriol. 78 (1995) 378−383.

[59] A. Deilamy, H. Abbasipour, Comparative bioassay of different isolates of *Bacillus thuringiensis* subsp. *kurstaki* on the third larval instars of diamondback moth, *Plutella xylostella* (L.) (Lep.: *Plutellidae*), Arch. Phytopathol. Plant Prot. 46 (2013) 1480−1487.

[60] S.K. Halder, C. Maity, C. Jana, A. Das, T. Paul, P.K.D. Mohapatra, B.R. Pati, B.K.C. Mondal, Proficient biodegradation of shrimp shell waste by *Aeromonas hydrophila* SBK1 for the concomitant production of antifungal chitinase and antioxidant chitosaccharides, Int. Biodeterior. Biodegrad. 79 (2013) 88−97.

[61] N. Sharma, K.P. Sharma, R.K. Gaur, V.K. Gupta, Role of chitinase in plant defense, Asian J. Biochem. 6 (2011) 29−37.

[62] J. Inbar, I. Che, Evidence that chitinase produced by *Aeromonas caviae* is involved in biological control of soil borne plant pathogen by this bacterium, Soil Biol. Biochem. 23 (1991) 973−978.

[63] R.J. Vaidya, I.M. Shah, P.R. Vyas, H.S. Chatpar, Production of chitinase and its optimization from a novel isolate Alcaligenes xylosoxydans: potential antifungal biocontrol, World J. Microbiol. Biotechnol. 1 (2001) 62−69.

[64] W.T. Chang, Y.C. Chen, C.L. Jao, Antifungal activity and enhancement of plant growth by Bacillus cereus grown on shellfish chitin wastes, Bioresour. Technol. 98 (2007) 1224−1230.

[65] W.T. Chang, M. Chen, S.L. Wang, An antifungal chitinase produced by Bacillus subtilis using chitin waste as a carbon source, World J. Microbiol. Biotechnol. 26 (2010) 945−950.

[66] G.K. Kishore, S. Pande, A.R. Podile, Biological control of late leaf spot of peanut (*Arachis hypogaea*) with chitinolytic bacteria, Phytopathology 95 (2005) 1157−1165.

[67] K. Suma, A.R. Podile, Chitinase A from *Stenotrophomonas maltophilia* shows transglycosylation and antifungal activities, Bioresour. Technol. 133 (2013) 213−220.

[68] S. Ghasemi, G. Ahmadian, N.B. Jelodar, H. Rahimian, S. Ghandili, A. Dehestani, P. Shariati, Antifungal chitinases from *Bacillus pumilus* SG2: preliminary report, World J. Microbiol. Biotechnol. 26 (2010) 1437−1443.

[69] M. Liu, Q.X. Cai, H.Z. Liu, B.H. Zhang, J.P. Yan, Z.M. Yuan, Chitinolytic activities in *Bacillus thuringiensis* and their synergistic effects on larvicidal activity, J. Appl. Microbiol. 93 (2002) 374−379.

[70] D. Liu, J. Cai, C.-C. Xie, C. Liu, Y.-H. Chen, Purification and partial characterization of a 36-kDa chitinase from *Bacillus thuringiensis* spp. *colmeri*, and its biocontrol potential, Enzym. Microb. Technol. 46 (2010) 252−256.

[71] C. Aggarwal, S. Paul, V. Tripathi, B. Paul, M.A. Khan, Chitinolytic activity in *Serratia marcescens* (strain SEN) and potency against different larval instars of *Spodoptera litura* with effect of sublethal doses on insect development, BioControl 60 (2015) 631−640.

[72] N. Dahiya, R. Tewari, R.P. Tiwari, G.S. Hoondal, Production of an antifungal chitinase from *Enterobacter* sp. NRG4 and its application in protoplast production, World J. Microbiol. 21 (2005) 1611−1616.

[73] N. Dahiya, R. Tewari, G.S. Hoondal, Biotechnological aspects of chitinolytic enzymes: a review, Appl. Microbiol. Biotechnol. 2 (2006) 1−10.

[74] N.S. Ajit, R. Verma, V. Shanmugam, Extracellular chitinases of fluorescent Pseudomonads antifungal to *Fusarium oxysporum* f. sp. *dianthi* causing carnation wilt, Curr. Microbiol. 52 (2006) 310−316.

[75] N.K. Arora, M.J. Kim, S.C. Kang, D.K. Maheshwari, Role of chitinase and β-1,3-glucanase activities produced by a fluorescent pseudomonad and in vitro inhibition of *Phytophthora capsici* and *Rhizoctonia solani*, Can. J. Microbiol. 53 (2007) 207−212.

[76] S.S. Sindhu, K.R. Dadarwal, Chitinolytic and cellulolytic pseudomonas antagonistic to fungal pathogens enhance nodulation by *Mesorhizobium* sp. in chickpea, Microbiol. Res. 156 (2001) 353−358.

[77] M. Sridevi, K.V. Mallaiah, Factors effecting chitinase activity of Rhizobium sp. from *Sesbania sesban*, Biologia 63 (2008) 307−312.

[78] N. Someya, M. Nakajima, K. Hirayae, T. Hibi, K. Akutsu, Synergistic antifungal activity of chitinolytic enzymes and prodigiosin produced by the biocontrol bacterium *Serratia marcescens* strain B2 against the gray mold

pathogen, *Botrytis cinerea*, J. Gen. Plant Pathol. 67 (2001) 312–317.

[79] S. Martínez-Absalón, D. Rojas-Solís, R. Hernández-León, M.,M. Orozco-Mosqueda, J.J. del, C. Peña-Cabriales, S. Sakuda, E. Valencia-Cantero, G. Santoyo, Potential use and mode of action of the new strain *Bacillus thuringiensis* UM96 for the biological control of the gray mould phytopathogen *Botrytis cinerea*, Biocontrol Sci. Technol. 24 (2014) 1349–1362.

[80] N.N. Nawani, B.P. Kapadnis, Production dynamics and characterization of chitinolytic system of Streptomyces sp. NK 1057, a well-equipped chitin degrader, World J. Microbiol. Biotechnol. 20 (2004) 487–494.

[81] K.H. Bao-qin, Y. Chang-ying, L. Wan-shun, D. Ji-Xun, Purification and inhibition fungal growth of chitinases from *Vibrio pacini*, Wuhan Univ. J. Nat. Sci. 9 (2004) 973–978.

[82] J.M. Payne, E.W. Davidson, Insecticidal activity of crystalline parasporal inclusions and other components of the *Bacillus sphaericus* 1593 spore complex, J. Invertebr. Pathol. 43 (1983) 383–388.

[83] V. Tamizhazhagan, K. Pugazhendy, V. Sakthidasan, C. Jayanthi, B. Sawicka, S. Gerlee, K. Ramarajan, P. Manikandan, The toxicity effect of pesticide Monocrotophos 36% E.C on the enzyme activity changes in Liver and Muscles of *Labeo rohita* (Hamilton, 1882), Int. J. Pharma Sci. Res. 8 (5) (2017) 60–67.

[84] K.C. Hoang, T.H. Lai, C.S. Lin, C.Y.T. Chen, C.Y. Liau, The chitinolytic activities of Streptomyces sp. TH-11, Int. J. Mol. Sci. 12 (2011) 56–65.

[85] K. Sakai, A. Yokota, H. Kurokawa, M. Wakayama, M. Moriguchi, Purification and characterization of three thermostable endochitinases of a noble Bacillus strain, MH-1, isolated from chitin-containing compost, Appl. Environ. Microbiol. 64 (1998) 3397–3402.

[86] S.N. Das, C. Neeraja, P.V.S.R.N. Sarma, J. Madhu Prakash, P. Purushotham, M. Kaur, S. Dutta, A.R. Podile, in: T. Satyanarayana, B.N. Johri, A. Prakash (Eds.), Microbial Chitinases for Chitin Waste Management (W) Microorganism in Environmental Management, Red, Springer, New York, 2012, pp. 135–150.

[87] A.S. Sahai, M.S. Manocha, Chitinases of fungi and plants: their involvement in morphogenesis and host parasite interaction, FEMS Microbiol. Rev. 11 (1993) 317–338.

[88] T.W. Liang, Y.J. Chen, Y.H. Yen, S.L. Wang, The antitumor activity of the hydrolysates of chitinous materials hydrolyzed by crude enzyme from *Bacillus amyloliquefaciens* V656, Process Biochem. 42 (2007) 527–534.

[89] Y. Cai, J. Yan, X. Hu, B. Han, Z. Yuan, Improving the insecticidal activity against resistant *Culex quinquefasciatus* mosquitoes by expression of chitinase gene chiAC in *Bacillus sphaericus*, Appl. Environ. Microbiol. 73 (2007) 7744–7746.

[90] Y.-T. Hu, Y. Ting, J.-Y. Hu, S.-H. Hsieh, Techniques and methods to study functional characteristics of emulsion systems, J. Food Drug Anal. (2017) 16–26. https://www.sciencedirect.com/science/article/pii/S1021948916301831.

[91] B. Krochmal-Marczak, B. Ślusarczyk, R. Tobiasz-Salach, I. Betlej, B. Sawicka, Chapter 3, in: B. Krochmal-Marczak (Ed.), Plants and Genetically Modified Food [in:] Selected Aspects of Food Safety, Publisher: The State Higher Vocational School named Stanisław Pigonia Edition I, Krosno, 2017, pp. 47–60. ISBN:973-83-64457-33-3 (in polish).

[92] E. Saks, U. Jankiewicz, Chitinolytic activity of bacteria, Postep. Biochem. 56 (2010) 427–434 (in polish).

[93] Anonymous, Global Trends in the Market of Chemicals Based on Biotechnology, Regulations and a Competitive Landscape. Perspectives up to 2020, 2019. https://www.transparencymarketresearch.com/biotechnology-based-chemicals-market.html.

[94] J. Clapp, The Political Economy of Food Aid in an Era of Agricultural Biotechnology, TIPEC Working Paper 04/6, 2004, p. 25, http://www.trentu.ca/org/tipec/4clapp6.pdf.

[95] B.L. Cantarel, P.M. Coutinho, C. Rancurel, T. Bernard, V. Lombard, B. Henrissat, The Carbohydrate-Active Enzymes database (CAZy): an expert resource for Glycogenomics, Nucleic Acids Res. 37 (2009) 233–238.

[96] S.L. Wang, W.T. Chang, Purification and characterization of two bifunctional chitinases/lysozymes extracellularly produced by *Pseudomonas aeruginosa* K-187 in shrimp and crab shell powder medium, Appl. Environ. Microbiol. 63 (1997) 380–386.

[97] X. Wang, Z. Lee, H. Hzeng, Q. Wei, J.S. Elledge, L. Li, Genomic instability and endoreduplication caused by RAD17 deletion 17, Genes Dev. 8 (2003) 965–970, https://doi.org/10.1101/gad.1065103.

[98] S.L. Wang, C.H. Chao, T.W. Liang, C.C. Chen, Purification and characterization of protease and chitinase from *Bacillus cereus* TKU006 and conversion of marine wastes by these enzymes, Mar. Biotechnol. 11 (2009) 334–344.

[99] S. Fukuda-Parrzai, D. McNeill, Post 2015: a new era of responsibility? J. Glob. Ethics 11 (1) (2015) 10–17. https://doi.org/10.1080/17449626.2015.

[100] S. Fukuda-Parrzai, A. Orr, GM Crops for Food Security in Africa – The Path Not yet Taken, WP 2012-018: 2, UNDP Regional Bureau of Africa, 2012, p. 51.

[101] Anonymous, Biotechnology Research in an Age of Terrorism Confronting the Dual Use Dilemma, Committee on Research Standards and Practices, 2003.

CHAPTER 15

Homemade Preparations of Natural Biopesticides and Applications

CHARLES OLUWASEUN ADETUNJI, PHD • CHUKWUEBUKA EGBUNA, BSC, MSC • HABIBU TIJJANI, PHD • DICKSON ADOM, PHD • LAITH KHALIL TAWFEEQ AL-ANI • KINGSLEY C. PATRICK-IWUANYANWU, PHD

INTRODUCTION

Medicinal plants, which are part of the large biodiversity available, are one of the natural gifts to mankind. Most farmers in developing countries depend on medicinal plants due to their easy accessibility, efficacy, reliability, cost-effectiveness, easy preparation, green and eco-friendly nature. The utilization of medicinal plants has formed a sustainable solution adopted by farmers for the preparation of homemade natural biopesticides. This is gradually forming a permanent replacement to synthetic agrochemicals which might be linked to various hazards experienced whenever they are utilized [1,2]. Homemade biopesticides have greater control for use over numerous pests and also come at a relatively cheaper cost when compared to synthetic pesticides. Farmers normally collect various parts of the medicinal plants which might include the leaves, barks, roots, flowers, and fruit peels. About 10 kg of the plant parts have to be thoroughly washed using clean water and dried using sunlight for some period of time. The air-dried/sun-dried plant parts can then be pulverized or ground into small pieces until it turns into powder. The powdered form is later soaked in a solvent that may be aqueous or organic in nature for a period of 3 or up to 7 days. The solvent containing the active ingredients are later evaporated until the aqueous solution turns into concentrated extract [3,4]. The obtained plant extract will then be used as insecticides, fungicides, herbicides, and disinfectant by farmers. Moreover, the farmer could also apply the powder obtained from the crude extract as dust for the management of pests and diseases affecting crops during preharvest as well as post harvest period. Also, the formulated botanical can also be applied by dipping them near the soil beside heaps so that the active ingredient could be diffused into the soil in a systematic way in order to prevent soil pests and diseases on the farm. This method is also effective against soil insect and termite soil pathogens.

EXTRACT DELIVERY FORMS

Powders

The powdered formulation is normally prepared by collecting various plant materials followed by grinding sun-dried medicinal plant materials into a powder. This method is more effective against stored product prone to fungi and insect attacks. The necessary amount of powder is admixed with a suitable amount of commodity prior to storage. The powder extracted with water is referred to as water-extractable powders and could be separated with muslin cloths and applied as a solution.

Aqueous Extracts

The crude extracts acquired by extracting the active ingredients using water may be obtained by pressing juice from the raw leaves of medicinal plants. The crude extract might later be diluted. Plant infusion may also be obtained by boiling the various herbal parts of medicinal plants. The extract obtained from the crude extract is more effective against field crop insect pests and diseases [2,5].

Mixed Formulations

Most farmers also use a synergetic formulation containing two different herbal mixtures as protection against agricultural pests. The efficacy of mixed formulation has been observed to be effective against stored grain pests. Also, it has been observed that a mixture of two essential oils might show better effectiveness against insect pests [6].

Natural Remedies for Pest, Disease and Weed Control. https://doi.org/10.1016/B978-0-12-819304-4.00015-4

Use of Adjuvants

Most farmers in the developing countries normally use some adjuvants like cassava, yam, and plantain as a diluent in the formulation of insecticidal dust from the dry fruits of *Piper guineense* and from buds of *Eugenia aromatica*. Moreover, some farmers attest that some cubes of sugar could be added to aqueous neem extracts which enhance the adherence capability of the active ingredient on the various plant parts [4,7,8].

SCIENTIFIC EXPERIMENTS ON THE USE OF PLANT EXTRACTS FOR PESTS, DISEASES, AND WEED CONTROL

Disinfectant from Plant Extract

Martins et al. [9] evaluate the efficacy of plant extracts and a commercially available disinfectant (Proxitane 1512 AL) on the fungus *Beauveria bassiana*. The alcoholic and aqueous plant extracts of guava (*Psidium guajava* (L.), jabuticaba (*Myrciaria cauliflora* (Mart.), and jambolan (*Syzygium cumini* (L.) were used as an insecticide against the insect pests. The following parameters were tested; colony-forming unit (CFU) counts, vegetative growth, conidia production, and insecticidal activity of the fungus. The result obtained shows that the aqueous extract from guava exhibited a higher activity on the tested insect pathogen.

Spentzouris [10] compared the antibacterial activities of *Aloe vera* and *Thymus vulgaris* with a commercial disinfectant and evaluated their efficacy against some nosocomial infections causing microorganisms like *Acinetobacter baumanii*, *Escherichia coli*, *Pseudomonas aeroginosa*, methicillin-resistant *Staphylococcus aureus*, and *Klebsiella pneumonia*. The result obtained show that the thyme essential oil exhibited an antibacterial effect on the six bacterial and reduced their number to 2 log reduction while that of ethanol extract of *Aloe vera* ($1 < DR < 2$) exhibited a lower antibacterial activity. Generally, it was observed that the plant extract shows a higher disinfectant activity when compared to commercial disinfectant used as positive control. Their study shows that plant extract could be used as a disinfectant but should be used at a higher concentration in order to achieve 5 log reductions against any tested pathogens. Also, their study also suggested that plant extracts may also be used in the food industry when consumed directly or used to flavor foods.

Urja et al. [11] prepared three plant-based formulations that could be utilized for household cleaning. The result obtained shows that all the three formulations exhibited a very high antimicrobial effect against all the tested bacterial pathogens like *E. coli* and *S. aureus*. Their study shows that herbal formulation

from *T. chebula* could form a permanent replacement to synthetic disinfectant commonly used as detergents, disinfectants, stain removers, and cosmetics releasing chemicals which have been observed to exhibit a hazardous effect to humans and the environment.

Saad et al. [12] evaluated a herbal formulation from *Matricaria chamomilla* flowers extracts. The antimicrobial efficacy of the herbal product was performed when compared to a synthetic product using disc diffusion method. The result obtained shows that the chamomile soap formulation exhibited more efficacy especially against inhibition against skin pathogens when compared to the commercial antiseptic soaps without any side effects.

Sekyere et al. [2] performed antifeeding and mortality tests on cowpea weevils using the soluble extracts of *Hyptis suaveolens* and *Hyptis spicigera*. The results of the experimental study showed that the extracts of the plants killed the cowpea weevils and prevented them from attacking the cowpea seeds.

Oparaeke et al. [13] used the crude extracts of West African black pepper, *Piper guineense*, on cowpea insect pests and noted that it prevented pod damage, thereby increasing the grain yield.

Antibacterial from Plant

Alavijeh et al. [14] utilized a methanolic extract of some medicinal plants from *Acacia nilotica*, *Azadirachta indica*, *Withania somnifera* against some strains of bacteria with a profound result. Again in their studies, they found that *A. tinolica* and *A. indica* exhibited significant antifungal activity against *Ziziphus mauritiana* and *Aspergillus flavus*.

Nisar et al. [15] evaluated the insecticidal, phytotoxic, antifungal, cytotoxic, antibacterial activity of *Indigofera gerardiana*. The result obtained shows that the crude extract demonstrated a more antimicrobial effect against *Salmonella typhi* and *Microsporum canis* with a minimum inhibitory concentration of 0.37 mg/mL and 0.09 mg/mL, respectively. Moreover, the insecticidal effect of the plant extract shows a minimum inhibitory effect on *Rhyzopertha dominica*, *Sitophilus oryzae*, *Callosobruchus analis* while there was no observable inhibition on *Tribolium castaneum*. The brine shrimp lethality assay reveals the absence of any cytotoxicity effect from the action and the crude extract of *Indigofera gerardiana*. Also, the fraction exhibited an herbicidal effect against *Lemna minor* when evaluated at the concentration of 1000 mg/mL.

Antifungal from Plant

Bluma et al. [16] validated the antifungal effect of 96 extracts from 41 Argentinian plant species against four strains of *Aspergillus* section *Flavi*. Their result revealed that essential oil from mountain thyme, poleo, and clove exhibited more antifungal activity when compared to others. Varma and Dubey [17] also validated the effect of essential oil from *Mentha arvensis* and *Caesulia axillaris* when tested at the concentration of 1000 and 1500 μL/L. Their result revealed that the oranges treated with essential oil from *Caesulia axillaris* and *Mentha arvensis* increased the shelf life to 3 and 7 days, respectively. Also, there was no observable symptom of spoilage on the orange peel.

Guynot et al. [18] performed antifungal studies of essential oils containing eugenol on *P. islandicum* and *A. flavus*. Their result showed that the essential oil has antimicrobial activity which might be linked to the presence of eugenol serving as an active ingredient. In another study, Guynot et al. [19] showed that the clove, thyme (*Thymus vulgaris*), lemongrass, and bay (*Laurus nobilis*) exhibited a higher antifungal effect against *A. niger* and *A. flavus*. Karthikeyan et al. [20] validated the antifungal efficacy of an aqueous extract of *Allium sativum* L. and *Allium cepa* L. when tested against fungus responsible for moldiness of sorghum grains. The tested fungus, *Alternaria alternate*, *Aspergillus flavus*, and *Fusarium moniliforme* had an inhibition growth values, respectively, of 71.1%, 74.4%, and 73.3%. Moreover, Dikbas et al. [21] showed that the methanol extract and the essential oil of *Satureja hortensis* exhibited an antifungal effect on *A. flavus* obtained from lemon fruit.

Shukla et al. [22] revealed that the aqueous extracts from the leaves of *A. indica* prevented the aflatoxin biosynthesis probably by inhibiting the fungus used during the in vitro *assay* (*A. flavus* 100% and *A. parasiticus* more than 95%, using extract concentrations at 10% v/v). Moreover, AFB1 synthesis and the toxic effect of aflatoxin were averted when treated with *Allium sativum* in a semisynthetic medium. Akpomeyade and Ejechi [23] evaluated the antifungal effect of the aqueous extract of *Zingiber officinale* and *Xylopia aetiopica* when tested at the concentration of 1%−3% against *A. niger*, *A. flavus*, or *Rhizopus stolonifer*. Their result revealed that the aqueous extract has the capability to reduce the growth of these spoilage fungi.

The high rate of postharvest losses encountered in developing countries as a result of lack of control and modified storage facilities has intensified and aggravated the high rate of wastage experience on harvested fruits and vegetable. Edible coatings have been identified as one of the sustainable techniques used for the postharvest management and extension of fruits and vegetables. They minimize microbial multiplying, delay dehydration, and prevent high rate of transpiration

from fruits and vegetables [5]. *Aloe vera* gel has been shown to contain active ingredient with proven medicinal and therapeutic properties. These exceptional features have made *Aloe vera* a candidate for edible coatings to prevent spoilage, reduce microorganism proliferation, and maintain the quality of fruit during storage. The *Aloe* gel coating has been also shown to enhance the postharvest quality of fruits [3,5,24,25].

In another study, Nie et al. [26] performed an in vitro assay with the crude extract of *Jatropha curcas* on *Aspergillus niger v. Tieghem* is responsible for the postharvest rot of onion bulb. The result obtained shows that the aqueous extract from *Jatropha curcas* inhibited the mycelial and growth of *Aspergillus niger v. Tieghem*. Also, Adetunji et al. [27] found that *Opuntia cactus* mucilage extract inhibited the activity of spoilage microorganisms during storage at evaporative coolant system and finally extend the shelf life of *Carica papaya* fruits.

Herbicidal Properties from Plants

Randhawa et al. [28] performed an herbicidal assay with the aqueous extract of *Wedelia chinensis*. The result obtained showed that the aqueous extract possesses the ability to inhibit the seedling growth, reduce the seed germination which consequently led to yellowing of leaves, and reduced resistance to disease in weeds. The concentration of 0.4 fresh weight ml^{-1} water exhibited the highest herbicidal activity against the following weeds: *Paspalum thunbergii*, *Alternanthera sessilis*, *Cyperus difformis*. In a nutshell, the study shows the efficacy of the crude extract from *W. chinensis* as a potential pre-emergence bioherbicides for the biological control of weed germination.

Kato et al. [29] also validate the effect of sorghum extract for the reduction of seedling growth and seed germination of *Trianthema portulacastrum* when tested at a very high concentration which ranges from 75% to 100% but stimulated shoot length of the weed when applied at a very low concentration of 25%.

In another study, Iqba et al. [30] established the bioherbicidal potential of acetone extract obtained from 30-day-old lemon balm (*Melissa officinalis*) shoots. The growth-inhibiting compound was obtained using silica gel column chromatography. Their result shows that 0.3 μg ml^{-1} exhibited the highest bioherbicidal effect on the target weeds.

Macias et al. [31] revealed that the aqueous extract obtained from root exudates of buckwheat from Aomori, Japan, showed a significant effect in reducing weed biomass in comparison to plots that does not have buckwheat. Their results show that the root

TABLE 15.1
Natural and Homemade Insecticides.

Type	Pests/Insects	Preparation	Effects on Pests/Insects	References
Oil spray insecticide	Aphids, mites, thrips	Mix 1 cup of vegetable oil with 1 tablespoon of soap (cover and shake thoroughly), and then when ready to apply, add 2 teaspoons of the oil spray mix with 1 quart of water, shake thoroughly, and spray directly on the surfaces of the plants which are being affected by the little pests.	The oil coats the bodies of the insects, effectively suffocating them, as it blocks the pores through which they breathe.	[32]
Soap spray insecticide	Soft-bodied arthropods (aphids, whiteflies, psyllids, mealybugs, spider mites)	Mix 1 1/2 teaspoons of a mild liquid soap (such as castile soap) with 1 quart of water, and spray the mixture directly on the infected surfaces of the plants. It is always recommended to NOT apply it during the hot sunny part of the day, but rather in the evenings or early mornings	Effective against varied soft-bodied arthropods, killing them to protect plants.	[34]
Neem oil insecticide	Capable of disrupting the life cycle of insects at all stages (adult, larvae, and egg). Mildew and other fungal infections	Mix 2 teaspoons of neem oil and 1 teaspoon of mild liquid soap and shake thoroughly with 1 quart of water, and then spray on the affected plant foliage	Neem oil acts as a hormone disruptor and as an "antifeedant" for insects that feed on leaves and other plant parts	[35]
Garlic insecticide spray	Insecticides/repellent of many especially the mealworm beetle, *Tenebrio molitor* Linnaeus	2 whole bulbs (not just 2 cloves) and puree them in a blender or food processor with a small amount of water. Quart of water. Let the mixture sit overnight, then strain it into a quart jar, adding 1/2 cup of vegetable oil (optional), 1 teaspoon of mild liquid soap, and enough water to fill the jar. To use this homemade insecticide, use 1 cup of mixture with 1 quart of water and spray liberally on infested plants.	Garlic essential oil causes lethal and sublethal effects on *T. molitor*	[36]

Ginger insecticide spray	Caterpillars of the diamond back moth (*Plutella xylostella*), the cabbage webworm (*Hellula undalis*), cabbage aphids (*Brevicoryne brassicae*), and cabbage looper (*Trichoplusia ni*)	The rhizomes were crushed and milled into a fine powder using an electric miller. The solution was sieved to remove any coarse materials and stored in a plastic container. Extract concentration of 20% (w/v) of ginger was prepared by 400g of ginger to 2 liters water in a plastic container and mixed well. It was again sieved to remove all solid particles before it was poured into a knapsack sprayer. The concentration was sprayed on sampled cabbage leaves infested with the four dangerous cabbage pests.	Effective against cabbage pests on caterpillars of the diamond back moth (*Plutella xylostella*), the cabbage webworm (*Hellula undalis*), cabbage aphids (*Brevicoryne brassicae*), and cabbage looper (*Trichoplusia ni*)	[37]
Hyptis suaveolens and Hyptis spicigera insecticide	Cowpea weevils	Masses of 400 and 500 g of pulverized samples of the two plants, Hyptis suaveolens and Hyptis spicigera were separately extracted by percolation with 2200 mL of 95% ethanol, respectively. The percolates were evaporated to dryness at room temperature to give crude extracts of both plants which were each subjected to a partition process. Soluble solvent extracts of the two plants are then applied on cowpea seeds infested with cowpea weevils	The chloroform soluble extracts were found to kill the cowpea weevils preventing them from attacking the cowpea seeds.	[2]
Basic oil spray	Insects, mites, aphids	2 tablespoons biodegradable dishwashing liquid 1 gallon of warm water Mix and use as a spray	Kills soft-bodied insects, mites, thrips, aphids by penetrating the cell membranes causing the insects to dry out	[38]

exudates reduced the shoot and root biomass of the weeds containing *Brassica juncea, Digitaria ciliaris, Echinochloa crus-galli, Trifolium repens, Amaranthus palmeri*. The result of the structural elucidation confirmed that the active compounds responsible for the bioherbicidal activity were Fagomine, 4-piperidone, and 2-piperidinemethanol.

Sarwar et al. [32] revealed that the crude extract obtained from *Helianthus annuus* possess an inhibitory effect on seedling growth and seed germination of different weeds containing *Flaveria australasica, Echinochloa crus-galli, Amaranthus albus, Digitaria sanguinalis, P. hysterophorus, Celosia cristata, Sida spinosa, Flaveria australasica, Ambrosia artemisiifolia, Avena fatua, Cynodon dactylon, Chloris Barbara, Dactyloctenium aegyptium, D. ciliaris, Agropyron repens, Veronica persica, Trianthema portulacastrum, Amaranthus viridis.* Salamci et al. [33] showed that essential oil obtained from *T. chiliophylllum* and *T. aucheranum* exhibited an inhibitory effect against seedling growth and seed germination of *A. retroflexus*. The active compounds responsible for the herbicidal activity in the essential oil includes borneol, 1, 8-cineole, α-terpineol, camphor. Some naturally prepared insecticides were presented in Table 15.1.

CONCLUSION AND FUTURE PROSPECTS

This chapter has provided a comprehensive and recent methodology commonly used for the manufacturing of homemade biopesticdes and their application as bioherbicides, biofungicides, bioinsecticides, antifungal, antibacterial and biodisinfectant in diverse field. Also, this book has provided sufficient detail on the modes of action through which these homemade biopesticides are normally applied for the management of pests and diseases. The various applications of these biopesticides is well-validated. Therefore, in order to increase the efficacy and mass production of the active ingredient responsible for their activity, there might be a need to introduce some biotechnological techniques that could improve the effectiveness of these plant extract through genetic engineering. This will go a long way in improving the actions of numerous metabolite containing biologically active ingredients in already established plants showing enhanced biological activities against pests, diseases, insects, and weeds. There is also a need for the government of the developing country to establish a policy that will support the utilization of all the highlighted homemade biopesticides due to their uncountable benefits when compared to the synthetic pesticides which comes with several environmental and health hazards on mankind. Also, it will reduce the financial stress on these farmers that might not have enough capital, subsidies, and accesses to government intervention, cooperatives, and bank loans for them to acquire these costly synthetic pesticides for the management of these agricultural pests and pathogens. This will also help most of the poor farmers in the developing countries that could not get access to agricultural inputs distributed by government and relevant agency. Hence, this chapter has listed some beneficial homemade biopesticides that are easily adaptable which are a sustainable, cost-effective, eco-friendly, easy mode of preparation, easy to prepare, as well as their durability and stability in nature.

REFERENCES

[1] C. Adetunji, J. Oloke, A. Kumar, S. Swaranjit, B. Akpor, Synergetic effect of rhamnolipid from *Pseudomonas aeruginosa* C1501 and phytotoxic metabolite from LasiodiplodiapseudotheobromaeC1136 on Amaranthushybridus L. and Echinochloa crus-galliweeds, Environ. Sci. Pollut. Control Ser. 24 (15) (2017) 13700–13709, https://doi.org/10.1007/s11356-017-8983-8.

[2] P.A. Sekyere, V.C. Mbatchou, D. Adom, F.A. Ayisi, Pesticidal effects of extracts from Hyptis suaveolens and Hyptis spicigera on cowpea weevils, Int. J. Environ. Agric. Biotechnol. 3 (5) (2018) 1691–1699.

[3] C.O. Adetunji, O.O. Olaniyi, A.T.J. Ogunkunle, Bacterial activity of crude extracts of Vernoniaamygdalinaon clinical isolates, J. Microbiol. Antimicrob. 5 (6) (2013) 60–64.

[4] T.I. Ofuya, Formulation of medicinal plants for crop protection in Nigeria, in: Proceeding of the Humboldt Kellog/5th SAAT Annual Conference of Formulations of Medicinal Plants in Plant and Animal Production in Nigeria, School of Agriculture and Agricultural Technology, The Federal University of Technology, Akure, Ondo State, Nigeria, 2009, pp. 1–6.

[5] C.O. Adetunji, F.S. Omojowo, E.S. Ajayi, Effects of Opuntia cactus mucilage extract and storage under evaporative coolant system on the shelf life of Carica papaya fruits, J. Agrobiotech 5 (2014b) 49–66.

[6] D.B. Olufolaji, Combination of plant extracts of Chromolaenaodorataand ocimumgratissimumin the control of pineapple disease of sugarcane in Nigeria, in: Y.-R. Li, M.I. Nasr, S. Solomon, G.P. Rao (Eds.), Proceedings of the InternationalConference of Professionals in Sugar and Integrated Technologies. InternationalAssociation of Professionals in Sugar and Integrated Technologies, Nanning, PR China, 2008, pp. 426–428.

[7] L.E.N. Jackai, The use of neem in controlling cowpea pests, Int. Inst. Tropical Agric. Res. 7 (1993) 5–11.

[8] C.O. Adetunji, O.J.K. Oluwaseun, O.O. Osemwegie, Environmental fate and effects of granular pesta formulation from strains of *Pseudomonas aeruginosa* C1501 and Lasiodiplodiapseudotheobromae C1136 on soil activity and weeds, Chemosphere 195 (2018) 98–107. https://doi.org/10.1016/j.chemosphere.2017.12.056.

[9] C.C. Martins, L.F.A. Alves, A.P. Mamprim, Effect of plant extracts and a disinfectant on biological parameters and pathogenicity of the fungus Beauveriabassiana(Bals.) Vuill. (Ascomycota: Cordycipitaceae), Braz. J. Biol. 76 (2) (2016) 420–427.

[10] N. Spentzouris, Comparative Study on Disinfection Efficacy of Thymus Vulgaris and Aloe Vera Extracts with Commercial Disinfectants, on Bacteria Isolated in Nosocomial Environment, MSc Thesis, 2015, pp. 1–43.

[11] P. Urja, D. Ankit, S.S. Nirmal, Development of herbal disinfectants formulation for mopping households and its antibacterial activity, Nat. Prod. Res. 31 (22) (2017) 2665–2668. https://doi.org/10.1080/14786419.2017.1283491.

[12] A.H. Saad, S.N. Gamil, R.B. Kadhim, R. Samour, Formulation and evaluation of herbal hand wash from Matricaria chamomilla flowers extracts, Int. J. Res. Ayurveda Pharm. 2 (6) (2011) 1811–1813.

[13] A.M. Oparaeke, Effect of crude extracts of mixtures of different plant species in the control of Megalurothrips sjostedti Trybom and pod setting of cowpea plants, Arch. Phytopathol. Plant Prot. 40 (3) (2007) 201–206.

[14] P.K. Alavijeh, K.A. Parisa, S.A. Devindra, Study of antimicrobial activity of few medicinal herbs, Asian J. Plant Sci. Res. 2 (4) (2012) 496–502.

[15] M. Nisar, S.A. Tariq, K.I. Marwat, M.R. Shah, I.A. Khan, Antibacterial, antifungal, insecticidal, cytotoxicity and phytotoxicity studies on Indigoferagerardiana, J. Enzym. Inhib. Med. Chem. 24 (1) (2009) 224–229, https://doi.org/10.1080/14756360802051313.

[16] R. Bluma, M.R. Amaiden, M. Etcheberry, Screening of Argentine plantsextracts: impact on growth parameters and aflatoxin B1 accumulation by Aspergillussection Flavi, Int. J. Food Microbiol. 122 (2008) 114–125.

[17] J. Varma, N.K. Dubey, Efficacy of essential oils of Caesuliaaxillaris and Menthaarvensis against some storage pests causing biodeterioration of food commodities, Int. J. Food Microbiol. 68 (2001) 207–210.

[18] P. Lopez, C. Sanchez, R. Batlle, C. Nerin, Solid- and vapor-phase antimicrobialactivities of six essential oils: susceptibility of selected food borne bacterial and fungal strains, J. Agric. Food Chem. 53 (2005) 6939–6946.

[19] M.E. Guynot, A.J. Ramos, L. Seto, P. Purroy, V. Sanchis, S. Marin, Antifungal activity of volatile compounds generated by essential oils against fungi commonly causing deterioration of bakery products, J. Appl. Microbiol. 94 (2003) 893–899.

[20] M. Karthikeyan, R. Sandosskumar, R. Radhajeyalakshmi, S. Mathiyazhagan, S.E. Khabbaz, K. Ganesamurthy, B. Selvi, R. Velazhahan, Effects of formulated zimmu (Allium cepa L. x Allium sativum L.) extract in the management of grain mold of sorghum, J. Sci. Food Agric. 87 (2007) 2495–2501.

[21] N. Dikbas, R. Kotan, F. Dadasoglu, F. Sahin, Control of Aspergillusflavus with essential and methanol extract of Saturejahortensis, Int. J. Food Microbiol. 124 (2008) 179–182.

[22] R. Shukla, A. Kumar, C.S. Prasad, B. Srivastava, N.K. Dubey, Antimycotic and antiaflatoxigenic potency of Adenocalymmaalliaceum miers on fungi causing biodeterioration of food commodities and raw herbal drugs, Int. Biodeterior. Biodegrad. 62 (2008) 348–351.

[23] D.E. Akpomeyade, B.O. Ejechi, The hurdle effect of mild heat and two tropical spice extracts on the growth of three fungi in fruit juices, Food Res. Int. 31 (1999) 339–341.

[24] C.O. Adetunji, O.B. Fawole, J.K. Oloke, J.B. Adetunji, O.R. Makanjuola, Effect of edible coatings from Aloe vera gel on Citrus sinensis during ambient storage, J. Agric. Res. Develop. 11 (1) (2012) 77–84.

[25] C.O. Adetunji, O.B. Fawole, K.A. Arowora, S.I. Nwaubani, E.S. Ajayi, J.K. Oloke, O.N. Majolagbe, B.A. Ogundele, J.A. Aina, J.B. Adetunji, Effects of edible coatings from Aloe vera gel on quality and postharvest physiology of Ananascomosus (L.) fruit during ambient storage, Global J. Sci. Frontier Res. Bio-tech Genetics 12 (5) (2012) 39–43.

[26] C. Nie, Y. Wen, H. Li, I. Chen, M. Hong, J. Huang, C. Nie, Y. Wen, H. Li, L.Q. Chen, M.Q. Hong, J.H. Huang, Study on allelopathic effects of Wedeliachinensison some weeds in South China, Weed Sci. China 2 (2002) 15.

[27] C.O. Adetunji, A.E. Fadiji, O.O. Aboyeji, Effect of chitosan coating combined Aloe vera gel on cucumber (Cucumis Sativa L.) post-harvest quality during ambient storage, J. Emerg. Trends Eng. Appl. Sci. 5 (6) (2014b) 391–397.

[28] M.A. Randhawa, Z.A. Cheema, M.A. Ali, Allelopathic effect of sorghumwater extract on the germination and seedling growth of Trianthemaportulacastrum, Int. J. Agric. Biol. 4 (2002) 383–384.

[29] N.H. Kato, K. Kawabata, Isolation of allelopathic substances in lemon balmshoots, Environ. Control Biol. 40 (2002) 389–393.

[30] Z. Iqba, S. Hiradate, A. Noda, S. Isojima, Y. Fujii, Allelopathy of buckwheat: assessment of allelopathic potential of extract of aerial parts of buckwheat and identification of fagomine and other related alkaloids as allelochemicals, Weed Biol. Manag. 2 (2002) 110–115.

[31] F.A. Macias, R.M. Varela, A. Torres, J.L.G. Galindo, J.M.G. Molinillo, Allelo chemicals from sunflowers: chemistry, bioactivity and applications, Chem. Ecol. Plants (2002) 73–87.

[32] M. Sarwar, M. Salman, Toxicity of oils formulation as a new useful tool in crop protection for insect pests control, Int. J. Chem. Biomol. Sci. 1 (4) (2015) 297–302.

[33] E. Salamci, S. Kordali, R. Kotan, A. Cakir, Y. Kaya, Chemical compositions,antimicrobial and herbicidal effects of essential oils isolated from Turkish Tanacetumaucheranumand Tanacetumchiliophyllumvar. chiliophyllum, Biochem. Syst. Ecol. 35 (2007) 569–581.

[34] W.S. Cranshaw, Insect Control: Soaps and Detergents. Insect Series, Fact Sheet No. 5-547, Colorado University, U.S., 1996.

[35] E.V.R. Campos, J.L. de Oliveira, de L.R. Pascoli, L.F. Fraceto, Neem oil and crop production: from now to the future, Front. Plant Sci. 7 (2016) 1–8.

[36] A. Plata-Rueda, L.C. Martinez, M.H. Dos Santos, F.L. Fernandes, C.F. Wilcken, M.A.A. Soares, J.E. Serrao, J.C. Zanuncio, Insecticidal activity of garlic essential oil and their Constituents against the mealworm beetle, Tenebrio molitor Linnaeus (Coleoptera: tenebrionidae), Sci. Rep. 7 (2017) 46406. http://www.nature.com/scientific reports/.

[37] P.A. Sekyere, D. Adom, A. Addo, Controlling Dangerous Cabbage Pests in Ghana: The Effects of Aqueous Extracts of Ginger (Zingiber Officinale Roscoe) as Organic Pesticides (Ongoing Study).

[38] C.E. Bogran, S. Ludwig, B. Metz, Using Oils as Pesticides, AgriLife Extension, 2006, p. E-419. http://AgriLife bookstore.org.

CHAPTER 16

The Use of Vermiwash and Vermicompost Extract in Plant Disease and Pest Control

INTAN SORAYA CHE SULAIMAN, BSC, MSC, PHD • AZHAM MOHAMAD, BSC, MS

INTRODUCTION

Historically, the early use of earthworms in organic wastes decomposition was first reported in Germany in 1980 and continued in the United States [1]. Further research and development of earthworms for vermiculture and vermicomposting as biofertilizers has led to their new potential application as biocontrol agents against pests and diseases [2–6]. By definition, vermiculture is the process of breeding earthworms, whereas the liquid filtered from the watery wash of earthworms is called vermiwash [7], while, vermicomposting is the transformation process of organic waste to compost or vermicompost by the use of earthworms. Vermicompost can be produced in a shorter time compared to conventional compost [8]. This owed to the humic compounds in vermicompost that increases the humification process, which fastens the conversion of organic matter [9].

Vermicomposting is one of the effective waste management techniques across the world. The technology employed combination action of earthworms with associated microbes in composting process of organic substrate. Substrate used for earthworms feeding in vermicomposting process can be sourced out from agricultural by-products and wastes. Poor animal waste management may cause public health risk due to discharge of unprocessed animal manures into agricultural fields that contaminated groundwater. Hence, vermicomposting is found to be an effective method to treat or converting agricultural by-product and waste into useful products.

Interestingly, this eco-friendly composting technique produces high nutritional quality of end products and the nonthermophilic conditions overcome the major drawbacks associated with conventional composting methods that volatilized ammonia during thermophilic process and losses about 55% of organic matter [1,10,11]. Healthy soil can be characterized as soil with high organic matter, balanced nutrient cycle, good structure, high water-holding and good drainage capacity, sufficient rooting depth, and having a diversity of beneficial microorganisms [12]. Hence, soil organic matter plays an important role in soil productivity as source of nutrients for crops. But, these nutrients may be removed from the system during crops harvesting process. In this manner, it restores soil organic matter by the addition of organic amendment from vermicompost or worm casting which is necessary to effectively increase or preserve adequate level of soil organic matter overtime and simultaneously improve soil biological activity [11]. Previous studies reported that, plants fortified with organic amendments possess a relatively greater pests and disease resistant ability than inorganic fertilizer amendments [13]. The reason behind this increment is addition of organic amendments into soil which alters the physicochemical and microbiological environment of soil, thereby facilitating to pest and disease suppression [14].

Vermicompost extracts also known as vermicompost teas are concentrated essence of vermicompost that are regularly prepared in the ratio of 5%–20% of solid vermicompost to water [15]. Application method for vermiwash and vermicompost extracts is mostly in the form of foliar sprays which can be directly applied to plant leaves or drench on soil surface. In that way, the beneficial nutrients from the aqueous solutions can be deposited effectively on target areas and thus only a small amount of material is needed [15,16]. Foliar spray and soil drench are more effective to protect against plant pest and disease compared with solid form of vermicompost because in liquid form, it is easier to reach the target area such as rhizosphere in soil where most nematodes live in.

The term vermiwash is often confused with vermi-compost tea. Basically, vermiwash is a product of ver-miculture, a liquid wash filtered through the body of earthworms, while vermicompost tea is the solution extracted from vermicompost [7]. Vermiwash contains mucus secretion of earthworms and their excretory products including micronutrients from the soil organic molecules [17]. Similarly, foliar application of vermi-wash exhibited significant effects of growth promoter and protector for crop plants [18]. The preparation of vermiwash using lukewarm water by mildly agitating may help to secrete higher amount of body fluids and mucus of earthworms [19]. Vermiwash is rich in vita-mins, amino acids, minerals (potassium, calcium, zinc, copper, nitrogen, iron and magnesium), beneficial microorganisms, and growth hormones of cytokinin's and auxins [4,20]. A combined use of vermicompost and vermiwash spray has been reported to reduce two serious pests of chilli, trips and mites [21].

PRINCIPLES OF VERMICOMPOSTING

Vermicomposting is a unique process that occurs in earthworm's gut to convert organic wastes into organic fertilizer or vermicompost by using joint action of earthworms and microorganisms [22]. Likewise, con-ventional composting, vermicomposting is an aerobic process that required the presence of oxygen. The pro-cess is natural, emits little or no odor [7], and only involved mesophilic phase as compared with conven-tional composting that requires thermophilic and mes-ophilic phases [10]. Briefly, when the earthworm consumes organic wastes, the substrate passes through earthworm's gut and gets digested in the intestine of earthworm with the aid of beneficial microbes. In the intestinal tract, mucus or chemical secretions, enzymes, and antibiotics help in the breakdown of substrate to finely divided peatlike material called vermicompost, which is readily available to plants [7,8,21,23]. Hence, vermicompost can be applied as plant growth media or as soil amendments [1].

The presence of vermicompost in soil may act as a soil conditioner by supplying nutrients to plants, lowering C to N ratio, improving the soil texture, increasing soil porosity and water holding capacity, thereby requiring less tillage and irrigation [10,24]. Fig. 16.1 illustrates the interaction between earthworm and microorganisms in earthworm's gut during vermicomposting.

Mesophilic phase in vermicomposting consists of two stages, active and maturation phases [21]. Active phase is an initial phase which involves the combined actions of earthworms and beneficial microorganisms to break down organic substrate by metabolic activities thus modifying their physical and soil biological activ-ities. While, maturation phase is the subsequent phase that involves microorganism action to continue the decomposition process, once the earthworm moves to a fresh layer of undigested wastes [10,24]. Earthworms play an important role in aerating, conditioning, frag-menting, and altering the biological activities of a sub-strate, while microorganisms are needed for biochemical degradation of organic matter [21,25]. Earthworms can consume organic wastes equal to their body weight per day [22], and the duration taken to break down organic substrates through their metabolic activities (digestion, ingestion, and assimilation) varies according to earthworm species and their density [21,24].

Among earthworm categories, epigeic earthworms are the most widely used earthworm species in vermi-composting, owing to their tolerance to a wide range of environmental factors (temperature and moisture) [17] and consumption of large amounts of organic wastes daily [1]. These species of earthworms are usu-ally found in topsoil, which are rich areas of organic matter and microorganisms. They have short life span, high reproductive rates, easy handling, and thus, serves as potential candidates for vermicomposting [1]. Two epigeic species of earthworms, *Eisenia fetida* (redworm) and *Eisenia andrei*, account for 80%—90% of the earth-worms commercially grown for vermicomposting [1].

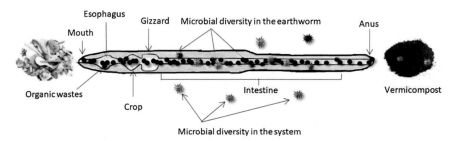

FIG. 16.1 Illustration of interaction between earthworm and microorganisms in earthworm's gut.

These common composting species do not burrow deeply into soil, and are therefore classified as nonburrowing type of earthworms. They are adapted to decaying organic material and pile of leaves or manure. Nonburrowing types of earthworms are mainly employed in vermicomposting than burrowing types due to their faster action in converting organic wastes into vermicompost [7]. A comparative study to investigate the effects of different earthworm species (*Metaphire guillemi* and *E. fetida*) on *Fusarium* wilt of strawberry was carried out by Bi et al. [26]. From the results, both tested earthworm species demonstrated a promising capability in reducing infestation of *Fusarium* wilt of strawberry. However, *M. guillemi* showed to be more effective than *E. fetida* in reducing the disease index, proving that different earthworm species may display different abilities in reducing plant diseases.

MECHANISM OF VERMICOMPOST EXTRACT AND VERMIWASH FOR SUPPRESSION OF PLANT PEST AND DISEASE

Previous scientific research found that chemical composition and autochthonous microbial communities in vermicompost are influenced by their parental wastes [27]. Moreover, Warman and AngLopez [28] reported the significant differences in plant growth when introduced to vermicompost produced from various types of parental wastes and vermicompost in varying concentrations. Due to the aforementioned factor, earthworm diet plays an essential role to shape the microbial community structure in vermicompost [15]. Besides its chemical and biological properties, physical appearance of vermicompost also may vary with regard to earthworm species used or their ecological group, production process, and age of vermicompost [10,29]. In addition to feedstock and earthworm species used, the variation in vermicompost or vermicast properties also may influence soil properties.

Soil type is shown to have great influence on vermicast properties (62%) as compared to earthworm species factor (10%) [30]. A similar observation has been reported by Nath and Singh [31] for vermiwash nutritional status in relation to different combinations of feed materials. From the experiment, pH, total organic carbon, C/N ratio, electrical conductivity, total Kjeldahl nitrogen, total calcium, and total phosphorus of vermiwash taken within before and after vermicomposting using different combinations of buffalo dung with water hyacinth and gram bran have shown a significant difference. The results confirm that feed materials play an essential role to develop nutritional values of end products (vermicompost). Furthermore, the quality of vermicompost derivatives produced, for instance, vermiwash and vermicompost extract, may also reflect the nutrient values of the vermicompost used [32]. Apart from recycling of organic waste into biofertilizer and biocontrol agent, vermicomposting also make benefits for bioremediation of contaminated soils [33].

Previous studies reported that vermicompost prepared with nonthermophilic approach comprises microbial diversity and activity compared with conventional thermophilic compost which only promotes selected microbes [1,21,34]. The presence of beneficial microbiota in vermicompost such as fungi, bacteria, and actinomycetes could benefit plant growth by producing plant growth–promoting enzymes and hormones, thereby indirectly suppressing plant pest and diseases [10,23]. For example, the presence of actinomycetes in the rhizosphere may boost the rivalry for iron with other rhizosphere phytopathogens [35]. Actinomycetes and proteobacteria are two major classes of microbials in vermicompost [36]. Principally, microbial antagonism resultant from microbial diversity and activity is a key factor that contributes to biological mechanism of disease suppression. As a result, the soil with higher level of active microbial biomass or total microbiological activity may possess higher level of disease suppressive ability [37]. This is because soil with greater microbiological activity may use up large amount of nutrients, carbon, and energy which may limit the nutrients accessibility to phytopathogens [38]. Several researchers reported that the addition of vermicompost which is a microbiologically active organic amendment to soil may also promote microbial proliferation and activity [1,21,39] through its binding agents of nutrients and calcium humate [34]. Some of the antagonistic effects reported in literatures for disease control were presented in Table 16.1.

Soil-borne pathogens or soil organisms may cause disease in crops and are responsible for great economic losses [39]. They share the same habitat with other soil organisms and compete for limited resources of food, space, and water [34]. When soil is ample with microbial diversity and activity, the massive number of beneficial microorganisms compete with pathogens for nutrients and act as antibiosis by producing antibiotics that reduce pathogens survival and growth [23,34]. Mechanisms involved in disease suppression are not restricted only to single mechanism but can be a combination of other mechanisms as well. Beneficial microorganisms may also act as a parasite by puncturing cell wall and consuming the plant pathogens [12,41].

TABLE 16.1
Antagonistic Effects Between Indigenous Beneficial Microbes and Phytopathogens for Disease Control.

Source	Parental Wastes	Earthworm Species	Beneficial Microbes	Phytopathogens	Plants	Disease Control	Reference
				DISEASE CONTROL			
Vermiwash	NAD	Eisenia fetida	Burkholderia sp.	Fusarium solani, Alternaria solani, Fusarium oxysporum, and Rhizoctonia solani	Brinjal Tomato Brinjal Mustard	Fusarium wilt Early leaf blight Fusarium wilt Damping off root	[40]
Vermiwash	NAD	E. fetida	Bacillus sp.	F. solani, A. solani, F. oxysporum, R. solani, and Xanthomonas campestris	Brinjal Tomato Brinjal Mustard Tomato	Fusarium wilt Early leaf blight Fusarium wilt Damping off root Bacterial leaf spot	[40]
Vermicompost extract	Cow dung	E. fetida	Bacillus subtilis	Botrytis cinerea	NAD	Controlling fungal phytopathogens	[3]
Vermicompost extract	NAD	NAD	B. subtilis	Pectobacterium carotovorum (soft rot bacterium)	Carrot	Soft rot disease	[41]
Vermicompost extract	Straw and goat manure	E. fetida	Pseudomonas sp. B. subtilis	Sarocladium oryzae, F. oxysporum, Pestalotia theae, Curvularia lunata, Colletotrichum gloeosporioides, Cylindrocladium floridanum, and Bipolaris oryzae F. oxysporum, Macrophomina phaseolina, Curvularia lunata, Colletotrichum gloeosporioides, Cylindrocladium floridanum, and Cylindrocladium scoparium	Rice Banana Tea Maize Mango Banana Rice Banana Groundnut Maize Mango Banana Banana	Sheath rot disease Fusarium wilt Leaf spot disease Leaf spot disease Anthracnose disease Root necrosis Brown spot disease Fusarium wilt Charcoal rot Leaf spot disease Anthracnose disease Root necrosis Root necrosis	[42] [42]

Vermicompost extract	Herbal	E. fetida	Streptomyces tsusimaensis, Streptomyces africanus, and Streptomyces S. africanus and Streptomyces	Rhizoctonia bataticola M. phaseolina	Chickpea Sorghum	Dry root rot Charcoal root rot	[35]
Vermicompost extract	Paper sludge and dairy sludge	E. fetida	Streptomyces microflavus Streptomyces coelicolor Streptomyces violascens Streptomyces somaliensis Streptomyces coelicolor Streptomyces felleus Streptomyces pulveraceus Streptomyces praecox	Colletotrichum coccodes, Phytophthora capsici, and Fusarium moniliforme C. coccodes and F. moniliforme C. coccodes, R. solani, and F. moniliforme C. coccodes and F. moniliforme C. coccodes, R. solani, F. moniliforme, and Pythium ultimum C. coccodes, R. solani, and F. moniliforme R. solani, P. capsici, and F. moniliforme P. capsici and F. moniliforme	NAD	Inhibit phytophogenic fungi	[43]
Vermicompost extract	Fruit and vegetable wastes	NAD	NAD	Fusarium moniliforme	Rice	Foot rot disease	[44]
Vermicompost	Lemongrass	Eudrilus eugeniae	Pseudomonas montelii and Glomus intraradices	Fusarium chlamydosporum, Ralstonia solanacearum, and Meloidogyne incognita	Coleus forskohlii	Complex root disease (root rot/wilt disease)	[45]
Vermicompost	Ipomoea leaves	E. fetida	NAD	Liriomyza spp. (leaf miner) and Alternaria alternate (fungal leaf spot)	Ladies finger	Leaf miner damage and leaf spot disease	[2]

Continued

TABLE 16.1

Antagonistic Effects Between Indigenous Beneficial Microbes and Phytopathogens for Disease Control.—cont'd

			DISEASE CONTROL				
Source	Parental Wastes	Earthworm Species	Beneficial Microbes	Phytopathogens	Plants	Disease Control	Reference
Vermicompost extract	Food wastes	NAD	NAD	*F. oxysporum*	Tomato and cucumber	Root rot disease	[15]
Vermicompost extract	Food wastes	NAD	NAD	*P. capsici*	Tomato and cucumber	Foliar phytopathogens	[15]
				R. solani	Tomato and cucumber		
				P. ultimum	Tomato and cucumber		
				Plectosporium tabacinum	Tomato and cucumber		
				B. cinerea	Tomato and cucumber		
				Scleretonia rolfsii	Tomato and cucumber		
				Verticillium wilt	Tomato		
Vermicompost extract	Cattle manure, tree bark, potato culls, and apples	*E. fetida*	NAD	*R. solani*	Cucumber	Damping-off disease	[46]
Vermicompost extract	Cow dung, *Parthenium hysterophorus*, neem, and *Lantana camara*	*E. eugeniae*	NAD	*X. campestris*	Tomato	Bacterial spot disease	[47]

NAD, not appropriately described.

Moreover, they may secrete lytic enzymes that could restrict spore germination and germ tube elongation [48].

Several bacterial volatiles that potentially serve as plant growth promoters are also capable of inducing systemic resistance in plants [3,37]. In this manner, plant's active defense can be triggered when the plants are exposed to phytopathogen threat [49]. Fig. 16.2 illustrates the possible mechanisms for plant disease suppression.

Meanwhile, the mechanisms by which liquid vermicompost derivatives suppress plant pest are numerous but speculative [15]. The combination of the various mechanisms of action may still not be fully understood or scientifically proven. Vermicompost products are very rich in microorganism diversity. Among others are organisms with predatory characteristic to pest. An increase in population of various predators will affect the population density of pest by directly killing the organism [50,51].

Vermicompost have a potential to provide a systemic approach for plants to defend against pest attack. Compared with inorganic fertilizer, vermicompost contributes not only important nutrients, for instance, organic nitrogen and potassium for plants growth, but there are many other components that are still not identified which contributed to plant systemic ability to defend against pest attack. Vermicompost may have almost complete nutrition required by plants to grow healthy and have resistance against pest infestation. A greenhouse experiment by Arancon et al. [13] suggested this possible mechanism by which, plants show increased resistance or become less palatable to be a target for pest. A similar response of vermicompost addition to planting media against pest attack was also demonstrated by Ramesh [52].

Plant pests, for instance, nematode and arthropod, can be killed by exposure to certain chemical substances. These substances such as ammonia, hydrogen sulfide, nitrites, and even minerals like zinc can be toxic to pest. Therefore, toxic substances released from organic amendment breakdown during vermicompost degradation in the soil [53] have a potential to eliminate plant pest by directly killing the organism. Harmful enzyme and toxic substances can also be produced by other microorganism live in vermicompost. Siddiqui and Mahmood [54] reported a similar activity of rhizobacteria from vermicompost, colonizing the roots and killing parasitic nematode by producing toxic substances. The mechanisms of possible pest suppression through application of vermicompost derivatives are illustrated in Fig. 16.3. Table 16.2 tabulated some of the antagonistic effects reported in literatures for pest control.

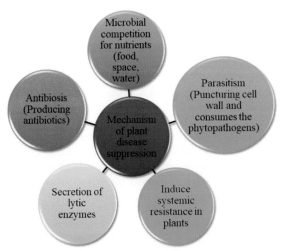

FIG. 16.2 Mechanisms of plant disease suppression through application of vermicompost derivatives. (Sources: Own study based on A.H.C. van Bruggen, M.R. Finckh, Plant diseases and management approaches in organic farming systems, Annu. Rev. Phytopathol. 54 (1) (2016) 25–54; J. Pathma, N. Sakthivel, Microbial and functional diversity of vermicompost bacteria, in Bacterial Diversity in Sustainable Agriculture, Springer, 2014, 205–225; M.K. Meghvansi, A. Varma, Organic Amendments and Soil Suppressiveness in Plant Disease Management, Springer vol. 46, 2015; M.S. Rao, M. Kamalnath, R. Umamaheswari, et al., *Bacillus subtilis* IIHR BS-2 enriched vermicompost controls root knot nematode and soft rot disease complex in carrot, Sci. Hortic. (Amst.) 218 (2017) 56–62; Y. Simsek-Ersahin, Suggested mechanisms involved in suppression of Fusarium by vermicompost products, in Organic Amendments and Soil Suppressiveness in Plant Disease Management, Springer International Publishing Switzerland, 2015, 331–352)

APPLICATION OF VERMIWASH AND VERMICOMPOST EXTRACT IN DISEASE AND PEST CONTROL

Many research studies in development of vermicompost derivatives, particularly liquefied form of vermiwash and vermicompost extracts, for plant protection were reported in the last two decades. Since then, interest in their role as biocontrol agents for inhibition of a variety of plant pathogens and pests has increased tremendously as indicated by the growing number of published literatures [1,7,16,34,40].

Disease Control

Application of vermicompost liquid derivatives on plants aboveground through foliar spray and underground through soil drench or mixed with soil has

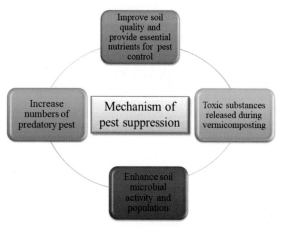

FIG. 16.3 Mechanisms of pest suppression through application of vermicompost derivatives. (Sources: Own study based on N.Q. Arancon, P.A. Galvis, C.A. Edwards, Suppression of insect pest populations and damage to plants by vermicomposts, Bioresour. Technol. 96 (10) (2005) 1137–1142; A. Edwards Clive, Earthworm Ecology, second ed. CRC press, Washington, 2004; M. Renco, N. Sasanelli, P. Kovacik, The effect of soil compost treatments on potato cyst nematodes Globodera rostochiensis and Globodera pallida, Helminthologia 48 (3) (2011) 184–194; R. Rodríguez-Kábana, Organic and inorganic nitrogen amendments to soil as nematode suppressants, J. Nematol. 18 (2) (1986) 129–134)

been investigated to understand the mechanisms underlying the potential to suppress plant pest and disease control. For instance, vermiwash contains enzymes, nutrients, and secretions of earthworms and helps in plant protection by killing or repelling pests prior to foliar spray application [4,65]. Unlike its solid counterpart, liquefied forms of vermiwash and vermicompost were found to be cost-effective and are efficient ways to deliver beneficial attributes of their biocide properties to counteract nutrient leaching, which is commonly encountered when performing soil amendments [66]. Biocidal activities of the essence may promote to their promising use as an alternative to synthetic pesticide with minimal or without environmental threats [40].

Kumar and Chandel [67] have reported the effect of vermiwash application on rose powdery mildew under greenhouse conditions. Powdery mildew is the most common fungal disease caused by Podosphaera pannosa which affects worldwide crops production, particularly roses. Currently, growers of cut roses for consumption depend on 40% of fungicide sprayed to control the respective disease which apparently caused an unpleasant looking, distortion, and death of leaves and shoots [68]. The results obtained revealed that vermiwash at

concentration of 20% exhibited a disease index of 35.02% as compared with control 61.39%. While in the case of germ tube growth, vermiwash at 5% concentration showed to reduce the germ tube elongation with 80.67 μm in comparison with control 114.60 μm.

Pattnaik et al. [40] investigated the effect of vermiwash on common fungal and bacterial phytopathogens. From the in vitro study, antagonistic activities of two bacterial species isolated from vermiwash which identified as Burkholderia sp. and Bacillus sp. V08 (accession no. KF543076) have shown a significant zone of inhibition against tested fungal phytopathogens. Those fungal phytopathogens, Fusarium solani (brinjal), Alternaria solani (tomato), Fusarium oxysporum (brinjal), and Rhizoctonia solani (mustard), were isolated from diseased leaves and seeds of various infected plants. Besides, antibiosis activity of Bacillus sp. V08 against bacterial phytopathogen of Xanthomonas campestris which was isolated from bacterial leaf spot of tomato has demonstrated a significant zone of inhibition. In addition, the researchers also investigated in vivo effect of storage period of vermiwash toward their antibiosis activity against phytopathogens. From the experiments, 2-month-old vermiwash displayed a maximum effectiveness in controlling bacterial and fungal phytopathogens than freshly prepared vermiwash. This suggested that bacteria present in vermiwash may produce or secrete some secondary metabolite or bioactive compounds during storage time [40]. This is in agreement with Esakkiammal et al. [69], when 60 days of vermicompost containing Vigna mungo leaf waste plus vermiwash showed an increased level of bacteria, fungi and actinomycetes concentrations as compared with initial day of vermicomposting and control. However, any preparation of liquid vermicompost is only recommended to be used within 24 h due to the availability of microorganism that will be significantly reduced after that.

Khan et al. [37] proposed the combination of vermiwash and Arbuscular mycorrhiza fungi (AM fungi) for plant disease suppression. It is known that vermiwash has the ability to inhibit pathogenic microbes through the antagonistic effect of microbial content [37], while AM fungi has the ability to suppress the growth of root-borne plant pathogens by anatomical changes in the root system [70]. Hence, the combined approaches of dual control measures may assure the efficiency of long-term disease controlled. A synergistic effect that was observed between AM fungi and foliar spray of vermiwash on C. tezpurensis — infected C. assamicum showed an early induction of nearly all targeting genes that are responsible for plant defense system and other physiological parameters [37].

TABLE 16.2
Antagonistic Effects Between Indigenous Beneficial Microbes and Phytopathogens for Pest Control.

Source	Parental Wastes	Earthworm Species	Beneficial Microbes	Phytopathogens	Plants	Pest Control	Reference
						PEST CONTROL	
Vermicompost extract	NAD	NAD	*Bacillus subtilis*	*Meloidogyne incognita*	Carrot	Root-knot nematode	[41]
Vermicompost extract	NAD	NAD	*B. subtilis*	*M. incognita*	Tomato	Root-knot nematode	[55]
Vermicompost	Cow manure	*Eisenia fetida*	NAD	*M. incognita*	Tomato	Root-knot nematode	[56]
Vermicompost and vermicompost extract	Municipal green wastes	NAD	NAD	*Globodera rostochiensis* and *Globodera pallida*	Potato	Potato-cyst nematodes	[16]
Vermicompost	Food wastes	NAD	NAD	*Acalymma vittatum* (striped cucumber beetle) and *Diabrotica undecimpunctata hovardii* (spotted cucumber beetle) *Manduca quinquemaculata* (hornworm caterpillars)	Cucumber Tomato	Cucumber beetle and hornworm populations	[57]
Vermicompost	Food wastes	NAD	NAD	*Myzus persicae* Sulz. (Aphids), *Pseudococcus* sp. (mealy bugs) and *Pieris brassicae* L. (cabbage white caterpillars)	Pepper Tomato Cabbage	Insect pest populations and plants damage	[13]
Vermicompost	Food wastes	NAD	NAD	*Tetranychus urticae* (spotted spider mite), *Pseudococcu s* sp. (mealy bug) and *M. persicae* (Aphid)	Bush bean and eggplant Cucumber and tomato Cabbage	Insect pest populations and plants damage	[58]
Vermicompost extract	Food wastes	NAD	NAD	*A. vittatum* (cucumber beetles) *Manduca sexta* (caterpillars of tobacco hornworm)	Cucumber Tomato	Cucumber beetle and hornworm populations	[15,59]

Continued

TABLE 16.2
Antagonistic Effects Between Indigenous Beneficial Microbes and Phytopathogens for Pest Control.—cont'd

			PEST CONTROL				
Source	Parental Wastes	Earthworm Species	Beneficial Microbes	Phytopathogens	Plants	Pest Control	Reference
Vermicompost	Animal wastes and agro wastes	*E. fetida*	NAD	*Pratylenchus* sp.	Tomato	Soil nematode	[60]
Vermicompost	Ipomoea leaves	*E. fetida*	NAD	*Earias vittella* (fruit borer)	Ladies finger	Fruit borer	[2]
Vermiwash	NAD	*Eudrilus eugeniae*	NAD	*Helicoverpa armigera* (pod borer)	Chickpea	Pod borer	[61]
Vermiwash	Platanus leaf and cow manure	*E. fetida*	NAD	*T. urticae* (two-spotted spider mite)	Bean	Damage to host plant	[4]
Vermiwash and vermicompost extract	NAD	*E. fetida*	NAD	*Meloidogyne javanica*	Cucumber	Root-knot nematode	[36]
Vermiwash	Buffalo dung, wheat bran and vegetable wastes	*E. fetida*	NAD	NAD	Soybean	Reduced pest's infestation	[62]
Vermiwash	Animal dung and agro-kitchen wastes	*E. fetida*	NAD	*Lipaphis erysimi* (aphid)	Mustard	Reduced pest's infestation	[63]
Vermicompost extract	Food wastes	NAD	NAD	*Meloidogyne hapla* *M. persicae* (aphids) and *Tetranychus* sp. (spider mites)	Tomato Tomato	Root-knot nematode Suppression of arthropod pests	[64]

NAD, not appropriately described.

According to Simsek-Ersahin (2015), the ability of vermicompost to effectively suppress plant pathogen was mainly attributed to its antagonistic effects [48]. This suggestion was proven by vermicompost which was subjected to sterilization or autoclaving that showed no effect to pathogen reduction because beneficial microbiota in vermicast was killed during high temperature [14,71]. Studies by Szczech (1999) concluded that vermicompost was responsible for the increase in the population of antagonistic bacteria and fungi that suppress the growth and infection of tomato *Fusarium* wilt [71]. Numerous scientific reports describe the application of vermicompost extract as suppressants of plant diseases [15]. Among various employing approaches of vermicompost extract for plant disease control, suppression of fungal disease caused by *Fusarium* sp. e.g., *Fusarium* wilt and foot rot disease, has been much reported in literatures [15,42–44].

A study conducted at Ohio State University has revealed the potential of vermicompost extract to suppress root and foliar pathogens of cucumber and tomato [15]. From the study, various concentrations of vermicompost extract drenched to growth medium were found to significantly suppress damage of root pathogens *F. oxysporum*, *Phytophthora capsici*, *R. solani*, and *Pythium ultimum* on tested vegetables. Meanwhile, the mean leaf area or mean plant height of cucumber and tomato that affected by *P. ultimum*, *R. solani*, and *P. capsici* showed to increase significantly as compared with control after treated with aqueous extract of vermicompost which demonstrated the pathogen suppression. In the same study, foliar spray of vermicompost extract onto crop foliage has shown to control foliar pathogens of tomato and cucumber caused by *Plectosporium tabacinum*, *Botrytis cinerea*, *Scleretonia rolfsii*, and *Verticillium* wilt. All the tested concentration of vermicompost extract (5%, 10%, and 20%) significantly suppressed foliar pathogen diseases. However, vermicompost extract prepared at concentration ranges of 10%–20% provide satisfactory suppression effects.

Studies by Jayakumar and Natarajan [42] investigated functional potential of bacteria isolated from straw- and goat manure–based vermicompost extract. *Bacillus* sp. (57%) was the most abundant groups of bacteria in the extract, followed by *Pseudomonas* sp. (15%) and *Microbacterium* sp. (12%). Most vermicompost bacteria particularly *Pseudomonas* sp. and *Bacillus* sp. showed plant growth–promoting traits and promising antagonistic effect. For example, *Bacillus subtilis* and *Pseudomonas* sp. produced indole-3-acetic acid (IAA) of 36.00 and 25.33 μg/mL, respectively. IAA is important for root growth, proliferation of plant tissue,

and has potential for plant pathogen suppression [72]. Furthermore, *B. subtilis* was reported to possess antifungal activity against *F. oxysporum*, *Colletotrichum gloeosporioides*, *Macrophomina phaseolina*, *Cylindrocladium scoparium*, *Cylindrocladium floridanum*, and *C. lunata*. While, *Pseudomonas* sp. demonstrated antagonistic effect against fungal pathogens viz. *Sarocladium oryzae*, *C. gloeosporioides*, *F. oxysporum*, *Pestalotia theae*, *C. floridanum*, *C. lunata*, and *Bipolaris oryzae*.

Rao et al. [41] reported the efficiency of seed treatment in conjunction with soil application of *B. subtilis*–enriched vermicompost for maximal yield production and controlling root-knot nematode and soft rot disease complex in carrot. Under combination of aforementioned treatment, crop yield progressively increased up to 28.8%, and a noticeable reduction of nematode population in roots and disease incidence were observed by 69.3% and 70.2%, respectively. Remarkably, antagonistic activity between *B. subtilis* against *Meloidogyne incognita* and *Pectobacterium carotovorum* has demonstrated a greater reduction in pest and disease incidence compared with application of chemicals carbofuran and streptocycline.

Mu et al. [3] evaluated the antagonistic activity of mixed volatile organic compounds (VOCs) released by biocontrol bacterium isolated from vermicompost extract, *B. subtilis* against *Botrytis cinerea*. A number of VOCs were identified and tested for their inhibitory activity against *B. cinerea*. Among them, acetic acid butyl ester, 3-methyl-3-hexanol, 1-butanol, and 1-heptylene-4-alcohol have entirely inhibited the mycelia growth at low concentration. This means that VOCs released by antifungal microbes in vermicompost may contribute to disease suppression.

In a study by Simsek-Ersahin et al. [46], a significant suppression of *R. solani* by aqueous extract of vermicompost was reported proportionate to amendment rate of vermicompost used. Pots with 20% and 30% of vermicompost showed to effectively control damping-off disease in cucumber. However, ratio of 20% vermicompost is suggested for optimum performance by considering the maximum disease inhibition and plant growth effect. Hence, plant growth effect such as leaf size, leaf number, and plant height slightly retarded at concentration of 30% vermicompost due to salinity stress.

Gopalakrishnan et al. [35] have evaluated actinomycetes isolated from herbal based vermicompost for managing *Fusarium* wilt in chickpea. By using dual-culture assay, actinomycetes *Streptomyces africanus* and an identified species of *Streptomyces* have shown a significant inhibition against *M. phaseolina* in sorghum and

Rhizoctonia bataticola in chickpea. In the same study, *Streptomyces tsusimaensis* also reported to possess antagonistic effect against *R. bataticola* in chickpea.

Reddy et al. [47] have proposed the treatment of seed with 10% vermicompost extract prior to soil application of vermicompost. Among various feedstock used, vermicompost derived from neem showed strong inhibitory effect against *X. campestris* with a highest yield. In addition, Yasir et al. [43] have examined the bacterial communities and chitinase gene diversity in vermicompost extract derived from paper sludge and dairy sludge. In the present study, Actinobacteria and chitinolytic bacteria being the prevailing bacterial communities in the vermicompost. These populations of bacteria were responsible for reducing the sporangium germination of *Fusarium moniliforme*.

Pest Control

Aghamohammadi et al. [4] assessed the bioefficacy of the combined use of vermiwash and acaricide azocyclotin for managing two-spotted spider mite, *Tetranychus urticae* in bean. *T. urticae* caused damage to host plant, and it is one among the most damaging pests worldwide. Their populations are hard to control because they developed resistance to insecticides and acaricides. Recently, *T. urticae* populations have been controlled using chemical pesticide such as pyridaben, azosyclotin, and spirodiclofen, but frequent use of these chemicals has consequences to pesticide resistance in them [4,73]. Thus, the incorporation of vermiwash into acaricide was found to cause increased mortality of *T. urticae* at 100% as compared to acaricide alone.

Rostami et al. [36] investigated the effect of vermicompost extract and vermiwash against root-knot nematode in cucumber. *In vitro* study demonstrated that highest rate of egg hatching inhibition was achieved at 10% and 100% concentration of vermicompost extract and also at 100% vermiwash. Under laboratory conditions, vermicompost extract and vermiwash at a concentration of 10% and 100% show a significant effect on mortality of *Meloidogyne javanica* larvae observed after 24, 48, and 72 h. Besides, the greenhouse experiments demonstrated reduction in the number of nematode juveniles and gall index for both products, and the best conditions for controlling disease was suggested by blending solid vermicompost with 10% vermicompost extract. Meanwhile, treatment with vermicompost plus 10% of vermiwash was recommended for plant growth. Application of vermicompost and its derivatives had been reported to attract much scientific interest for control of *Meloidogyne* sp. in various crops, including carrot [41], tomato [55,56], and cucumber [36].

Nath et al. [62] reported that a mixture of vermiwash and neem oil sprayed onto soybean crop exhibited significant growth and productivity, increased number of healthy pod and flower in soybean, and concurrently reduced the pest infestation. In this manner, possible mechanisms for suppression of pest attack is through improving soil quality and providing essential nutrients for plant defense [2]. In another study by Nath and Singh (2012), the researchers have reported the efficiency of sprayed vermiwash in combination used with neem oil and custard apple for reduction of aphid population [63]. The productivity of the crop, mustard, significantly increased three and a half times than control after 30 days of treatment.

In a previous study, Edwards et al. [64] examined the effect of vermicompost extract on plant parasitic nematodes and arthropod pests in tomato. The experimental results showed that drenched treatment of vermicompost extract onto seedling transplanting significantly reduced the number of root-knot galls. While, treatment of vermicompost extract at all tested concentration significantly declined the number of aphid populations.

An experimental study by Renčo and Kováčik [16] demonstrated the efficiency of liquefied form of vermicompost than its solid form. Sole application of vermicompost extract or in mixing with urea exhibited strong suppression effect against potato-cyst nematodes: *Globodera rostochiensis* and *Globodera pallida*. All treated samples have shown a positive response compared with control at all doses. The number of eggs and juveniles of both species were significantly reduced when soil treated with vermicompost.

Vermicomposts can increase the number of predatory or omnivorous nematodes or arthropods such as mites that selectively prey on plant parasitic nematodes [51]. Edwards et al. [59] investigated the efficiency of vermicompost in the form of liquid extract as a mode of protection for plants against pest. Survival rates and reproduction of cucumber beetles (*Acalymma vittatum*) and tobacco hornworms (*Manduca sexta*) attacks in cucumber and tomatoes, respectively, found to be significantly suppressed with application of foliar vermicompost extract. The mode of protection increased with the rate of application. The experiments also suggested that the solution not actually kill the pest but instead reduce the pest interest to attack due to the availability of an unattractive phenolic compound derived from vermicompost extract.

Arancon et al. [13] in a closed system greenhouse experiment demonstrated the ability of vermicompost to produce a plant with resistance to pest invasion. In the experiment, solid vermicompost has been used as a replacement over soil as planting media. The results demonstrated a significant suppression against aphid (*Myzus persicae*), mealy bug (*Pseudococcus* spp.), and cabbage caterpillar (*Peiris brassicae*) attacks on tomato and peppers. Plant's resistance against the pest was developed systematically because there was no direct contact of pest and vermicompost in the experiment.

CONCLUSION

Aqueous forms of vermicompost derivatives, which are vermiwash and vermicompost extracts, have been scientifically demonstrated to be beneficial in pest and disease control in plant. Foliar application of liquid vermicompost derivatives protects the aboveground part of plants from pest and disease, while the underground part is protected when the solution drenched or mixed with solid vermicompost, fertilizer, or any organic material in soil. Vermicompost and its derivatives, particularly vermiwash and vermicompost extracts, can be used in combination to achieve the best result not only for pest and disease control but more importantly for soil health and plant growth.

REFERENCES

[1] S.K. Kiyasudeen, M.H. Ibrahim, S. Quaik, S. Ahmed Ismail, Prospects of Organic Waste Management and the Significance of Earthworms, Springer International Publishing, 2016.

[2] N. Hussain, T. Abbasi, S.A. Abbasi, Evaluating the fertilizer and pesticidal value of vermicompost generated from a toxic and allelopathic weed ipomoea, J. Saudi Soc. Agric. Sci. (2018) (in press).

[3] J. Mu, X. Li, J. Jiao, G. Ji, J. Wu, F. Hu, et al., Biocontrol potential of vermicompost through antifungal volatiles produced by indigenous bacteria, Biol. Control 112 (2017) 49−54.

[4] Z. Aghamohammadi, H. Etesami, H.A. Alikhani, Vermiwash allows reduced application rates of acaricide azocyclotin for the control of two spotted spider mite, *Tetranychus urticae* Koch, on bean plant (*Phaseolus vulgaris* L.), Ecol. Eng. 93 (2016) 234−241.

[5] N. Hussain, T. Abbasi, S.A. Abbasi, Enhancement in the productivity of ladies finger (*Abelmoschus esculentus*) with concomitant pest control by the vermicompost of the weed salvinia (*Salvinia molesta*, Mitchell), Int. J. Recycl. Org. Waste Agric. 6 (4) (2017) 335−343.

[6] N. Soobhany, Preliminary evaluation of pathogenic bacteria loading on organic Municipal Solid Waste compost and vermicompost, J. Environ. Manag. 206 (2018) 763−767.

[7] K. Naidoo, H. Swatson, K.S. Yobo, G.D. Arthur, Boosting our soil with green technology: conversion of organic waste into 'Black Gold', in: Food Bioconversion, Elsevier Inc., 2017, pp. 491−510.

[8] S. Adhikary, Vermicompost, the story of organic gold: a review, Agric. Sci. 3 (7) (2012) 905−917.

[9] A. Singh, G.S. Singh, Vermicomposting: a sustainable tool for environmental equilibria, Environ. Qual. Manag. 27 (1) (2017) 23−40.

[10] S. Datta, J.J.J. Singh, S. Singh, J.J.J. Singh, Earthworms, pesticides and sustainable agriculture: a review, Environ. Sci. Pollut. Res. 23 (9) (2016) 8227−8243.

[11] M. Tejada, C. Benítez, Organic amendment based on vermicompost and compost: differences on soil properties and maize yield, Waste Manag. Res. 29 (11) (2011) 1185−1196.

[12] A.H.C. van Bruggen, M.R. Finckh, Plant diseases and management approaches in organic farming systems, Annu. Rev. Phytopathol. 54 (1) (2016) 25−54.

[13] N.Q. Arancon, P.A. Galvis, C.A. Edwards, Suppression of insect pest populations and damage to plants by vermicomposts, Bioresour. Technol. 96 (10) (2005) 1137−1142.

[14] R. Joshi, J. Singh, A.P. Vig, Vermicompost as an effective organic fertilizer and biocontrol agent: effect on growth, yield and quality of plants, Rev. Environ. Sci. Biotechnol. 14 (1) (2014) 137−159.

[15] C.A. Edwards, N.Q. Arancon, R. Sherman, Use of aqueous extracts from vermicomposts or teas in suppression of plant pathogens, in: C. Edwards, N. Arancon, R. Sherman (Eds.), Vermiculture Technology, CRC Press, USA, 2011, pp. 183−207.

[16] M. Renčo, P. Kováčik, Assessment of the nematicidal potential of vermicompost, vermicompost tea, and urea application on the potato-cyst nematodes *Globodera rostochiensis* and *Globodera pallida*, J. Plant Prot. Res. 55 (2) (2015) 187−192.

[17] J. Domínguez, J.C. Sanchez-Hernandez, M. Lores, Vermicomposting of winemaking by-products, in: Handbook of Grape Processing By-Products: Sustainable Solutions, Elsevier Inc., 2017, pp. 55−78.

[18] R.K. Sinha, S. Agarwal, K. Chauhan, D. Valani, The wonders of earthworms & its vermicompost in farm production: Charles Darwin's 'friends of farmers', with potential to replace destructive chemical fertilizers from agriculture, in: Agricultural Sciences, vol. 1 (2), Scientific Research Publishing, Inc., USA, 2010, pp. 76−94.

[19] M. Gopal, A. Gupta, C. Palaniswami, R. Dhanapal Thomas, Coconut leaf vermiwash: a bio-liquid from coconut leaf vermicompost for improving the crop production capacities of soil, Curr. Sci. 98 (9) (2010) 1202−1210.

[20] T. Sivananthi, J.A.J. Paul, Fungicidal activity of vermicompost water extract, Scrut. Int. Res. J. Microbiol. Bio Technol. 1 (1) (2014) 7−11.

[21] J. Pathma, N. Sakthivel, Microbial diversity of vermicompost bacteria that exhibit useful agricultural traits and waste management potential, SpringerPlus 1 (1) (2012) 1−19.

[22] M.E. El-Haddad, M.S. Zayed, G.A.M. El-Sayed, M.K. Hassanein, A.M. Abd El-Satar, Evaluation of compost, vermicompost and their teas produced from rice straw as affected by addition of different supplements, Ann. Agric. Sci. 59 (2) (2014) 243–251.

[23] J. Pathma, N. Sakthivel, Microbial and functional diversity of vermicompost bacteria, in: Bacterial Diversity in Sustainable Agriculture, Springer, 2014, pp. 205–225.

[24] K. Sharma, V.K. Garg, Solid-state fermentation for vermicomposting, in: Current Developments in Biotechnology and Bioengineering, Elsevier B.V., 2018, pp. 373–413.

[25] U. Ali, N. Sajid, A. Khalid, L. Riaz, M.M. Rabbani, J.H. Syed, et al., A review on vermicomposting of organic wastes, Environ. Prog. Sustain. Energy 34 (4) (2015) 1050–1062.

[26] Y. Bi, G. Tian, C. Wang, Y. Zhang, D. Wang, F. Zhang, et al., Differential effects of two earthworm species on Fusarium wilt of strawberry, Appl. Soil Ecol. J. 126 (2018) 174–181.

[27] M.J. Fernández-Gómez, R. Nogales, H. Insam, E. Romero, M. Goberna, Role of vermicompost chemical composition, microbial functional diversity, and fungal community structure in their microbial respiratory response to three pesticides, Bioresour. Technol. 102 (20) (2011) 9638–9645.

[28] P.R. Warman, M.J. AngLopez, Vermicompost derived from different feedstocks as a plant growth medium, Bioresour. Technol. 101 (12) (2010) 4479–4483.

[29] J. Singh, Role of earthworm in sustainable agriculture, in: Sustainable Food Systems from Agriculture to Industry, Elsevier Inc., 2018, pp. 83–122.

[30] J. Clause, S. Barot, B. Richard, T. Decaëns, E. Forey, The interactions between soil type and earthworm species determine the properties of earthworm casts, Appl. Soil Ecol. 83 (2014) 149–158.

[31] S. Nath, K. Singh, Analysis of different nutrient status of liquid bio-fertilizer of different combinations of buffalo dung with gram bran and water hyacinth through vermicomposting by Eisenia fetida, Environ. Dev. Sustain. 18 (3) (2016) 645–656.

[32] M. Zarei, V.A. Jahandideh Mahjen Abadi, A. Moridi, Comparison of vermiwash and vermicompost tea properties produced from different organic beds under greenhouse conditions, Int. J. Recycl. Org. Waste Agric. 7 (1) (2018) 25–32.

[33] J.D. Juan, C. Sanchez-Hernandez, Vermicompost derived from spent coffee grounds: assessing the potential for enzymatic bioremediation, in: Handbook of Coffee Processing By-Products, Elsevier, 2017, pp. 369–397.

[34] M.K. Meghvansi, A. Varma, Organic Amendments and Soil Suppressiveness in Plant Disease Management, vol. 46, Springer, 2015.

[35] S. Gopalakrishnan, S. Pande, M. Sharma, P. Humayun, B.K. Kiran, D. Sandeep, et al., Evaluation of actinomycete isolates obtained from herbal vermicompost for the biological control of Fusarium wilt of chickpea, Crop Protect. 30 (8) (2011) 1070–1078.

[36] M. Rostami, M. Olia, M. Arabi, Evaluation of the effects of earthworm Eisenia fetida-based products on the pathogenicity of root-knot nematode (Meloidogyne javanica) infecting cucumber, Int. J. Recycl. Org. Waste Agric. 3 (2014) 58.

[37] M.H. Khan, M.K. Meghvansi, R. Gupta, et al., Combining application of vermiwash and Arbuscular Mycorrhizal fungi for effective plant disease suppression, in: Organic Amendments and Soil Suppressiveness in Plant Disease Management, Springer International Publishing, Switzerland, 2015, pp. 479–494.

[38] Y. Simsek-Ershahin, The use of vermicompost products to control plant diseases and pests, in: Biology of Earthworms, vol. 24, Springer, 2011, pp. 191–213.

[39] C.M. Mehta, U. Palni, I.H. Franke-Whittle, A.K. Sharma, Compost: its role, mechanism and impact on reducing soil-borne plant diseases, Waste Manag. 34 (3) (2014) 607–622.

[40] S. Pattnaik, S. Parida, S.P. Mishra, J. Dash, S.M. Samantray, Control of phytopathogens with application of vermiwash, J. Pure Appl. Microbiol. 9 (2) (2015) 1697–1701.

[41] M.S. Rao, M. Kamalnath, R. Umamaheswari, et al., Bacillus subtilis IIHR BS-2 enriched vermicompost controls root knot nematode and soft rot disease complex in carrot, Sci. Hortic. (Amst.) 218 (2017) 56–62.

[42] P. Jayakumar, S. Natarajan, Molecular and functional characterization of bacteria isolated from straw and goat manure based vermicompost, Appl. Soil Ecol. 70 (2013) 33–47.

[43] M. Yasir, Z. Aslam, S.W. Kim, et al., Bacterial community composition and chitinase gene diversity of vermicompost with antifungal activity, Bioresour. Technol. 100 (19) (2009) 4396–4403.

[44] T. Manandhar, K. Yami, Biological control of foot rot disease of rice using fermented products of compost and vermicompost, Sci. World 6 (6) (2008) 52–57.

[45] R. Singh, S. Tiwari, R.P. Patel, S.K. Soni, A. Kalra, Bioinoculants and AM fungus colonized nursery improved management of complex root disease of Coleus forskohlii Briq. under field conditions, Biol. Control 122 (2018) 11–17.

[46] Y. Simsek-Ershahin, K. Haktanir, Y. Yanar, Vermicompost suppresses Rhizoctonia solani Kühn in cucumber seedlings, J. Plant Dis. Prot. 116 (4) (2009) 182–188.

[47] S.A. Reddy, D.J. Bagyaraj, R.D. Kale, Management of tomato bacterial spot caused by Xanthomonas campestris using vermicompost, J. Biopestic. 5 (1) (2012) 10–13.

[48] Y. Simsek-Ershahin, Suggested mechanisms involved in suppression of Fusarium by vermicompost products, in: Organic Amendments and Soil Suppressiveness in Plant Disease Management, Springer International Publishing, Switzerland, 2015, pp. 331–352.

[49] P.K. Singhai, B.K. Sarma, J.S. Srivastava, Biological management of common scab of potato through Pseudomonas species and vermicompost, Biol. Control 57 (2) (2011) 150–157.

[50] A. Edwards Clive, Earthworm Ecology, second ed., CRC press, Washington, 2004.

[51] M. Renco, N. Sasanelli, P. Kovacik, The effect of soil compost treatments on potato cyst nematodes *Globodera rostochiensis* and *Globodera pallida*, Helminthologia 48 (3) (2011) 184–194.

[52] Ramesh, Effects of vermicomposts and vermicomposting on damage by sucking pests to ground nut (*Arachis hypogea*), Indian J. Agric. Sci. 70 (5) (2000) 334.

[53] R. Rodríguez-Kábana, Organic and inorganic nitrogen amendments to soil as nematode suppressants, J. Nematol. 18 (2) (1986) 129–134.

[54] Z.A. Siddiqui, I. Mahmood, Role of bacteria in the management of plant parasitic nematodes: a review, Bioresour. Technol. 69 (2) (1999) 167–179.

[55] M.S. Rao, R. Umamaheswari, A.K. Chakravarthy, et al., A frontier area of research on liquid biopesticides: the way forward for sustainable agriculture in India, Curr. Sci. 108 (9) (2015) 1590–1592.

[56] Z. Xiao, M. Liu, L. Jiang, et al., Vermicompost increases defense against root-knot nematode (*Meloidogyne incognita*) in tomato plants, Appl. Soil Ecol. 105 (2016) 177–186.

[57] E.N. Yardim, N.Q. Arancon, C.A. Edwards, T.J. Oliver, R.J. Byrne, Suppression of tomato hornworm (*Manduca quinquemaculata*) and cucumber beetles (*Acalymma vittatum* and *Diabotrica undecimpunctata*) populations and damage by vermicomposts, Pedobiologia 50 (1) (2006) 23–29.

[58] N.Q. Arancon, C.A. Edwards, E.N. Yardim, et al., Suppression of two-spotted spider mite (*Tetranychus urticae*), mealy bug (*Pseudococcus* sp) and aphid (*Myzus persicae*) populations and damage by vermicomposts, Crop Protect. 26 (1) (2007) 29–39.

[59] C.A. Edwards, N.Q. Arancon, M. Vasko-Bennett, A. Askar, G. Keeney, Effect of aqueous extracts from vermicomposts on attacks by cucumber beetles (*Acalymna vittatum*) (Fabr.) on cucumbers and tobacco hornworm (*Manduca sexta*) (L.) on tomatoes, Pedobiologia 53 (2) (2009) 141–148.

[60] G. Nath, K. Singh, Combination of vermicomposts and biopesticides against nematode (*Pratylenchus* sp.) and their effect on growth and yield of tomato (*Lycopersicon esculentum*), IIOAB J. 2 (5) (2011) 27–35.

[61] S. Haralu, S.S. Karabhantanal, S.B. Jagginavar, G.K. Naidu, Utilization of vermiwash as biopesticide in the management of pod borer, *Helicoverpa armigera* (Hubner), in chickpea (*Cicer arietinum* L.), Appl. Biol. Res. 20 (1) (2018) 37–45.

[62] G. Nath, K. Singh, Effect of foliar spray of biopesticides and vermiwash of animal, agro and kitchen wastes on soybean (*Glycine max* L.) crop, Bot. Res. Int. 4 (3) (2011) 52–57.

[63] G. Nath, K. Singh, Combination of vermiwash and biopesticides against Aphid (*Lipaphis erysimi*) infestation and their effect on growth and yield of Mustard (*Brassica campestris*), Dyn. Soil Dyn. Plant 1 (1) (2012) 96–102.

[64] C.A. Edwards, N.Q. Arancon, E. Emerson, R. Pulliam, Suppressing plant parasitic nematodes and arthropod pests with vermicompost teas, Biocycle 48 (12) (2007) 38–39.

[65] A. Chattopadhyay, Effect of vermiwash of Eisenia foetida produced by different methods on seed germination of green mung, *Vigna radiate*, Int. J. Recycl. Org. Waste Agric. 4 (4) (2015) 233–237.

[66] M.H. Khan, M.K. Meghvansi, R. Gupta, et al., Foliar spray with vermiwash modifies the Arbuscular mycorrhizal dependency and nutrient stoichiometry of bhut jolokia (*Capsicum assamicum*), PLoS One 9 (3) (2014) e92318.

[67] V. Kumar, S. Chandel, Management of rose powdery mildew (*Podosphaera pannosa*) through ecofriendly approaches, Indian Phytopathol. 71 (3) (2018) 393–397.

[68] M. Linde, N. Shishkoff, Powdery mildew, in: Reference Module in Life Sciences, vol. 23 (4), Elsevier, 2017, pp. 235–301.

[69] B. Esakkiammal, C. Esaivani, K. Vasanthi, L. Lakshmibai, N.S. Preya, Microbial diversity of vermicompost and vermiwash prepared from *Eudrilus euginae*, Int. J. Curr. Microbiol. Appl. Sci. 4 (9) (2015) 873–883.

[70] J. Wehner, P.M. Antunes, J.R. Powell, J. Mazukatow, M.C. Rillig, Pedobiologia Plant pathogen protection by arbuscular mycorrhizas : a role for fungal diversity? Pedobiologia 53 (3) (2009) 197–201.

[71] M.M. Szczech, Suppressiveness of vermicompost against fusarium wilt of tomato, J. Phytopathol. 147 (3) (1999) 155–161.

[72] E. Khare, N.K. Arora, Effect of indole-3-acetic acid (IAA) produced by *Pseudomonas aeruginosa* in suppression of charcoal rot disease of chickpea, Curr. Microbiol. 61 (2010) 64–68.

[73] S. Wang, Q. Wu, Y. He, et al., Status of pesticide resistance and associated mutations in the two-spotted spider mite, *Tetranychus urticae*, in China, Pestic. Biochem. Physiol. 150 (2018) 89–96.

CHAPTER 17

Genetic Modification as a Control Mechanism to Plant Pest Attack

ONYEKA KINGSLEY NWOSU, BSC, MSC, PGDE •
KINGSLEY IKECHUKWU UBAOJI, PHD

INTRODUCTION

Genetic modification encompasses the introduction into an organism of new genes, from related or an unrelated organism, using artificial laboratory techniques. It can also be regarded as genetic manipulation or genetic engineering. Genetic modification is in essence a process done to manipulate the genome of an organism in order to produce desired traits. This modern genetic manipulation involves scientific procedures to add new DNA to or silence gene sequences in an organism (both plants and animals) [1].

As regards to plants in the past, this was achieved by selective breeding. Selective breeding can be described in such that a plant would be born with a desired trait and a farmer would breed the plant to produce more desirable plants with that trait. Selective breeding is why we have cabbage, broccoli, cauliflower, kale, among others [2].

Selective breeding may be described as inefficient because it is left to chance. In order to cultivate a plant with new feature, you have to wait for the feature to occur spontaneously. Modern genetic manipulation makes breeding plants with desired traits more efficient. It uses genetic engineering to build the genes to give an organism the desired traits and uses modern biotechnology to introduce the traits into the genome [1].

The first step toward being able to engineer new genes was the discovery of DNA ligase and restriction enzymes that lead to the laboratory process known as Cut and Paste. Both serve as scissors and glue. Restriction enzymes are like the molecular scissors that can cut the DNA while DNA ligases are like the molecular glue that can be used to glue DNA sequences back together. The process allows the transfer of useful characteristics (such as resistance to a disease/pest) into a plant, animal, or microorganism by inserting genes from one organism to another. When the transfer is between closely related organisms, it is referred to as **Cisgenesis**. When the transfer is between nonclosely related organisms, it is referred to as **Transgenesis.** All crops improved with transferred gene/DNA is referred to as genetically modified organisms (GMOs) [3].

Reasons of Modification or Altering Plant Genes

Genes are altered to introduce newer or improved desirable traits or even eliminate undesirable ones. The major reasons include the following:

1) To accelerate the introduction of a gene to provide a better characteristic into a crop. Examples include the following:
   a) Improved resistance to pest and disease, e.g., Bt cotton, Bt maize.
   b) Improved nutritive value, e.g., β-carotene introduction in to the "Golden rice."
   c) Improved adaptation to extreme environmental stress such as drought condition and soil chemical regime. Flavr Savr tomato was engineered to delay decay.
2) To protect a crop against an herbicide that is used to control weeds: Through the introduction of a new gene or modifying an existing gene into a crop, scientists now provide herbicide tolerance to the crops. It is therefore possible to control weeds in fields without affecting the crop yield.
3) To be used as quick way to prove that a gene which is supposed to provide a new characteristic to a plant is actually doing so: If a gene located on a chromosome in a plant is potentially resistant to

virus, to be sure that the gene is responsible for the resistance, there may be an introduction of this gene in a variety that is susceptible to virus. If the variety becomes resistant, then the gene is said to be a resistant gene and can be used to improve resistance of crops to the virus in the field.

Genetic Modification Resulting to Pest-Resistant Crops

Pest-resistant GM crops have been genetically modified, so they are toxic to certain insects. They are often called Bt crops because the introduced genes were originally identified in a bacterial species called *Bacillus thuringiensis* (Bt). These bacteria produce a group of toxins called Cry toxins.

Bacillus thuringiensis (Bt) is a Gram-positive bacterium that produces insecticidal proteins as crystal inclusions during its sporulation phase of growth, known as Cry or Cyt toxins, which have been proven to be effective against important crop pests and also against mosquitoes that are vectors of human diseases such as dengue and malaria [4]. For decades, Bt has been sprayed on fields as an organic pesticide; several major pests of some crops that are difficult and expensive to control with chemical insecticide are susceptible to Bt when sprayed on the surface of crops; however, Bt toxins break

down quickly when exposed to ultraviolet light, and they also wash off in a heavy rain. To address these problems, several varieties of crops have been genetically engineered to incorporate Bt genes which are specific to various insect pests. Some strains of Bt produce proteins that are selectively toxic to caterpillars, while others target mosquitoes, rootworms, or beetles [2].

To create a Bt crop variety, plant scientists select the gene for a particular Bt toxin and insert it into the cells of the plant at the embryo stage (Fig. 17.1). The resulting mature plants have the Bt gene in all its cells and express the insecticidal protein in its leaves. Insect pests like caterpillars ingest the toxins, which fatally damage the lining of the gut. Because the Bt plant produces an insecticide within its tissues, the toxic proteins are protected from the sun and thus persist longer. Moreover, these Bt plants make the toxin continually over a season, extending its protective effects. Since Bt plants offer an alternative to spraying chemical insecticides, they offer environmental and economic benefits to farmers.

Bt was first discovered in 1901 in Japan by Shigetane Dishwaters when the causal agent of wilt disease in silkworm (*Bombyx mori*) was isolated. Few years later Bt was rediscovered in Germany by Ernst Berliner from a Mediterranean flour moth (*Ephestia kuehniella*) [2]. The first Bt formulation was developed using the Bt strain

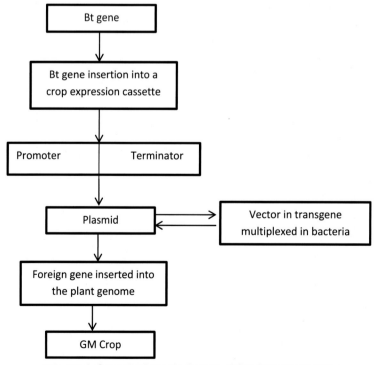

FIG. 17.1 General schematic diagram of GM Crop Production.

isolated by Berliner in 1938. However, the success of Bt as a bioinsecticide came with the development of Bt crops that express the cry gene resulting in crops that resist insect attack including borers that were difficult to control with topical Bt formulations leading to the commercial release of Bt crops in 1995 [2]. Bt toxins are specific to a limited number of insect species with no toxicity against humans or other organisms [4]. In other words, there are no known effects to mammals, fish, or birds, and they appear safe for consumers. In 2010, more than 58 million hectares were grown worldwide with Bt Maize or Bt cotton [5].

Mechanism of Bt Toxicity

As earlier said, Bt has two classes of toxins: Cytolysins (Cyt) and Crystal delta-endotoxins (Cry). While Cyt proteins are toxic toward the insect orders like *Coleoptera* (beetles) and *Diptera* (flies), Cry proteins selectively target *Lepidopterans* (moths and butterflies). The Cry proteins bind to specific receptors on the membranes of midgut (epithelia) cells resulting in rupture of those cells [6]. If a Cry protein cannot find a specific receptor on the epithelial cell to which it can bind, then the Cry protein is not toxic. Bt strains will have different complements of Cyt and Cry proteins, thus defining their host ranges [7]. The genes encoding many Cry proteins have been identified providing biotechnologists with genetic building blocks to great GM crops that express a particular Cry protein in particular crops that are toxic to particular pests and yet potentially safe for human consumption.

Insect/pest resistance to Bt technology

A major threat for the use of Cry toxins in transgenic plants is the appearance of insect resistance. Bt crops are supposed to be grown with refuges of non-GM crops and plants to reduce the likelihood of the targeted pests developing resistance (meaning that they are no longer killed by the toxin produced in the plant). Also, the need to introduce Bt crops that produce two or more relatively dissimilar toxins in the field is another resistance management tactic. Presumably, it is less likely that any one insect will be simultaneously resistant to more than one toxin. Despite this, resistant pests have been found in the United States and in India. In China, there have been reports of surges in other types of pest that are not affected by the toxin produced by Bt cotton. Therefore, an alternative for the screening and isolation of novel Cry toxin protein in nature is the in vitro genetic evolution of Cry toxins with the aim of enhancing toxicity against specific pests, to kill novel targets, or to recover toxicity in the case of the appearance of resistance in the field [8].

Genetically modified insect/pest resistant crops

Bt cotton. Bt cotton has been genetically modified by the insertion of one or more genes from a common soil bacterium, *Bacillus thuringiensis*. These genes that have been inserted into cotton produce toxins that are limited in activity almost exclusively to caterpillar pests (*Lepidoptera*). Bollgard cotton (a trade mark of Monsanto) was the first Bt cotton to be marketed in the United States in 1996. This cotton produces a toxin called cry1Ac that has excellent activity on tobacco budworm and pink bollworm. These two insects are extremely important caterpillar pests of cotton and both are difficult and expensive to control with traditional insecticides. Bollgard toxin also has moderate activity on loopers, fall armyworm, and beet armyworm [9].

Bollgard II representing the next generation of Bt cottons was introduced in 2003 and contains a second gene from the Bt bacteria, which encodes the production of Cry 2Ab. Later in 2004, WideStrike (a trademark of Dow Agrosciences) was registered for use. WideStrike cotton expresses two Bt toxins (Cry1Ac and Cry 1F). Both Bollgard II and WideStrike have better activity on a wider range of caterpillar pests than the original Bollgard technology. Currently, Bt cottons from many other companies have been developed where some have been commercially introduced and others yet to be commercially released [5].

Bt Maize. Bt maize or corn which can as well be regarded as transgenic maize is the maize that has been modified to produce the insecticidal proteins that naturally occur in *Bacillus thuringiensis*. Most of Bt maize hybrids marketed worldwide is targeted either against the European corn borer (ECB) (i.e., Cry1Ab or Cry1F protein) or the corn rootworm (CRW) (i.e., Cry3Bb, Cry34Ab1/Cry35Ab1, or mCry 3A protein) or both (i.e., stacked hybrids containing Cry proteins active against ECB and Cry proteins active against CRW). On caterpillars like the ECB, the expressed Bt proteins are only lethal to a specific group of insects that are caterpillars in their juvenile stage (*Lepidoptera*). On the beetle larvae like the CRW, the Bt proteins expressed are only lethal to a specific group of insects that are beetles as adults. The larvae are targeted for control.

Bt soybean. Early efforts to obtain Bt soybean cultivars showed to be promising alternatives to control defoliating caterpillars in soybeans. The first report of the successful introgression and expression of a native

Cry1Ab Bt gene in soybeans was published in 1994 [10], and several more studies on the subject have since been published. Stewart et al. [11] produced a transgenic soybean plant with synthetic Bt cry1Ac (tic 107) and described the tissue culture and transformation procedures and the molecular and biotic characteristics of the Bt soybean that was produced. Walker et al. [12] demonstrated a transgenic lineage of the soybean "Jack," *Glycine max* (L.) Merrill, expressing a synthetic Cry1Ac gene from (B) *thuringiensis* variety kurstaki (Jack-Bt), which was evaluated for resistance on four *lepidopteran* pests in the field: *Helicoverpa zea* (corn earworm) *Boddie*, (A) *gemmatalis* (soybeans caterpillar) Hubner, (C) *includens* (soybean looper) Walker, and *Elasmopalpus lignosellus* (lesser cornstalk borer) Zeller. Data from these experiments suggested that the expression of this cry1Ac construct in the soybean should provide adequate levels of resistance to several *lepidopteran* pests under field conditions.

Bt soybean has recently been developed by combining the transformation events MON 87,701 (expressing Cry1Ac protein) and MON 89,788 (glyphosate tolerance). This soybean cultivar is characterized by having the cry1Ac Bt gene encoding the Cry1Ac protein and the protein 5-enolpyruvylshikimate-3-phosphate synthase (EPSPS) from Agrobacterium spp. that confers tolerance to the herbicide glyphosate. This product was commercially released in Brazil in 2014. Based on the experiences with Bt maize and cotton, Bt soybean technology is expected to control major *Lepidoptera* pests.

Bt potato. Potato is one of the world's most important staple foods, as it is part of the diet for 60 to 70% of humans globally. The effort by the society to improve potato is ancient, and many of these efforts have produced cultivars resistant to insect pest. In the United States, potato has comparatively simple pest complex, including several species of pathogens and insects. The Colorado potato beetle (CPB), *Leptinotarsa decemlineata* (Say), is the most important insect pest over much of Northern America and Northern Europe. Other insect pests include leafhoppers, aphids, and numerous sporadic species. Almost all commercial transgenic potatoes express Cry3A gene that kill the CPB.

Possible Risks Related to the Use of Genetically Modified Plants

The application of genetic modification allows genetic material to be transferred from any species into plants or other organisms. In the process of this modification, copies of a gene may be integrated, additional fragments inserted, and gene sequences rearranged and deleted—which may result in lack of operation of the genes' instability or interference with other gene functions possibly cause some potential risks. Based on this, there could be a number of predictable and unpredictable risks related to release of GMOs in the open environment. These possible risks include the following:

a) *Genetic contamination*: Introduced GM plant may interbreed with the wild-type or the sexually compatible relatives of the plant. The novel trait may disappear in wild types unless it confers a selective advantage to the recipient. However, tolerance abilities of wild types may also develop, thus altering the native species' ecological relationship and behavior.

b) *Ecosystem impacts:* The effects of changes in a single species may extend well beyond to the ecosystem. Single impacts are always joined by the risk of ecosystem damage and destruction.

c) *Competition with natural species:* Faster growth of GM plants may facilitate them to have a competitive advantage over the native organisms. This may allow them to become invasive, to blow out into new habitats, and cause ecological and economic damage.

d) *Impossibility of follow-up:* There is a probability that once GM plant has been introduced into the environment and some problems arise; it becomes impossible to eliminate them. Many of these risks are identical to those incurred with regards to the introduction of naturally or conventionally bred species. But still this does not suggest that GM plants should be less scrutinized.

e) *Horizontal transfer of transgenes to other organisms:* One particular risk with great concern relating to GM plants is the risk of horizontal gene transfer (HGT). HGT is the acquisition of foreign genes (via transformation, transduction, and conjugation) by organisms in a variety of environmental situations. It occurs especially in response to changing environments and provides organisms, especially prokaryotes, with access to genes other than those that can be inherited [13]. HGT of an introduced gene from a GM plant may confer a novel trait in another organism, which could be a source of potential harm to the health of people or the environment. Recent evidence from the HGT technology confirms that transgenes in GM crops and products can spread by being taken up directly by viruses and bacteria as well as plant and animals' cells. Very recently, Yoshida et al. [14] reported that HGT also moved from a nuclear monocot gene into the genome of the eudicot parasite witchweed, which infects many grass species in Africa.

Regulation of Genetic Engineering/ Modification

The deliberate modification of crops and the resulting products thereof have become a concern of controversy all over the world. Benefits aside, genetically modified organisms (GMOs) have always been considered a threat to environment and human health. In the act of addressing those threats, any GMO that will be commercially released must be comprehensively assessed. This makes GMOs the most extensively regulated products all over the world. The regulation of these GMOs is referred to as "Biosafety Regulation". It has been considered necessary by biosafety regulations of individual countries to test the feasibility of GMOs in contained and controlled environments for any potential risks they may pose. Biosafety regulations have become mandatory in countries using modern biotechnology and/or its products (GMOs). Biosafety regulation is intended to do the following:

a) Minimize or eliminate the possible harmful effects of modern biotechnology genetic engineering on the environment, , and human health using policies, laws, and guidelines.

b) Determine in advance when hazards to human health and natural systems will result, if any particular GMO is released into the environment.

c) Detect whether GMO will actually yield benefits it was designed to provide.

d) Anticipate when a given GMO or any of its product(s) will be harmful, if it becomes part of human or animal food

e) Make as certain as possible that hazards will not arise when GMOs are transported intentionally, between different ecosystems and countries among many others.

Risk assessments and management are the quintessence of biosafety regulations and protocols in the modern biotechnology (genetic engineering/modification) arena. Risk assessment intends to quantify risks and evaluate the probabilities of possible outcomes on the basis of scientific data. It is a fundamental part of improving quality, being the quality of products or the quality of life, and plays a central role in the innovation required to maximize benefits. A critical step in risk assessment is identification of circumstances that may give rise to an adverse effect(s) (risk identification or "what could go wrong") [15]. The level of risk is then estimated from both the likelihood ("how likely is it to happen") and severity/consequences ("would it be a problem") associated with the circumstances of concern. This is then followed by characterization of the risk based on evaluation of likelihood and consequences of the identified adverse effects being realized ("what is the risk" step) [1].

Environmental risk assessment (ERA) considers the impact of introducing a GM plant into a given environment. ERA is concerned with assessing the potential for harm to ecosystem components given that there is exposure to the GM plant. It is noteworthy to highlight that the focus and degree of emphasis on elements of the ERA will change during the development process for the GM plant as the scope of environmental release ranges from confined field trials of limited extent through to larger-scale trials and seed increases in more environments, and to the final unconfined commercial release. The assessment of any potential risk of GM plant to the environment is conducted on a case-by-case basis, comparative, and uses lines of evidence to reach an all-inclusive understanding of the nature and degree of risk posed by the particular type of environmental release being analyzed [16]. In addition, a stepwise approach of data generation and analysis is used in order that the focus is directed to important concerns within the universe of possibilities.

Once a risk is assessed, it must be managed. The management of risk is an exclusively political action, resulting in a decision regarding whether to accept or not the risk previously estimated. It can take additional aspects (e.g., socioeconomic or ethical) into consideration and concerns methods used to reduce the scientifically identified risk. Risk management process also serves as a second focus of the economic/political component of the GMO biosafety issue. Whereas a risk/benefit analysis concludes that risks exist with regard to a GMO introduction or other activity but are sufficiently outweighed by the benefits of that action, it will probably still be required both practically and legally to take steps to manage the risk and to ensure that damage will be minimized [17].

The international agreements such as Cartagena Protocol on Biosafety, Convention on Biological Diversity (CBD), and the International Plant Protection Convention (IPPC) address the environmental aspects of GMOs. Biosafety regulatory frameworks are expected to serve as mechanisms for ensuring the safe use of modern biotechnology/genetic modification products without imposing unintended constraints to technology transfer.

REFERENCES

[1] G.T. Tzotzos, G.P. Head, R. Hull, Principles of risk assessment, Genet. Modif. Plants (2009) 33–63.

[2] G. Sanahuja, R. Banakar, R.M. Twyman, T. Capell, P. Christou, Bacillus thuringiensis: a century of research,

development and commercial applications, Plant Biotechnol. J. 9 (3) (2011) 283–300.

[3] R. Raman, The impact of genetically modified (GM) crops in modern agriculture: a review, GM Crops Food 8 (4) (2017) 195–208.

[4] A. Bravo, S. Likitvivatanavong, S.S. Gill, M. Soberón, Bacillus thuringiensis: a story of a successful bio-insecticide, Insect Biochem. Mol. Biol. 41 (2011) 423–431.

[5] C. James, Global Status of Commercialized Biotech/GM Crops, ISAAA Brief No. 42, ISAAA, Ithaca, NY, 2010.

[6] J.A. Dorch, M. Candas, N. Griko, W. Maaty, E. Midboe, R. Vadlamudi, L. Bulla, Cry1A toxins of Bacillus thuringiensis bind specifically to a region adjacent to the membrane-proximal extracellular domain of bt-R-1 in manduca sexta: involvement of cadherin in the entomopathogenicity of bacillius thuringiensis, Insect Biochem. Mol. Biol. 32 (2002) 1025–1036.

[7] R.A. De Maagd, A. Bravo, N. Crickmore, How Bacillus thuringiensis has evolved specific toxins to colonize the insect world, Trends Genet. 17 (2001) 193–199.

[8] L. Pardo-López, C. Muñoz-Garay, H. Porta, C. Rodríguez-Almazán, M. Soberón, A. Bravo, Strategies to improve the insecticidal activity of cry toxins from Bacillus thuringiensis, Peptides 30 (2009) 589–595.

[9] S. Downes, T. Parker, R. Mahon, Incipient resistance of Helicoverpa punctigera to the Cry2Ab Bt toxin in bollgard II cotton, Pls ONE 5 (9) (2010) e12567.

[10] W.A. Parrott, J.N. All, M.J. Adang, M.A. Bailey, H.R. Boerma, C.N. Stewart Jr., Recovery and evaluation of soybean plants transgenic for a Bacillus thuringiensis var. kurstaki insecticidal gene, in: Vitro Cellular & Developmental Biology Plant vol. 30, 1994, pp. 144–149.

[11] C.N. Stewart, M.J. Adang, J.N. All, H.R. Boerma, G. Cardineau, D. Tucker, Genetic transformation, recovery, and characterization of fertile soybean transgenic for a synthetic Bacillus thuringiensis CryIAc gene, Plant Physiol. 112 (1996) 121–129.

[12] D.R. Walker, J.N. All, R.M. McPherson, H.R. Boerma, W.A. Parrott, Field evaluation of soybean engineered with a synthetic Cry1Ac transgene for resistance to corn earthworm, soybean looper, velvet-bean caterpillar (Lepidoptera: Noctuidae), and lesser cornstalk borer (Lepidoptera: Pyralidae), J. Econ. Entomol. 93 (2000) 613–622.

[13] H. Ochman, J.G. Lawrence, E.A. Grolsman, Lateral gene transfer and the nature of bacterial innovation, Nature 405 (6784) (2000) 299–304.

[14] S. Yoshida, S. Maruyama, H. Nozaki, K. Shirasu, Horizontal gene transfer by the parasitic plant striga hermonthica, Science 328 (5982) (2010) 1128.

[15] European Food Safety Authority (EFSA), Statement of the Scientific Panel on GMO on the Safe Use of the Notion Antibiotic Resistance Marker Gene in Genetically Modified Plants, 2007. http://www.efsa.europa.eu/etc/medialib.

[16] European Food Safety Authority (EFSA), Guidance Document of the Scientific Panel on Genetically Modified Organisms for the Risk Assessment of Genetically Modified Plants and Derived Food and Feed, 2004. http://www.efsa.europa.eu/en/science/gmo/gmo_guidance.html.

[17] Advisory committee on releases to the environment (ACRE), managing the footprint of agriculture: towards a comparative assessment of risks and benefits for novel agricultural systems, in: Proceedings of the Advisory Committee on Releases to the Environment, 2006 (London, UK).

CHAPTER 18

Biopesticides: Formulations and Delivery Techniques

SMRITI KALA, PHD • NISHA SOGAN, PHD • AMRISH AGARWAL • S.N. NAIK •
P.K. PATANJALI • JITENDRA KUMAR

INTRODUCTION

Plants have prolonged history of usages as traditional folk medicines, modern drugs, pest repellents, etc. The presence of various phytochemicals in plant provided mankind the natural solution to target undesirable pests, responsible for loss in agricultural production. Plant-derived products alias herbal products are in trend these days in every sector; from pharma, cosmetics to pesticides. The pesticide sector is seeking it in more capacity because of several advantages like biodegradability, nonpersistency, and eco-friendly compared to synthetic pesticides [1,2].

The biopesticides derived from plant origin refers to "botanical pesticides" and can be sourced from different parts of plants. The botanical pesticides can reduce pest populations and provide sustainable solution against the damage caused by pests [3]. The biodegradability, nonpersistence, and user's safety are the root causes of their preference [1]. Plant-derived pesticides have also gained tremendous importance in pest management due to other benefits. The plant products are indigenous and abundantly available at low cost and have a wide spectrum of activities such as insecticidal, antifeedant, repellent, larvicidal, and ovicidal [4]. The damage causing agents, which affects agricultural production, includes nematodes, fungus, field pests, pests of stored grains, etc. The potential of biopesticides have been reported through crude extracts, oils, de-oiled cakes against all these wide range of pests. This chapter describes the potential of botanical pesticide and their formulations for agricultural pest management.

Review of Plant-Derived BioPesticides
Insecticidal potential of botanicals

Crop losses due to insects is a major threat to the incomes of farmers and to food security worldwide [5]. Crop loss can occur in the field (preharvest) or during the storage (postharvest); due to biotic or abiotic factors [6]. Among botanicals, neem (*Azadirachta indica*) is registered under Insecticide Act, 1968 [7]. The neem-based pesticides are extensively studied and found very effective in protecting agricultural crops [8]. *Azadirachtin* (tetranortriterpenoids) is responsible for insecticidal activity occurs in the seeds and leaves of neem tree [9,10]. The insecticidal activity of the seed and leaf extract of castor (*Ricinus communis*) against *Spodoptera frugiperda* reported that ricinine is a major ingredient responsible for the activity [10]. Crude extract of *Alpinia galanga* (Linnaeus), a medicinal plant belonging to family Zingiberaceae, was found to be effective against *Spodoptera litura*; a polyphagous pest causing economic losses to a large number of agriculturally important crops [11]. The diamondback moth (*Plutella xylostella*) (Lepidoptera: *Plutellidae*) and the aphid (*Brevicoryne brassicae*) (Hemiptera: Aphididae) are the most important pests of cultivated brassicas responsible for more than 90% of crop loss. Crude aqueous extracts of plants, *Maerua edulis* and *Bobgunnia madagascariensis*, were found effective in controlling pests of cabbage [12]. The essential oil hemp (*Cannabis sativa*) was found effective against phytophagous insects (*Myzus persicae*) aphids, which are key pests responsible for loss of agriculture production [13].

Efficacy of botanicals against stored grains pest

Global annual losses in the stored products due to stored grain pests are estimated as 43% in developing countries [3]. In recent years, studies have been focused on the use of plants and their phytochemicals as possible alternatives to synthetic insecticides. Powdered leaves of *Olax zeylanica* were found to have repellent potential against *Sitophilus oryzae*, a rice weevil [14]. Aqueous extract-based formulation of Karanja (*Pongamia glabra*) and Jatropha (*Jatropha curcas*) seedcakes

Natural Remedies for Pest, Disease and Weed Control. https://doi.org/10.1016/B978-0-12-819304-4.00018-X

had a significant effect on the *Tribolium castaneum* (red flour beetle), a pest of stored grains. The main constituents of Karanja and Jatropha responsible for insecticidal activity are Karanjin and Phorbol esters, respectively [3]. Saponins are class of steroidal or triterpenoidal compounds found in plants with diverse range of biological activities and are characterized by high molecular weight in the presence of nonpolar aglycone along with polar sugar molecules. Saponins have received attention as insecticidal compounds due to their toxic nature against many serious pests of crops and stored grains [15]. Castor oil had been reported to have insecticidal activity against *Zabrotes subfasciatus* [16].

Biopesticides as nematicide, fungicide, and weedicide

Castor oil, obtained from the seed of castor (*R. communis*), is found widely grown in India, East Africa, tropical and warm temperate regions throughout the world and have insecticidal properties against a variety of pests [17]. Root-knot nematode is one of the most destructive pests that cause significant loss in yield of rice crop [18], the castor plant reported to have nematicidal activity [19,20]. De-oiled cake of Karanja seeds (*P. glabra*) also possesses insecticidal and nematicidal activities [21]. The plant-derived pesticides such as hydroalcoholic crude extracts from different parts (leaves, stems, rhizomes) of hop plant (*Humulus lupulus* L.), as well as hops essential oil, were reported for their activity against *Zymoseptoria tritici* fungus [22]. Extracts contained some polyphenols (apigenin, gallic acid, catechin, quercetin, and tannic acid), which are well-known compounds possessing antifungal activity, were reported [23]. The efficacy of a bark extract of *Magnolia officinalis* to control pathogens including *Plasmopara viticola* (causing grapevine downy mildew) and *Venturia inaequalis* (causing apple scab) was described [24]. Weeds are considered a serious enemy crop production, suppressing their growth and development. Essential oil extracted from *Tagetes erecta* was evaluated for its herbicidal activities (pre- and postemergence) against *Echinochloa cruss-galli* [25].

Obstacles and Challenges in using BioPesticides

The broad efficacy of the botanicals, as biopesticides, cannot be overlooked. However, for pest management practices, the common trend is being followed that has been reported in several studies, which typically involves the usage of biobotanical in their crude or raw state. The promising results expected are usually not observed under realistic conditions during field trials.

The degradation and volatile natures of bioactive compounds are majorly responsible for reduced efficacy of plant-based products under field conditions [26]. Further, the botanical is required in larger concentration to achieve desired efficacy. The potential of biobotanicals for use in pest management is not well accepted due to the aforesaid drawbacks. To avoid these drawbacks and to attain best efficacy, the formulation of biobotanicals is desired.

Overview of Formulation Technology

The processing of pesticide with inert materials (adjuvants) to induce certain advantageous properties makes up its "formulation." The processes of formulation include grinding; mixing with inert materials like talc, silica, diatomaceous earth, etc.; and adding adjuvants like wetting agents, spreaders, emulsifiers, stickers, and stabilizers to deliver unique properties [7].

1. Formulation technology involves usage of different adjuvants (polymers, emulsifying agents, surfactants, solvents, stabilizers, defoamers, etc.) to provide safe and effective products, which are convenient to use and also provide enhanced efficacy.
2. Formulation of pesticide improves storage, handling, safety, application, and effectiveness of biopesticide.
3. It also help provide adherence to the target sites and controlled release of the bioactive compounds, depending on the type of formulation [26].
4. The formulation technology can also improve quality of existing formulations, or new formulation of an existing pesticide.

Different Types of BioPesticide Formulations
Conventional formulations

In the past, most of the agrochemical formulation technologies were based on simple solvent-based solutions or powder mix. The presence of petroleum-based solvents and dust in these conventional formulations has a negative impact on the environment and mankind [27]. These are difficult to mix in spray tanks and have poor compatibility and may be phytotoxic to crops. The conventional formulation usually includes wettable powders (WPs) and emulsifiable concentrates (ECs).

Wettable powders

The initial formulation under conventional category is WP. It is a pesticide in a dry form mixed with wetting, dispersing agent (surfactants) and a fine solid carrier as fillers (Fig. 18.1). The commonly employed fillers

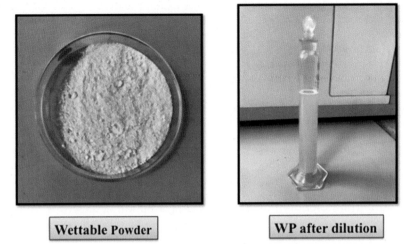

FIG. 18.1 Wettable Powder (WP) and its dilution.

for WP are china clay and silica. The WP disperses in water to form a homogenous and stable suspension. While formulating WP of solid active ingredient (AI), it is gradually mixed with surfactants and added in china clay. In case of liquid pesticide, first it is absorbed on the carrier silica, followed by the addition of other ingredients.

After complete addition of the AI it is mixed with wetting/ dispersing agents in a mixer to get a uniform composition. The steps involved in preparation are the following:

- **Step I:** Blending of ingredients (Simple mixing using kitchen mixer at laboratory scale and Ribbon Blender used for large production)
- **Step II:** Grinding (Through air jet milling, air classifier miller to bring average particle size of 4–5 microns) [28].
- **Step III:** Packaging. The composition, advantages, and drawbacks of WP are summarized in Table 18.1.

TABLE 18.1
Composition, Advantages, and Drawbacks of Wettable Powders.

Composition	Advantages	Drawbacks
Pesticides, wetting/ dispersing agents, fillers [29]	Economic (production and packaging), easy to handle	Dust inhalation risk, lower stability of suspension [27]

Emulsifiable concentrate

ECs are formulated by dissolving the pesticide with emulsifiers, surfactants in a solvent (Fig. 18.2). The solvents used in EC preparations are generally petroleum distillates. The typically used solvents include C-9, Aromax, and Solvesso. Due to environmental concern with EC formulation a new trend is being followed in the EC formulations, which involves using biodiesel [30]. Currently, vegetable oil–based EC, replacing petroleum distillates, are also trending. Ethylene glycol di-acetate has been reported as an

FIG. 18.2 Emulsifiable concentrate (EC) and its dilution.

alternative carrier solvent in EC [31]. EC formulations when diluted in water give a stable "milky" emulsion (Fig. 18.2). In an ideal EC formulation, emulsion after diluting with water should be stable initially and after 24 h, creaming and oil separation should not be observed. The selection of surfactant plays an important role in emulsion stability. Phenolic compounds (phlorizin, resveratrol, alkylresorcinols, considered natural alternatives) were reported as antifungal ECs [32]. Neem oil EC has been reported [33]. The composition and other details of EC formulation are summarized in Table 18.2.

New-Generation Formulations: Recent Advancement Incorporating Biobotanicals

The new generation formulations are advancement over conventional formulation technologies and these are free from dust and solvent. The developments in formulation technology include innovative formulation types which can give products a competitive advantage, of extended self-life, and improved safety to consumers [7,27]. The said types of formulations are referred to as new-generation formulations or user and environment friendly pesticide formulations.

O/W emulsions

The first category among new-generation formulation is oil-in-water emulsion (EW). In EW formulation, a solid active dissolved in water-immiscible solvent and dispersed into continuous water phase with the high shear mixing the two phases are stabilized by selection of suitable emulsifiers [27]. Detailed steps of preparation of EW are shown in Fig. 18.3. The average particle size of EW typically ranges in 5–6 microns. EW formulations are very prone to creaming, sedimentation through the process of coagulation, flocculation, and

Ostwald ripening. The factors, which can govern stability of EW, are summarized in Table 18.3.

Abamectin is a macrocyclic lactone derived from the soil bacterium *Streptomyces avermitilis* was reported as EW [34]. Tobacco (*Nicotiana tabacum*) extract was formulated as a concentrated emulsion (EW) formulation containing palm oil, emulsifiers (Tween and Span), giving a more physically stable product [35].

Microemulsions

Microemulsions (MEs) are homogeneous and isotropic dispersions with low viscosity, optical transparency, thermodynamic stability, and an internal (dispersed) phase having typical sizes of 10–200 nm (Fig. 18.4). Essential oil ME of aniseed is reported as insecticide [36]. Neem oil ME have been reported [37]. The common composition, commonly used emulsifiers in ME preparation along with its advantages and disadvantages are given in Table 18.4.

Nanoemulsions

Nanoemulsions (NEs) are fine oil-in-water dispersions having droplet size ranging from 1 to 200 nm [40]. They are kinetically stable with a natural oil and water in combination with a surfactant [41]. Eucalyptus oil NE was found effective against *Tribolium castaneum*, red flour beetle. Castor oil NE (Fig. 18.5) was reported as an insecticide [42]. Table 18.5 presents details of NE.

Controlled Release Formulations. The controlled release technology intended for the controlled release of pesticides which can be modulated as per requirement. Controlled release formulation (CRF) technology is based on encapsulation of pesticide in any carrier material.

TABLE 18.2
Composition and Commonly Used Emulsifiers in Emulsifiable Concentrates.

Composition	Advantages	Disadvantages	Emulsifiers Used in EC
Pesticides Emulsifiers Solvent [28,29]	Low cost Physicochemical stability Prepared by simple mixing	Large amount of solvent [27]	Emulsifiers with low HLB Anionic: Calcium alkyl dodecyl benzene sulfonate Nonionic: Alkyl phenols with 3–6 ethylene oxide (Eo) units and fatty alcohols with 2–5 Eo Emulsifiers with high HLB [33] Anionic: Sodium dodecyl benzene sulfonate Nonionic: Alkyl phenols with 8–30 Eo, castor oil with 20–60 Eo

Preparation of EW

Water Phase + Surfactant

Oil phase containing active

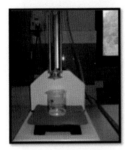

High Shear mixer

EW

Emulsification of oil phase into water

FIG. 18.3 Preparation of EW.

TABLE 18.3
Composition, Advantages, and Disadvantages of Oil-in-Water Emulsion (EW).

Composition	Advantages	Disadvantages	Factors Affecting Stability of EW
Pesticide, emulsifier, solvent (minimum), antifreeze, defoamer, water, viscosity modifiers	Solvent free No phytotoxicity Low dermal toxicity Safer in transport and storage [7]	Are prone to creaming, sedimentation through the process of coagulation, flocculation, and Ostwald ripening.	• Proper selection of Emulsifier • high shear emulsification • Optimum viscosity

Advantages of controlled release formulation

1. Prolonged effective duration of nonpersistent pesticides [48].
2. Less dose as compared with conventional formulation, pesticide used for the same period of activity, resulting in less waste and fewer applications.
3. Reduced environmental contamination, particularly reduced contamination of surface water and groundwater.
4. Reduced losses due to environmental factors (evaporation, photolysis, leaching with water, and degradation due to chemical and microbial factors), resulting in savings in the cost of the AI [49].

5. Reduced toxicity to nontarget species of plants, mammals, birds, fish, and other organisms; Improved efficacy of pesticides due to better targeting.
6. Greater safety for the users and those who come in contact with the pesticide formulations. The different types of CRFs were described in Table 18.6.

A formulation based on microencapsulated Boldo *(Peumus boldus)* essential oil (EO) was evaluated on in-pod stored peanut *(Arachis hipogaea)* to preserve the seed quality [57]. Encapsulation of mustard essential oil and microencapsulation of Neem and Karanja oil in alginate beads was reported [58,59].

Tablet formulations. Tablets are, "preformed solids of uniform shape and dimensions, usually circular, with either flat or convex face, the distance between faces being less than the diameter" [60]. Tablet is an optimized mixture of active ingredients (AIs) with adjuvants, which is compressed into a solid mass of uniform shape and size (Fig. 18.7).

Advantages of tablets formulations

1. Due to compressed form, these are easy in handling and free from dust.

2. Premeasured dose rates: The tablet formulation offers premeasured dose rate. There is no need of weighing and making other measurement for does preparation, as does can be simply decided by counting number of tablets [61].

3. Tablet is suitable in case where there is low dose requirement. The different types of tablet formulations were summarized in Table 18.7.

Preparation of ME

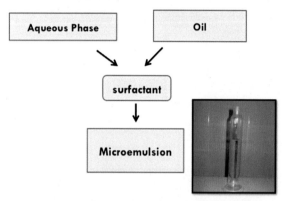

FIG. 18.4 Preparation of Microemulsion (ME).

Castor oil nanoemulsion

FIG. 18.5 Castor oil nanoemulsion.

TABLE 18.4
Composition, Commonly Used Emulsifiers in Microemulsion (ME) Preparation.

Composition	Advantages	Disadvantages	Surfactant (Emulsifiers) for ME
Pesticides Surfactants Cosurfactants Water	• Extended shelf life • Ease of preparation and scalability with reduced external energy input [38] • Improved solubilization of poorly water-soluble compounds • High possibility of enhanced absorption due to small size	• Higher amount of surfactant used. • Loading of pesticide is less	• Polyoxyethylene sorbitan monostearate, nonylphenol ethoxylate [37] • POE (20) sorbitan monooleate [36] • Lauryl alcohol ethoxylated with 6 mol of ethylene oxide; CABS: Calcium salt of dodecyl benzene sulfonic acid • Unitop 100: Nonylphenol ethoxylated with 9.5 mol of ethylene oxide • Unitop FFT 40: Castor oil ethoxylated with 40 mol of ethylene oxide [39]

TABLE 18.5
Details of Nanoemulsion (NE) Formulation.

Composition	Advantages	Disadvantages	Emulsifiers for NE
• Pesticides • Surfactants • Cosurfactants • Water	• Enhanced shelf life [43] • Increased rate of absorption due to smaller droplet size • Enhanced solubility of poorly water-soluble actives [44]	• Higher amount of surfactant used • Loading of pesticide content is less	• Castor oil ethoxylate-40, polysorbate-20 [45] • Polyoxyethylene 10 oleoyl ether [46] • Propylene glycol monocaprylate (Capryol 90) and caprylocaproyl macrogol-8-glyceride (Labrasol) [47]

TABLE 18.6
Details of Controlled Release Formulation (CRF) Technologies.

Types of CRF	Description	Composition
Coated pesticide granules (comes under the category of conventional formulation)	Pesticide is adsorbed on a solid carrier and sticking agents are used to provide controlled release of pesticide. These are for direct application in soil and used against soil borne pests.	Carrier (bentonite Fig. 18.6), Pesticide Coating Polymer (polyethylene glycol, low-density polyethylene [50])
Matrix containing physically trapped pesticide	The pesticide get physically entrapped into biodegradable polymeric matrix (e.g., alginate matrix) and cross-linked by suitable polymer to provide control for the release of pesticide [51]	Polymer, cross-linker and Pesticide [51]
Microencapsulation/ Nanoencapsulation (Fig. 18.6)	Encapsulation is the process of coating of small solid particles, liquid droplets, or with a thin coating (capsules Fig. 18.6). These can be further classified into two types, capsules smaller than 1 μm as nanocapsules and greater than 1000 μm as microcapsules [52–55].	Pesticide [56] Solvent (minimum to dissolve pesticide), emulsifier, defoamer, antifreeze, monomer, cross-linker

Multifunctional emulsion based on biopesticide
The new concept in the formulation technology is a multifunctional plant-based emulsion [65]. The formulation is basically a W/O emulsion, which consists of neem oil as continuous phase and metallic nanoparticle (low percentage) as dispersed water phase along with combination of other biopesticides, which impart multifunctional efficacy and synergistic efficacy in one single formulation (Fig. 18.8).

Features of multifunctional formulation:
• The emulsion provided insecticidal activity as an antifeedant against *Spodoptera latera*).
• The emulsion was effective against plant parasitic nematodes, which reduce crop yield through their feeding and movement through root (*Meloidogyne* sp.).
• It also possesses antifungal properties against soil-borne pathogens which cause damage to a

Capsules containing pesticide

Scanning microscope (SEM) image of
Capsules in microcapsulated suspension

Microencapsulated   Dilution of
suspension   M.encap

Bentonite carrier

FIG. 18.6 Capsulated suspension, SEM images of capsules, Bentonite carrier for coated granules.

Tablet formulation

Floating tablets

Effervescent action in dispersible tablets

FIG. 18.7 Tablets, Floating tables, and effervescent tablets.

TABLE 18.7
Types of Tablets.

Types of Tablets	Description	Composition	Application	Advantages
Water dispersible/ effervescent tablet (Fig. 18.7)	Disperses in water very fast, produces homogenous and stable suspension	Pesticide [62], wetting and dispersing agent, binder, effervescent agents optional (sodium bicarbonate and citric acid), filler (clay)	For spray application. The tablet is dropped in water producing suspension.	Disintegrates fast and produces stable suspension
Controlled release floating tablets (neem oil) Fig. 18.7 [63]	Tablets for direct application in aquatic environment provide controlled release. After application in water bodies, these float on the surface of water releasing active on slow rate	Pesticide, carrier, wetting and dispersing agent, binder, and lubricant	Aquatic pest/weed control. Applied directly to water bodies.	Controlled release of active Reduces frequency and dose of application Safe to nontarget organisms
Waste biomass-based bait tablet for household pest control (Fig. 18.7)	Applied in houses to control undesirable pests [64]	Waste biomass with pesticide property, binder, lubricant, attractant, fillers	Applied directly to control the house pest.	Safe to user and biodegradable

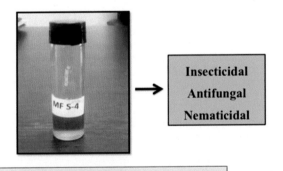

Neem oil based multifunctional emulsion

FIG. 18.8 Neem-based multifunctional formulation.

formulation technology are in front area of investigation. The main focus of formulation technology is cost reduction and environmental impact. The formulation technology incorporating biobotanical can effectively combat the alarming effects of synthetic pesticides and provide the cost-effective and efficacious solution to pest management. The new-generation biobotanical pesticide formulation can provide enhanced efficacy along with the user's safety. The demand of product performance, efficacy, environmental and user safety are the key concerns of new formulation technology, and the biopesticide formulation needs to be optimized considering these factors.

wide range of agricultural crops including variety of fruits and vegetables like tomato, potato, apple, peanut, etc.

Conclusion

The utilization of biopesticides can reduce the loads of the synthetic pesticides in the environment and can provide sustainable pest control. Besides numerous formulation technologies available to target pest, the demands of new safer, effective possibilities in

REFERENCES

[1] M. Govindarajan, Larvicidal and repellent properties of some essential oils against *Culex tritaeniorhynchus* Giles and *Anopheles subpictus* Grassi (Diptera: Culicidae), Asian Pac. J. Trop. Med. 4 (2) (2011) 106—111.

[2] P. Stevenson, M. Isman, S. Belmain, Pesticidal plants in Africa: a global vision of new biological control products from local uses, Ind. Crop. Prod. 110 (30) (2017) 2—9.

[3] M. Pant, S. Dubey, P. K Patanjali, S. N Naik, S. Sharma, Insecticidal activity of eucalyptus oil nanoemulsion with karanja and jatropha aqueous filtrates, Int. Biodeterior. Biodegrad. 91 (2014) 119—127.

[4] G.G.- Pirasanna-Pandi, T. Adak, B. Gowda, N. Patil, M. Annamalai, M. Jena, Toxicological effect of underutilized plant, Cleistanthus collinus leaf extracts against two major stored grain pests, the rice weevil, *Sitophilus oryzae* and red flour beetle, *Tribolium castaneum*, Ecotoxicol. Environ. Saf. 154 (2018) 92–99.

[5] J. Avelino, M. Cristancho, S. Georgiou, P. Imbach, L. Aguilar, G. Bornemann, The coffee rust crises in Colombia and Central America (2008–2013): impacts, plausible causes and proposed solutions, Food Secur. 7 (2) (2015) 303–321.

[6] R. Cerda, J. Avelino, C. Gary, P. Tixier, E. Lechevallier, C. Allinn, Primary and secondary yield losses caused by pests and diseases: assessment and modeling in coffee, PLoS One 12 (1) (2017) 0169133, https://doi.org/10.1371/journal.pone.0169133.

[7] P. Bhandari, M. Pant, P. Patanjali, S. Raza, Advances in bio-botanicals formulations with incorporation of nanotechnology in intensive crop management, in: R. Prasad (Ed.), Advances and Applications through Fungal Nanobiotechnology. Fungal Biology, Springer, Cham, 2016, pp. 291–305.

[8] T. Chermenskayaa, E. Stepanychevaa, A. Shchenikovaa, A. Chakaeva, Insectoacaricidal and deterrent activities of extracts of Kyrgyzstan plants against three agricultural pests, Ind. Crop. Prod. 32 (2) (2010) 157–163.

[9] N. Ahmad, M. Shafiq Ansari, F. Hasan, Effects of neem based insecticides on *Plutella xylostella* (Linn.), Crop Protect. 34 (2012) 18–24.

[10] M. Ramos-López, G. Pérez, C. Rodríguez-Hernández, P. Guevara-Fefer, M. Zavala-Sánchez, Activity of *Ricinus communis* (Euphorbiaceae) against *Spodoptera frugiperda* (Lepidoptera: *noctuidae*), Afr. J. Biotechnol. 9 (2010) 1359–1365.

[11] R. Datta, A. Kaur, I. Saraf, I. Singh, S. Kaur, Effect of crude extracts and purified compounds of *Alpinia galanga* on nutritional physiology of a polyphagous lepidopteran pest, *Spodoptera litura* (Fabricius), Ecotoxicol. Environ. Saf. 168 (30) (2019) 324–329.

[12] E. Mazhawidza, B. Mvumi, Field evaluation of aqueous indigenous plant extracts against the diamondback moth, *Plutella xylostella* L. and the rape aphid, *Brevicoryne brassicae* L. in *Brassica* production, Ind. Crop. Prod. 110 (30) (2017) 36–44.

[13] G. Benellia, R. Pavelac, R. Petrellid, The essential oil from industrial hemp (Cannabis sativa L.) by-products as an effective tool for insect pest management in organic crops, Ind. Crop. Prod. 122 (2018) 308–315.

[14] M. Kłys, N. Malejky, M. Nowak-Chmura, The repellent effect of plants and their active substances against the beetle storage pests, J. Stored Prod. Res. 74 (2017) 66–77.

[15] B. Singh, A. Kaur, Control of insect pests in crop plants and stored food grains using plant saponins: a review, LWT - Food Sci. Technol. (Lebensmittel-Wissenschaft-Technol.) 87 (2018) 93–101.

[16] D.M.K. Mushobozy, G. Nganilevanu, S. Ruheza, G. Swella, Plant oils as common bean (*Phaseolus vulgaris* L.) seed protectants against infestations by the Mexican bean weevil *Zabrotes subfascistus* (Boh.), J. Plant Prot. Res. 49 (1) (2009) 35–39.

[17] M. Ali, A.M. A Ibharim, Castor and camphor essential oils alter hemocyte populations and induce biochemical changes in larvae of *Spodoptera littoralis* (Boisduval) (Lepidoptera: noctuidae), J. Asia Pac. Entomol. 21 (2) (2018) 631–637.

[18] P.G. Kavitha, M. Umadevi, S. Suresh, V. Ravi, The rice root-knot nematode (*Meloidogyne graminicola*) — life cycle and histopathology, Int. J. Sci. Nat. 7 (3) (2016) 483–486.

[19] W. Mohamed Abd-Elhameed El-Nagdi, M. Ahmed Youssef, Comparative efficacy of garlic clove and castor seed aqueous extracts against the root-knot nematode, meloidogyne incognita infecting tomato plants, J. Plant Prot. Res. 53 (3) (2013) 285–288.

[20] J.S. Prasad, K.S. Varaprasad, Y.R. Rao, E. Srinivasa Rao, M. Sankar, Comparative efficacy of some oil seed cakes and extracts against root-knot nematode (*Meloidogyne graminicola*) infection in rice, Nematol. Medit 33 (2005) 191–194.

[21] S. Sharma, M. Verma, R.P.D. Yadav, Efficacy of non-edible oil seedcakes against termite (Odontotermes obesus), J. Sci. Ind. Res. 70 (2011) 1037–1041.

[22] L. Bocqueta, C. Rivièrea, C. Dermontb, J. Samailliea, J. Hilberta, P. Halamab, A. Siahb, S. Sahpaz, Antifungal activity of hop extracts and compounds against the wheat pathogen Zymoseptoria tritici, Ind. Crop. Prod. 122 (2018) 290–297.

[23] P. Sittisart, S. Yossan, P. Prasertsanb, Antifungal property of chili, shallot and garlic extracts against pathogenic fungi, Phomopsis spp., isolated from infected leaves of para rubber (Hevea brasiliensis Muell. Arg.), Agriculture Nat. Resourc. 51 (2017) 485–491.

[24] B. Thueriga, J. Ramseyerb, M. Hamburgerb, Efficacy of a *Magnolia officinalis* bark extract against grapevine downy mildew and apple scab under controlled and field conditions, Crop Protect. 114 (2018) 97–105.

[25] C. Laosinwattana, P. Wichittrakarn, M. Teerarak, Chemical composition and herbicidal action of essential oil from *Tagetes erecta* L. leaves, Ind. Crop. Prod. 126 (2018) 129–134.

[26] D. Borges, E. Lopes, A. Moraes, Formulation of botanicals for the control of plant-pathogens: a review, Crop Protect. 110 (2018) 135–140.

[27] D. Hazara, Recent advancement in pesticide formulations for user and environment friendly pest management, Int J Res Rev 2 (2) (2018) 35–39.

[28] A. Knowles, Recent developments of safer formulations of agrochemicals, Environmentalist 28 (1) (2007) 35–44, https://doi.org/10.1007/s10669-007-9045-4.

[29] M. Sarwar, Commonly available commercial insecticide formulations and their applications in the field, Int. J. Mat. Chem. Phys. 1 (2) (2015) 116–123.

[30] C. Chin, C. Lan, Ho-ShingWu, Application of biodiesel as carrier for insecticide emulsifiable concentrate formulation, J. Taiwan Inst. Chem. E 43 (4) (2012) 578–584.

[31] X. Zhang, T. Jing, D. Zhang, J. Luo, F. Liu, Assessment of ethylene glycol diacetate as an alternative carrier for use in agrochemical emulsifiable concentrate formulation, Ecotoxocol Enviorn Saf 163 (15) (2018) 349–355.

[32] H. Patzke, A. Schieber, Growth-inhibitory activity of phenolic compounds applied in an emulsifiable concentrate - ferulic acid as a natural pesticide against *Botrytis cinerea*, Food Int. Res. 113 (2018) 18–23.

[33] J. Waghmare, A. Ware, S. Momin, Neem oil as pesticide, J. Dispersion Sci. Technol. 28 (2007) 323–328.

[34] B. Zhang, Development of 5% abamectin EW formulation, J. Chem. Pharm. Res. 6 (6) (2014) 28–32.

[35] J. Puripattanavong, C. Songkram, L. Lomlim, T. Amnuaikit, Development of concentrated emulsion containing nicotiana tabacum extract for use as pesticide, J. Appl. Pharm. Sci. 3 (11) (2013) 16–21.

[36] R. Pavelaa, G. Benelli, L. Pavonid, G. Bonacucinad, M. Cespid, K. Cianfaglionee, I. Bajalang, M. Morshedlooh, G. Lupidid, D. Romanoi, A. Canalec, F. Maggi, Microemulsions for delivery of Apiaceae essential oils – towards highly effective and eco-friendly mosquito larvicides, Ind. Crop. Prod. 129 (2019) 631–640.

[37] M. Singla, P. Patanjali, Phase behaviour of neem oil based microemulsion formulations, Ind. Crop. Prod. 44 (2013) 421–426.

[38] D. McClements, Nanoemulsions versus microemulsions: terminology, differences, and similarities, Soft Matter 8 (6) (2012) 1719–1729.

[39] A. Pratap, D. Bhowmick, Pesticides as microemulsion formulations, J. Dispersion Sci. Technol. 29 (2008) 1325–1330.

[40] C. Solans, J. Esquena, A. Forgiarini, D. Morales, N. Uson, P. Izquierdo, Nanoemulsion: Formulation and Properties, 2002, pp. 525–554. New York.

[41] K. Bouchemal, S. Briançon, E. Perrier, H. Fessi, Nanoemulsion formulation using spontaneous emulsification: solvent, oil and surfactant optimization, Int. J. Pharma. 280 (2004) 241–251.

[42] N. Sogan, N. Kapoor, S. Kala, P. Patanjali, B. Nagpal, K. Vikram, N. Valech, Larvicidal activity of castor oil Nanoemulsion against malaria vector Anopheles culicifacies, Int. J. Mos. Res. 5 (3) (2018) 1–6.

[43] A. Gupta, H. Eral, T. Hattona, P. Doyle, Nanoemulsions: formation, properties and applications, Soft Matter 12 (2016) 2826–2841, https://doi.org/10.1039/c5sm02958a.

[44] C. Fernandes, F. Almeida, A. Nunes Silveira, M. Gonzalez, C. Brasileiro Mello, D. Feder, R. Apolinário, M. Guerra Santos, J. Carvalho, L. Tietbohl, L. Rocha, D. Quintanilha Falcão, Development of an insecticidal nanoemulsion with *Manilkara subsericea* (*Sapotaceae*) extract, J. Nanobiotechnol. (2014) 12–22.

[45] A. Sharma, N. Sharma, A. Srivastavab, A. Katariab, S. Dubey, S. Sharma, B. Kundu, Clove and lemongrass oil based non-ionic nanoemulsion for suppressing the growth of plant pathogenic *Fusarium oxysporum* f. sp. *Lycopersici*, Ind. Crop. Prod. 123 (2018) 353–362.

[46] W. Chien Lu, D. Huang, C. Wang, C. Hua Yeh, P. Hsien Li, Preparation, characterization, and antimicrobial activity of nanoemulsion incorporating citral essential oil, J. Food Drug Anal. 26 (1) (2018) 82–89.

[47] A. Azeem, M. Rizwan, F.J. Ahmad, Z. Iqbal, K. Roop Khar, M. Aqil, S. Talegaonkar, Nanoemulsion components screening and selection: a technical note, AAPS PharmSciTech 10 (1) (2009) 69–76, https://doi.org/10.1208/s12249-008-9178-x.

[48] S. Dubey, V. Jhelum, P. Patanjali, Controlled release of agrochemical formulation: a review, J. Sci. Ind. Res. 70 (2011) 105–112.

[49] P. Mulqueen, Recent advances in agrochemical formulation, Adv. Colloid Interface Sci. 106 (2003) 83–107.

[50] N. Kimato, A. Takahashi, K. Inubushi, Design and release profile of timed-release coated granules of systemic insecticide, J. Pest. Sci. 32 (4) (2007) 402–406, https://doi.org/10.1584/j pestics.G07-16.

[51] H. Scher, Pesticide Formulations. Innovations and Developments in Pesticide Formulations: An Overview ACS Symposium Series, vol. 371, American Chemical Society, Washington, DC, 1998, pp. 1–5.

[52] W. Hsieh, C. Chang, Y. Gao, Controlled release properties of chitosan encapsulated volatile citronella oil microcapsules by thermal treatments, Colloids Surf., B 53 (2006) 209–214.

[53] T.K. Maji, I. Baruah, S. Dube, M.R. Hussain, Microencapsulation of Zanthoxylum limonella oil (ZLO) in glutaraldehyde crosslinked gelatin for mosquito repellent application, Bioresour. Technol. 98 (2007) 840–844.

[54] M. Moretti, G. Sanna-Passino, S. Demontis, E. Bazzoni, Essential oil formulations useful as a new tool for insect pest control, AAPS Pharm. Sci. Technol. 3 (2004) 64–74.

[55] M.R. Hussain, T.K. Maji, Preparation of genipin crosslinked chitosan–gelatin microcapsules for encapsulation of *Zanthoxylum limonella* oil (ZLO) using salting-out method, J. Microencapsul. 25 (2008) 414–420.

[56] H. Secher, Microencapsulation of pesticides by interfacial polymerization: process and performance considerations, Pest. Residues Formul. Chem. (1983) 295–300.

[57] N. Girardia, M. Alejandra Passonea, D. Garcíaa, A. Nescia, M. Etcheverry, Microencapsulation of Peumus boldus essential oil and its impact on peanut seed quality preservation, Ind. Crop. Prod. 114 (2018) 108–114.

[58] C. Peng, S. Zhao, J. Zhang, G. Huang, L. Chen, F. Zhao, Chemical composition, antimicrobial property and microencapsulation of mustard (*Sinapis alba*) seed essential oil by complex coacervation, Food Chem. 165 (2014) 560–568, https://doi.org/10.1016/j.foodchem.2014.05.126.

[59] M. Pant, S. Dubey, S. Raza, P. Patanjali, Encapsulation of neem and karanja oil mixture for synergistic as well as larvicidal activity for mosquito control, J. Sci. Ind. Res. 71 (2012) 348–352.

[60] F.A.O. Specification, Manual on Development & Use of FAO & WHO Specifications for Pesticides November 2nd Revision, first ed., 2010. & 3rd revision.

[61] S. Sharma, A. Upadhyay, M. Haque, K. Padhan, P. Tyagi, C. Batra, T. Adak, A.D.S. K. Subb, Village-Scale evaluation of mosquitoe nets treated with a tablet formulation of deltamethrin against malaria vector, Med. Vet. Entomol. 19 (2005) 286–292.

[62] T. Seenivasagan, K. Sharma, S. Prakash, Electroantennogram, flight orientation and oviposition responses of *Anopheles stephensi* and *Aedes aegypti* to a fatty acid ester-propyl octadecanoate, Acta Trop. 124 (2012) 54–61.

[63] P. Patanjali, S. Kala, A. Agarwal, S. Raza, A Composition for Preparing Controlled Release Floating Tablets, 2012. Indian Patent, Number: 880/DEL/2012.

[64] P. Patanjali, S. Dubey, M. Pant, S. Raza, S. Naik, Insecticidal Compositions for Control- Ling Household Pests, 2014. Indian patent Application No: 2705/DEL/2012.

[65] P. Patanjali, S. Kala, Madhuri. Multifunctional and Efficicous Bio-Botanical Emulsion Based Pesticide Formulation, 2017. Indian Patent, Number: 201611018872.

CHAPTER 19

Safe Storage and Preservation Techniques in Commercialized Agriculture

NARASHANS ALOK SAGAR, PHD • SUNIL PAREEK, PHD

INTRODUCTION

A rapid growth in technological innovations and mechanization has increased the production of various agricultural and horticultural crops in recent years. However, a major part of this produce gets lost during various processes from farm to fork [1]. Based on shelf life, there are two types of agricultural produce: perishables and durables. Perishable materials are fruits and vegetables (F&V) which can be stored for a longer period after preservation or processing. Durables like cereals, pulses, and other grains have shelf life of several years only when preserved and stored properly [2]. Food preservation is a procedure of handling and treating agricultural produce in such a way that minimum loss and maximum shelf life could be achieved. Preservation also stops or retards the growth of microorganisms which are responsible for loss of nutritive value, quality, and edibility. However, some methods include benign bacteria, fungi, or yeasts are used for preserving foods and adding qualities. Food preservation inhibits discoloration and natural aging of fresh produce that can take place during food preparation (enzymatic browning of F&V). Some preservation methods such as canning need a proper seal of the food after treatment to avoid recontamination of microbes while others like drying do not need sealing after processing for storage [3]. Preservation of agricultural produce has been a great challenge since ages because microbial spoilage and pest contamination are the cause of huge losses of foods during storage, transportation, and marketing. Among different food commodities, fruits and vegetables (F&V) are highly perishable and exhibit higher postharvest losses up to 35%–40%, may vary crop to crop. The reason behind the loss is higher water activity (a_w), microbial spoilage, mechanical damage due to delicacy, and environmental conditions of F&V [4].

Water is the main and crucial component of foods. It affects flavor and texture, microbiological growth, and fat oxidation of dried food items [5]. Dehydration is the oldest and most common processing method used to improve food stability by decreasing water activity and microbiological spoilage of the foods. It also minimizes physicochemical changes in the produce [6,7]. Dehydration can be classified into mechanical dewatering, thermal drying, and osmotic drying and as per the water-removing method [8]. Although many preservation techniques have been developed so far for F&V preservation and most of them are efficient for processed food products, there is still a chance for more efficient, innovative, and greener techniques for the preservation of fresh F&V and grains [9]. However, there are various innovative techniques such as high-pressure processing, irradiation, and plasma technology which are being used for the preservation of agricultural produce [2]. Chemical method of preservation was also explored as an alternative, but excessive amount of chemicals during processing and storage in agricultural produce may arise various health problems for the consumers and may be consequently resulted in decrement in consumer acceptance and export [10].

Grains are sensitive products against various biological agents like mites, insects, and fungi. Additionally, quality and flavor of grains may be reduced due to oxidative processes during handling and storage. The biological pathogens cause discoloration and contamination of dangerous mycotoxins in grains. Insects can create favorable conditions for fungal development and their dispersion. Insects and fungus can cause allergy to the workers and consumers. After harvesting, the grain is at high risk of insect infestation due to higher moisture content [11]. Hence, processing industries dry and take necessary chemical measures before storage of grains, but sometimes chemical methods become harmful for workers, consumers, and environment due to the residues which ultimately increase the losses. Therefore, to reduce losses and maintain the quality of agriculture produces, integrated and innovative technologies are required with sustainable approach to mitigate the environment impact [12].

Preservation Techniques for Fruits and Vegetables

Food preservation is an important part of food processing which is carried out to minimize the spoilage and increase the shelf life of the foods [13,14]. There are various preservation techniques and processing methods which are used in the food processing industries to enhance the shelf life of F&V (Fig. 19.1).

Thermal preservation methods

Following are the thermal methods of preservation.

Freezing. Freezing is a process to slow down the biochemical and physiochemical reactions by formation of ice from available water. The temperature goes below freezing line and retards the growth of pathogens which deteriorate the F&V [15,16]. It decreases the water activity and available water of the food items [17]. Freezing creates a complex reaction in a food material during heat transfer. It includes phase change and alteration of thermal attributes [18]. Nucleation (ice crystal formation) followed by growth (crystal size increment) are two main consecutive actions of freezing [19].

Individual quick freezing (IQF) is a quick-freezing technique to preserve solid F&V like cut beans, green peas, and cauliflower pieces. Inversely, quick freezing is related to freezing of pulpy, semiliquid, and liquid items such as fruit juices, papaya pulps, and mango pulp. Quick freezing causes less injury to the food texture and structure of the cell because it forms very small ice crystals. IQF also provides higher capability and cost cutting to the freezing plants (commercial). However, a quick-freezing plant needs big investment for setting up [20]. There are several other quick-freezing methods, i.e., air-blast freezing, contact plate freezing, and cryogenic freezing, which are utilized for preserving F&V [2].

Chilling. Chilling process decreases the onset temperature and sustains the final temperature of foods for a longer period. The temperature range (-1 to $8°C$) is maintained for food products in chilling procedure [21]. It retards the microbiological and biochemical

activities, resulting in longer shelf life of F&V [22]. When cooling is done at $<15°C$, it is known as chilling [23]. Superchilling decreases the rate of ice formation and enhances the shelf life of fresh produce in the industries. Sometimes, superchilling is referred as partial freezing [24]. Various instruments can be utilized for chilling process such as ice implementation system, vacuum attribution system, continuous air cooler, plate heat exchanger, cryogenic chamber, ice bank cooler, and jacketed heat exchanger [25]. Rate of chilling depends on different factors like initial temperature of F&V, thermal conductivity, density, moisture content, and size and weight of the food item [26].

Drying. Drying (dehydration) is one of the ancient techniques of preservation. It is a water-removing process from foods through evaporation which resulted in a solid or hard product with very low quantity of water [27]. Water is responsible for enzyme activation and growth of microorganisms which ultimately triggers food spoilage. Almost all the microorganisms can grow at a_w 0.95, but they cannot flourish below a_w 0.88. Drying drags down the water level at a point where microorganisms cannot grow [28,29]. Drying is considered as a cheapest method for preserving foods, and it has several advantages like it decreases the weight and volume of fruits, vegetables, and other foods and thus helps in better packaging, storage, and transportation [30]. Inversely, it is responsible for the destruction of aroma and flavor, polyphenols, vitamin C, lipids, protein, and thiamine [31,32].

Drying is broadly classified into two types: natural drying or sun drying (oldest method of preservation) and artificial drying. In sun drying, controlling of the temperature is a challenge, whereas temperature can be easily controlled in artificial drying. Artificial drying is the best suited method for commercial purpose. Commercial drying includes mainly conductive and convective methods. Convective method is mostly used for dehydration of foods up to 90%. Based on the way of operation, dryers can be categorized as batch or continuous. Batch dryers are best suited for small-scale processes while continuous dryers are preferred for drying operations of longer period [33]. F&V are the best suited agricultural produce for drying. Drying time and temperature of different F&V is given in Table 19.1.

Retorting. It is a kind of thermal sterilization in which food is packaged in a vessel and sterilized [38]. Food products having pH more than 4.5 need $100°C$

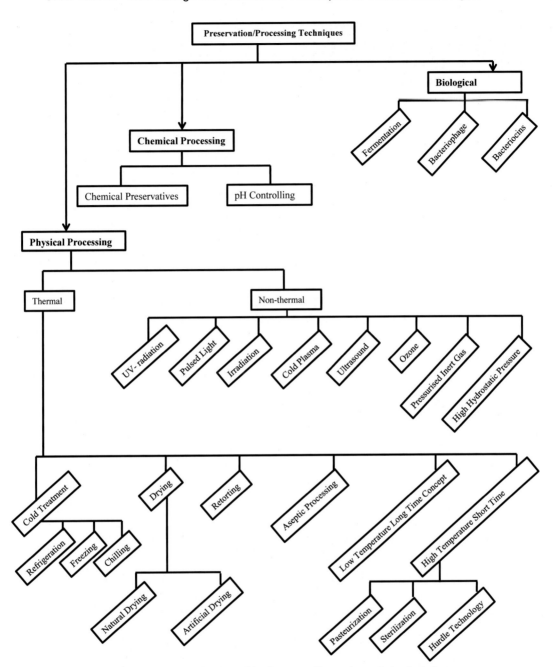

FIG. 19.1 Schematic diagram of food preservation and processing techniques.

temperature for sterilization. This range of temperature is possible with continuous or batch retort process. Continuous operation has gradually replaced batch retort process [39]. In food industries, rotary cookers and hydrostatic retorts are the popular continuous systems.

Aseptic packaging. Aseptic packaging is also a type of thermal sterilization. It is process in which commercially sterilized food item is placed in a sterilized package followed by sealing under aseptic environment. It is highly recommended for the preservation of F&V slices, tomato paste, and fruit juices [39]. Heat treatment,

TABLE 19.1
Drying Temperature and Time for Different Fruits and Vegetables [7,34—37].

Foods	Drying Time (hour)	Processing Temperature (°C)
FRUITS		
Apple	Until dry	70
Cherries	2—3	70
Coconuts	Until dry	45
Longan (cv. *Daw*)	1—2	60—75
Pineapples	1—2	70
Persimmons	1—2	60
Pears (Asian)	Until dry	60
VEGETABLES		
Asparagus	2—3	60
Beetroots (red)	1—2	30—100
Carrot	Until dry	70
Green beans	2	60
Mushrooms	2—3	25—30
Onions	1—2	70
Parsley	Until dry	30—50
Potatoes (*Agata*)	1—2	30—70

[43] used partial dehydration, antimicrobials, and packaging in polymeric bag for carrots in combination and extended the shelf life up to 6 months in ambient condition. When pineapple slices underwent hurdle process (osmotic dehydration, gamma irradiation, and infrared drying), microbial load reduced and shelf life was enhanced for 40 days [44]. Shelf life of fresh scraped coconut was increased by 3 months in refrigerated condition ($5 \pm 2°C$) and up to 1 month in ambient condition when coconut was preserved using additives (acidulants, humectants) and preservatives [45]. Moreover, combined use of a_w, antimicrobials, and heat treatment (hurdle technology) was applied to preserve banana, pineapple, peach, mango, and papaya, and this combination extended the shelf life of the fruits [46].

Nonthermal and novel techniques of preservation and shelf life extension

Nonthermal techniques such as ultraviolet radiation (UV-radiation), pulse light, irradiation, cold plasma, ultrasound, ozone, pressurized inert gas, and high hydrostatic pressure can change inner temperature of F&V but does not affect the quality of foods [47]. Moreover, nonthermal preservation methods are not under temperature boundary and heat transfer limit. Being green technologies, nonthermal techniques not only increase food quality and efficiency of processing but also decrease drying time [48]. Some important nonthermal processing techniques are discussed below.

chemical application, or both can be utilized for sterilization in this process. Paper and plastic are generally used in conventional aseptic packaging. Apart from this, foil, metal cans, and various metal containers can be used in aseptic packaging [40]. This wide range of packaging materials increase proficiency and reduce the cost of the process. Basically, there are two approaches of aseptic packaging: direct approach (composed of steam infusion and steam injunction) and indirect approach (comprises scrapped surface heat exchanger, plate heat exchanger, and tubular heat exchanger) [41]. Steam injection is a faster process hence sometimes volatile is lost while steam infusion gives better control on the processing conditions thus decreases the chance of overheating of foods [41].

Hurdle technology. Hurdle technology is a technique of combined application of existing and novel techniques for the preservation of foods. It is used to improve the quality of food items and inhibit the growth of microbial pathogens [42]. Vibhakara et al.

Ultraviolet radiation. UV light is a kind of nonionizing radiation within the wavelength range of 100 —400 nm. UV-A (315—400 nm), UV-B (280—315 nm), and UV-C (100—280 nm) are the three basic categories of UV irradiation [49]. UV-C has the highest germicidal capacity at 254 nm. UV irradiation directly hits the DNA of living microorganisms which consequently produces dimers like pyrimidine 6-4 pyrimidone and cyclobutane pyrimidine. These photoproducts initiate mutagenesis followed by cell death [50]. UV-C is best suited for surface sterilization of F&V because enzymatic degradation and microbial spoilage take place on the surface and UV irradiation covers wide range of microorganisms [51].

Various studies showed a significant reduction in microbial load in F&V when subjected to UV. When UV-C dose was given to fresh-cut apples with 1.2 kJ m^{-2}, total viable counts were reduced to 2 log units lower compare with untreated samples for 8 days in storage at $6°C$ [51]. Similarly, reduction of microorganisms

(*Escherichia coli*, *Listeria monocytogenes*, and *Salmonella enteritidis*) was found when kailan-hybrid broccoli was treated with UV-C and stored [52]. Fresh-cut carambola was treated with UV radiation with 12.5 kJ m^{-2} dose, it remained fresh up to 21 storage days. Firmness is directly related to freshness of the F&V. It was observed that when peppers were treated with UV-C, the firmness was 50% better than the untreated one during storage for 12 days at 4°C [49]. Apart from this, UV-C treatment showed good concentration of vitamin C and total phenolic content in mandarin and paprika [53,54].

Pulse light. Pulse light (PL) is a nonthermal and novel technique of food preservation. Its high pulse power and short duration work as surface decontaminant of F&V or packaged items and kill microorganisms [55]. A flash lamp of an inert gas (mostly xenon) is used to generate the pulses and broad-spectrum white light which includes ultraviolet to near-infrared wavelength. The mechanism of action of PL is to destroy the cell and DNA of microorganisms by photochemical effect [56]. The main merits of this innovative technique are flexibility, no residues, and lower energy cost [57]. It has already been proven to be a novel preservation technique for various food products like meat, fruit and vegetable juices, and milk [58−60]. In recent years, PL has taken good attention as a preservation method for F&V. It not only conserves nutritional aspects and sensorial attributes of foods but also inactivates pathogens. Accurate conditions and optimized parameters are required to give best results otherwise undesirable loss may occur [61]. In mangoes treated with PL, color, carotenoids, and firmness were found preserved up to 7 days at 6°C, while loss in carotenoids, color, and firmness was reported in untreated samples after 3 days [62]. In addition to this, when avocado underwent PL dose with 6 kJ m^{-2}, maintained level of chlorophyll *a* and chlorophyll *b* and decreased level of yeasts and molds were reported [63]. Likewise, when mushroom (fresh-cut) was treated with higher dose (12 kJ m^{-2}) of PL, less counts of *Listeria innocua* and *E. coli* were reported [64]. Further, it was confirmed by transmission electron microscopy (TEM) that notable destruction was observed in cell membrane and cytoplasm of pathogens. Moreover, exposure of apple to high dose of PL showed a significant decrease in *L. innocua* [65].

Irradiation. Irradiation is also known as cold-sterilization in which selected dose of ionizing radiation (IR) is applied on food products [66]. There are two modes of IR, i.e., natural (gamma rays, high-energy UV and X-rays) and artificial (secondary radiation and electron beam) [67,68]. The advantages of IR include decontamination of F&V and grains, extension in shelf life of F&V by slowing the maturation or inhibiting the sprouting, and safety of foods from microorganisms [69,70]. IR dose is calculated in kilo grays (kGy) means 1 gray is equal to absorbed ionizing dose by 1 kg of irradiated food [67]. Irradiation does not affect nutritional properties such as protein, carbohydrates, lipids, most vitamins, and minerals of foods but few reports showed some impact on vitamin A, B1, and E [71].

Cold plasma. Cold plasma (CP) is green and novel nonthermal preservation technology. After solid, liquid and gas, plasma is considered as forth state of matter. It is known as quasi-neutral ionized gas which comprises different particles like free electrons, photons, excited or nonexcited atoms and positive or negative ions [72]. Usually, oxygen, air, nitrogen, or noble gases mixture (neon, argon, and helium) are used in CP. Various energy sources such as laser, electricity, microwaves, magnetic field, and radiofrequency are utilized to produce plasma [73,74]. The combination of these active particles release UV and visible light. Active particles of the plasma break the cell and DNA structure of bacteria or viruses and provide higher shelf life to the food product [73].

Application of cold plasma has been successfully tested as an innovative sterilization on potato, cherry tomato, lettuce, strawberry, cabbage, and milk to inhibit microbes. Treatment of CP on cabbage and lettuce showed 0.3−2.1 log CFU g^{-1} reduction in *L. monocytogenes* and 1.5 log CFU g^{-1} reduction in *Salmonella typhimurium*, respectively [75]. Likewise, Fernandez et al. [76] reported 2.72, 0.94, and 1.76 log decrement in *S. typhimurium*, respectively, for lettuce, potato, and strawberry when subjected to CP. Misran et al. [77] found total mesophilic count reduction (12%−85%) and yeast and mold count reduction (44%−95%) in strawberries treated with CP. There was no notable change observed in color, rate of respiration, and firmness of F&V by CP. CP treatment was applied on cherry tomato for short duration, i.e., 10, 60, and 120 s, and an undetectable drop was observed in *Salmonella*, *E. coli*, and *L. monocytogenes* count from initial population 6.3, 3.1, and 6.7 log10 CFU/sample, respectively [78]. Similarly, when atmospheric CP was applied on blueberries for 0, 15, 30, 45, 60, 90, or 120 s, a notable drop in yeast and molds (from 1.5 to

2.0 log CFU g^{-1}) and total aerobic count (from 0.8 to 1.6 log CFU g^{-1}) were reported in comparison with control up to 7 days [79].

Ozone. In 1997, the U.S. FDA approved ozone as Generally Recognized as Safe (GRAS) sterilizing agent for foods [80]. Ozone destroys nucleic acid, important enzymes, spore coats, and viral capsid of microbes and does not leave any toxic residue. The main by-product is oxygen after ozone oxidation process [81]. Various previous studies have been carried out for F&V to inhibit microbial growth. Gaseous ozone treatment (950 mL L^{-1}, 20 min) on lettuce and spinach showed 1.0−1.5 log decrement in *L. innocua* and *E. coli* [78]. Similarly, Alwi and Ali [82] reported reduction of 3.06, 2.89, and 2.56 log in the counts of *L. monocytogenes*, *S. typhimurium*, and *E. coli* O157 in bell pepper (fresh-cut) when subjected to ozone (9 ppm, 6 h). Liu et al. [83] applied aqueous ozone on apple for 5−10 min at 1.4 mg L^{-1} and found reduced ethylene production and total bacteria counts. Same results were reported in the case of papaya [84] and melon [85] when treated with ozone.

Pressurized inert gas. Pressurized inert gas is a green and novel preservation technique in which high pressure is applied with the help of inert gases like argon (Ar), krypton (Kr), neon (Ne), nitrogen (N$_2$), and xenon (Xe). Ice-like crystals (clathrate hydrate) are formed when these gases are dissolved in water with high pressure [86,87]. Clathrate crystals are stable above 0°C temperature under a particular pressure and can be seen visually [88]. In previous years, this technique has been successfully utilized to preserve various F&V [89,90]. The shelf life of green asparagus was found to be extended up to 12 days at 4°C when preserved with a mixture of xenon (Xe) and argon (Ar) gases under 1.1 MPa for 24 h [91]. Likewise, when green peppers (fresh-cut) was treated with argon at 4 MPa for 1 h, reduction in water loss, growth of yeast and molds, and enzyme activities were reported which consequently extended the shelf life 4 days longer (up to 12 days) than untreated sample [92]. Cucumbers (fresh-cut) showed a longer shelf life (3−4 days) than control when subjected to argon treatment with 1 MPa pressure at 20°C for 60 min [93]. Additionally, when pineapples were treated with argon treatment (1.8 MPa for 60 min), shelf life was extended from 9 days to 15 days [94]. Same parameters were taken for apple treatment with argon and extension in shelf life was reported from 7 days to 12 days [95]. Apart from this, reduction in the growth

of *Saccharomyces cerevisiae* and *E. coli* was observed in pineapples and apples than untreated samples when treated with xenon and argon gases with 1.8 MPa pressure (2:9 v:v Ar and Xe in partial pressure) [96].

High hydrostatic pressure. High hydrostatic pressure or high-pressure processing (HPP) is a novel nonthermal preservation technique that deals with pressure up to 900 MPa for disinfestation of microbes in food materials. HPP retard food spoilage, inactivates enzymatic and chemical process responsible for deterioration, and maintains crucial physicochemical properties of F&V. HPP preserves important vitamins, color, and flavors of foods [97,98]. Being environment friendly, it is a green technique and consumes less energy without leaving any residue [99]. HPP is applicable for almost all fruits and vegetables. Vegetables are preserved at approximately 3000 bar pressure at ambient temperature. Bulk processing and container processing are two main methods of HPP technology.

Chemical methods of preservation
Chemical method of food preservation is one of the oldest methods [100]. Its effectiveness depends on physicochemical properties of foods, type of microbes, and selectivity and amount of the chemical compounds [101]. Food preservatives and additives are the main component of chemical preservation. Food additives and chemical preservatives are strictly regulated through different rules and acts in various countries because of health issues [102−104]. Food preservatives are responsible for retarding, inhibiting, and arresting the microbial growth and extending the shelf life of F&V. They also work as flavoring agent, antioxidants, and antimicrobials [105,106]. Food preservatives are of two types: natural and artificial. Sugar, salt, honey, and oil are considered as natural preservatives, whereas artificial preservatives, for example, butylated hydroxyl anisole (BHA) and vitamin E are used as antioxidants; sodium benzoate, potassium sorbet, and sorbic acid (2,4-hexadienoic acid) are used as antimicrobial agents; and polyphosphates are utilized as antienzymatic agents during preservation of F&V [107,108].

Some side effects of artificial preservatives gave space for acidic electrolyzed water (AEW) technique to be evolved. AEW technique was first developed in Japan and called functional water [109]. In AEW, diluted sodium salts (potassium and sodium chloride) are used and passed via an electrolytic cell (anode and cathode divided by a membrane) and different chlorine ions/compounds such as Cl$_2$, OCl, and HOCl are produced

[110]. AEW was reported as a potential killer of microbes without affecting the nutritional and organoleptic attributes of foods. In recent years, it has been utilized as an alternative to chlorine disinfectant for safety concerns and to enhance the shelf life of F&V [111]. AEW has showed an effective role against microbes in apples [112]. Baby leaves of mizuna revealed a significant inhibition of microorganisms when treated with AEW (available chlorine 260 mg/L) for 2 min than control samples up to 11 days at 5°C [113]. When broccoli (fresh-cut) was treated with AEW (available chlorine 70 and 100 mg L^{-1}, 2 min), more concentration of total phenols was reported than NaOCl-treated samples [12].

Biological methods of preservation

Bacteriophages. Bacteriophages are biological agents, used as an innovative, effective, and eco-friendly method of food preservation. Their specific mechanism of action is infection and multiplication in the host cells [114]. Therefore, they are safe to animals, humans, and plants. Significant results have been recorded using bacteriophages against foodborne pathogens like *E. coli* O157:H7, *L. monocytogenes*, and *Salmonella* spp. [115,116]. Leverentz et al. [117] carried out the first experiment with bacteriophages in honeydew melon and found noticeable decrement in *Salmonella* when stored at 5, 10, and 20°C. Sharma et al. [118] treated lettuce and cantaloupes with ECP-100 against *E. coli* and found significant reduction in the counts. Likewise, Viazis et al. [119] used bacteriophage mixture with baby lettuce and baby spinach and found reduction in *E. coli* 0157:H7 counts. Moreover, when *Salmonella* was mixed with lettuce, a reduction was observed in *Salmonella* spp. [120].When bacteriophage (Listex P100) was applied on apple, pear, and melon, *L. monocytogenes* population decreased [121].

Bacteriocins. Bacteriocins are antimicrobial proteins or peptides which inhibit spoilage microbes [122]. Many bacteria produce bacteriocins which are utilized for shelf life extension, antimicrobial reaction, and biopreservation. They have vast application such as food sector and medical field. Bacteriocins obtained from lactic acid bacteria (LAB) gained big attention because LAB is known to be beneficial to food industry and humans as well [121]. As of now, only two bacteriocins, i.e., nisin and carbocyclic A, are utilized in food products, and they are produced by *Lactococcus lactis* and *Corynebacterium maltaromaticum* UAL307 [120]. Leverentz et al. [123] recorded a noticeable decrement (2.7 log units) in *L. monocytogenes* on melon and up to 2.3 log units on apple when treated with mixture of nisin and bacteriophage. Similarly, reduced units (2.7 log) of *L. monocytogenes* was reported in lettuce (fresh-cut) after 7 days of storage at 4°C when subjected to bacteriocin than control sample. The results suggested bacteriocins as safe sanitizer for F&V [124]. When antimicrobial film (cellulose + 25% nisin) was applied on mangoes and stored for 12 days at 5°C, all physicochemical properties were found to be maintained and a noticeable reduction of *L. monocytogenes* was recorded compared with untreated sample [125]. Similarly, Narsaiah et al. [126] applied alginate coating having bacteriocin to preserve papaya and observed inhibition of microbes and better stability of physical and chemical characters than the control one.

Preservation and Storage of Grains

Grains such as rice, wheat, corn, oats, and pulses are important source of our diet and energy. Their preservation and storage is very crucial to meet our future needs. Preservation (drying, gamma irradiation, and ozone treatment) and storage (short term or long term) may vary according to the commodity and purpose. Long-term storage is done at commercial level [127,128]. Global grain loss is about 20% due to improper facilities of storage. The demand of grains is increasing everyday with the population [129]. Microbes, insects, mechanical damage, enzymatic activities, and problem of heat result in loss of quality in grains at the time of storage. Right method of storage and selection of efficient machinery can prevent this loss [128]. Proper harvesting and storage of grain are the main factors to protect grains from beetles, moths, weevils, and rodents [130]. Before storing the grains, cleaning and drying of grains, selection of suitable storage site and structure, and proper fumigation of store are important factors [128]. Apart from this, there is a need to utilize integrated pest management strategies due to limitation of regulating agency and consumer concerns over residues of chemicals [131].

Thermal disinfestation

This technique is used in grain processing units to protect grains from insects by high temperature range (54−60°C). The duration of the operation may vary from few hours to 24 or more hours, depending upon the area of storage house [132]. Thermal disinfestation is becoming popular in flour mills and about to replace methyl bromide (MB) fumigation, because in chemical treatment, manufacturing process should be stopped, whereas thermal disinfestation can be carried out in running manufacturing condition [131].

Aeration systems

Mechanical aeration by fans is popular practice to decrease temperature of grains. Aeration refers movement of forced ambient air via bulk of grain to revamp storage quality [131]. Aeration is one of the popular methods of grain preservation and to maintain abiotic factors such as humidity, temperature, and atmospheric condition of stored grain bulk. Force aeration is used in commercial practice which gets advantage of porosity and thermal insulation property of grains. As per the experiment for wheat, it will take 220 h to cool down the temperature from 35°C to 20°C at 3.49 $m^3 h^{-1} t^{-1}$ airflow rate and 200 h at 3.88 $m^3 h^{-1} t^{-1}$ aeration rate. If airflow rate is increased to 5.63 $m^3 h^{-1} t^{-1}$, the cooling time would be 135h [133].

Refrigerated air

Ambient air is not sufficient in the case of fungi, mites, and insects; therefore, chill air is applied to preserve stored grains from insects, germination, and self-heating and to maintain quality during warm season. Chilling of grains was successfully applied on sensitive grains such as maize, soybean, seeds, edible beans, and dry gains for the protection against insects and mites [133].

Storage technologies for grains

Basically, there are two ways for storage of the grains: temporary method and long-term storage methods [130]. Temporary storage is done at field level, and aerial storage, storage on drying floor or ground, and open timber platforms are the temporary storage methods. Long-term storage is classified into solid wall bins, storage baskets, calabashes, gourds, jar, pots, and underground storage [128]. In modern way, warehouses are the best structures constructed for quality grain storage. Aeration and sealing are the main components to be taken care of during bulk storage. Based on requirement, ambient or refrigerated aeration can be applied to stop storage loss. Following are the popular storage techniques.

Storage in bag. Storage in bag is a long-term storage method. Rice and legumes are packed into bags (sacks) and then stored to maintain the quality of grains. Numbering and sampling of sacks are easy in this method but controlling of the grains in bags is tough. Apart from this, limited amount of grains is stored per unit area in this method compare to bulk

storage method. It is an expensive method of storage because it is time consuming and involves higher labor cost.

Underground storage. This method of storage is considered effective to keep the grain safe for a longer time. Sometimes pits are made airtight, and they keep grain cool [134]. Before soil covering, stalk, polyethylene, hay, and apron are used at base and on the grains in this technique. This method does not allow the air to come in contact with grains during whole storage. Due to unavoidable weather conditions, this method has been obsolete for storing grains [128].

Storage in silo. This is most preferred method of grain storage in plants. The merits of this method are less time consuming, less cost of labor, easy discharge and hygiene maintenance during processing. More quantity of grains can be stored in silo due to its vertical orientation. Silos can be made up of concrete, wood, and steel. Wood silos are prone to fire, and so they are generally ignored for preservation. Steel, concrete, and galvanized silos are the common and ideal options for grain storage. They are easy to control and resistant to insects and fire. Barley, wheat, rye, and oats can be stored in concrete and steel silo as well [134].

Bulk storage. Bulk storage offers both horizontal and vertical warehouses for grain storage. Leveling of the bulk stack surface is important in this technique. It gives more unit area for storing wheat, rye, oat, barley, corn, lentil, and chickpea. It is time saving and requires less labor cost [135].

Modified atmosphere technology

Modified atmosphere (MA) comprises CO_2, O_2, and N_2 gases in different ratio that provides protection to food materials [136]. From past decades, atmospheric manipulation has been carried out throughout the world for the protection of grains from adverse conditions and pathogens [137–139]. Hermetic and airtight storage methods have been extensively used to protect the quality of grains from chemical residues, attacking insect, and fungi. MA and controlled atmosphere (CA) technologies offer a replacement for conventional chemical methods which has negative effects on food and environment. However, the commercial use of MA and CA still needs to be explored because it is limited to some countries for grain storage [139].

Hermetic storage. Hermetic storage is a kind of MA technology which is also called as airtight storage, sealed storage, and sacrificial sealed storage. This method seals the available atmosphere by increasing CO_2 and decreasing O_2 which is responsible for the growth of aerobic pathogens and insects [140]. Hermetic environment keeps the grain quality intact for longer period and protects from mites, insects, and other microorganisms [11]. Hermetic storage has been proven to be a better technique for insect control (up to 99.9% killing) than conventional fumigants. When grains were stored under such condition, only 0.15% loss was observed during 15 months of storage [141]. Protective storage of pulses, cereals, oilseeds, coffee, and cocoa can be done by this technique for many years. This method has been successfully used in plastic structures and bulk storage for long-term storage of grains. Additionally, there are various storage systems based on hermetic technique—(1) bunker gastight storage: for conserving huge bulks from 10,000 to 15,000 tons; (2) flexible gastight silo: for preserving up to 1000 tons of bulk in bags and weld-mesh container; (3) gas-tight liners: suitable for storage cubes of 5–1000 tons of corn or rice paddy. These systems provide space to MA technology for storage of grains throughout the world [131].

Hermetic storage of paddy and rice seeds has been carried out from a long time in different countries such as India, Indonesia, Pakistan, Cambodia, East Timor, Vietnam, and Sri Lanka [142]. Coconut are multicapacity (5–1000 tons) storage structure which is used to store grains in a protective environment. Coconuts are widely used in African and Asian countries to store shelled corn and corn seeds. The flexible storage units (hermetic) are used at village as well as district level to reserve the corn [142]. In Cyprus, hermetic bunkers (10,000–15,000 tons) are utilized to store barley up to 3 years with germination (above 88%) and total losses up to 0.98% [143]. In the case of wheat, grains are stored at 12.5% of moisture content in hermetic unit to prevent quality degradation [132]. Pulses are susceptible to *Callosobruchus chinensis* and *Callosobruchus macula* which can be inhibited by hermetic storage. In many African countries, pulses are stored in coconut which provides storage capacity (20–50 tons) to the farmers during off season [144].

Conclusions

The demand for agriculture commodities (fruits, vegetables, and grains) is increasing with the rapid growth of population. Higher production often leads to greater losses in the agriculture produce. Therefore, it is necessary to preserve agriculture produce to meet the demand of the generation and to reduce the postharvest losses as well. Along with preservation, storage is also an important aspect for cereals, pulses, and other grains. There are physical, chemical, and biological measures for the preservation of agriculture produce. Thermal preservation methods are sometimes providing good results and sometimes show negative outcomes in terms of nutritional characteristics, but they are economically efficient, whereas nonthermal techniques are better but require high monetary inputs for set up. Among various storing techniques, hermetic storage methods appear best for all types of grains. There is a hope of further research toward integrated technologies and storage systems so that preservation and grain storing could be more efficient and better.

REFERENCES

[1] H.A. Mostafavi, S.M. Mirmajlessi, H. Fathollahi, The potential of food irradiation: benefits and limitations, in: A.A. Eissa (Ed.), Trends in Vital Food and Control Engineering, IntechOpen, Rijeka, Croatia, 2012, pp. 43–68.

[2] S.K. Amit, M.M. Uddin, R. Rahman, S.R. Islam, M.S. Khan, A review on mechanisms and commercial aspects of food preservation and processing, Agric. Food Security 6 (2017) 51.

[3] A.A. Olunike, Storage, preservation and processing of farm produce, Food Sci. Quality Manag. 27 (2014) 28–33.

[4] L.L. Huang, M. Zhang, Trends in development of dried vegetable products as snacks, Drying Technol. 30 (2012) 448–461.

[5] L. Fan, M. Zhang, Q. Tao, G. Xiao, Sorption isotherms of vacuum-fried carrot chips, Drying Technol. 23 (2005) 1569–1579.

[6] L. Mayor, A.M. Sereno, Modelling shrinkage during convective drying of food materials: a review, J. Food Eng. 61 (2004) 373–386.

[7] D.I. Onwude, N. Hashim, R. Janius, K. Abdan, G. Chen, A.O. Oladejo, Non-thermal hybrid drying of fruits and vegetables: a review of current technologies, Innov. Food Sci. Emerg. Technol. 43 (2017) 223–238.

[8] Z. Duan, M. Zhang, Q. Hu, J. Sun, Characteristics of microwave drying of bighead carp, Drying Technol. 23 (2005) 637–643.

[9] R.A. Molins (Ed.), Food Irradiation: Principles and Applications, John Wiley and Sons, New York, USA, 2001.

[10] S.V.R. Reddy, R.R. Sharma, G. Gundewadi, Use of irradiation for postharvest disinfection of fruits and vegetables, in: M.W. Sidiqqui (Ed.), Postharvest Disinfection of Fruits and Vegetables, Academic Press, Amsterdam, 2018, pp. 121–136.

[11] Z.G. Weinberg, Y. Yan, Y. Chen, S. Finkelman, G. Ashbell, S. Navarro, The effect of moisture level on high-moisture maize (*Zea mays* L.) under hermetic

storage conditions—in vitro studies, J. Stored Products Res. 44 (2008) 136—144.

[12] J. Navarro-Rico, F. Artés-Hernández, P.A. Gómez, M.A. Núñez-Sánchez, F. Artés, G.B. Martínez-Hernández, Neutral and acidic electrolysed water kept microbial quality and health promoting compounds of fresh-cut broccoli throughout shelf life, Innov. Food Sci. Emerg. Technol. 21 (2014) 74—81.

[13] M.S. Rahman (Ed.), Handbook of Food Preservation, second ed., Taylor and Francis, Boca Raton, 2007.

[14] O. Rodriguez-Gonzalez, R. Buckow, T. Koutchma, V.M. Balasubramaniam, Energy requirements for alternative food processing technologies - principles, assumptions, and evaluation of efficiency, Comp. Rev. Food Sci. Food Saf. 14 (2015) 536—554.

[15] M.S. Rahman, C.O. Perera, Drying and food preservation, in: M.S. Rehman (Ed.), Handbook of Food Preservation, 9Ed, Marcel Dekker, New York, 1999, pp. 173—216.

[16] M. George, Freezing, in: G.S. Tucker (Ed.), Food Biodeterioration and Preservation, Blackwell Publisher, Singapore, 2008, pp. 117—136.

[17] J.G. Brennan, A.S. Grandison, M.J. Lewis, Separations in food processing, in: J.E. Brennan, A.S. Gradins (Eds.), Food Processing Handbook, Wiley-VCH Verlag & Co, Weinheim, Germany, 2006, pp. 281—329.

[18] H.S. Ramaswamy, M.A. Tung, A review on predicting freezing times of foods, J. Food Process. Eng. 7 (1984) 169—203.

[19] R. Bhat, A.K. Alias, G. Jaliyah (Eds.), Progress in Food Preservation, John Wiley & Sons, London, UK, 2012.

[20] J.S. Prutah, Quick Freezing Preservation of Foods: Foods of Plant Origin, vol. 2, Allied Publishers, New Delhi, India, 1999.

[21] G.D. Saravacos, A.E. Kostaropoulos, Handbook of Food Processing Equipment, Kluwer Academic/Plenum, Amsterdam, Netherlands, 2002.

[22] K.P. Sudheer, V. Indira, Postharvest Technology of Horticultural Crops, New India Publishing Agency, New Delhi, India, 2007.

[23] M. Karel, D.B. Lund, Physical Principles of Food Preservation, CRC Press, Taylore and Francis, Boca Raton, FL, 2003.

[24] O.M. Magnussen, A. Haugland, A.K.T. Hemmingsen, S. Johansen, T.S. Nordtvedt, Advances in super chilling of food—Process characteristics and product quality, Trends Food Sci. Technol. 19 (2008) 418—424.

[25] G.S. Tucker (Ed.), Food Biodeterioration and Preservation, John Wiley & Sons, London, UK, 2008.

[26] N. Light, A. Walker, Cook-chill Catering: Technology and Management, Springer Science & Business Media, Germany, 1990.

[27] Z. Berk, Food Process Engineering and Technology, second ed., Academic Press, Amsterdam, Netherlands, 2018.

[28] J.M. Jay, Modern Food Microbiology, Aspen Publ. Inc, Gaithersburg, MD, 2000, p. 410.

[29] K. Rayaguru, W. Routray, Effect of drying conditions on drying kinetics and quality of aromatic *Pandanus amaryllifolius* leaves, J. Food Sci. Technol. 47 (2010) 668—673.

[30] D. Agrahar-Murugkar, K. Jha, Effect of drying on nutritional and functional quality and electrophoretic pattern of soy flour from sprouted soybean (*Glycine max*), J. Food Sci. Technol. 47 (2010) 482—487.

[31] M. Kutz (Ed.), Handbook of Farm, Dairy and Food Machinery, first ed., Academic Press, New York, 2007.

[32] J.A. Salvato, N.L. Nemerow, F.J. Agardy, Environmental Engineering, fifth ed., John Wiley & Sons, London, UK, 2003.

[33] C.G. Baker (Ed.), Industrial Drying of Foods, Springer Science & Business Media, Germany, 1997.

[34] S. Mizuta, Y. Yamada, T. Miyagi, R. Yoshinaka, Histological changes in collagen related to textural development of prawn meat during heat processing, J. Food Sci. 64 (1999) 991—995.

[35] L. Kristensen, P.P. Partlow, The effect of processing temperature and addition of mono-and di-valent salts on the hemi-nonheme-iron ratio in meat, Food Chem. 73 (2001) 433—439.

[36] A. Sequeira-Munoz, D. Chevalier, A. LeBail, H.S. Ramaswamy, B.K. Simpson, Physicochemical changes induced in carp (*Cyprinus carpio*) fillets by high pressure processing at low temperature, Innov. Food Sci. Emerg. Technol. 7 (2006) 13—18.

[37] V.R. Sagar, P.S. Kumar, Recent advances in drying and dehydration of fruits and vegetables: a review, J. Food Sci. Technol. 47 (2010) 15—26.

[38] P.L. Knechtges, Food Safety: Theory and Practice, Jones & Bartlett Publishers, 2011.

[39] Wiley, Kirk-Othmer Food and Feed Technology, vol. 1, John Wiley & Sons, London, UK, 2007.

[40] G.D. Miller, J.K. Jarvis, L.D. McBean, Handbook of Dairy Foods and Nutrition, CRC Press, Taylor and Francis, Boca Raton, FL, 2006.

[41] T. Ohlsson, N. Bengtsson, Minimal Processing Technologies in the Food Industry, Woodhead Publication, Cambridge, UK, 2002.

[42] A. Pundhir, N. Murtaza, Hurdle technology - an approach towards food preservation, Int. J. Current Microbiol. Applied Sci. 4 (2015) 802—809.

[43] H.S.J. Vibhakara, D.K.D. Gupta, K.S. Jayaraman, M.S. Mohan, Development of a high-moisture shelf-stable grated carrot product using hurdle technology, J. Food Process. Preserv. 30 (2006) 134—144.

[44] S. Saxena, B.B. Mishra, R. Chander, A. Sharma, Shelf stable intermediate moisture pineapple (*Ananas comosus*) slices using hurdle technology, Food Sci. Technol. 42 (2009) 1681—1687.

[45] K.D.P.P. Gunathilake, Application of hurdle technique to preserve fresh scraped coconut at ambient and refrigerated storage, J. Nat. Sci. Foundation Sri Lanka 33 (2005) 265—268.

[46] S.M. Alzamora, M.S. Tapia, A. Argaiz, J. Welli, Application of combined methods technology in minimally processed fruits, Food Res. Int. 26 (1993) 125—130.

[47] J. Raso, G.V. Barbosa-Cánovas, Nonthermal preservation of foods using combined processing techniques, Crit. Rev. Food Sci. Nutr. 43 (2003) 265−285.

[48] P.J. Cullen, B.K. Tiwari, V.P. Valdramidis, Status and trends of novel thermal and non-thermal technologies for fluid foods, in: P.J. Cullen, B.K. Tiwari, V.P. Valdramidis (Eds.), Novel Thermal and Non-Thermal Technologies for Fluid Foods, Academic Press, Elsevier Inc., Cambridge, Massachusetts, United States, 2012, pp. 1−6.

[49] G. Gonzalez-Aguilar, J.F. Ayala-Zavala, G.I. Olivas, L.A. de la Rosa, E. Alvarez- Parrilla, Preserving quality of fresh-cut products using safe technologies, J. für Verbraucherschutz und Lebensmittelsicherheit 5 (2010) 65−72.

[50] E. Gayan, S. Condon, I. Alvarez, Biological aspects in food preservation by ultraviolet light: a review, Food Bioprocess Technol. 7 (2014) 1−20.

[51] L. Manzocco, S. Da Pieve, A. Bertolini, I. Bartolomeoli, M. Maifreni, A. Vianello, et al., Surface decontamination of fresh-cut apple by UV-C light exposure: effects on structure, colour and sensory properties, Postharvest Biol. Technol. 61 (2011) 165−171.

[52] G.B. Martínez-Hernández, J.P. Huertas, J. Navarro-Rico, P.A. Gómez, F. Artés, A. Palop, F. Artés-Hernández, Inactivation kinetics of foodborne pathogens by UV-C radiation and its subsequent growth in fresh-cut kalian-hybrid broccoli, Food Microbiol. 46 (2015) 263−271.

[53] Y. Shen, Y. Sun, L. Qiao, J. Chen, D. Liu, X. Ye, Effect of UV-C treatments on phenolic compounds and antioxidant capacity of minimally processed Satsuma Mandarin during refrigerated storage, Postharvest Biol. Technol. 76 (2013) 50−57.

[54] I.L. Choi, T.J. Yoo, H.M. Kang, UV-C treatments enhance antioxidant activity, retain quality and microbial safety of fresh-cut paprika in MA storage, Hort. Environ. Biotechnol. 56 (2015) 324−329.

[55] G.V. Barbosa-Canovas, D.W. Schaffner, M.D. Pierson, Q.H. Zhang, Pulsed light technology, J. Food Sci. 65 (2000) 82−85.

[56] V. Heinrich, M. Zunabovic, J. Bergmair, W. Kneifel, H. Jaeger, Post-packaging application of pulsed light for microbial decontamination of solid foods: a review, Innov. Food Sci. Emerg. Technol. 30 (2015) 145−156.

[57] G. Oms-Oliu, O. Martín-Belloso, R. Soliva-Fortuny, Pulsed light treatments for food preservation: a review, Food Bioprocess Technol. 3 (2010) 13−23.

[58] M. Ferrario, S.M. Alzamora, S. Guerrero, Study of the inactivation of spoilage microorganisms in apple juice by pulsed light and ultrasound, Food Microbiol. 46 (2015) 635−642.

[59] M. Ganan, E. Hierro, X.F. Hospital, E. Barroso, M. Fernández, Use of pulsed light to increase the safety of ready-to-eat cured meat products, Food Control 32 (2013) 512−517.

[60] N. Innocente, A. Segat, L. Manzocco, M. Marino, M. Maifreni, I. Bortolomeoli, M.C. Nicoli, Effect of pulsed light on total microbial count and alkaline phosphatase activity of raw milk, Int. Dairy J. 39 (2014) 108−112.

[61] L. Ma, M. Zhang, B. Bhandari, Z. Gao, Recent developments in novel shelf life extension technologies of fresh-cut fruits and vegetables, Trends Food Sci. Technol. 64 (2017) 23−38.

[62] F. Charles, V. Vidal, F. Olive, H. Filgueiras, H. Sallanon, Pulsed light treatment as new method to maintain physical and nutritional quality of fresh-cut mangoes, Innov. Food Sci. Emerg. Technol. 18 (2013) 190−195.

[63] I. Aguiló-Aguayo, G. Oms-Oliu, O. Martín-Belloso, R. Soliva-Fortuny, Impact of pulsed light treatments on quality characteristics and oxidative stability of fresh-cut avocado, Food Sci. Technol. 59 (2014) 320−326.

[64] A.Y. Ramos-Villarroel, N. Aron-Maftei, O. Martín-Belloso, R. Soliva-Fortuny, The role of pulsed light spectral distribution in the inactivation of *Escherichia coli* and *Listeria innocuous* on fresh-cut mushrooms, Food Control 24 (2012) 206−213.

[65] P.L. Gómez, D.M. Salvatori, A. García-Loredo, S.M. Alzamora, Pulsed light treatment of cut apple: dose effect on colour, structure, and microbiological stability, Food Bioprocess Technol. 5 (2012) 2311−2322.

[66] I.S. Arvanitoyannis (Ed.), Irradiation of Food Commodities: Techniques, Applications, Detection, Legislation, Safety and Consumer Opinion, Academic Press, Netherlands, 2010.

[67] M.M. Islam, G.S. Uddin, Irradiation to insure safety and quality of fruit salads consumed in Bangladesh, J. Food Nutr. Res. 4 (2016) 40−45.

[68] B.A. Niemira, C.H. Sommers, New applications in food irradiation, in: D.R. Heldman, C.I. Moraru (Eds.), Encyclopaedia of Agricultural, Food, and Biological Engineering, CRC Press, Taylor and Francis, Boca Raton, FL, 2010, pp. 864−868.

[69] D.R. Heldman, C.I. Moraru (Eds.), Encyclopaedia of Agricultural, Food, and Biological Engineering, CRC Press, Taylor and Francis, Boca Raton, FL, 2014, pp. 869−872.

[70] S.R. Kanatt, R. Chander, A. Sharma, Effect of radiation processing of lamb meat on its lipids, Food Chem. 97 (2006) 80−86.

[71] J.S. Smith, S. Pillai, Irradiation and food safety, Food Technol. 58 (2004) 48−55.

[72] A. Fernandez, N. Shearer, D.R. Wilson, A. Thompson, Effect of microbial loading on the efficiency of cold atmospheric gas plasma inactivation of *Salmonella enterica serovar Typhimurium*, Int. J. Food Microbiol. 152 (2012) 175−180.

[73] B.A. Niemira, Cold plasma decontamination of foods, Ann. Rev. Food Sci. Technol. 3 (2012) 125−142.

[74] R. Thirumdas, C. Sarangapani, U.S. Annapure, Cold plasma: a novel non-thermal technology for food processing, Food Biophys. 10 (2015) 1−11.

[75] H. Lee, J.E. Kim, M.S. Chung, S.C. Min, Cold plasma treatment for the microbiological safety of cabbage, lettuce, and dried figs, Food Microbiol. 51 (2015) 74–80.

[76] A. Fernandez, E. Noriega, A. Thompson, Inactivation of *Salmonella enterica serovar Typhimurium* on fresh produce by cold atmospheric gas plasma technology, Food Microbiol. 33 (2013) 24–29.

[77] N.N. Misra, S. Patil, T. Moiseev, P. Bourke, J.P. Mosnier, K.M. Keener, P.J. Cullen, In-package atmospheric pressure cold plasma treatment of strawberries, J. Food Eng. 125 (2014) 131–138.

[78] D. Ziuzina, S. Patil, P.J. Cullen, K.M. Keener, P. Bourke, Atmospheric cold plasma inactivation of *Escherichia coli, Salmonella enterica serovar Typhimurium* and *Listeria monocytogenes* inoculated on fresh produce, Food Microbiol. 42 (2014) 109–116.

[79] A. Lacombe, B.A. Niemira, J.B. Gurtler, X. Fan, J. Sites, G. Boyd, H. Chen, Atmospheric cold plasma inactivation of aerobic microorganisms on blueberries and effects on quality attributes, Food Microbiol. 46 (2015) 479–484.

[80] N. Tzortzakis, I. Singleton, J. Barnes, Deployment of low-level ozone-enrichment for the preservation of chilled fresh produce, Postharv. Biol. Technol. 43 (2007) 261–270.

[81] W. Krasaekoopt, B. Bhandari, Fresh-cut vegetables, in: M. Siddiq, M.A. Uebersax (Eds.), Handbook of Vegetables and Vegetable Processing, Wiley-Blackwell, Oxford, UK, 2010, pp. 219–242.

[82] N.A. Alwi, A. Ali, Reduction of *Escherichia coli O157, Listeria monocytogenes* and *Salmonella enterica* cv. *typhimurium* populations on fresh-cut bell pepper using gaseous ozone, Food Control 46 (2014) 304–311.

[83] C. Liu, T. Ma, W. Hu, M. Tian, L. Sun, Effects of aqueous ozone treatments on microbial load reduction and shelf life extension of fresh-cut apple, Int. J. Food Sci. Technol. 51 (2016) 1099–1109.

[84] W.K. Yeoh, A. Ali, C.F. Forney, Effects of ozone on major antioxidants and microbial populations of fresh-cut papaya, Postharvest Biol. Technol. 89 (2014) 56–58.

[85] R. Botondi, R. Moscetti, R. Massantini, A comparative study on the effectiveness of ozonated water and peracetic acid in the storability of packaged fresh-cut melon, J. Food Sci. Technol. 53 (2016) 2352–2360.

[86] H. Ando, S. Takeya, Y. Kawagoe, Y. Makino, T. Suzuki, S. Oshita, In situ observation of xenon hydrate formation in onion tissue by using NMR and powder X-ray diffraction measurement, Cryobiology 3 (59) (2009) 405.

[87] Y.A. Purwanto, S. Oshita, Y. Seo, Y. Kawagoe, Concentration of liquid foods by the use of gas hydrate, J. Food Eng. 47 (2001) 133–138.

[88] D.S. Reid, O.R. Fennema, Water and ice, in: O.R. Fenema (Ed.), Fenneman's Food Chemistry, fourth ed., CRC Press, Boca Raton, FL, 2008, pp. 17–82.

[89] F. Artés, P. Gómez, E. Aguayo, V. Escalon, F. Artés-Hernández, Sustainable sanitation techniques for keeping quality and safety of fresh-cut plant commodities, Postharvest Biol. Technol. 51 (2009) 287–296.

[90] D. Zhang, P.C. Quantick, J.M. Grigor, R. Wiktorowicz, J. Irven, A comparative study of effects of nitrogen and argon on tyrosinase and malic dehydrogenase activities, Food Chem. 72 (2001) 45–49.

[91] M. Zhang, Z.G. Zhan, S.J. Wang, J.M. Tang, Extending the shelf-life of asparagus spears with a compressed mix of argon and xenon gases, Food Sci. Technol. 41 (2008) 686–691.

[92] X. Meng, M. Zhang, B. Adhikari, Extending shelf-life of fresh-cut green peppers using pressurized argon treatment, Postharvest Biol. Technol. 71 (2012) 13–20.

[93] X. Meng, M. Zhang, Z. Zhan, B. Adhikari, Changes in quality characteristics of fresh-cut cucumbers as affected by pressurized argon treatment, Food Bioprocess Technol. 7 (2014) 693–701.

[94] Z.S. Wu, M. Zhang, B. Adhikari, Application of high pressure argon treatment to maintain quality of fresh-cut pineapples during cold storage, J. Food Eng. 110 (2012) 395–404.

[95] Z.S. Wu, M. Zhang, S. Wang, Effects of high pressure argon treatments on the quality of fresh-cut apples at cold storage, Food Control 23 (2012) 120–127.

[96] Z.S. Wu, M. Zhang, S. Wang, Effects of high-pressure argon and nitrogen treatments on respiration, browning and antioxidant potential of minimally processed pineapples during shelf life, J. Sci. Food Agric. 92 (2012) 2250–2259.

[97] C.P. Dunne (Ed.), High Pressure Processing of Foods, vol. 12, John Wiley & Sons, Oxford, UK, 2008.

[98] T. Koutchma, V. Popović, V. Ros-Polski, A. Popielarz, Effects of ultraviolet light and high-pressure processing on quality and health-related constituents of fresh juice products, Comp. Rev. Food Sci. Food Safety 15 (2016) 844–867.

[99] H.B. Nielsen, A.M. Sonne, K.G. Grunert, D. Banati, A. Pollák-Tóth, Z. Lakner, et al., Consumer perception of the use of high-pressure processing and pulsed electric field technologies in food production, Appetite 52 (2009) 115–126.

[100] P.M. Davidson, J.N. Sofos, A.L. Branen (Eds.), Antimicrobials in Food, CRC Press, Taylor and Francis, Boca Raton, FL, 2005.

[101] A. Frank, H.Y.P. Paine, A Handbook of Food Packaging, Springer, New York, 1993.

[102] M.N. Islam, M. Mursalat, M.S. Khan, A review on the legislative aspect of artificial fruit ripening, Agric. Food Security 5 (2016) 1–10.

[103] M.N. Islam, A.H.M.S. Rahman, M. Mursalat, A.H. Rony, M.S. Khan, A legislative aspect of artificial fruit ripening in a developing country like Bangladesh, Chem. Eng. Res. Bull. 18 (2015) 30–37.

[104] M. Mursalat, A.H. Rony, A.H.M.S. Rahman, M.N. Islam, M.S. Khan, A critical analysis of artificial fruit ripening: Scientific, legislative and socio-economic aspects, ChE Thoughts 3 (2013) 1–7.

[105] M.R. Adams, M.O. Moses, Food Microbiology, third ed., The Royal Society of Chemistry, Cambridge, UK, 2008, pp. 98–99.

[106] T.A.M. Msagati, Chemistry of Food Additives and Preservatives, Wiley-Blackwell, New York, 2012.

[107] N. Garg, K.L. Garg, K.G. Mukerji, Laboratory Manual of Food Microbiology, I.K. International Pvt Ltd., New Delhi, India, 2010.

[108] J. Smith, Technology of Reduced Additive Foods, second ed., Wiley-Blackwell, New Jersey, 2004.

[109] Y.R. Huang, Y.C. Hung, S.Y. Hsu, Y.W. Huang, D.F. Hwang, Application of electrolyzed water in the food industry, Food Control 19 (2008) 329–345.

[110] M.I. Gil, V.M. Gómez-López, Y.C. Hung, A. Allende, Potential of electrolyzed water as an alternative disinfectant agent in the fresh-cut industry, Food Bioprocess Technol. 8 (2015) 1336–1348.

[111] E.J. Park, E. Alexander, G.A. Taylor, R. Costa, D.H. Kang, The decontaminative effects of acidic electrolyzed water for *Escherichia coli* O157: H7, *Salmonella typhimurium*, and *Listeria monocytogenes* on green onions and tomatoes with differing organic demands, Food Microbiol. 26 (2009) 386–390.

[112] A. Graça, M. Abadias, M. Salazar, C. Nunes, The use of electrolyzed water as a disinfectant for minimally processed apples, Postharvest Biol. Technol. 61 (2011) 172–177.

[113] A. Tomás-Callejas, G.B. Martínez-Hernández, F. Artés, F. Artés-Hernández, Neutral and acidic electrolyzed water as emergent sanitizers for fresh-cut mizuna baby leaves, Postharvest Biol. Technol. 59 (2011) 298–306.

[114] A. Meireles, E. Giaouris, M. Simões, Alternative disinfection methods to chlorine for use in the fresh-cut industry, Food Res. Int. 82 (2016) 71–85.

[115] J.A. Hudson, C. Billington, A.J. Cornelius, T. Wilson, S.L.W. On, A. Premaratne, N.J. King, Use of a bacteriophage to inactivate *Escherichia coli* O157: H7 on beef, Food Microbiol. 36 (2013) 14–21.

[116] D.A. Spricigo, C. Bardina, P. Cortés, M. Llagostera, Use of a bacteriophage cocktail to control *Salmonella* in food and the food industry, Int. J. Food Microbiol. 165 (2013) 169–174.

[117] B. Leverentz, W.S. Conway, Z. Alavidze, W.J. Janisiewicz, Y. Fuchs, M.J. Camp, A. Sulakvelidze, Examination of bacteriophage as a biocontrol method for *Salmonella* on fresh-cut fruit: a model study, J. Food Prot. 64 (2001) 1116–1121.

[118] M. Sharma, J.R. Patel, W.S. Conway, S. Ferguson, A. Sulakvelidze, Effectiveness of bacteriophages in reducing *Escherichia coli* O157: H7 on fresh-cut cantaloupes and lettuce, J. Food Prot. 72 (2009) 1481–1485.

[119] S. Viazis, M. Akhtar, J. Feirtag, F. Diez-Gonzalez, Reduction of *Escherichia coli* O157: H7 viability on leafy green vegetables by treatment with a bacteriophage mixture and trans-cinnamaldehyde, Food Microbiol. 28 (2011) 149–157.

[120] M. Oliveira, M. Abadias, P. Colás-Medà, J. Usall, I. Viñas, Biopreservative methods to control the growth of foodborne pathogens on fresh-cut lettuce, Int. J. Food Microbiol. 214 (2015) 4–11.

[121] M. Oliveira, I. Vinas, P. Colas, M. Anguera, J. Usall, M. Abadias, Effectiveness of a bacteriophage in reducing *Listeria monocytogenes* on fresh-cut fruits and fruit juices, Food Microbiol. 38 (2014) 137–142.

[122] E.M. Balciunas, F.A.C. Martinez, S.D. Todorov, B.D.G. de Melo Franco, A. Converti, R.P. de Souza Oliveira, Novel biotechnological applications of bacteriocins: a review, Food Control 32 (2013) 134–142.

[123] B. Leverentz, W.S. Conway, M.J. Camp, W.J. Janisiewicz, T. Abuladze, M. Yang, A. Sulakvelidze, Biocontrol of *Listeria monocytogenes* on fresh-cut produce by treatment with lytic bacteriophages and a bacteriocin, Appl. Environ. Microbiol. 69 (2003) 4519–4526.

[124] C.L. Randazzo, I. Pitino, G.O. Scifò, C. Caggia, Biopreservation of minimally processed iceberg lettuces using a bacteriocin produced by *Lactococcus lactis* wild strain, Food Control 20 (2009) 756–763.

[125] A.A.T. Barbosa, H.G.S. de Araújo, P.N. Matos, M.A.G. Carnelossi, A.A. de Castro, Effects of nisin-incorporated films on the microbiological and physico-chemical quality of minimally processed mangoes, Int. J. Food Microbiol. 164 (2013) 135–140.

[126] K. Narsaiah, R.A. Wilson, K. Gokul, H.M. Mandge, S.N. Jha, S. Bhadwal, S. Vij, Effect of bacteriocin-incorporated alginate coating on shelf-life of minimally processed papaya (*Carica papaya* L.), Postharvest Biol. Technol. 100 (2015) 212–218.

[127] R. Bucklin, S. Thompson, M. Montross, A. Abdel-Hadi, Grain storage systems designs, in: M. Kutz (Ed.), Handbook of Farm, Dairy, Food Machinery Engineering, Elsevier, New York, US, 2013, pp. 123–175.

[128] H. Pekmez, Cereal storage techniques: a review, J. Agric. Sci. Technol. B 6 (2016) 67–71.

[129] P.P. Said, R. Pradhan, Food grain storage practices - a review, J. Grain Process. Storage 1 (2014) 1–5.

[130] A. Mishra, P. Prabuthas, H.N. Mishra, Grain storage: methods and measurements, Qual. Assur. Saf. Crops Foods 4 (2012) 136–158.

[131] S. Navarro, Advanced grain storage methods for quality preservation and insect control based on aerated or hermetic storage and IPM, J. Agric. Eng. 49 (2012) 13–20.

[132] S. Navarro, A. Varnava, E. Donahaye, Preservation of grain in hermetically sealed plastic liners with particular reference to storage of barley in Cyprus, in: Proceedings of International Conference on Controlled Atmosphere and Fumigation in Grain Storages, Winnipeg, Canada, 1993, pp. 223–234.

[133] S. Navarro, R. Noyes (Eds.), The Mechanics and Physics of Modern Grain Aeration Management, CRC Press, Boca Raton, FL, 2002, p. 647.

[134] S. Bhardwaj, Grain storage structures for farmers, Popular Kheti. 2 (2014) 202–205.

[135] I. Kemaloğlu, B. Baran, Wheat in milling, in: Handbook of Miller, İstanbul, Turkey: Parantez, 2011, pp. 23–47.

[136] S. Bodbodak, M. Moshfeghifar, Advances in modified atmosphere packaging of fruits and vegetables, in:

M.W. Siddiqui (Ed.), Eco-Friendly Technology for Post-harvest Produce Quality, Academic Press, Netherlands, 2016, pp. 127–183.

[137] C. Adler, H.G. Corinth, C. Reichmuth, Modified atmospheres, in: B. Subramanyam, D.W. Hagstrum (Eds.), Alternatives to Pesticides in Stored Product IPM, Kluwer Academic Publishing, Norwell, MA, 2000, pp. 105–146.

[138] M. Calderon, R. Barkai-Golan, Food Preservation by Modified Atmospheres, CRC Press, Boca Raton, FL, 1990.

[139] S. Navarro, Modified atmospheres for the control of stored-product insects and mites, in: J.W. Heaps (Ed.), Insect Management for Food Storage and Processing, American Association of Cereal Chemists International, 2006, pp. 105–146.

[140] N.D.G. White, D.S. Jayas, Controlled atmosphere storage of grain, in: A. Chakravarty, A.S. Majumdar, H.S. Ramaswamy (Eds.), Handbook of Postharvest Technology: Cereals, Fruits, Vegetables, Tea, and Spices, Marcel Dekker Inc., New York, 2003, pp. 235–251.

[141] A. Varnava, Hermetic storage of grain in Cyprus, in: T. Batchelor, J. Bolivar (Eds.), Proceedings of International Conference on Alternatives to Methyl Bromide, Sevilla, 2002, pp. 163–168.

[142] R. Montemayor, World Grain Magazine, November, 2004 Issue, Better Rice in Store. IRRI, 2004, pp. 43–45.

[143] A. Varnava, C. Mouskos, 7-Year results of hermetic storage of barley under PVC liners: losses and justification for further implementation of this method of grain storage, in: Proceedings of International Conference of Controlled Atmosphere and Fumigation in Stored Products, 1996, pp. 21–26.

[144] MINAGRI, Systems de Stockage au Rwanda: 36 moist experience, RSwandan Ministry of Agriculture. MINAGRI, Kigali, Rwanda, 2006.

CHAPTER 20

Advances in Application of ICT in Crop Pest and Disease Management

CHANDAN KUMAR PANDA, PHD

INTRODUCTION

Crop-protection chemicals are used during presowing, sowing, and postsowing stages of farming. To multiply their savings, it is essential that farmers use crop-protection chemicals judiciously across these stages. The use of crop-protection chemicals across the value chain can increase the overall yield of crops, not only resulting in rise in incomes for the farmers but also boosting their profitability with significant cut-down in crop losses [1]. Controlling pest outbreaks through chemical means is not always cost-effective [2]. Internet has immense scope to impact research, agricultural extension, and farmers teaching in entomology domain. Extension entomologists may use information on pest management and can disseminate to the deserving farmers. Farmers may communicate directly with concern experts for remedies of their specific problem. Internet supportive communication is gradually becoming well acquainted among extension workers and farmers, as it allows broader options for the farmers and permits sufficient time and space for the subject expert to reply a query [3]. In the management of insect pest and disease, the web-based models and decision support systems (DSS) are getting their special attention because minimal or no client software is essential; in consequence, it reduces software management and distribution costs [4], as information technology had strengthened the efficacy of data collection and its analysis; pest recognition, biocontrol agent selection [5], and epidemiological information can be used with weather data to anticipate disease onset, and disease advisories can be applied to reduce the unnecessary use and improve the timing of fungicide sprays [6]. Integrated pest management failed to replicate in sub-Saharan Africa in their major smallholder food crops due to inadequate extension, underinvestment in agricultural research, and faulty government policies [2]. However, realizing the scope

of information and communication technology (ICT) in plant protection with multifold possibilities of centralization and decentralization and considering the roles and responsibilities of the stakeholders involved, the National Research Center for Integrated Pest Management (NCIPM) in India with its mandate of eliciting national pest scenario across crops *vis-á-vis* dissemination of integrated pest management (IPM) practices to the growers revolutionized the ICT-driven pest surveillance [7].

Increasing Use of Pesticide

Food and Agricultural Organization of United Nation estimates that the use of different types of pesticide in the developing countries was found that collectively pests, weeds, and disease cause the destruction of around 40% of crops. The damage to crops is caused mainly by insects, followed by pathogens and weeds. There is indiscriminate use of pesticide throughout the world. The use of pesticide was more in Asia (3.62 kg a.i. ha^{-1}) and America (3.39 kg a.i. ha^{-1}) as compared to Africa (0.31 kg a.i. ha^{-1}), Oceania (1.17 kg a.i. ha^{-1}), and Europe (1.67 kg a.i. ha^{-1}) as per available latest data from FAOSTAT. The world average use of pesticide was 1.98 kg a.i. ha^{-1} in the year 2000 and now it is around 2.57 kg a.i. ha^{-1}, so there is an increase use of pesticide, 29.80%. The top 10 pesticide used countries in the world are Bahamas (32.22 kg a.i.ha^{-1}), Costa Rica (18.78 kg a.i.ha^{-1}), Barbados (18.06 kg a.i.ha^{-1}), Malta (16.77 kg a.i.ha^{-1}), Colombia (14.57 kg a.i.ha^{-1}), Saint Lucia (14.33 kg a.i.ha^{-1}), Japan (14.18 kg a.i.ha^{-1}), Republic of Korea (12.74 kg a.i.ha^{-1}), Maldives (12.29 kg a.i.ha^{-1}), and Israel (11.60 kg a.i.ha^{-1}). The aforesaid data are compiled from FAOSTAT latest update in their website.

Smallholder farmers immensely contribute in agriculture. In Tanzania, the distribution of pesticides used by smallholder farmers in insecticides use was

59%, followed by fungicides (29%), herbicides (10%), and the remaining being rodenticides (2%) [8]. In a study about vegetable production in Red River Delta of Vietnam, it was noted that farmers' expenditure on pesticide application is around one-fourth of their total cost of cultivation [9]. The average costs of cultivation for 1179 farms in case of cultivation of corn and soybeans was $98 per acre in the United States. The costs of cultivation include fertilizer ($37 per acre), pesticides ($32 per acre), and seed costs ($29 per acre) [10].

Intelligence in Crop Pest and Disease Management

Agriculture practice is a dynamic domain where conditions cannot be generalized to suggest a common solution. With the techniques of artificial intelligence (AI), the complex details of each situation can be analyzed and a best fit solution of the problem can be provided. With the advancement of computational technology in AI, the complex problems are being solved more precisely [11]. DSSs in crop pest and disease management are easy-to-use techniques that execute intricate tasks effectively and efficiently. The Internet-based DSSs delivery increases user accessibility, allows the DSSs to be updated easily and continuously (for rapid transfer of knowledge efficiently), and allows farmers to maintain get in touch with service providers. For diagnosis of potato diseases, the prototyping an expert system, Boyd and Sun [12] divulged excellent potential for expert system (ES) use in managing the disease component of potato seed production. This ES was ruling-based systems, and its 127 rules diagnose 11 pathogenic diseases and 6 nonpathogenic diseases, and it utilizes 8 knowledge bases accessed via an agenda in a Pascal environment. Most of the fungal and bacterial plant pathogens need free moisture for infection and reproduction; accordingly, disease forecasting model requires for estimating wetness duration on plant surfaces if direct measurement is not available. Artificial neural network models with a backpropagation architecture can predict wetness on wheat flag leaves by using environmental variables recorded at 0.5 h intervals with an electronic data logger [13]. Various meteorological data such as temperature, humidity, and leaf wetness duration played a vital role in the multiplication of microorganism responsible for disease. Precise forecasting of such plant diseases based on climate data must help the farmers to take timely actions to control the diseases. Weather-based forecasting system was considered a part of agricultural decision support system, which is a knowledge-based system. Keeping this in consideration, Tilva et al. [14] used fuzzy logic-based structure for plant disease forecasting system. With minimum weather data, that is, temperature and humidity, the proposed method can be implemented.

Farmers Information Access in Crop Pest and Disease Management

The most frequently mentioned source of information on pesticide usage is from commercial media, government agricultural extension officers, village leaders, and the opinions of other community leaders [15]. mASK—building an Agricultural Advisory System that would leverage the power of two-way communication over mobile phones would help bridge the information gaps existing between farmers and experts, and carry small group of farmers along by supporting them with customized agricultural advisories. Customized advisories would play a crucial role in revitalizing agriculture [16]. The use of ICT for pest surveillance comprises e-pest surveillance that is basically an internet-based system for capturing of pest-related information from fields and send information to the plant protection experts to provide advisory services to the state agricultural extension agencies who further advise the concerned farmers for crop protection [17] and plant disease forecasting was valuable when the systems give timely information to the farmers for their judicious pest management decisions. A Case Study of USDA's Soybean Rust Coordinated Framework showed that soybean producers' profits by a total of $11–$299 million, or between 16 cents per acre, depending on the quality of information and other factors supported through the forecast framework [18].

A study was conducted among 120 farmers of India between March and August 2018 with the objective to know the roles of different information sources on plant pest and disease management advisory service to the farmers. Data obtained were presented in Tables 20.1, 20.2 and Fig. 20.1.

It was noted that farmers received plant protection measure-related advisory information from 15 sources. The major sources of information were progressive farmer (ranked I), neighbor farmer (ranked II), agri-input dealers (ranked III), Kisan Salahkar (ranked IV), and Bihar Krishi apps (mobile apps) (ranked V), and other information sources as shown in Table 20.1. It was also noted that farmers accessed on an average eight information sources, out of which five sources (average) was tried. The mean frequency of use of sources was 3.4, that is, the respondents used "tried information sources" from fortnightly to monthly (Table 20.2).

Farmers received plant protection measure-related advisory information from aforesaid 15 sources. Considering types of information disseminated by sources, modus operand of sources, and ease to access,

TABLE 20.1
Rank Position of Different Sources of Information (Agriculture-Related Activities).

Order No.	Information Sources (Agriculture-Related Activities)	Weighted Mean	Rank Order
1.	Progressive farmer (FIS1)	4.87	I
2.	Neighbor farmer (FIS2)	4.67	II
3.	Kisan Salahkar (FIS3)	4.25	IV
4.	Assistant technical manager (FIS4)	3.21	XII
5.	Block technological manager (FIS5)	3.12	XIII
6.	Agriculture coordinator (FIS6)	3.95	VII
7.	Scientists of Bihar Agricultural University (FIS7)	4.06	VI
8.	Kisan call center (FIS8)	2.89	XIV
9.	BAU, Kisan help line (FIS9)	3.86	VIII
10.	Bihar Krishi apps (mobile apps) (FIS10)	4.12	V
11.	Radio (FIS11)	2.17	XV
12.	Television (FIS12)	3.23	XI
13.	Newspaper (FIS13)	3.78	IX
14.	Bihar Agricultural University, Kisan diary (FIS14)	3.67	X
15.	Agri-input dealers (FIS15)	4.45	III

Data Source: Primary data collected through field survey [19] (FIS = farmers' information source. This code is used in dendrogram); $n = 120$.

TABLE 20.2
Summary Statistics for Variables Used in Information Search.

Order No.	Variables	Mean	SD	Maximum	Minimum
1.	Number of sources accessed	8.34	2.34	10.00	3.00
2.	Number of sources from which information tried	5.43	2.12	7.00	2.00
3.	Mean of frequency of use (6 = daily, 5 = weekly, 4 = fortnight, 3 = monthly, 2 = seasonal, 1 = yearly, 0 = none)	3.40	1.30	6.00	0

Data Source: Primary data collected through field survey [19].

it became imperative to group or club or cluster similar type of sources. Hence, clustering of the sources was done through the dendrogram using single linkage. It was noted that the information sources agriculture coordinator (FIS6), agri-input dealers (FIS15), Kisan Salahkar (FIS3), BAU Kisan diary (FIS14), and BAU, Kisan help line (FIS9) have similar type of implication and these were in one cluster. Most interesting thing to note that Bihar Krishi apps (mobile apps) (FIS10) is stand-alone in dendrogram, it implied that these mobile apps are strong enough to support farmers all kinds

of advisory in plant protection timely measures (Fig. 20.1).

Public and Private Extension System ICT Endeavor

Throughout the world, national government and private agencies support farmers through advisory service in crop pest and disease management. However, in developed nation, the role of public extension system in general advisory service is less as compared to the private extension service. Although, this scenario is reverse

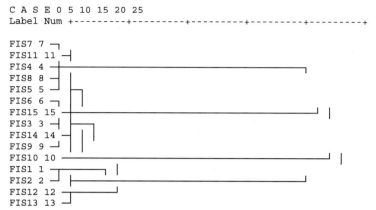

FIG. 20.1 Dendrogram of Cluster Analysis on Information Sources used by the farmer (FIS = farmers' information source). (Data Source: Primary data collected through field survey [19].)

in developing and low-income nations. In developed nation, the mechanization in pest and disease management is well recognized and established; however, it is less in low-income nations. In developing nation, the agri-input dealers play a pivotal role in crop pest and disease advisory. In a study conducted in India by National Sample Survey Organization [20], it was noted that agri-input dealers' contribution in farmers' advisory service is 13%. In India, Bangladesh, Nepal, and Sri-lanka, under the Training and Visit (T&V) System, the farmers were supported with advisory service by village level extension workers. For the last 2 decades, the Agricultural Technology Management Agency (ATMA) has strong extension services throughout India. This ATMA network is working in close contact of farmers in plant protection advisory services. Apart from this, some group approaches are working in South East Asia, viz. Farmer Club, Commodity Interest Group, Farmer Interest Group, Farmers Producers Organization, Farmer Producer Company, Farmers Club, etc. These groups are formed in close association with farmers, and in most cases government provides some financial support to the group as boost up fund. These group approaches support farmers with crop pest and disease management advisory services through some ICT tools viz. Mobile SMS, WhatsApp, Facebook, and Twitter.

ICT Tools in Crop Pest and Disease Monitoring and Management

With rapid scientific progress in information and communication technology, the use of ICT in

agriculture is progressing and government extension wings, private extension service providers, companies, international agencies are applying it for better plant pest and disease advisory services to the farmers. The ICT tools that are used in pest and disease advisory services are website and web portals, mobile apps (iOS & android), mobile SMS (text & voice), sensors, drones, wireless sensor network, etc.

Mobile apps

The agricultural sector has developed mobile applications (or "apps") for agricultural consultants and producers with educational materials and support tools. These "Ag-Apps" received increasing interest among farmers, and the use of this "Ag-Apps" technology increases the efficiency in communicating and decision-making of farmers [21] and these are developed with number of purposes viz. identification purpose (weeds, insect pest, crop disease, nutrient deficiency) and diagnosis and recommendation (controlling the weed, crop insect and disease), market price, weather information and advisory, fertilizer calculation, etc. The use of "Ag-Apps" proves its worthiness, to bridge the gap between "Information-haves" and "information-have-nots" and reducing the "digital divide." Agricultural extension services (viz. weather-related information, crop advisory service, market price, buyers, and seller's information) is provided to farmers through mobile SMS either through basic mobile, feature mobile, and android mobiles. However, potentiality of Android mobile in agricultural extension is more as compared to basic mobile [22]. The following are some important Mobile Apps used in agricultural extension.

I. **Plantix:** It supports farmers, extension workers, and gardeners in diagnosing plant diseases, pest damages, and nutrient deficiencies affecting crops and offers corresponding treatment measures In free of cost. This mobile app is worldwide popular, and it was downloaded more than one million among 150 nations. The Plantix also hosts online community to network with other farmers, discuss plant health issues, and access their local weather reports.

II. **Kisan Suvidha:** It provides current weather information and forecast weather for the next 5 days, market prices of agricultural commodities in the nearest town of the users, knowledge on use of fertilizers, seeds, farm machinery, etc. The farmers can use this app in major Indian languages. which facilitates its more widely accessible.

III. **Pusa Krishi:** Farmers receive information on new release and old varieties of crops, resource conservation cultivation, farm machineries, and agricultural implements, release plant protection measures through this mobile app. This app is developed by Indian Council of Agriculture Research (ICAR), New Delhi.

IV. **Field Check App:** Through this mobile app the pesticide applicators can easily locate speciality crop and also beehive location from their smart phone or tablet. The special features of this app were ease of use; larger icons and choosing of desired search location through GPS support.

V. **Agri Sync:** AgriSync makes it possible for farmers and advisors to connect and resolve problems related to issues on the farm viz., insect pest and disease management. Farmers can connect with number of advisors from different companies to submit problems and receive solution in real-time via video. Advisors can manage multiple services through a dashboard and remote videos that allow advisors to see what the farmer sees in real-time basis.

VI. **Climate Basic:** Climate Basic support farmers with track up-to-the-minute, field-level information on current and future weather data, soil, and crop growth stages information. It can set alerts for each field of the farmers.

Website and portal

In agricultural extension advisory services, website and portal provide technical knowledge to the farmers and also extension advisors. Some important websites in these regards are as follows:

http://www.knowledge bank.irri.org
http://www.rkmp.co.in
http://www.eXtension.org
http://www.digitalgreen .org
http://www.agmarknet. nic.in
http://www.vikaspedia.in
http://www.nafis.go.ke

http://www.kilimo.go.ke
http://www.aesa-gfras.net
http://www.meas-extension.org

http://www.agritech.tnau.ac.in
http://www.e-agriculture.gov.gh
http://www.accessagriculture. org
https://www.digitalgreen.org/
https://www.agtube.org/en
http://www.ncipm.res.in
http://www.celkau.in/ ecropdoctor.aspx
https://www.mkisan.gov.in/
https://www.farmer.gov.in/

Use of drone

The technological breakthrough in the use of unmanned aerial vehicles or drones in field crop management, livestock monitoring, fish farming, forests resource management, and other natural resource-based management activities represents a new technological edge and opens a range of thrilling opportunities [23], as the raw data collected through drone provide the basis for analytical models for agriculture advisory to the farmers and also conducting research. Use of drone in precision farming for soil health scans, crop health monitor, irrigation schedules, fertilizers application, yield data generation, and data for weather analysis is well recognized through the world [24]. Use of drone in agriculture measures plant health details, gives plant counts and estimates yield, optimizes input timing, identifies plant health patterns before these are visible, measures water conditions in fields, detects resistant weeds or invasive weed species, regulates nitrogen fertilization application in row crops, quickly estimates storm damage to settle crop insurance claims rapidly and assesses areas for replanting, optimizes water usage by monitoring drought stress at different growth stages, records plants density, plant size, and condition, and provides guidance for optimal harvest timing of crops [25]. The advantages of using drone are as follows:

1. Drones are affordable, requiring a very modest capital investment when compared to most farm equipment.
2. Drones can pay for themselves and start saving money within a single growing season.

3. Operation is relatively simple and getting easier with every new generation of flight hardware.

4. Drones are easy to incorporate into the regular crop-scouting workflow; instead of visiting a crop field to check for pests' infestation or other ground issues, the drone can provide ground data through aerial data capture to server.

5. The real advantages of drones are not about the hardware; the value is in the convenience, quality, and utility of the final data product.

Wireless sensor network in crop pest and disease management

Wireless sensor network (WSN) is considered as a group of spatially dispersed and dedicated sensors for monitoring and recording the physical environmental conditions. The data and images collected through the wireless sensor are stored in central location for subsequent interpretation. WSNs measure environmental parameters viz. temperature, humidity, wind, fog, rainfall, etc. [26] those are used in crop pest disease forecasting. An acoustic sensor device can monitor the noise level of the pests and give an indication to the farmer through an alarm when the noise crosses a threshold [27]. The use of WSN in crop pest and disease management and also advisory services to the farmers are newer in ICT intervention in agriculture.

Social medias in crop advisory

Social media is an internet-based digital tool for information sharing among people. It serves as a platform for sharing of text messages, video, audio, image, multimedia, and other relevant contents. This platform also gives opportunity for expression of views [28]. Social media play effective role in agricultural crisis during pest and disease outbreaks as well facilitating quick communication among scientists, advisory experts, farmers, and other actors helping in countering situations [29]; in a research program conducted by Extension wing of Iowa State University in association with other universities in United States, it was noted that by using social media (Twitter) it was possible to map and identify the spread of pests and pathogens in major crops [30]. Stellenbosch University has started IPM service to help growers and industry players identify unknown pests in the field through WhatsApp, which acts as a basic information sharing hub, as opposed to the more comprehensive reports provided by the Insect ID Service when samples are submitted [31].

CONCLUSIONS

From the aforesaid discussion, the following conclusions can be drawn on Advances in Application of Information and Communication Technology in Crop Pest and Disease Management:

1. Information technology has strengthened the efficiency of data collection and analysis, pest identification, control agent selection, and pesticide field applications.

2. DSS and web-based models in pest and disease management are drawing attention of experts as it requires little or no client software. The cloud server and cloud computing are the basis for it. Ultimately, it reduces the software management and distribution costs in pest management advisory.

3. As internet-based communication provides broader choices for the farmers and permits enough time to experts to answer a query, hence it is becoming more accepted among farmers in pest and disease management.

4. With rapid development of computational technology, especially artificial intelligence, there is a shift to automated expert system in crops' pest and disease management and precise and localized crop advisory service to the farmers.

5. Drone in crop disease and pest monitoring and pesticide application, more accuracy of different sensors in studying different parameters of weathers and soil, changes the way of pest and disease management of crops.

6. Social medias connected farmers and experts in real-time basis in crop advisory and feedback console.

7. Judicious application of weather data along epidemiological information to anticipate disease onset, accordingly the disease advisories to farmers from the experts can reduce unnecessary use and improve the timing of fungicide sprays in the crop fields.

8. e-pest surveillance that is basically an ICT-based system of integrating pest information from fields and producing instant and customized pest advisory reports to the plant protection experts for advisory services to the farmers. Use of e-pest surveillance can have reduced the cost of pest and disease management.

9. The Mobile "Ag-Apps" are able to identify insect pest and crop disease in situ and provide advisory service to the farmers.

10. WSN can capture environment factors (temperature, humidity, sunshine, etc.) and organize the collected data at a central server and predict the occurrence of crop pest and diseases appearance.

REFERENCES

[1] Anonymous 2019 Indian Agrochemical Industry, Doubling Farmers, Income: Role of Crop Protection Chemicals & Solutions, 2018. http://ficci.in/spdocument/23002/Knowledge%20Paper%20Agrochemicals%202018.pdf.

[2] A. Orr, Integrated pest management for resource-poor African farmers: is the emperor naked? World Dev. 31 (5) (2003) 831–845.

[3] A.K. Dhawan, S. Kumar, R.K. Saini, Role of information and communication technology (ICT) in entomology, J. Agricultural Technol. 7 (4) (2011) 879–894.

[4] D.J. Power, S. Kaparthi, The changing technological context of decision support systems, in: D. Berkeley, G. Widmeyer, P. Brazilian, V. Rajkovic (Eds.), Context-sensitive Decision Support Systems, 1998, pp. 41–54.

[5] R.E. Stinner, Information management: past, present, and future, in: G.C. Kennedy, T.B. Sutton (Eds.), Emerging Technologies for Integrated Pest Management — Concepts, Research, and Implementation, Conference Proceedings, March 8–10, Raleigh, NC, 1999, pp. 474–481.

[6] J.E. Bailey, Integrated method of organizing, computing and deploying weather-based decision advisories for selected peanut diseases, Peanut Sci. 26 (1999) 74–80.

[7] S. Vennila, A. Birah, V. Kanwar, C. Chattopadhyay, Success Stories of Integrated Pest Management in India, ICAR-National Research Centre for Integrated Pest Management, New Delhi, 2016. Available from. http://www.ncipm.res.in/NCIPMPDFs/folders/Success%20stories.pdf.

[8] A.V.F. Ngowi, T.J. Mbise, A.S.M. Ijani, L. London, O.C. Ajayi, Pesticides use by smallholder farmers in vegetable production in Northern Tanzania, Crop Protection (Guildford, Surrey) 26 (11) (2007) 1617–1624.

[9] H.P. Van, P.J. Arthur, P. Oosterveer, V.P. Brink, Pesticide distribution and use in vegetable production in the Red River Delta of Vietnam, Renew. Agric. Food Syst. 24 (2009) 174–185.

[10] G. Schnitkey, Farm Economics Facts & Opinions, Farm Business Management, March, University of Illinois Extension, 2004, pp. 1–3. http://www.farmdoc.illinois.edu/manage/newsletters/fefo04_04/fefo04_04.pdf.

[11] G. Bannerjee, U. Sarkar, S. Das, I. Ghosh, Artificial intelligence in agriculture: a literature survey, Int. J. Sci. Res. Compu. Sci. Appli. Manag. Stud. 7 (3) (2018) 112–118.

[12] D.W. Boyd, M.K. Sun, Prototyping an expert system for diagnosis of potato diseases, Comput. Electron. Agric. 10 (3) (1994) 259–267.

[13] L.J. Francl, S. Panigrahi, Artificial neural network models of wheat leaf wetness, Agric. For. Meteorol. 88 (1997) 57–65.

[14] V. Silva, J. Patel, C. Bhatt, Weather based plant diseases forecasting using fuzzy logic, [In:] Nirma University International Conference on Engineering, 28–30 Nov. 2013, Ahmedabad, India, Institute of Electrical and Electronics Engineers Print ISSN: 2375-1282, Available from: https://ieeexplore.ieee.org/abstract/document/6780173.

[15] W.O. Nyakundi, G. Muruga, J. Ocharaandand, A.B. Nyenda, A Study of Pesticide Use and Application Patterns Among Farmers: A Case Study from Selected Horticultural Farms in Rift Valley and Central Provinces, Kenya, 2010. Available from: https://eleaning.jkuat.ac.ke/journals/ojs/indexphp/jscp/Article/view/file/744/686.

[16] J. Umadikar, U. Sangeetha, M. Kalpana, M. Soundarapandian, S. Prashant, A. Jhunjhunwala, mASK: a functioning personalized ICT-based agriculture advisory system, in: IEEE Region 10 Humanitarian Technology Conference (R10 HTC). Held at Chennai, India, from 6-9 August 2014, 2014. https://ieeexplore.ieee.org/document/7026313.

[17] Haryana State Horticulture Development Agency and ICAR- National Research Centre for Integrated Pest Management, ICT Based Pest Surveillance and Advisory Services for Horticultural Crops in Haryana, 2018. Available from: http://www.ncipm.res.in/HaryanaWebReport/PDF/AboutUs.pdf.

[18] M.J. Roberts, D.E. Schimmelpfennig, E. Ashley, M.J. Livingston, M.S. Ash, U. Vasavada, The Value of Plant Disease Early-Warning Systems: A Case Study of USDA's Soybean Rust Coordinated Framework, 2006. https://www.ers.usda.gov/webdocs/publications/45312/28407_err18fm_1_.pdf?v=41063.

[19] AgriOrbit, 2019. https://www.agriorbit.com/new-whatsapp-service-insect-identification.

[20] National Sample Survey Organisation, Situation Assessment Survey of Farmers: Access to Modern Technology for Farming, ministry of statistics and programme implementation government of India, 2005. Report 499(59/33/2).

[21] K-State, Agricultural Mobile Apps: A Review and Update of ID Apps, Agronomy eUpdate. Issue 549, February 8, 2016. https://webapp.agron.ksu.edu/agr_social/m_eu_article.throck?article_id=820.

[22] C.K. Panda, Mobile phone usage in agricultural extension in India: the current and future perspective, in: F.J. Mtenzi, G.S. Oreku, D.M. Lupiana, J.J. Yonazi (Eds.), Mobile Technologies and Socio-Economic Development in Emerging Nations, IGI Global, Hershey PA, USA, 2018, ISBN 9781522540298, pp. 1–21.

[23] Anonymous 18 March, Unmanned Aerial Vehicles (UAV) for Agriculture [online], 2019. http://www.uav4ag.org.

[24] G. Sylvester, E-agriculture in Action: Drones for Agriculture, Food and Agriculture Organization of the United Nations and International Telecommunication Union, Bangkok, 2018. http://www.fao.org/3/i8494en/i8494en.pdf.

[25] Anonymous, Drone Data Collection and Analytics for Agriculture 2019. Available from: https://www.precisionhawk.com/agriculture.

[26] https://en.wikipedia.org/wiki/Wireless_sensor_network.

[27] N. Srivastav, G. Chopra, P. Jain, B. Khatter, Pest Monitor and control system using WSN with special reference to Acoustic Device; International Conference on Electrical and Electronics Engineering 27th Jan 2013, Goa, ISBN: 978-93-82208-58-7.

[28] D. Andres, J. Woodard, Social Media Handbook for Agricultural Development Practitioners, Publication by FHI360 of USAID.

[29] S. Bhattacharjee, S. Raj, Social Media: Shaping the Future of Agricultural Extension and Advisory Services, GFRAS Interest Group on ICT4RAS, 2016. Available from: https://www.g-fras.org/en/knowledge/gfras-publications.html?download=414:social-media-shaping-the-future-of-agricultural-extension-and-advisory-services.

[30] G. Lucht, Farmers Follow Social Media to Track Crop Threats, 2018. Available from: https://www.agupdate.com/iowafarmertoday/news/crop/farmers-follow-social-media-to-track-crop-threats/article_25164bea-7a1b-11e8-bbb0-13943abfe8f4.html.

[31] Anonymous, New Whatsapp™ insect ID service launched by IPM Initiative.

CHAPTER 21

Applications of Computational Methods in Plant Pathology

KESHAVI NALLA • SESHU VARDHAN POTHABATHULA, PHD •
SHASHANK KUMAR, PHD, MSC, BSC

INTRODUCTION

The application of different computational methods in plant pathology has made disease detection and plant pathogenic studies easier. Analyzing the diseased conditions of plants early helps the farmers to take necessary precautions to control the spread of the disease. These computational methods involve different theoretical methods and data analytic methods to determine and analyze different systems. Computational methods like digital image processing, color space models, feature extraction, etc. indulge different mathematical models and algorithms in such a way to extract the features on a plant and their location [1]. The use of these different computational techniques makes accurate data processing and disease monitoring to be done with high accuracy and without delay. Applying these computational techniques to traditional practices increases the protection and early detection of the disease so that the spread of the disease can be controlled. Apart from disease detection, there are many applications of computational methods in agriculture. These applications mainly involve the control of the weed, estimation of usage of fertilizer, pest control, etc. This computer vision concept is nothing but developing the technology for analyzing and understanding the real-time information by using digital systems. This improves the visualization of different organisms and helps in analyzing them in a clear aspect and in a digital way. This helps the computers to attain higher computational intelligence and brings good interface with this real complex world. These techniques enhance the practical visualization of a problem and its solution in an algorithmic way by converting the data from analog to digital and vice versa. This analog-to-digital conversion of the data involves the sampling of the signal (i.e., data collected in the signal form). This signal containing information is sampled and the signal is allotted with 0 and 1 for the conversion purpose and then the signal is converted to digital. This information is analyzed by applying different algorithms. This kind of computational approach helps in identifying the disease, its effects, and its classification. It also helps in identifying the species of the organism that affects the plant.

Digital Signal Processing

Digital image processing is one of the improvised techniques that can be used to analyze the data of a system without destructing it [2]. It allows the involvement of different algorithms and apply them to the input data and is used as one of the solutions to the different complex problems without bisecting or destructing the material. Generally, this process requires the equipment consisting of a high-resolution digital camera, supercomputer, and software tools to perform digital image processing. Though the images captured by the devices are taken in two-dimensional, i.e., in x and y, this technique can also be applied to the multidimensional systems. Digital image processing is very advantageous as it provides the application of different algorithms to avoid problems like noise and signal distortion. As the digital image is a composition finite number of elements, every element of it has the specified location and value. The images are segmented into objects and background, and these data are processed using the algorithms. Infrared images can also be used in this technique but there should be no need for segmentation. This means that the infrared images have a fixed threshold value for segmentation in many systems [3]. The digital image processing involves the techniques of image acquisition, feature extraction, object detection and extraction, image preprocessing (Fig. 21.1). The image segmentation is the crucial stage in observing the diseased plant surface in a practical way. Though

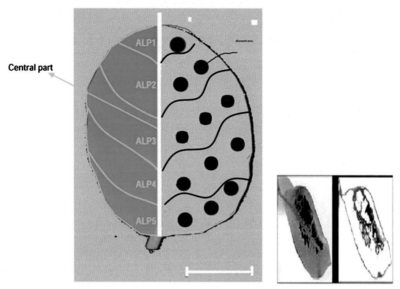

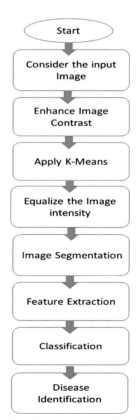

FIG. 21.1 Digital image processing in plant pathology [2].

there are many segmentation algorithms, this stage is based on analyzing the discontinuities and similarities in the data of the image [4]. We already know that most of the symptoms of the plant diseases can be seen on their surface tissues. At the stage of image segmentation, the similarities of the image are determined, so the diseased tissues of the plant can easily be identified. A region-based similarity algorithm can be used to identify and categorize the diseases when the image of the surface of the plants are analyzed. The color information and local basis of analysis basically depend upon the camera resolution which helps in detecting and categorizing the disease at the stage of image segmentation by applying different segmentation and other algorithms.

Image Segmentation

Image segmentation is nothing but dividing the elements of the image into different clusters or parts and analyzing each one (Figs. 21.2 and 21.3). This helps in analyzing it in a clear way. The segmentation is the key approach in many of the computational techniques. The key concept of this technique is to make the image easier to analyze and simplify the image. The image segmentation is a technique that results in establishing definition as a set of segments that cover the entire image. Each pixel in an image is assigned with a label which determines certain characteristics. Even every detail like lines and curves are identified more precisely by using this technique. Genetic algorithms are also

Start

Consider the input Image

Enhance Image Contrast

Apply K-Means

Equalize the Image intensity

Image Segmentation

Feature Extraction

Classification

Disease Identification

FIG. 21.2 Image Segmentation steps.

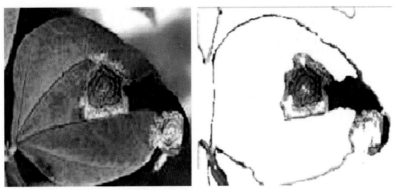

FIG. 21.3 Image Segmentation [5].

used in image segmentation. They are the search algo-rithms which process large number of variables at a sin-gle time. They are efficient and have large sampling cost. It gives a number of solutions so that we get different segmentation results easily. As genetic algorithms are efficient, they generate various results at the same time. The basic steps involve the generation of popula-tion of chromosomes, i.e., suitable solution for the problem: Observing the fitness of each chromosome generated and creating a new population by selecting the parent chromosomes according to their fitness. Ac-cording to the probability to crossover, a new offspring is formed by crossing the selected parents. Now mutate the offspring at each location in chromosome, and insert the new offspring in the new population. Replace this new generation, and run the algorithms and test whether the end statement is satisfied or not. If the state-ment is satisfied, stop the algorithms and return to the best solution in the present population, else continue the loop [5]. There are many other methods that are adopted at the phase of image segmentation. They are K-nearest-neighbor for class prediction and color space models.

Color Space Models

The color space model involves the organization of colors in a specific way and in the combination of a physical device which allows different reproducible rep-resentations of colors in both analog and digital format. This may contain specific color collection or can be structured mathematically [6]. Basically, a color model is an abstract mathematical model of colors describing the way of colors. Color space models are basic mathe-matical structures that are used in digital image process-ing for the stage of feature extraction and color processing. Color models generally explain how the colors are represented and specify the component of

color space as per the image extracted [7]. There are different color models, namely device-oriented color models, user-oriented color models, device-independent color models. Device-oriented color models are device-dependent models that are affected by the color, signal of the device, and the tools used for displaying. User-oriented models are related to the observer and the device regarding the color informa-tion. Device-independent color models are not related to the device, but the color representation will have a set of parameters without considering the parameters and configurations of the device. There are different co-lor models, Munsell, RGB, CMY(K), YIQ, YUV, YCbCr, HSI, HSV, HSL, CIE XYZ, CIEL*U*V*, CIEL*a*b. Actu-ally RGB, HIS, and CIEL*U*V are the basic models used in different techniques for plant disease detection. RGB is a basic color space model which is used in sensing and displaying in different electronic systems. This is a combination of red, blue, and green that produces different array of colors. The three colors are superim-posed to get different array of colors. Basically, the RGB image is an image represented by different color intensities brought up by combination of three different colors (Fig. 21.4). These do not require any color map in imaging an object. This color space model uses 8-bit monochrome standard and have 24 bit/pixel [5]. During segmentation, the RGB images undergo color mapping and clustering. Later on, the RGB color space models can be converted to different images namely gray scale and other color space models like CIEL*U*V, HSI, HSV, etc. CIEL*U*V color space generally follows uniformity, device independence, linearity, intuition. The storage of color is in UV components [8]. The CIE-L*U*V color space is considered as the best method for identification and segmentation of diseased plant leaves. The use of this color space model in a fuzzy-based approach gives better efficiency in identifying

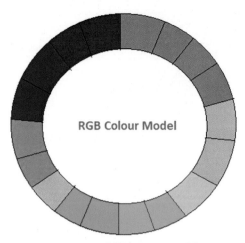

FIG. 21.4 RGB color model [5].

the diseases. The basic method in involving the application of color space model in disease identification has basic steps as explained. First of all, the images are collected using a digital camera as well as unnecessary information and RGB color space, and they are transformed from RGB to CIELuv and modified FCM is applied on the segment into clusters and the affected portion is extracted. This is a fine concept in developing better and efficient techniques in disease monitoring. Sometimes the HSV and HSI models are also used instead of CIEL*U*V. HSI is nothing but hue saturation intensity where the colors represent exact senses of our human eye (Fig. 21.5).

Feature to Feature Extraction

Feature to feature extraction is finding a subset that maximizes the ability of the users to classify different patterns on the leaves. This feature extraction can also

be done by applying different techniques such as machine learning, computation, deep learning, and artificial neural networks, etc. The basic methodology for feature extraction indulges image acquisition as primary step where the image is captured by the digital camera to extract the features in JPEG format, and the infected fruits are placed in a white background where the light source is placed at 45° to the fruit to eliminate reflection and to make the picture clear enough to extract the features. Later on, image preprocessing is done which involves the conversion of the input image from RGB to HSV color spaces and have histogram processing. After masking the green pixels, they are removed and it enters into the stage of image segmentation to obtain necessary segments. Later on, the features are computed using a color-co-occurrence methodology, and the texture statistics is evaluated to obtain the desired features [9].

Low-level feature extraction
Basically, the process of low-level feature extraction is just the same as the feature extraction of the leaf using image acquisition, image processing, and feature extraction and so on. But the features that are selected and are to be extracted might be quite different and optional. The low-level features involve minor details such as lines, dots that can be picked up, or SIFT which are abstract things like edges. So the low-level feature extraction involves the extraction of minor details like dots or small marks on the leaves such that it will be easy to find the minute diseased points on the leaf surfaces.

High-level feature extraction
The process will be the same for the high-level feature extraction just as the feature extraction. The high-level features require multilayering technique in deep neural networks where the edges, curves, and the whole object

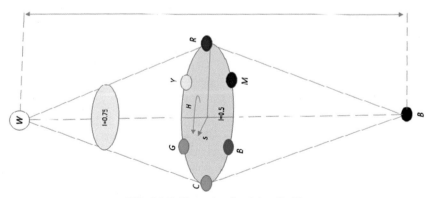

FIG. 21.5 Hue saturation intensity [6].

are detected in different layers such that the entire object will be detected and analyzed. The entire leaf or the surface of the plant is analyzed in a better way to get the accurate result regarding the diseased conditions of the plant.

Support Vector Machine

Support vector machine is one of the machine learning techniques. It is also called the support vector networks. They are the supervised learning algorithms that analyze the classification and other characteristics of an object (Fig. 21.6). They perform nonlinear classification which is also called kernel trick, and this technique has two different datasets: one is training and the other is train dataset [10]. It is the latest classifying technique where the algorithms are developed to identify or detect the disease. This technique is developed to implement better classification. First of all, the original image of the diseased plant or the plants are captured using the digital camera and the image undergoes the image preprocessing and it gives us black and the background is analyzed. The background of the image is taken into pixels, and the image is segmented to get the real classification of the hue part and the segmented part of the image so that disease detection is made where the diseased and the healthy portion of the plant is segregated by using this computational technique. There will be different algorithms adopted under this technique to detect different plant diseases. Support vector machine is a machine learning technique which is mainly adopted for the purpose of segregation and the classification of different objects.

K-Means

K-means clustering is an unsupervised technique dealing with the unlabeled data. This technique is a method of vector quantization that is originally

FIG. 21.6 Classification and Processed Image using Support Vector Machine.

expelled from signal processing. This is a key concept in data mining. K-means clustering is nothing but dividing the given data into groups and clusters and applying different algorithms to get the disease detected. The number of groups or clusters is organized in the unlabeled data. K-means clustering is the simplest way to analyze the data. In this technique, the image is captured through a digital camera, and it undergoes RGB image acquisition. The color transformation structure is created and the color value of RGB is converted to HSI. Then it undergoes k-means clustering where the features of the image denoted as $X = \{x1, x2, x3,... xn\}$ are taken as data and there are some centroids, and now the cluster is initialized as per the number of clusters taken and "c" centers are selected randomly. Euclidean distance is calculated for each point. The data points are assigned nearer to the centers and new centers are calculated and the process will be repeated until no centroid position changes or proceed the calculation on Euclidean distance. Then the infected cluster is detected. Then it undergoes masking, feature extraction, and finally by using neural networks the disease is identified [11].

Neural Networks

Neural networks are artificial models of interconnecting networks parallel to the network of neurons (Fig. 21.7). The nodes are assigned to represent the neuron in the actual human brain, and a neuron connects to an individual processing element which is known as the perceptron. The first model in artificial neural networks is given by Mc. Gulloch Pitts in 1943. There are many types of artificial neural networks that can be used in plant pathology, each one having their own importance. The basic types are single-layer perceptron (SLP) network, radial basis function (RBF) network, multilayer perceptron (MLP) network, probabilistic neural network (PNN), Kohonen's self-organizing map (SOM) network, and convolutional neural network. SLP is based on the activation function where this function transforms linear combination into a nonlinear function, and it is called as a simple discriminant. The extension of this method is linear discriminant where the input variables are directly transformed to the nonlinear combination. This SLP architecture provides simple and comprehensive idea to identify the disease when applied with hyperspectral data. MLP is used as SLP is not practical these days. This technique is designed for two layers, one having number of neurons that will be equal to the number of spectral bands processed and other consisting of two neurons that are varying from 5 to 25. Here four

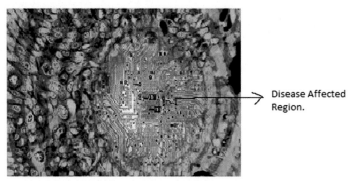

Disease Affected Region.

FIG. 21.7 Detection of infected cluster by neural networks [11].

optimal spectral bands are selected and tested, most efficient neurons are selected such that it provides 98% classification accuracy for healthy plants and 99% classification basis for diseased plants [12]. They are also called back-propagation algorithms [12]. RBF basically has input and output, and hidden layers and input layers control the entering of data. The hidden layer connects the output node linearly and analyzes the number of input variables that are being processed via input layer and output layer, and this technique is not more efficient than MLP, but it is advantageous as per its parametric modeling, nonlinear interpolation, and function approximation. Kohonen's SOM is an unsupervised technique which follows two layer feed-forward technique. Here the neurons are arranged in a grid form, either rectangular or hexagonal. This is the representation of high-dimensional data in low-dimensional view. This is first used in nematode detection [13]. PNN is a statistical approach in representing the data which is called a Bayesian classifier. This generally enhances the predictability of different objects or clusters likely with the relative and some other prior information into consideration. PNN is a nonparametric classifier obtaining the multivariate probability and estimating density. Nowadays, PNN is the most appropriate tool in the case of classifications. CNN acts as an interface for biological systems and neural systems. It acts as an application of deep learning in neural networks. It is inspired by the retina in the vision systems. It achieved an accuracy of 99.35% which is quite near to the MLP. It is used to detect 2 diseases over 14 crop species. This technique is considered as an effective tool for the identification and classification of the crop diseases using the hyperspectral images. Now it is used to detect the arctic vegetation using the most effective sensor data and following the approach of interfacing the hyperspectral data of the images [14].

Hyperspectral Imaging

Hyperspectral imaging is also one of the spectral imaging techniques that are used to process the information from the electromagnetic spectrum. This technique is used to analyze the spectrum for each pixel in every image of a pixel. The images are analyzed to detect plant diseases, identifying materials and finding objects. Technologies involved in hyperspectral data acquisition are spatial scanning, spectral scanning, nonscanning, spatiospectral scanning. The devices used for spatial scanning obtain slit spectra by projecting the strip of the image or scene onto a slit. The image is analyzed as per the lines that disperse the slit image by using a prism or a grating. This kind of line scan devices is used to scan and analyze different materials and surfaces which can be used for leaf disease detection. Spectral scanning is a 2D sensor output which represents a monochromatic and spatial line map of the image or the scene. The devices used for the spectral scanning are based on the optical bandpass filters. The image is scanned by exchanging different filters one after the other. Nonscanning is the 2D sensor output that consists both spatial and spectral data. This is a sort of snapshot system where the snapshot of the scene is analyzed. Many systems are designed for spectral imaging namely computed tomographic image spectrometry (CTIS), fiber reformatting imaging spectrometry (FRIS), integral field spectrometry (IRIS), filter stack spectral decomposition (FSSD), coded aperture snapshot spectral imaging (CASSI), etc. Spatiospectral scanning is a 2D sensor output which represents a rainbow colored and spatial map of the image. The system for this technique can be obtained by placing a camera behind the spatial scanning system using a combination of hyperspectral imaging where visible and near-infrared (VNIR) portions and short wavelength infrared (SWIR) and some machine learning techniques are used to identify the diseased plant. The features that are discriminatory are extracted using the full spectrum and a variety of vegetation indices and probabilistic top models and vegetation domain and are used to detect the tomato spotted wilt virus in capsicum [15].

Smart or Precision Control in Agriculture

Precision agriculture is also known as satellite crop management or site-specific crop management. This is used to develop a decision support system for complete farm management with the goal of optimizing the inputs and outputs of the different systems at regular intervals. This is a sort of phytogeomorphological approach. The practice of this precision control agriculture is GPS- and GNSS-based advent where the crop yield, moisture levels, pH levels, and many other variables regarding the field are considered. With the array of multiple sensors, multispectral imagery is also used to detect different specifications and diseased conditions in plants. Near-range and remote sensing techniques have demonstrated high potential in detecting different plant diseases using sophisticated technologies to detect different factors like epidemiological, environmental, etc. [16].

Deep Learning in Plant Pathology

Deep learning is the complex of different machine learning methods with the application of artificial networks. The image is captured, and the data are given for image annotation. The annotated data undergo augmentation, and these data are given for testing, training, and validation. These data are given for disease and pest detection and training parameters. This result undergoes performance verification and then finally the disease would be detected.

CONCLUSION

The application of computational techniques such as image segmentation, color space model, support vector machines, K-means, and neural networks in plant disease diagnosis will not only enhance disease diagnosis accuracy but lead to early detection of diseases or pests before they get to the stage where they cause damages to agricultural crops.

REFERENCES

[1] S.S. Chouhan, U.P. Singh, S. Jain, Applications of Computer Vision in Plant Pathology. Archives of Computational Methods in Engineering. Published in Springer. DOI:https://doi.org/10.1007/s11831-019-09324-0.

[2] J.A. Pandurang, S.S. Lomte, Digital image processing applications in agriculture: a survey, Int. J. Adv. Res. Comput. Sci. Softw. Eng. 5 (3) (March 2015). https://www.researchgate.net/publication/274418841. ISSN: 2277128X.

[3] J. Pan, Y. He. Recognition of plants by leaves digital image and neural networks. Published in 2008 International Conference on Computer Science and Software Engineering. DOI:10.1109/CSSE.2008.918.

[4] J.W. Funck, Y. Zhong, D.A. Butler, C.C. Brunner, J.B. Forrer, Image Segmentation algorithms applied to wood defect detection, Comput. Electron. Agric. 41 (2003) 157–179, https://doi.org/10.1016/S0168-1699(03)00049-8.

[5] K. Padmavathi, K. Thangadurai, Implementation of RGB and gray scale images in plant leaves diseases detection-comparative study in Indian, J. Sci. Technol. 9 (6) (2016), https://doi.org/10.17485/ijst/2016/v9i6/77739.

[6] R. Nayyer, B. Sharma, Use of color models in image processing, Int. J. Adv. Sci. Res. (2015), https://doi.org/10.7439/ijasr. ISSN:2395-3616(Online).

[7] N.A. Ibraheem, M.M. Hasan, R.Z. Khan, P.K. Mishra, Understanding color models: a review, ARPN J. Sci. Technol. (2011-2012). ISSN 2225-7217.

[8] P. Ganesan, G. Sanjiv, M.L. Leo, CIELuv color space for identification and segmentation of disease affected plant leaves using fuzzy based approach. Published in 2017 Third International Conference on Science Technology Engineering and Management (ICONSTEM).978-1-5090-4855-7/17.

[9] P. Priya, D.A. D'Souza, Study of feature extraction techniques for the detection of diseases of agricultural products. Published at National Conference on Advanced Innovation in Engineering and Technology 2015 in International Journal of Innovative Research in Electrical, Electronics, Instrumentation and Control Engineering. ISSN (Online) 2321 – 2004 ISSN (Print) 2321 – 5526.

[10] R. Kaur, S.S. Kang, An enhancement in classifier support vector machine to improve plant disease detection. Published in: 2015 IEEE 3rd International Conference on MOOCs, Innovation and Technology in Education (MITE). On 1–2 October 2015, INSPEC Accession Number: 15701262. DOI: 10.1109/MITE.2015.7375303.

[11] S.P. Patel, A.K. Dewangan, Automatic detection of plant leaf disease using K-means clustering and segmentation, Int. J. Sci. Res. Eng. Technol. 6 (7) (July 2017). ISSN: 2278-0882.

[12] S.T. Monteiro, Y. Kosugi, K. Uto, Watanabe E. Towards applying hyperspectral imagery as an intraoperative visual aid tool. In: Proceeding of the Fourth IASTED International Conference on Visualization, Imaging and Image Processing. Marbella, Spain, p. 483–488.

[13] G.W. Lawrence, R.A. Doshi, R.L. King, K.S. Lawrence, J. Caceres, Nematode management using remote sensing technology, self-organized maps and variable rate nematicide applications, in: World Cotton Research Conference. Texas USA, 2007.

[14] Z.L. Langford, J. Kumar, F.M. Hoffman. A convolutional neural network approach for mapping arctic vegetation using multi-sensor remote sensing fusion. In: International Conference on Data Mining Workshops IEEE. p. 322–331.

[15] P. Moghadam, D. Ward, E. Goan, S. Jayawardena, P. Sikka, E. Hernandez, Plant disease detection using hyperspectral imaging, in: International Conference on Digital Image Computing: Techniques and Applications (DICTA), 2017, https://doi.org/10.1109/DICTA.2017.8227476.

[16] A. Mahlein, E. Oerke, U.S. Heinz, W. Dehne, Recent advances in sensing plant diseases for precision crop protection, Eur. J. Plant Pathol. (2012), https://doi.org/10.1007/s10658-011-9878-z.

Index

A

Acetogenins, 38–39
Actinobacteria, 198
Aesthetic injury level (AET), 7
Aflatoxins, 2–3
African cassava mosaic virus, 136
Agricultural Advisory System, 236
Agricultural decision support system, 236
Agricultural pathogens
  classification
    bacteria, 135
    fungi, 135–136
    infectious disease–causing agent, 135
    nematodes, 136
    protozoa, 136
    viruses, 136
  detection methods, 135
    biosensors and quantum dots, 137
    colony-forming unit (CFU), 137
    enzyme-linked immunosorbent assay (ELISA). See Enzyme-linked immunosorbent assay (ELISA)
    nanotechnology-based technology, 137
    polymerase chain reaction (PCR). See Polymerase chain reaction (PCR)
  disease diagnosis, 135
    microscopy methods, 136
    molecular methods, 136
    spectroscopy methods, 136–137
  Food and Agriculture Organization (FAO), 135
  matrix-assisted laser desorption/ionization-time-of-flight mass spectrometry (MALDI-TOF), 139
  nanopore sequencing, 139
Agrobacterium tumefaciens, 121
Airborne infection, 19
Alarm pheromones, 83–84
Aldicarb, 29
Aldrin, 34f
Alkaloids, 38–39, 39f, 70–72
Allelochemicals, 82
Allelopathy
  aqueous/hydro alcoholic extracts, 112
  bialaphos, 111
  bioassay method, 112

Allelopathy (*Continued*)
  Cinch, 111
  compounds, 110–111
  definition, 110–111
  donor crop, 112
  mulches/crop residues, 112
  organic farming, 111
  phytotoxic compounds, 110
  plant communications, 111
  postemergence herbicides, 110
  properties, 110
  weed management, 112–113
Allicin, 49t–50t, 57–61
Allomones, 84
Aloe vera, 180–181
Amino acid synthesis inhibitors, 109
Anisomeles indica, 112
Antagonism, 4
Arabidopsis thaliana, 111
Araujo mosaic virus, 122–123
Arbuscular mycorrhiza fungi (AM fungi), 194
Artemisia annua, 111
Artificial insemination (AI), 171–172
Artificial intelligence (AI), 236
Aspergillus flavus, 2–3
Atropa belladonna, 38
Atropine, 38, 39f
Attract and kill technique, 81–82, 85
Augmentative biological control, 7
Avenacins, 37
Azadirachtin, 37

B

Bacillus subtilis, 197
Bacillus thuringiensis (Bt), 4
  cotton, 205
  Cry protein, 205
  insect/pest resistance, 205
  maize, 205
  potato, 206
  soybean, 205–206
  toxins, 205
Back-propagation algorithms, 247–248
Bacterial chitinases
  N-acetylglucosamine, 166
  gram-positive actinomycetes, 166
  in green biotechnology, 166–168, 167t
Beauveria bassiana, 44–45, 180
Beetles and storage butterflies, 1–2

Bialaphos, 111
Biocontrol, 3–4
  aesthetic injury level (AET), 7
  antagonism, 4–5
  antagonistic reactions, 5–6
  augmentative method, 7
  biofertilizer, 5
  biological control agent (BCA), 4–5
  biological factors, 4–5
  biological plant protection preparations, 7, 8t–9t
  classic biological method, 4–5
  conservation method, 7
  control action threshold (CAT), 7
  dissemination, 5
  economic injury level (EIL), 7
  environmental risk assessment (ERA), 9–12
  indirect mechanisms, 6–7
  macroorganisms, 9, 10t–11t
  microbial preparations, 7–9
  natural enemies, 9–12
  plant growth-promoting microorganisms (PGPRs), 5–6
  soil manipulation, 5
Biodynamic preparations, 45
Biofertilizers, 187
Biofungicides, 166, 167t
BioHerbicide approach, 119–120
  agricultural production, 107–108
  agro ecosystem, 107
  allelopathy, 113. See also Allelopathy
  biodegradation, 107
  chemical structure, 107
  crop production, 107
  environmental risks, 110
  weeds control methods. See Weeds control methods
Bioinsecticides, 168
Biological control agent (BCA), 4–5, 108–109
Biologically active ingredients, 46
Biological storage methods
  bacteriocins, 227
  bacteriophages, 227
Biological warfare, 3–4
Biological wars, 3–4
Biopesticides, 29. See also Natural biopesticides
  agricultural economy, 123–124
  agricultural production, 209
  agrochemical formulation, 210

Note: Page numbers followed by "t" indicate tables, "f" indicate figures.

Biopesticides (*Continued*)
  antifungal activity, 210
  biochemical pesticides, 120
  biodegradability, 209
  botanicals efficacy, 209–210
  characteristics and benefits
    biodynamic preparations, 45
    biologically active ingredients, 46
    biological protection organisms,
      43–44
    "cocktail effect", 43–44
    entomopathogenic mushrooms,
      44–45
    fungicides, 45–46
    gibberellins, 46
    herbicides, 46
    insecticides, 46
    integrated pest management, 44
    maximum residue limits (MRLs),
      46–47
    microbiological pesticides, 45
    natural insecticides, 45
    pheromones, 47
    plant protection products (PPP),
      43–44
    semichemicals, 43–44
    yellow zucchini mosaic virus, 46
  classification, 120
  crop loss, 209
  definition, 120
  ecoproducts, 43
  emulsifiable concentrate (EC),
    211–212, 211f, 212t
  EU-based regulations, 123–124
  formulation technology, 210
    controlled release formulation
      (CRF) technology. *See*
      Controlled release formulation
      (CRF) technology
    microemulsions (MEs), 212, 214f,
      214t
    nanoemulsions (NEs), 212–214,
      214f, 215t
    neem-based multifunctional
      formulation., 215–217, 217f
    O/W emulsions, 212, 213f, 213t
    types, 212
  fungal biopesticides, 122t
    Beauveria bassiana controls,
      121–122
    bioherbicides, 122, 122t
    entomopathogenic fungi,
      121–122
  health issues and legal requirements,
    47–48
  microbial pesticides
    bacterial bioherbicides, 121, 121t
    bacterial biopesticides, 121, 121t
    Bacillus. thuringiensis, 120–121
  natural bioactive compounds,
    49t–50t
  nematodes, 123, 123t
  pest management, 210
  plant-incorporated-protectants, 123

Biopesticides (*Continued*)
  root-knot nematode, 210
  "substitute candidates", 43
  trends and market demand, 48–49
  viral, 122, 122t
    bioherbicides, 122–123, 123t
  wettable powder (WP), 210–211,
    211f, 211t
Biorational pesticides, 36
Biosafety Regulation, 207
Biosensors, 137
Blue biotechnology, 164
Botrytis cinerea, 61
Brome mosaic virus Potato virus, 136
BTG 504, 39f
BTG 505, 39f
Bursaphelenchus xylophilus, 69–70

C
Caffeine, 38, 39f
Calendula micrantha, 66–67
Camouflage, mating disruption, 82
Campylobacter detection, 143–144
Candida oleophila, 46
Carbamates, 35
Cardamom oil
  antifeedant activity, 69
  ovipositional deterrence property, 69
Cauli-flower mosaic virus, 136
Ceiba pentandra essential oil, 67
Cell membrane disruptors, 109
Ceratitis capitata, 66–67
Ceratocystis ulmi, 22
Chaetomium globosum YY-11, 128
Chemical pesticides, 29
Chemical storage methods, 226–227
Chitin, 169
  hydrolysis, 166
  sources and structure, 165
Chitinases
  bacterial chitinases
    N-acetylglucosamine, 166
    Gram-positive actinomycetes, 166
    in green biotechnology, 166–168,
      167t
  endochitinases, 165–166
  glycoside hydrolases (GH), 165–166
  β-glycosidic bonds, 165–166
  in red biotechnology, 168
  in white biotechnology, 168
Chitinolytic bacteria, 198
Chitosan, 49t–50t, 61
Chlordane, 34f
Cinch, 111
1,8-Cineole, 66–67
Cinnamaldehyde, 66–67
Cisgenesis, 203
Citrus psorosis virus, 138
Citrus tristeza virus, 138
Classic biological method, 4–5
Club root disease, 18
Cluster analysis, 236, 238f
"Cocktail effect", 43–44
Colletotrichum gloeosporioides, 122

Colonization, 4
Colony-forming unit (CFU), 137
Color-co-occurrence methodology,
  246
Color space models, 245–246, 246f
Competition, 4
Competitive attraction, mating
  disruption, 82
Computational methods, plant
  pathology
  analog-to-digital conversion, 243
  color space models, 245–246, 246f
  data processing, 243
  deep learning, 249
  digital signal processing, 243–244,
    244f
  feature to feature extraction
    color-co-occurrence methodology,
      246
    high-level feature extraction,
      246–247
    image acquisition, 246
    low-level feature extraction, 246
    pattern classification, 246
  hyperspectral imaging, 248–249
  image segmentation, 244–245,
    244f–245f
  K-means clustering, 247
  neural networks, 247–248, 248f
  smart/precision control, 249
  support vector machine, 247, 247f
Coniothyrium minitans, 45–46
Control action threshold (CAT), 7
Controlled release formulation (CRF)
  technology, 215t
  advantages, 213
  essential oil (EO), 213
  tablet formulations
    active ingredients, 214
    advantages, 214, 217t
    shape and size, 214, 216f
Coptotermes formosanus, 66–67
Cryphonectria parasitica, 130
Cryptosporidium parvum, 146
Cry toxins, 204
Cucumber mosaic virus, 136
"Cut-off criteria", 43
Cyanogenic glycosides, 70
Cymbopogon citratus, 37, 92
Cymbopogon winterianus, 66–67
Cypermethrin, 36f

D
Decision support systems (DSSs),
  235–236
Deep learning, 249
Defense proteins, 39–40
Dermatophagoides pteronyssinus, 69
Desensitization, mating disruption,
  82
Dianthus caryophyllus, 21
Diazinon, 35f
Dichlorodiphenyl-trichloroethane
  (DDT), 30, 34f

Dichlorvos, 35f
Dieldrin, 34f
Digital polymerase chain reaction (dPCR)
    absolute quantification, 152
    cycle threshold (CT), 152
    digital droplet PCR (ddPCR), 152−153
    Poisson's law, 152
Digital signal processing, 243−244, 244f
Direct tissue blot immunoassay (DTBIA), 138
Disease cycle, 21
Disease tolerance, 24
Diterpenoids, 37
DNA microarray, 155
DNA polymerase enzyme synthesis, 144−145
Drechslera oryzae, 92
Drosophila melanogaster, 66−67

E

Economic injury level (EIL), 3, 7
EcoPCOR, 74−75
Ecoproducts, 43
Electron microscopy, viral pathogen detection, 139
Emulsifiable concentrate (EC), 211−212, 211f, 212t, 213f, 213t
Endemic disease, 19
Endophytic microorganisms
    antibiosis, 127
    bacteria, 129
        host plant, 130
    benefits, 131
    bioactive compounds, 127
    biocontrol agent, 128
    Chaetomium globosum YY-11, 128
    chemical fertilizers, 127
    colonies and biofilms, 127
    diseases/pest control
        Cryphonectria parasitica, 130
        genetic and environmental modifications, 130
        in-vitro called kakadumycins, 130
        Peppermint growth, 129−130
        plant resistance response, 129
        terpene production, 129−130
        volatile compounds, 130
    epiphytes, 127−128
    fungal and bacteria genera, 128
    host-plant resistance, 127
    insect-pests control, 128
    integrated pest management, 131
    molecular techniques, 131
    nitrogen-fixing bacteria, 127−128
    parasites, 129
    pathogenesis-related proteins (PR-protein), 130−131
    pathogens, 129
    phytohormones, 127−128
    phytopathogens, 127−128
    plant defense mechanisms, 130−131

Endophytic microorganisms (*Continued*)
    recombinant endophytes, 128
    soil contamination, 132
Endosulfan, 34f
Entomopathogenic fungi, 121−122
Entomopathogenic mushrooms, 44−45
Entomopathogenic nematodes, 123, 123t
Environmental risk assessment (ERA), 9−12, 207
Enzyme-linked immunosorbent assay (ELISA)
    direct tissue blot immunoassay (DTBIA), 138
    electron microscopy, viral pathogen detection, 139
    flow cytometry (FCM), 138−139
    fluorescence in-situ hybridization (FISH), 138
    gas chromatography-mass spectrometry (GC-MS), 139
    immunofluorescence (IF), 138
    lateral flow microarrays (LFM), 138
    pathogen strains, 138
Epidemic or epiphytotic disease, 19
Erwinia amylovora, 25−26
Essential oils (EOs), 213
    antifungal and antibacterial mechanism, 74
    anti-property, 69
    bioactive compounds, 65−66
    flavonoids and tannins, 72
    fumigant property, 67−69
    fungicide and bactericide activity, 72−74, 73t−74t
    insecticidal property, 66−67
    miticidal property, 69
    nematicidal activity. *See* Nematicidal plant compounds
    ovipositional deterrence propert, 69
    pesticidal properties, 74−75
    phenolics, 72
    repellence property, 67, 68t
Ethidium monoazide (EMA), 150, 152
Etiological symptoms, 25−26

F

Facultative parasites, 20
Facultative saprophytes, 20
Fe (III) complexing compounds, 5
Fenevelerate, 36f
Final host, 21
Flavonoids, 37
Flow cytometry (FCM), 138−139
Fluorescence imaging, 136
Fluorescence in-situ hybridization (FISH), 136, 138, 154−155
Fluorescent detection systems, 147
Fluvalinate, 36f
Foliage diseases, 19

Food and Agriculture Organization (FAO), 135
Fruit diseases, 19
Fruits and vegetables
    biological methods
        bacteriocins, 227
        bacteriophages, 227
    chemical methods, 226−227
    nonthermal techniques
        cold plasma (CP), 225−226
        drying time, 224
        high-pressure processing (HPP), 226
        irradiation, 225
        ozone, 226
        pressurized inert gas, 226
        pulse light (PL), 225
        ultraviolet (UV) radiation, 224−225
    thermal preservation methods
        aseptic packaging, 223−224
        chilling, 222
        drying, 222, 224t
        freezing, 222
        hurdle technology, 224
        retorting, 222−223
Fumigants, 30
Fungal biopesticides, 122t
    Beauveria bassiana controls, 121−122
    bioherbicides, 122, 122t
    entomopathogenic fungi, 121−122
Fungicides, 45−46
Furanocoumarin, 37−38, 38f

G

Gas chromatography-mass spectrometry (GC-MS), 139
Generally Recognized as Safe (GRAS) sterilizing agent, 226
Genetically modified organisms (GMOs)
    artificial laboratory techniques, 203
    Bacillus thuringiensis (Bt), 204. *See also Bacillus thuringiensis* (Bt)
    bioinsecticide, 204−205
    Biosafety Regulation, 207
    Cisgenesis, 203
    contamination, 206
    crop production, 204, 204f
    Cry toxins, 204
    DNA ligase, 203
    ecosystem impacts, 206
    Environmental risk assessment (ERA), 207
    horizontal gene transfer (HGT), 206
    international agreements, 207
    plant gene alteration, 203−204
    risks, 206
    selective breeding, 203
    transgenesis, 203
Genetic struggle and resistance, 4
"Genomic reproduction" technologies, 169
Gibberellins, 46

Ginkgo, 37
Glucosinolates (GLSs), 70
Glycoside hydrolases (GH), 165
Glyphosate, 35f
Grains, storage methods
    aeration systems, 228
    bulk storage, 228
    demand of, 227
    integrated pest management
        strategies, 227
    modified atmosphere (MA)
        technology, 228–229
    refrigerated air, 228
    storage in bag, 228
    storage in silo, 228
    thermal disinfestation, 227–228
    underground storage, 228
Grain weevil, 2
Graphium ulmi, 22
Grayanoid diterpenes, 37
Green biotechnology, 164
    biofungicides, 166, 167t
    bioinsecticides, 168
    chitin hydrolysis, 166

H

Haematobia irritans, 129
Half-parasites, 20–21
"Heads" tapeworms, 21
Helianthus annuus, 183–184
Heptachlor, 34f
Herbicide-resistant crops, 109
Herbicides, 46
Horizontal gene transfer (HGT), 206
Horizontal resistance, 23
Hosts, 21
Houttuynia cordata, 112
Hyperspectral imaging, 248–249
Hyptis spicigera, 180
Hyptis suaveolens, 180

I

Image segmentation, 244–245,
    244f–245f
Immunofluorescence (IF), 138
Inclusion viruses (IV), 122
Induced mutation-assisted breeding
    (IMAB), 171–172
Infection, 21
Information and communication
    technology (ICT)
    Agricultural Advisory System, 236
    agricultural decision support system,
        236
    artificial intelligence (AI), 236
    cluster analysis, 236, 238f
    crop-protection chemicals, 235
    decision support systems (DSSs),
        235–236
    drone, 239–240
    mobile apps
        "Ag-Apps" technology, 238–239
        Agri Sync, 239
        Climate Basic, 239

Information and communication tech-
    nology (ICT) (*Continued*)
    Field Check App, 239
    Kisan Suvidha, 239
    Plantix, 239
    Pusa Krishi, 239
    National Research Center for
        Integrated Pest Management
        (NCIPM), 235
    pest and disease advisory services,
        238
    pesticide, 235–236
    public and private extension system,
        237–238
    rank position, 236, 237t
    social medias, 240
    statistics, 236, 237t
    web-based models, 235
    website and portal, 239
    wireless sensor network (WSN), 240
Infrared (IR) spectroscopy, 136
Initiative of the International
    Organization of Biological Control
    (IOBC), 7
Inorganic fertilizer, 193
Inorganic pesticides, 33
Insecticides, 29, 46
Insect parasitoid kairomones, 82–83
Insect pheromones, 83
Integrated pest management (IPM), 4,
    55–56, 119
Integrated weed management (IWM),
    109–110
Intellectual property rights, 171–172
Intermediate host, 21
Ipomoea cairica essential oil, 67–69
Isobutylamides, 37
Isothermal-based amplification
    methods, 153

K

Kairomones, 82–84
K-means clustering, 247

L

Laboratory-based biotechnology,
    171–172
Laminaria digitata, 45–46
Lateral flow microarrays (LFM), 138
Laurus nobilis, 67
Lectin-containing extracts
    agglutination, 94, 97t
    biological activity
        agrimony, 96, 98f
        biotest, 95
        common yarrow, 97, 100f
        dwarf everlast, 96, 98f
        inhibitory effect, 96
        marigold, 96, 99f
        sea buckthorn, 96, 99f
        St. John's wort, 97f
    fungal spore germination, 94
    plant extracts, 96t
    research methodology, 94–95

Lectin-containing extracts (*Continued*)
    sources, 95
    Ustilago nuda teliospores
        germination
        common sea buckthorn, 102, 102f
        common yarrow, 100, 102
        dwarf everlast, 100–102, 101f
        marigold, 102, 102f
        St. John's wort, 100, 101f
        wheat extract, 94
        wheat germ agglutinin, 94
Leptinotarsa decemlineata larvae,
    44–45
Leveillula taurica, 92
Lignins, 38
Limonoids, 37, 70
Lindane, 34f
Lipid synthesis inhibitors, 109
Local infection, 22
Loop-mediated isothermal
    amplification (LAMP), 153–154

M

Macroscopic symptoms, 25–26
Malathion, 35f
Mandatory parasites, 20
Marker systems, 26
Mark-recapture approach, 86–87
Mass trapping, 82, 86–87
Mating disruption, 82, 85–86
Matricaria chamomilla, 180
Matrix-assisted laser desorption/
    ionization-time-of-flight mass
    spectrometry (MALDI-TOF), 139
Medicago polymorpha, 112–113
Melaleuca leucadendron essential oil, 67
Melia volkensii, 37
Microbial control
    antagonism, 120
    biological agents, 119
    biopesticides
        biochemical pesticides, 120
        classification, 120
        definition, 120
        fungal biopesticides, 121–122,
            122t
        microbial pesticides, 120–121,
            121t
        nematodes biopesticides, 123, 123t
        viral biopesticides, 122, 122t–123t
    chemical pesticides, 124
    crop production, 119
    parasitism, 120
    pests, 119
    population control, 119
    predation, 120
    resources limitations, 120
    weeds, 119–120
Microbiological insecticides, 4
Microemulsions (MEs), 212, 214f,
    214t
Microscopy methods, 136
Modified atmosphere (MA)
    technology, 228–229

Mold fungi, 2—3
Molecular methods, 136
Monocyclic pathogens, 22
Monophagous pathogen, 21
Multiple ovulation/embryo transfer
  (MOET), 171—172
Mycotoxins, 2—3

N

Nanoemulsions (NEs), 212—214,
  214f, 215t
Nanopore sequencing, 139
Nanotechnology-based technology,
  137
Napthoquinone, 39, 39f
Naringin, 49t—50t, 59—60
National Research Center for
  Integrated Pest Management
  (NCIPM), 235
Natural bioactive compounds,
  49t—50t
Natural biopesticides, 182t—183t
  adjuvants, 180
  agricultural pests, 179
  aqueous extracts, 179
  medicinal plants, 179
  plant extract
    antibacteria, 180
    antifungal, 180—181
    disinfectant, 180
    herbicidal properties, 181—184
  powders, 179
  soil insect and termite soil pathogens,
    179
Natural compounds
  antibacterial and antifungal activity
    allicin, 57—61
    animal origin compound, 61
    naringin, 59—60
    terpenes, 60
    thymol, 60—61
  pyrethrin, *Chrysanthemum
    cinerariaefolium*
    advantages and disadvantages, 57
    chemical composition, 56—57
    industrial production, 56—57
    mode of action, 56
Natural insecticides, 45
Natural organic pesticides, 33
Natural plant compounds
  antifungal components, 91—93
  antimicrobial properties, 93
  phytosanitary properties, 93
  resistance and susceptibility, 93—94
Near-infrared resonance (NIR), 136
Neem-based multifunctional
  formulation., 215—217, 217f
Neem oil, 46, 74—75
Nematicidal plant compounds, 123,
  123t
  alkaloids, 70—72
  Bursaphelenchus xylophilus, 69—70
  cyanogenic glycosides, 70
  glucosinolates (GLSs), 70

Nematicidal plant compounds
(*Continued*)
  limonoid triterpenes, 70
  Meloidogyne spp, 69—70
  parasitic natures, 69—70
  phytochemicals, 69—70
  polythienyls, 70
  saponins, 70
Neural networks, 247—248, 248f
Nitrogen-fixing bacteria, 127—128
Nitrogen metabolism inhibitors, 109
Nonchemical methods, 44
Nonculturable microorganisms,
  143—144
Noninclusion viruses (NIV), 122
Nonparasitic disease, 19
Nonsystemic contact pesticides, 30
Nonthermal storage techniques
  cold plasma (CP), 225—226
  drying time, 224
  high-pressure processing (HPP), 226
  irradiation, 225
  ozone, 226
  pressurized inert gas, 226
  pulse light (PL), 225
  ultraviolet (UV) radiation, 224—225
Nuclear magnetic resonance (NMR),
  136—137
Nucleic acid—based methods,
  foodborne pathogens
  amplification methods
    diagnostic tools, 145
    DNA extraction procedures, 145
    DNA polymerase enzyme
      synthesis, 144—145
    DNA/RNA sequences, 144
    isothermal-based amplification
      methods, 153
    loop-mediated isothermal
      amplification (LAMP), 153—154
    nested PCR, 145—146
    nucleic acid sequence—based
      amplification (NASBA), 154
    PCR inhibitors, 145
    real-time PCR. *See* Quantitative
      PCR (qPCR) technique
    temperature control, 144—145
    thermal cycling amplification, 144
    Thermus aquaticus (Taq)
      polymerase enzyme, 144—145
    in vitro amplification and
      hybridization, 144
  Campylobacter detection, 143—144
  cultural methods, 143—144
  "gold standard" methods, 155
  hybridization methods
    DNA microarray, 155
    double-stranded DNA probes, 154
    fluorescent in situ hybridization
      (FISH) method, 154—155
    radioactive/nonradioactive marker,
      154
    rRNA molecules, 154
  incidence, 143

Nucleic acid—based methods, foodborne pathogens (*Continued*)
  microbial food safety, 143—144
  nonculturable microorganisms,
    143—144
  US Department of Agriculture, 143
  World Health Organization (WHO)
    report, 143
Nucleic acid hybridization, 136
Nucleic acid sequence—based
  amplification (NASBA), 154

O

Obligatory parasites, 20
Occasional parasites, 20
Odor-baited traps, 86—87
Omni phages, 21
Optional parasites, 20
Organic farming, 111
Organic fertilizer, 188
Organochlorine pesticides, 34, 34f
Organophosphorus, 34—35, 35f
Ostericum grosseserratum, 66—67
Oxyproline-rich glycoproteins (ORG),
  94

P

Pandemic disease, 19
Para-pesticides, 30
Parasitic affinity, 20
Parasitic disease, 19
Parasitism, 120
Parathion, 35f
Parthenium hysterophorus, 112
Pathogenesis-related proteins (PR-
  protein), 130—131
Pathogenic agents
  Clostridium botulinum, 19
  facultative parasites, 20
  infectious agents, 20
  noninfectious agents, 19—20
  obligatory parasites, 20
  occasional parasites, 20
  parasitism and pathogenicity
    aggressiveness, 20
    facultative parasites and
      saprophytes, 20
    half-parasites, 21
    hosts, 21
    infection, 21—22
    mandatory parasites, 20
    nonpathogenic parasites, 20
    optional parasites, 20
    parasite, 20
    parasitic affinity, 20
    parasitic specialization, 21
    pathogen, 20
    pathogen life cycle and disease
      cycle, 21
    symbiotes, 20
  pathogen spread
    air currents, 22
    animals, 22
    contamination, 22

Pathogenic agents (*Continued*)
  flowing water and raindrops, 22
  humans, 22
  monocyclic and polycyclic
    pathogens, 22
  polycyclic pathogens, 23
  seeds, 22
  vegetative organ, 22
PCR. *See* Polymerase chain reaction
  (PCR)
Penicillium verrucosum, 2–3
Peptaibol, 5
Permethrin, 36f
Pesticides. *See also* Biopesticides
  alcohols and oils, 29
  biological, 29
  chemical, 29
  classification
    biorational pesticides, 36
    function, 33t
    inorganic pesticides, 33
    natural organic pesticides, 33
    origin, 33t
    plant origin, 36–40
    synthetic pesticides, 33–35
    target pest species, 33t
  commercialized agriculture, 30–31
  definition, 65
  demerits
    continuous usage, 31
    environmental pollution, 31–32
    erroneous dosage, 32
    health challenges and food safety
      issues, 31
    inadequate safety measures, 32
    low government intervention, 32
    nontarget effect, 31
    poor extension services, 32
    soil fertility attenuation, 32
    soil impoverishment, 32
    water bodies pollution, 32
  history, 29–30
  insecticides, 29
  mechanisms, 29
  nonsystemic contact pesticides, 30
  systemic contact pesticides, 30
Pests
  bacterial interaction, 4
  beetles and storage butterflies, 1–2
  biocontrol. *See* Biocontrol
  biological conservation strategy, 3–4
  biological warfare, 3–4
  characteristics, 1
  chemical fight, 3
  crops, 2
  cultivation methods, 3
  definition, 1
  integrated pest management (IPM), 4
  losses caused, 1
  Prostephanus truncates, 2
  Sitophilus Zea mays, 2
  storage pests
    bacteria and fungi, 2–3
    grain weevil, 2

Pests (*Continued*)
  mites, 2
  rodents, 3
  types, 1
Phenolics, 37–38
Pheromones, 45, 47, 81–82
  alarm pheromones, 83–84
  insect pheromones, 83
  sex pheromone, 83
  synthesis, 83
Phoma exigua, 4
Photosynthesis inhibitors, 109
Physocnemum brevilineum, 128
Phytoalexins, 37
Phytoecdysones, 37
Phytohormones, 127–128
Phytophthora fungus, 45–46
Pigment inhibitors, 109
Plant biotechnology, food security
  breeding tools, 163–164
  business support, 172–173
  climate change, 164–165
  crop intensification, 173–174
  environmental protection, 163–164
  food supply, 171
  genetically modified plants (GMOs),
    163–164
  genetic applications, 170–171
  genetic engineering
    artificial insemination (AI),
      171–172
    endoreduplication, 172
    induced mutation-assisted
      breeding (IMAB), 171–172
    intellectual property rights,
      171–172
    multiple ovulation/embryo
      transfer (MOET), 171–172
  history of
    classical biotechnology, 164
    degree of, 164
    division, 164
    modern biotechnology, 164
  Mendelian genetics, 164–165
  microorganisms
    chitinases. *See* Chitinases
    chitin, sources and structure, 165
    enzymatic hydrolysis, 165
    renewable biopolymers, 165
  molecular breeding approach,
    163–164
  preservation
    applications, 170
    bioactive compounds, 169–170
    plant productivity performance,
      170
  sustainable food production, 169
Plant defense mechanisms, 130–131
Plant disease
  affected plant organ, 18–19
  classification scheme, 18t
  diagnosis
    automatic disease detection, 26
    disease symptoms, 25–26

Plant disease (*Continued*)
    marker systems, 26
    pathogen threat assessment, 26
  epidemics
    crop type, 25
    genetic resistance, 24–25
    genetic uniformity, 25
    new pathogens, 24
    pathogen's ecology, 24
    pathogen spreading ways, 25
    soil type and preparation,
      25
    virulent pathogen, 25
  external environment, 17
  horizontal resistance, 23
  internal environment, 17
  localized disease, 19
  manifestations, 17
  natural perpetuation and infection
    mode, 19
  nonparasitic disease, 19
  occurrence and distribution, 19
  parasitic disease, 19
  pathogenic agents. *See* Pathogenic
    agents
  physiological imbalance, 17
  quality losses, 17
  quantitative losses, 17
  symptoms, 17–18
  systemic disease, 19
  tolerance, 24
  vertical resistance, 23–24
Plant extracts
  antibacteria, 180
  antifungal, 180–181
  disinfectant, 180
  herbicidal properties, 181–184
Plant growth-promoting
  microorganisms (PGPRs), 5–6
Plant-incorporated-protectants, 123
Plant protection process, 128
Plant protection products (PPP),
  43–44
Plum pox virus, 136
Plutella xylostella, 129
Podosphaera pannosa, 194
Poisson's law, 152
Polycyclic pathogens, 22–23
Polymerase chain reaction (PCR), 136
  , 144–145
  digital polymerase chain reaction
    (dPCR)
    absolute quantification, 152
    cycle threshold (CT), 152
    digital droplet PCR (ddPCR),
      152–153
    Poisson's law, 152
  multiplex nested polymerase chain
    reaction, 137
  multiplex polymerase chain reaction,
    137
  nested PCR, 145–146
  nested polymerase chain reaction,
    137

Polymerase chain reaction (PCR) (*Continued*)
  quantitative PCR (qPCR) technique. *See* Quantitative PCR (qPCR) technique
  real-time polymerase chain reaction, 137–138
  target sequences, 137
Polyphagia, 21
Polythienyls, 70
Pongamia pinnata essential oil, 67
Pore forming antibiotics, 5
Potato cyst nematodes, 136
Potato tubers, 92
Potato virus Y, 136
Potentilla erecta, 92
Predation, 120
Primary symptom, 25–26
Probabilistic neural network (PNN), 247–248
Propidium monoazide (PMA), 150
Prostephanus truncates, 2
Pseudomonas, 121
Push–pull strategy, 87
Pyrethroids, 35, 36f

Q

Quantitative PCR (qPCR) technique, 151t
  amplification curve, 146, 146f
  complementary DNA (cDNA), 150
  cultural-based methods, 150
  definition, 146
  DNA isolation, 150–152
  DNA sequence, 146–147
  ethidium monoazide (EMA), 150, 152
  fluorescent detection systems, 147
  FRET probes, 149
  molecular beacons, 149
  post-PCR analysis, 146
  propidium monoazide (PMA), 150
  reverse-transcriptase (RT)-qPCR, 150
  scorpion primers, 149–150
  standard curve method, 146–147, 147f
  SYBR Green I fluorescence mechanism, 147, 148f
  TaqMan probe, 148, 149f
  target specific probes, 148
  threshold cycle, 146
Quantum dots, 137

R

Red biotechnology, 164, 168
Repellents, 30
  semiochemicals, 82, 86
Reverse-transcriptase (RT)-qPCR, 150
Rhizosphere microorganisms, 5
Rocaglamides, 37
Rodents, 3
Root-colonizing microorganisms, 129
Root-knot nematode, 210

Rosemary oil based formulation, 74–75
Rotenone, 38–39

S

Salmonella typhimurium, 146
Salvia officinalis, 92
Saponins, 37, 70
Secondary symptom, 25–26
Seedborne infection, 19
Seedling diseases, 19
Seedling growth inhibitors, 109
Selected ion flow tube mass spectroscopy (SIFT-MS), 136
Selective breeding, 203
Self-destructive method, 4
Semiochemicals
  attract and kill technique, 81–82, 85
  biological control, 87
  characterization, 81
  insect parasitoid kairomones, 82–83
  isolation, 82
  mass trapping, 82, 86–87
  mating disruption, 82, 85–86
  pest control strategies, 81
  pheromones, 81–82
    alarm pheromones, 83–84
    insect pheromones, 83
    sex pheromone, 83
    synthesis, 83
  pros and cons, 84–85
  push–pull strategy, 87
  repellency, 82, 86
Sex pheromones, 84
Short-wave infrared wavelength (SWIR), 136
Sitophilus Zea mays, 2
Soilborne infection, 19
Solanum vivarium, 122–123
Specialized pheromone and lure application technology (SPLAT), 85
Spectroscopy methods, 136–137
Sporadic disease, 19
Squamocin, 38–39, 39f
Stem diseases, 19
Stomach poisoning pesticide, 30
Storage and preservation techniques, 223f
  biological pathogens, 221–222
  chemical method, 221
  fruits and vegetables. See Fruits and vegetables
  grains
    aeration systems, 228
    bulk storage, 228
    demand of, 227
    integrated pest management strategies, 227
    modified atmosphere (MA) technology, 228–229
    refrigerated air, 228
    storage in bag, 228
    storage in silo, 228
    thermal disinfestation, 227–228

Storage and preservation techniques (*Continued*)
    underground storage, 228
  microbial spoilage, 221
  pest contamination, 221
"Substitute candidates", 43
Support vector machine, 247, 247f
SYBR Green I fluorescence mechanism, 147, 148f
Symbiosis, 24
Symbiotes, 20
Synthetic pesticides
  carbamates, 35
  organochlorine pesticides, 34, 34f
  organophosphorus, 34–35
  pyrethroids, 35
Systemic contact pesticides, 30
Systemic infection, 22

T

Tablet formulations
  active ingredients, 214
  advantages, 214, 217t
  shape and size, 214, 216f
Taiwania cryptomerioides heartwood, 69
Tannins, 38
TaqMan probe, 148, 149f
Terpenes, 49t–50t, 60
Terpenoids, 36–37
Tetranychus urticae, 198
Thermal cycling amplification, 144
Thermal preservation methods
  aseptic packaging, 223–224
  chilling, 222
  drying, 222, 224t
  freezing, 222
  hurdle technology, 224
  retorting, 222–223
Thermus aquaticus (Taq) polymerase enzyme, 144–145
Thymol, 37, 49t–50t, 60–61, 67–69
Thymus vulgaris, 180
Tobacco mosaic virus, 136
Tomato spotted wilt virus, 136
Tomato yellow leaf curl virus, 136
Toxaphene, 34f
Transgenesis, 203
Trilactone terpenes, 37
Triterpenoids, 37

U

US Department of Agriculture, 143
US Environmental Protection Agency (EPA), 120

V

Vegetative organ, 22
Vermiwash and vermicompost extract
  biochemical degradation, organic matter, 188
  biocontrol agents, 193
  biofertilizers, 187
  crops harvesting process, 187

Vermiwash and vermicompost extract (*Continued*)
cytokinin and auxins, 188
decomposition process, 188
definition, 187
disease control
Actinobacteria, 198
Arbuscular mycorrhiza fungi (AM fungi), 194
Bacillus subtilis, 197
chitinolytic bacteria, 198
Podosphaera pannosa, 194
root-borne plant pathogens, 194
vermicompost liquid derivatives, 193—194
in vivo effect, 194
volatile organic compounds (VOCs), 197
earthworm *vs.* earthworm's gut microorganisms, 188, 188f
environmental factors, 188—189
Fusarium, 188—189, 197
maturation phase, 188
nonthermophilic conditions, 187
organic fertilizer, 188
organic wastes decomposition, 187
pest control, 198—199
plant pest and disease suppression bacterial volatiles, 193

Vermiwash and vermicompost extract (*Continued*)
derivatives, 193, 194f
inorganic fertilizer, 193
mechanisms, 193, 193f
microbial antagonism, 189, 190t—192t, 193, 195t—196t
nonthermophilic approach, 189
population density, 193
soil-borne pathogens, 189—193
soil properties, 189
toxic substances, 193
public health risk, 187
soil amendments, 188
Vertical resistance, 23—24
Violet biotechnology, 164
Viral pesticides, 122—123, 122t—123t
Virulence, 20
Visible (VIS) spectroscopy, 136
Visible wavelength spectroscopy (VIS), 136
Volatile isoprene, 36
Volatile organic compounds (VOCs), 139, 197

W
Web-based models, 235
Wedelia chinensis, 181

Weeds control methods
biological method, 108—109
chemical method, 109
cultural method, 108
integrated weed management (IWM), 109—110
management, 108
mechanical method, 108
physical method, 108
Wettable powder (WP), 210—211, 211f, 211t
Wheat germ agglutinin, 94
White biotechnology, 164, 168
Wireless sensor network (WSN), 240
World Health Organization (WHO) report, 143

X
Xanthomonas, 121

Y
Yellow zucchini mosaic virus, 46

Z
Zanthoxylum alatum essential oil, 69
Zanthoxylum limonella, 112

Printed in the United States
By Bookmasters